NUTRITION AND NUTRITIONAL PHYSIOLOGY OF THE MINK

A HISTORICAL PERSPECTIVE

WILLIAM L. LEOSCHKE, Ph.D.

Order this book online at www.trafford.com
or email orders@trafford.com

Most Trafford titles are also available at major online book retailers.

Printed in the United States of America.

ISBN: 978-1-4251-5098-3 (sc)
ISBN: 978-1-4251-5100-3 (e)

Trafford rev. 04/04/2011

 www.trafford.com

North America & international
toll-free: 1 888 232 4444 (USA & Canada)
phone: 250 383 6864 ♦ fax: 812 355 4082

Nutrition and Nutritional Physiology of the Mink

A Historical Perspective

William L. Leoschke, Ph.D.

Nutrition and Nutritional Physiology of the Mink
A Historical Perspective

William L. Leoschke, Ph.D.
Director of Fur Animal Nutrition Research
National Fur Foods Company
New Holstein, Wisconsin 53061
Emeritus Professor of Chemistry
Valparaiso University
Valparaiso, Indiana 46385

Edited by Kathleen Mullen, Ph.D.
Professor of English Emerita
Valparaiso University

Book Design by Robert Sirko, MFA
Associate Professor of Graphic Design
Valparaiso University

Foreword

Nutrition and Nutritional Physiology of The Mink – A Historical Perspective is written by Dr. William L. Leoschke, a fur animal nutrition scientist who is recognized worldwide for his basic research studies on mink and chinchilla nutrition at the University of Wisconsin, 1950-1959, and his numerous articles on mink nutrition in fur animal trade journals over a period of almost three decades, including the years 1960-1965 in association with Dr. R. Shackelford, geneticist, and Dr. G. Hartsought, veterinarian, in a monthly column in *American Fur Breeder* entitled "Ask The Experts—About Diseases—About Genetics—About Nutrition." This fur animal education program was followed by a similar column entitled "Dr. Leoschke On Mink Nutrition" from 1965-1973 in *American Fur Breeder* and later in *U.S. Fur Rancher.* In addition, each year of "The Blue Book of Fur Farming" from 1973 to 1988 featured a significant mink nutrition article by Dr. Leoschke.

Relative to the worldwide fur industry, Dr. Leoschke provided articles for the *British Fur Farmers Gazette* from 1966-1967 as well as a "Question and Answer" column. He has served as a member of both the 1968 and 1982 National Academy of Science National Research Committee that prepared the publication on "Nutrient Requirements of Mink and Foxes." In addition, Dr. Leoschke has provided scientific presentations at the International Congresses In Fur Animal Production initially in 1976 in Finland and every four years following: 1980, Denmark;1984, France; 1988, Canada; 1992, Norway; 1996, Poland; 2000 Greece; and 2004, The Netherlands.

In addition to these significant activities, Dr. Leoschke has been a consultant to National Fur Foods Company, New Holstein, Wisconsin, as Director of Fur Animal Nutrition Research since 1955. The basic goal of the experimental studies was the development of top performance pellet formulations for mink, fox, ferrets, and even opossums. After many years of intensive and extensive research, the goal has been achieved wherein the performance of fur animals on a commercial pellet has equaled that of common fresh/frozen/fortified cereal programs in terms of reproduction, lactation, and kit growth. Relative to fur development it has provided unique qualities of light leather, silkiness, and a sharp color unequalled on any practical fresh/frozen/ fortified cereal program within the worldwide fur industry.

The book is documented with more than one thousand references, including multiple tables, figures, and graphs. The scientific observations are presented in a manner useful for mink nutrition scientists, commercial mink cereal and pellet manufacturers, veterinarians, and practical mink ranchers throughout the world.

Preface

From my elementary school years, I had always been very sympathetic to the role of teachers in the classroom. By the age of 12, I knew that I wanted to be a teacher and the choice of a university position as a professor of chemistry was related, in part, to a chemistry set I received for Christmas, 1940. After a year in the U.S. Navy, I became a freshman student at Valparaiso University with a major in chemistry and a minor in mathematics. At the end of my junior year in 1949, I was essentially "bored" with chemistry and was giving serious consideration to a major in mathematics, a field of science that had fascinated me since high school courses in plane and solid geometry. Fortunately, in the summer of 1949, Professor Buls' course in conservational geography, which I took reluctantly to fulfill the social science course requirements for a B.A. degree, introduced me to the wonderful world of agricultural chemistry. This new fascination with biochemistry has never left my mind. At one point in the course, Professor Buls mentioned a brilliant approach to certain economic problems of Appalachia. Why not simply plant oak acorns? These would produce trees to minimize erosion and provide acorns as food for pigs which were an economic resource for the farmers. Only one problem: the pork tasted like acorns from the tannins deposited in the flesh, causing the farmers to receive minimal compensation for their product. The basic concept that an animal's diet could have a direct bearing of the taste of the flesh was one of the most exciting chemical concepts I had ever heard. To explore agricultural chemistry in depth, I resolved to become a biochemist.

For graduate studies, I applied to seven schools, with the primary goal of becoming a student in the Biochemistry Department, University of Wisconsin, the oldest and largest department of biochemistry in the world. As part of the application, the university required a simple essay on why the individual student wanted to pursue graduate studies. I diligently wrote of my interest in teaching at the university level so that I could combine two lifetime goals, teaching and biochemical research studies.

Then with great enthusiasm, I naively wrote of a childhood experience related to my rabbit colony. Men from the University of Rochester were visiting a local farm purchasing calves for medical research studies and asked the farmer about the presence of any small 4-H project rabbit colonies in the area where they could obtain some healthy rabbits for experimental studies. The net result was that some of my rabbits ended up at a university research laboratory.

When Dr. C. A. Elvehjem, Head of the Biochemistry Department, University of Wisconsin, read about Leoschke and his rabbits, he simply said, "This is the man we need for the mink project." The previous graduate student on the mink nutrition research studies was Sam Tove from Brooklyn, New York. The quality of his experimental work was undercut by the basic fact that he did not particularly like animals and especially disliked the mink ranchers supporting his work. As he told me in our first conversation, "Keep away from those 'dumb' mink farmers." As a young, eager graduate student, I did the exact opposite and attended local mink rancher meetings and the International Mink Show in Milwaukee each year. As a direct result of my sincere concern for the welfare of Wisconsin mink farmers, I obtained the basic data for my Ph.D. thesis, not at the University of Wisconsin fur animal research facility, but at the Elmer Christiansen mink ranch, 30 miles from the university at Cambridge, Wisconsin.

The offer of a $100/month research associate stipend made me even more enthusiastic about pursuing graduate studies at the University of Wisconsin. With that vital financial support, I would have been happy to work on any project, even studies on the nutrition of cockroaches; I was seeking a Ph.D. degree, essential for college teaching, and was not actually interested in a career as a mink nutrition scientist. At Valparaiso University, one of my fellow students said, "Bill, you will work with mink the rest of your life." I immediately replied, "No way. As soon as I obtain the coveted Ph.D. degree from the university, I will be leaving town an hour later." Fortunately, it took me five years to leave the University of Wisconsin after completing my graduate studies. I spent those years working on projects sponsored by the Mink Farmers Research Foundation and waiting the opportunity to join the faculty of Valparaiso University as an assistant professor of biochemistry.

Within a year of the completing my graduate studies, I was hired as a mink nutrition consultant for National Fur Foods. The primary reason for my hiring was simple public relations. As Paul Langenfeld said years later, "The mink ranchers trust Bill Leoschke." The primary reason for my acceptance was also simple. It gave me the opportunity to have available hundreds of mink for experimental nutrition studies. This work would lead to the development of a mink pellet providing performance of the mink superior to that of current conventional mink ranch diets. In the '50s most ranch diets contained a fortified cereal base at 15% and 85% fresh/frozen feedstuffs, horsemeat, fish, tripe, liver, and rabbits as well as some poultry by-products. Although it took many years to achieve that goal, today in the 21st century, high quality pellet formulations are providing superior performance of the mink in all phases of the mink ranch year. Performance is especially unique in fur production with the marketing of pelts with sharper color, lighter leather, and silkier fur than that attainable on any conventional mink ranch diet anywhere within the worldwide fur industry.

Why A Book On Mink Nutrition?

After 60 years of involvement in fur animal nutrition experimental studies including mink, fox, ferrets, and chinchillas, why set aside more than a decade of one's life to write this book?

The answer is simple in terms of my dual lifetime goals of being both a college professor and active researcher in biochemistry. I was not interested in a career as a graduate professor in fur animal nutrition research at the University of Wisconsin, Cornell University, or Michigan State University. In these schools, the emphasis on guiding graduate students left very limited time for actual teaching and working with young men and women interested in future careers as chemists, dietitians, nurses, or physicians. Thus I had to find an alternative opportunity for research studies toward achieving the goal of developing pellets for the modern nutrition of the mink. That opportunity arose in spring, 1955 with my appointment as mink nutrition consultant to National Fur Foods Co., Fond du Lac, Wisconsin. My employment within the structure of a commercial mink food company, obviously, provided very limited opportunity for publication of experimental data; however, at the same time, full credit must be given to National Fur Foods Co. management for their support in allowing me to present significant scientific reports at International Congresses in Fur Animal Production from 1976 through 2004 in Finland, Denmark, France, Canada, Norway, Poland, Greece, and The Netherlands.

In 1959, I accepted an appointment as Assistant Professor of Chemistry and Home Economics at Valparaiso University, Valparaiso, Indiana, and enjoyed 36 wonderful years teaching students about the world of chemistry, biochemistry, and human nutrition as well as advising pre-medical and pre-dental students. The appointment as a professor at Valparaiso University allowed me the opportunity for "free" summer months which could be devoted to mink nutrition research. In the summers of 1960-61, I worked at a small Mink Metabolism Laboratory in a barn at my first residence working on Wet-Belly Disease of the mink and then on the digestibility of carbohydrates by the mink. Since 1962, I have had the privilege of spending each summer as "Biochemist In Residence" at National Fur Foods mink research facilities in Illinois and Wisconsin.

As early as 1964, I was asked to leave my position as Associate Professor of Chemistry and Home Economics, at Valparaiso University, to work full time for National Fur Foods Co. at more than double my university stipend. However, I was not in any way interested, inasmuch as my primary goal in life has always been teaching. In addition, my wonderful wife, Marjorie, said, "Take the job, and I will divorce you," knowing that only 20% of my work would involve actual experimental studies with mink with 80% of my efforts as a "Salesman On The Road" for National Fur Foods.

However, I continued in the position of Director of Fur Animal Nutrition Research, National Fur Foods Company. There I have had wonderful opportunities for mink, fox, and ferret nutrition research studies with as many as 2,000 mink a year on experimental studies in recent years. Still, inasmuch as these studies on the development of pellets for modern mink nutrition were sponsored by a commercial enterprise, the opportunity for publishing the experimental data was obviously limited.

Scientists over the centuries have contributed to knowledge of chemistry, biochemistry, and animal nutrition by publishing original research papers year after year and/or compiling extensive reviews of research endeavors and even books covering as many as 50-100 years of experimental studies on a specific scientific area. My experimental labors of the last 60 years on fur animal nutrition research have provided multiple articles on modern nutrition of the mink and fox in trade journals such as *The American Fur Breeder, U.S. Fur Rancher, Fur Journal of Canada* and other journals. However, my actual publication of specific scientific data has been limited to fewer than 10 significant articles in the period of the last four decades. Thus, I decided to contribute to the world's knowledge of mink nutrition and nutritional physiology via the publication of a book covering more than 1,000 scientific articles from more than 100 journals throughout the world fur industry.

Obviously, I am not alone in valuing this method of authorship, for I join Edward Gibbon in his perspective: *"Having declined the fame and envy of original composition, we can only require at his hands method, choice, and fidelity, the humble though indispensable virtues of a compiler."*

Acknowledgements

This book is dedicatd to the hundreds of scientists and their associates, including graduate students too numerous to list, throughout the world, who have made significant contributions to our basic understanding of the nutrition and nutritional physiology of the mink. Without question they deserve recognition for their efforts.

In terms of specific individuals, this book is dedicated to my father, William C. Leoschke, who not only set a high standard for diligence in work, often reminding me, "Take a job: Do it small, Do it right, Or not at all," but also constructed the original rabbit colony building which project led to my future with mink. I also dedicate it to Gunnar Jorgensen, Denmark, who has provided more to experimental studies on mink nutrition and to mink rancher education via the publication *Scientifur* than any other mink nutrition scientist with the world fur animal research community. Without question, Gunnar Jorgensen is the "Mink Scientist/Educator" of the 20th century.

Special credit is also well deserved by the workers at the National Research Ranch facilities which led in the period from 1955 to 2006 to the development of a mink nutrition pellet program unequalled within the world fur industry in terms of being supportive of the top performance of the mink in all phases of the mink ranch year. Full credit to these individuals including Mark Michels, who has held the position of mink research ranch manager for the longest period, is provided in the chapter in this book on modern mink nutrition-pellets.

xv

Chapter 5

Disorders Related to Nutrition Mismanagement

Chapter 6
Toxic Substances In The Feed Supply

Chapter 7
Modern Mink Nutrition Management

Chapter 8
Modern Mink Nutrition—Pellets

Chapter 9
Nutrition Management—21st Century

1

Natural Diet of the Mink

Natural Diet of the Mink

The mink, *Mustelidae vison*, is a model carnivore. The family *Mustelidae* includes weasels, otters, skunks, wolverines, martins, and old world badgers. The *Mustelidae* along with the family of the foxes (*Canidae*) belongs to *Vulpavines*, the New World Carnivores, which is also known as the "dog branch" of the Carnivore family. The "cat branch", *Viverravines*, includes among others the domestic cat. These branches have evolved separately for about 60 million years. During their evolution since the Paleocene Miacids, carnivores have adapted to diets low in carbohydrates (Macdonald, 1992). The *Vulpavines* are distinct from the *Viverravines* inasmuch as the "dog branch" are omnivores as adults while the "cat branch" remain carnivores as adults (Buddington et. al., 1991).

Field Studies

The mink in the wild is an efficient semi-aquatic opportunistic catholic predator with significantly different dietary regimens related to seasonal availability of feedstuffs. During the summer months the intake of easily caught crayfish may be very high with a lower proportion of frogs, fish, and birds and their available eggs and a minimal proportion of mammals. In the fall and winter months with reduced availability of birds, mammals, especially muskrats, are a major food resource (Akande, 1972; Day and Linn, 1972; Dearfor, 1932; Dunstone and Birks, 1987; Hamilton, 1936; La Due, 1930; Sealander, 1943).

Aside from man, the mink is the worst predator with which the muskrat must contend. Wild mink exhibit the capacity to store muskrats, fish, birds, and red squirrels for future use as feed resources (Yeager, 1943).

Mammals consumed by the mink, in addition to the muskrat, include mice, moles, shrews, voles, wood mice, harvest mice, dormice, common rats, weasels, grey squirrels, rabbits, hares, and gophers (Anon, 1936; Chanin and Linn, 1980; Gilbert and Nancekivell, 1982; Lode, 1993). Birds are part of the wild mink's dietary program including duck, teal, ringnecked pheasant, mallard, Hungarian pheasant, blackbird, mudhen and bittern (La Due, 1930). Other feedstuffs of the wild mink include snakes, insects, and earth worms. Grass and bits of vegetables are frequently found in the stomachs of wild caught mink, indicating that this matter is not taken fortuitously, but rather by design (Hamilton, 1936).

A study by Birks and Dunstone (1985) is the only experimental report in the fur animal literature wherein the sexual dietary preferences of the mink have been noted. Their studies indicated that the male animals were involved more with larger animals such as rabbits while the smaller female mink in the same geographical area preyed more heavily on fish and crustaceans.

It is of interest to record the different dietary patterns of mink in Estonia. The European mink (*Mustela lutreola*), a relative of the North American mink (*Mustela vison*), preferred fast flowing water, eating mostly fish and crustaceans and frogs, while the North American mink preferred mammals near slow flowing rivers. Very likely the different feeding regimens were related to differences in habitat selection (Maran et al., 1998).

In Finland, Tolonen (1982) and Niemimaa and Polk (1990) note that although the ecological fitness of the mink is largely based on opportunistic feeding behavior, during the winter the mink of Northern Finland mainly catch fish.

Natural Diet of the Mink

2

Ranch Mink Nutrition History
North America

In terms of the world history of fur as human apparel, North America played a major role. The Chinese have used fur from about 500 B.C. and fur garments are recorded as being used in both ancient Greece and Rome. In medieval times, the use of fur was strictly controlled and only royalty, nobility, and certain clerics and layman were allowed to wear fur. Ermine could be used only for royal robes. In the middle ages and in Chaucer's time (1400 AD) beaver was used for making headwear and was the only fur worn by the common people. The English custom of using ermine as a trimming for judges' robes has survived from the middle ages.

Furs became available to the general population for the first time following the discovery of North America because the limited supply in Northern Europe restricted their use to the wealthy. The Hudson's Bay Company was founded in 1670 and spread across the continent of North America in search of furs. Fur animals are part of the North American heritage. Exploration of the Northeastern part of the continent was initiated by the French fur traders. The fur industry (fox, mink, and chinchilla) is still an important part of the North American economy.

Commercial fur farming in North America began in the early 1860s when a pair of young silver foxes was dug out of a den on Prince Edward Island. In terms of the North American mink industry in Canada, the first mink were bred in captivity in the 1880s and by 1907 ranch raised mink were beginning to achieve popularity. The mink industry did not come into prominence until about 1920 when it was found that the wearing qualities of mink pelts were superior to those of silver foxes. In the beginning the mink ranchers were ridiculed by the fox ranchers who felt that there was only one great fur, namely fox. However, by 1946, that industry had all but disappeared from North America and today mink is the fur of the world fur industry.

The original stock for commercial mink ranching was secured by trapping mink from all over the continent. The greatest contributions to current North American ranch mink were made by a little Eastern dark mink and the larger, more prolific Alaskan mink. The color of wild dark mink ranges from dark brown to almost black. The present day ranch dark mink is considerably darker than the original wild mink. This has been accomplished by selective breeding over a period of many years. Not until the early 1940s did mink other than dark mink appear on the market as the result of mutations. The first mutation to appear was that of the silver mink called Platinum. Since then over 60 color phases of mink have appeared.

Early Ranch Diets

As early as 1927, significant commentary on practical mink ranch diets was provided in the commercial trade journals (Palmer, 1927). Palmer commented on the biological value of protein resources with the statement that milk, glandular tissue (liver, kidney, and heart), and eggs were better sources of protein than muscle meats, a viewpoint that North American mink ranchers in 2008 still do not fully appreciate. Palmer appreciated the value of iodized salt for mid-Western mink ranch operations. However, his insights on practical mink nutrition were not entirely valid, as illustrated by his statement, "There is no evidence that vitamin E will ever be deficient in the diet of fur bearing animals fed mixtures of natural foods."

Obviously, Palmer at that time was not familiar with potential deficiencies of vitamin E in mink ranch diets employing rancid horsemeat or salmon waste products with high levels of polyunsaturated fatty acids sensitive to oxidative rancidity.

In 1937, Martin made several significant comments on differences in practical ranch diets for mink and fox required for top performance of the animals with a very interesting observation relative to pelt color of fox and mink: "We must bear in mind that while the fox starts growing its winter coat sometime in July, the mink does not usually start until about October 1st. So that anything you have read regarding the influence of summer feed on the color of the fox would not necessarily be applied to the mink, so early in the season.

"It has been definitely proven that the feeding of cod liver oil during the summer months tends to make fox fur brown in color and it seems reasonable to suppose that it might have the same influence on mink if given in the fall."

Later he comments on specific constituents of the ranch diet: Initial mink feed schedules were worked out from the results obtained by Smith (1927) in his work on fox feeding. However, it has been modified in different places to suit some of the known seasonal requirements of the mink and I do not believe that at any time the mink can handle as high a percentage of cereal as the fox."

First attempts to provide satisfactory rations for the mink were based on replicating natural feedstuffs using small mammals, fish, frogs, and eggs. Later, items such as horsemeat comprised the backbone of mink rations; fish also became a staple feed in the mink industry, especially in coastal North America and the Great Lakes areas.

By 1939 the ready mix feeding regimen of the Midvale Co-Op, Utah, consisted of horsemeat, 45%; beef tripe, 21%; fortified cereals, 17%; liver, 9%; tomatoes, 5% and green bone, 3% (Erikson, 1960). Within a few years, the shortage of horses required the employment of other high protein resources including whole fish, fish scrap, whale meat, and poultry by-products (heads, entrails and feet). By the 1960s, high protein fortified cereals employing quality fish meals and poultry meals were marketed that comprised as much as 40-60% of the ranch diet. By 1970, Paul Iams and Bill Kelly in Ohio marketed the first successful mink pellet which was soon followed by quality pellets from Ross-Wells, National Fur Foods, and XK Foods.

The nutrient composition of four typical ranch diets in September, 1998 in the midwest is of interest (Heger Co., 1998). See Table 2.1.

TABLE 2.1 Nutrient Composition—Modern Ranch

Nutrient	Average	Range
Protein,%	35	28-41
Fat, %	25	24-27
Ash, %	10	6-13
Calcium, %	1.5	1-2.2
Phosphorus, %	1.4	1-2.1
Ca/P Ratio	1.0	1.4-0.7
Salt, %	1.7	1.6-1.9
Sodium, %	0.7	0.6-0.9
Potassium, %	0.8	0.4-1.5
Sulfur, %	0.5	0.4-0.6
Magnesium, %	0.2	0.1-0.5
Manganese, ppm	70	50-90
Copper, ppm	40	10-110
Zinc, ppm	160	110-200
Iron, ppm	2,400	360-7,400
Vitamin A, IU/kg	7,500	4,000-10,000
Vitamin D, IU/kg	6,000	5,900-6,200
Vitamin E, IU/kg	80	23-150

Mink Nutrition Research Studies—North America

Fur animal research studies were initiated at Cornell University and Oregon State University in 1936 and at the University of Wisconsin in 1939 with initial studies involving fox and mink (Bernard and Smith, 1941; Coombes, 1941).

3

Anatomy and Physiology

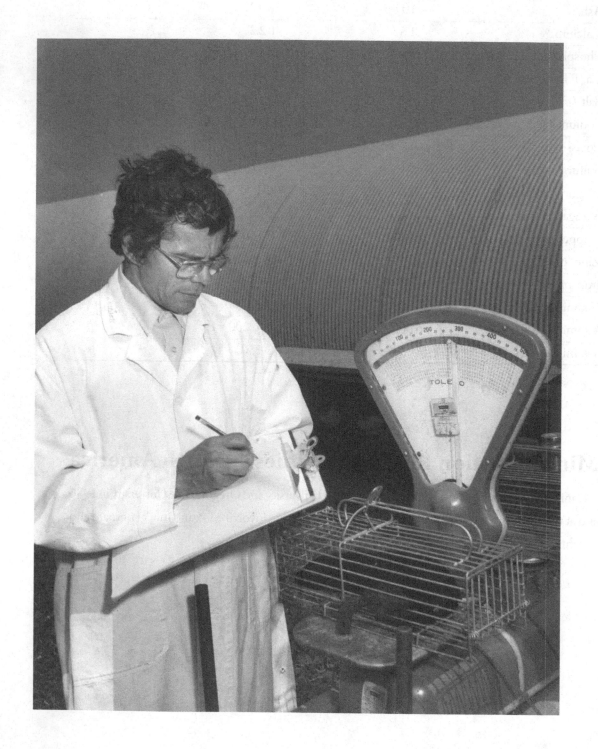

Anatomy

The mink, *Mustela vison*, is a mammal. The taxonomic position of the animal is as follows:

Phylum—Chordata

Superclass—Tetrapods

Class—Mammalia

Order—Carnivora

Family—Mustelidae

Genus—*Mustela*

Species—*M. vison*

The Family *Mustelidae* comprises 25 genera of carnivores including mink, weasels, ferrets, martens, sables, wolverines, badgers, skunks, otters and some others. The genus *Mustela* includes the weasels and ferrets in addition to mink.

Detailed discussion and illustrations of the anatomy of the mink may be found in books published by Klinger (1998 and 2000) and Schlough (1971), and additional references in the scientific literature include Aulerich (1968) and Kainer (1954a and 1954b).

Body Composition

Proximate Analysis Data

The data provided on the composition of male kits is based on experimental reports totaling 14 mink on each date (Charlet-Lery et. al., 1979; Charlet-Lery et. al., 1980; Enggaard-Hansen, N. and Glem-Hansen, N., 1980). The data on female kits involved 2 kits at each age as reported by Enggaard Hansen, N. and Glem-Hansen, N. (1980).

TABLE 3.1 Proximate Analysis Data—Mink Kits (males)

Kit Age	Body grams	Solids %	Protein %	Fat %	Ash %	Ca %	P %	Mg %
0	9	16	12	2	1.8			
3	120	24	12	9	1.7			
8	350	26	18	4	2.8			
10	730	33	20	10	3.2	0.8	0.5	0.3
13	1040	38	21	15	3.2	0.9	0.6	0.3
15	1250	43	19	22	2.9	0.9	0.5	0.3
22	1540	49	18	30	2.7	0.8	0.5	0.3
26	1780	51	18	31	2.7	0.8	0.5	0.3
28	1840	55	18	35	2.5	0.7	0.4	0.2

TABLE 3.1 Proximate Analysis Data—Mink Kits (females)

Kit Age	Body grams	Solids %	Protein %	Fat %	Ca %	P %	Mg %
12	750	21	14		1.0	0.6	0.3
14	860	20	21		0.9	0.6	0.3
17	1010	19	24		1.0	0.6	0.3
20	1050	18	32		0.8	0.5	0.3
24	1230	18	28		0.8	0.5	0.3
32	1140	19	31		0.7	0.4	0.2

Amino Acid Data

TABLE 3.2 Amino Acid Data—Mink Milk, Fetus, Kits and Fur

Amino Acid	Milk*	Fetus*	Days			Mature***	Fur*****
			4**	56***	92*		
Alanine			5.5	5.9		6.0	1.5
Arginine	6.1	6.7	6.4	6.4	6.4	6.4	7.5
Aspartate			7.3	7.4		8.0	2.2
Cystine			2.0	3.2		2.8	12.6
Glutamate			12.7	13.4		14.0	11.5
Glycine			8.4	10.0		8.6	5.6
Histidine	1.7	2.6	1.9	1.8	2.1	2.4	1.0
HO-Proline				3.2			
Isoleucine	3.7	3.5	3.0	2.9	3.5	3.2	2.6
Leucine	11.3	10.1	6.7	6.5	7.3	7.2	7.5
Lysine	5.5	7.8	5.7	5.7	6.6	6.6	3.5
Methionine	1.3	1.5	1.6	1.5	1.8	2.6	1.0
Phenylalanine	3.4	6.3	3.4	3.3	4.2	3.6	2.6
Proline			5.9	6.7		7.2	17.9
Serine			3.8	4.6		5.0	8.0
Threonine	3.9	4.9	3.7	3.8	3.9	4.3	5.6
Tryptophan	1.2	0.5	1.0	0.7	0.4	1.0	0.4
Tyrosine	2.1	2.8	2.9	2.5	2.9	2.8	5.2
Valine	5.6	6.9	4.4	4.4	5.0	4.8	5.4

   * Moustgaard, J. and Riis, P.M. (1957). **Glem-Hansen, N. (1979).
  *** Skrede, A. (1981).
 **** Jorgensen, G. and Eggum, B.O. (1971), Glem-Hansen, N. and Enggaard Hansen, N. (1981).
***** Moustgaard, J. and Riis, P.M. (1957) and Disse, P.H. (1957).

Fatty Acid Data

TABLE 3.3 Fatty Acid Composition Mink Milk and Body Fat

Fatty Acid	Milk Fat*	Body Fat**
Lauric Acid, C-12:0	0.2	trace
Myristic Acid, C-14:0	2.9	2.6
Myristoleic Acid, C-14:1	0.5	0.6
Other - saturated, C-15:0	0.4	0.2
Monoene C-15:1	0.1	0.1
Palmitic Acid, C-16:0	23.5	12.5
Palmitoleic Acid, C-16	14.1	13.6
Margaric Acid, C-17:0	0.8	0.6
Other monoene,	0.5	0.8
Stearic Acid, C-18:0	10.3	3.1
Oleic Acid, C-18:1	30.0	49.2
Linoleic Acid, C-18:2	13.7	14.1
Linolenic Acid, C-18:3	1.2	0.2
Arachidic Acid, 20:0	—	0.3
Gadoleic Acid, C-20:1	1.8	2.1
Other; diene, C-20:2	0.1	—
Unidentified (highly unsaturated C-20,C-22)	10.0	—

* Garcia-Mata, R. (1980), mixed milk from five females, wild type, 40 days post whelping.

** Fat taken from a nursing female.

Mineral Data

TABLE 3.4 Mineral Composition—Mink Body and Fur

Mineral	Dry Matter-%	Body ppm***	Fur	
	*	**	Darks***	Pastels***
Aluminum	3.2	4.6		
Calcium	4.0	5	920	240
Copper	0.001	5	8.1	7.9
Iron	0.04	28	57	62
Magnesium	0.13	2000	148	42
Manganese	0.0002	0.8	0.4	0.4
Phosphorus	2.4	5	575	400
Potassium	0.8	10	1,640	1,010
Sodium	0.6	20	910	740
Sulfur	—	400	—	—
Zinc	0.2	200	280	260

* Hansen, N.E. (1982), **Kangas, J. (1976), ***Hornshsaw, T. C. et al. (1990).

Milk Composition

Collection

The following technique was used by Jones et. al. (1980).

Vacuum was produced by a Duoseal pump which developed 12 mm mercury vacuum when the T-tube was unstoppered and a maximum of 322 mm mercury vacuum when the T-tube was stoppered. Prior to milk collection, mink were separated from their kits a minimum of two hours but no longer than four hours. Ten minutes prior to collection, 0.06 ml of 50 mg/ml telazol and 0.5 ml (10 units) of oxytocin were administrated intramuscularly to each female to be milked to permit ease of handling and to induce milk "let-down" respectively.

A greater quantity of milk was obtained when milking was initiated 10-15 minutes after the injection of oxytocin, as opposed to five minutes recommended by Conant (1962). The quantity of milk obtained was not related to either the number of teats or the milking time. The average volume of milk obtained was 0.5 to 1.0 ml, rarely less than 0.5 ml. However, up to 4.0 ml of milk was obtained from some animals and 2 ml was not uncommon. Additional injections of oxytocin did not increase the volume of milk yield. Maximum milk volumes were obtained during the 4th week of lactation and no differences were noted in milk yield between first year and second and third year breeders.

14

Nutritional Value

The following table from Sherman (1941) and Brody (1945) modified to include mink is of major significance in understanding the uniqueness of mink milk composition and mink kit growth.

TABLE 3.5 Double Birth Weight and Milk Composition-%

Animal	Days	Protein	Fat	Lactose	Ash
Man	180	1.6	3.7	6.9	0.2
Horse	60	2.3	1.5	6.1	0.5
Cow	47	3.5	3.6	4.5	0.7
Goat	22	3.7	4.1	4.2	0.8
Ewe	15	5.4	6.2	4.3	0.8
Pig	14	6.2	6.8	5.0	1.0
Cat	10	7.0	3.0	5.0	1.0
Dog	9	7.5	12.5	3.5	
Rabbit	6	13.0	10.0	1.0	2.4*
Mink	4**	7.0	8.0	7.0	0.7***

* Labas (1971), **Hartsough (1955) and Tauson and England (1989), ***Jorgensen (1960)

Proximate Analysis Data

At birth, the body of the mink kit is almost devoid of fat and no energy stores are available for growth except liver glycogen which will support survival for about one day (Tauson and England, 1989). It has been shown that the composition of mink milk is quite stable regardless of the plane of nutrition (Glem-Hansen et al., 1973). This point is well illustrated by the fact that with sub-optimum protein nutrition in terms of protein quantity or quality, the content of the milk will remain constant while the quantity may be significantly reduced. During the first three to four weeks of lactation the lactating mother is the sole source of nutrition for the rapidly growing kits. At the point of top lactation (25 days postpartum) the female sustains a litter biomass sometimes exceeding her own.

The composition of mink milk on the average may be considered to be percentage as solids, 24; protein, 8; fat, 7; carbohydrates, 8; and ash, 1. These data are based on studies by Jorgensen, 1960; Conant, 1962; Howell, 1975; Glem-Hansen and Jorgensen, 1973; Glem-Hansen et. al., 1973. It is of interest to note that the data of Glem-Hansen and Jorgensen (1973) and Glem-Hansen, et. al. (1973) indicated a mink milk composition with an average gross energy (GE) of 160 kcal/100 grams and that 35% of Metabolic Energy (ME) is represented by protein. These experimental data support the recommendations of mink nutrition scientists for 35-40% M.E. as protein for the lactation period (Leoschke, 2001).

15

Mink Milk Composition—Danish Data

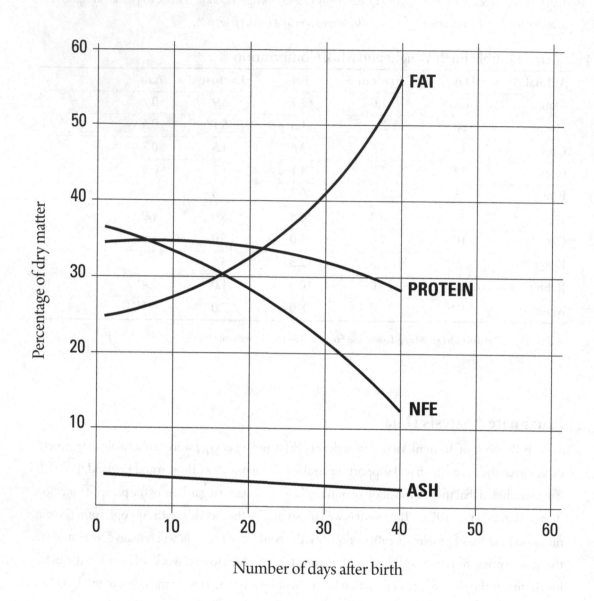

Obviously, these figures represent a basic estimate of the composition of mink milk over the 4 week lactation period inasmuch as mink milk composition varies very significantly over the period of milk production as seen in the following graph provided by Glem-Hansen and Jorgensen (1975). From the graph it is readily seen that the content of crude fat as percentage of dry matter solids is increasing throughout the lactation period at the expense of protein and carbohydrate content. Their graphical data are the basis of the common mink ranch practice of providing increasing levels of fat in the mink's ration as the lactation phase proceeds throughout May. The relatively high fat content of mink milk at six weeks post-whelping is confirmed by Kinsella (1971) with a percentage analysis of immediate post-weaning milk as protein, 7.2; fat, 13.4; carbohydrate, 6.3; and ash 1.3. More recent experimental data on the composition of mink milk over the period of lactation are presented in the following table:

TABLE 3.6 Variation in the Composition of Mink Milk—Lactation*

Days Post Partum	%-Nutrients-Solids Basis		
	Protein	Fat	Lactose
3	38	38	11
11	38	33	2
18	32	42	1
25	30	41	1
32	35	45	1
36	35	46	<1

* Olesen et al. (1992).

These more recent experimental data are consistent with the earlier studies on the composition of mink milk as presented in the graph provided.

Studies by Glem-Hansen et al. (1973) have indicated that neither the protein content nor the biological value of the protein provided the mink had any significant influence on the content of protein, fat, or carbohydrates in mink milk; that is, the quality of mink milk composition is not influenced by the feeding regimen. However, with kit growth as a parameter, the studies indicated that the quantity of milk provided the kits was to a great extent influenced by both the quantity and quality of protein provided the lactating animal. Experimental studies by Fink (2001) indicated that four weeks post-partum, dams nursing 9 kits did not produce more milk than dams nursing 6 kits, indicating a maximum capacity for mink milk production and that milk production limits the growth rate of kits raised in large litters. Newborn mink have no detectable gamma-globulin. It has been demonstrated by Porter (1965) that the kits acquire gamma-globulin from the mother via the colostrum and milk. Gamma-globulin concentration in the colostrum was lower than that in the maternal serum.

Protein Content

The protein content of mink milk is approximately seven times higher than that of most primates, but levels are consistent with most other carnivores (Oftedal, 1984). Mink do not require a specific level of protein intake for top performance throughout all phases of the mink ranch year but a unique combination of amino acids provided by the protein intake for each period of the calendar year.

TABLE 3.7 Amino Acid Composition of Mink Milk*

Amino Acid	%Protein	
	Moustgaard & Riis	Howell
Arginine	6.1	—
Cystine	—	1.0
Histidine	1.7	2.6
Isoleucine	3.7	3.7
Leucine	11.3	10.1
Lysine	5.5	6.3
Methionine	1.3	2.0
Phenylalanine	3.4	3.6
Threonine	3.9	3.9
Tryptophan	1.2	0.5
Tyrosine	2.1	3.4
Valine	5.6	6.4

* Moustgaard and Riis (1957), Howell (1975).

For mink ranchers and mink nutrition scientists interested in a more detailed examination of the protein and polypeptide content of the colostrums of mink milk and the changes that take place in the following nursing period, a valuable reference is Bjergegaard et al., 1999.

Lipid Content

The very high fat levels provided by mink milk yield a high caloric density which is unique and very supportive of top growth of the mink kits during the lactation period (Leoschke, 1966 and Tauson, 1988). High milk fat is usually seen in animals that have a short nursing period and high postnatal growth rate (Oftedal, 1984). The high fat: carbohydrate ratio is supportive of superior kit weights at 21 days of age (Skrede, 1981b).

18

Triglycerides

TABLE 3.8 Triglyceride Content of Mink Milk*

Carbon Number	%	Carbon Number	%
C-34	0.1	C-46	1.9
C-36	0.1	C-50	18.6
C-38	0.2	C-52	35.8
C-40	0.9	C-54	28.4
C-42	1.3	C-56	trace
C-44	1.2	C-58	trace

* Kinsella (1971).

Most of the triglycerides (96%) contained 2-3 fatty acids with 18-C; Free Fatty Acids, 1.5%, some monoglycerides, cholesterol, and cholesterol esters. 60% of all lipid fatty acids were 18-C with all fractions containing a high concentration of linoleic acid. The preponderance of long-chain fatty acids in the mink milk indicates that much of the lipids secreted in mink milk are derived from dietary or adipose tissue depot sources, and relatively little de novo synthesis of fatty acids occurs in the mammary tissue as compared to ruminant species.

Fatty Acids

TABLE 3.9 Fatty Acid Composition of Mink Milk*

Fatty Acid	%	Fatty Acid	%
Lauric acid, C-12:0	0.2	Stearic acid, C-18:0	10.3
Myristic acid, C-14:0	2.9	Oleic acid, C-18:1	30.0
Myristoleic acid, C-14:1	0.5	Linoleic acid, C-18:2	13.7
Other - saturated, C-15:0	0.4	Linolenic acid, C-18:3	1.2
Palmitic acid, C-16:0	23.5	Gadoleic acid, C-20:1	1.8
Palmitoleic acid, C-16:1	4.1	Unidentified (highly	
Margaric acid, C-17:0	0.8	unsaturated, C-20, C-22)	10.0
Other monoene, C-17:1	0.5		

* Garcia Mata (1980). Mixed milk from five females, wild type, 40 days post whelping.

It is obvious that the fatty acid profile of mink milk is very dependent on the fatty acid content of the diets provided to the lactating animal. Studies by Clausen et al. (2004) and Hansen et al. (2004) indicated that ranch diets providing relatively high levels of n-3 fatty acids, such as docosahexaenoic acid found in fish oils, yielded mink milk with high levels of n-3 fatty acids and lower levels of linoleic acid (n-6). Mink diets with high levels of n-6 fatty acids such as linoleic acid (vegetable oils and cereal grains) and arachidonic acid (animal fats) yielded mink milk with higher levels of n-6 fatty acids.

Phospholipids

TABLE 3.10 Phospholipic Content of Mink Milk*

Phospholipid	%Total
Lecithin	53
Lysophosphatidyl Choline	8.3
Phosphatidyl Ethanolamine	10.0
Sphingomyelin	15.5
Phosphatidyl Inositol	6.6
Phosphatidyl Serine	3.6
Cerebrosides	3.2

* Kinsella (1971). The level of phospholipids is unusually high. Lysophosphatidyl choline may conceivably enhance the rate of lipid absorption from the small intestine of the young nursing mink.

Carbohydrates

Milk from mink has a very special composition and amount of different types of glycosides/ carbohydrates. Lactose is not the quantitatively dominating carbohydrate, as in the case of milk from several other animals, especially cows. Up to 16 different glycosides have been demonstrated in mink milk, generally oligosaccharides of 4-6 monosaccharide units, with acetylglucosamine and 2-alpha-L-fucose as typical constituents. 2-alpha-L-fucosyl-D-galactopyranosyl-beta-(1-4)-D-glucopyranos (2-alpha-fucosyllactose) is among the quantitatively dominating glycosides identified. The carbohydrates in mink milk also call for attention due to an apparent structural similarity with the glycosidic parts of membrane components, glycolipids, and glycoproteins, including immunoglobulins and glycoproteins that are the structural determinants of blood type. Mink milk has, moreover, "colostrum appearance" for a relatively long part of the lactation period (Andersen et al., 1999). Studies of Olesen et al. (1992) showed that wherein the mean concentration of calcium increased during lactation from 22 to 40 nmol/l and the mean concentration of phosphorus from 35 to 48 nmol/l, sodium and chloride varied considerably (range 35-60 mmol/l). Magnesium remained fairly constant at 3 mmol/l while potassium decreased from 33 to 26 mmol/l.

Minerals

TABL.3.11 Mineral Content of Mink Milk

Major Minerals	% Solids*	Trace Minerals	ppm Solids**
Calcium	0.75	Boron	11
Phosphorus	0.71	Barium	1.4
Magnesium	0.03	Copper	5.3
Sodium	0.40	Iron	52
Potassium	0.50	Manganese	0.6
		Zinc	83

*Aulerich, et. al., 1985; Aulerich, et. al., 1989; Conant, 1962; Howell, 1975; Jorgensen, 1960; **Aulerich, et.al., 1985; Aulerich, et.al., 1989)

Vitamins

TABLE 3.12 Vitamin Content of Mink Milk*

Vitamin	pp solids	Vitamin	pp solids
Thiamine	6	Pyridoxine	0.7
Riboflavin	11	Vitamin B12	0.01
Niacin	72	Folic Acid	3.4
Pantothenic Acid	36	Vitamin C	48
Biotin	0.1	Vitamin A, USPU	130, 000

* Conant, 1962; Howell, 1975; Jorgensen, 1960

Milk Volume

Newborn mink kits lack mobilizable energy stores and hence totally depend on mother's milk for nourishment during the first 24-26 days of life. Using water isotope dilution technique, Wamberg and Tauson (1998a, 1998b, 1998c, and 1998e) have provided some wonderful data on the milk production of the lactating mink involving 42 kits (Table 3.13) Their experimental data is in close agreement with studies conducted by Tauson et al. (1998).

TABLE 3.13 Volume of Mink Milk

Milk Production - 2 Year Old Mothers - 6-8 Kits	
Week	**Milk Intake/Kit**
1	11.1+/- 0.7
2	18.0+/- 0.8
3	27.0+/- 1.4
4	27.7+/- 1.2

The kits move from milk to solid feed at about 3 weeks of age, when the nutrient requirements are only partially met by lactation. The researchers noted that the male kits showed an extra 10% growth compared to the female kits, growth directly related to significantly higher milk intake. Daily milk yield increased from about 90 grams/day Week One to 190 grams/day Week Four. This increase is equivalent to 110 to 240 Calories/day which corresponds well with calculated daily energy requirements of the kits for growth and maintenance.

The high nutrient demand on the lactating mother is remarkable: the data indicate a female weighing 1,100 grams is producing more than 3,000 grams of milk during the four week lactation period. This very high energetic demand of lactation may under specific circumstances of sub-optimum mink nutrition management and/or mink ranch management lead to a very severe negative energy balance of the dam which may lead to the metabolic disorder termed "nursing sickness."

Digestive Tract

Anatomy

The alimentary canal of animals is a tubular structure which extends from the mouth to the anus and includes the esophagus, stomach, small intestine, large intestine, colon, and rectum. As with other carnivores, the mink has a simply built digestive tract consisting of a simple stomach, a relatively short intestinal tract, and no cecum. There is no clear distinction between the small and large intestine. The passage of digesta from the small intestines to the colon of mink is not slowed by the presence of an ileocecal valve, and the very short unsacculated colon does not prolong retention of digesta time required for the development of micro-organisms with significant carbohydrate digestive action. The length of the small intestine and the large intestine in relation to the total length of the intestine is approximately 95% and 5% (Szymeczko and Skrede, 1990). The gastrointestinal tract shows anatomical peculiarities as compared to that of other monogastric animals (Laplace and Rougeot, 1976), including a relatively short intestine/body length as seen in Table 3.14.

22

TABLE 3.14 Digestive Tracts of Animals

Animal	Intestine/Body Lentgth Ratio
Pig	14/1*
Rabbit	10/1*
Dog	6/1*
Cat	4/1*
Mink	4/1**

* Colin (1854) and Stevens (1977). **(Neseni, 1935; Perel'dik et al., 1972; Elnif, 1987; Elnif and Enggaard Hansen, 1988; Szymeczko and Skrede, 1990)

More recent experimental work of Buddington et al. (2000) is of interest. Their studies indicate that with the mink kit, the intestinal dimensions increase up to eight weeks and then remain constant (length) or decrease (mass) into maturity, despite gains in body mass.

Physiology
Mouth

There is minimal mastication to yield smaller food particles which are mixed with saliva to yield a bolus which is delivered to the stomach. The saliva contains a salivary amylase enzyme for partial starch breakdown.

Stomach

A pepsin enzyme is provided for initial protein breakdown as well as HCl (hydrochloric acid) from parietal cells which enhances protein hydrolysis to yield smaller polypeptide structures. In addition, there is a lipase enzyme for fat hydrolysis.

Small Intestine

Enzymes are provided from the pancreas and intestinal wall for the breakdown of protein (proteinases and peptidases), fats (lipases), and carbohydrates (amylases, lactase, maltase, and sucrase).

Large Intestine

This is the major site of water and mineral absorption.

Digestive Enzymes

Studies by Roberts (1988) showed that the carbohydrate digesting enzymes in the mucosa of the intestinal wall of adult mink are the same as those of other monogastric animals. The development of hydrolase activity in the intestinal brush border membrane is important for the maturation of digestive function in early life. Intestinal hydrolases develop later in the mink and are sensitive to glucocorticoid induction for a longer period in postnatal life than in species such as rats, pigs, or humans, (Sangild and Elnif, 1996). The activity of protein and lipid hydrolases develops in the mink intestine at an earlier age than that of the carbohydratases (Oleinik and Svetchkina, 1992).

Salivary Alpha Amylase

Relative to the digestion of starch (a mixture of amylose and amylopectin) and glycogen (animal starch), the saliva of most monogastric animals contains an alpha amylase which hydrolyzes the large polymer to maltose (disaccharide) units.

Gastric Lipase

Gastric lipase activity increased significantly in both the fundus and the corpus from three weeks to nine weeks of age with equal activity in the fundus at 14 weeks but lower activity in the corpus. The lowest activity was found in the antrum region (Layton, 1998).

Intestinal Invertase

The activity of invertase became more or less high only by seven weeks. Very likely before that time there was not enough substrate for it (Oleinik and Svetchkina, 1992).

Intestinal Gamma Amylase

Local peak amylase activity is observed at three weeks, that is, at the transition to mixed feed (Oleinik and Svetchkina, 1992).

Intestinal Lipase

There is an increase in activity from birth to 30 weeks of age (Oleinik and Svetchkina, 1992).

Intestinal Peptidases

Aminopeptidase N and A undergo their major development increases in activity at 4-6 weeks with enzyme development stimulated by cortisol (Sangild and Elnif, 1996). Dipeptidase activity was high throughout the period from birth to 30 weeks (Oleinik and Svetchkina, 1992). Other researchers observed a slight decrease in the activity of dipeptidylpeptase IV with advancing age from 2-10 weeks (Sangild and Elnif, 1996).

Intestinal TPA

Intestinal Total Proteolytic Activity, TPA, that is, trypsinogen-trypsin and chymotrypsinogen-chymotrypsin, increased over the period from birth to 30 weeks of age (Oleinik and Svetchkina, 1992).

Lactase

The activity of lactase is maximal at 4 weeks and decreases to about 5% of this level during the following two weeks. Cortisol treatment stimulated total lactase activity at two weeks (70% of control) and reduced this activity at 4 weeks to 20% of control (Sangild and Elnif, 1996). These observations provide an enzymatic basis for the sensitivity of post-weaning kits to lactose as observed by Leoschke (1970) wherein pellet program diets with 9% skim milk powder provided normal fecal structure with pastel mink but major diarrhea with dark kits.

Maltase

There is a gradual increase from 2-10 weeks with activity highest in cortisol-treated kits (Sangild and Elnif, 1996).

Pancreatic Alpha Amylase

There is low or no detectable activity in the new-born; high activity occurs in the next four weeks; and during the following six weeks a declining activity is seen towards the levels found in the pancreas of adult mink (Elnif and Enggaard Hansen, 1988; Oleinik and Svetchkina, 1992). Studies indicate that the pancreatic amylase does not seem to be sensitive to variations in the dietary supply of carbohydrates and this might perhaps explain why this carnivorous animal does not perform very well on diets in which the starch level exceeds a certain threshold (Simoes-Nunes et al., 1984).

Pancreatic Lipase

A considerable level of pancreatric lipase was detected at birth with an increasing activity and a later powerful rise of lipase activity taking place at mixed feeding (Oleinik and Svetchkina, 1992). Studies by Simoes-Nunes et al. (1984) indicate that pancreatic lipase exosecretion of the mink does adapt to the dietary lipid content.

Pancreatic TPA

The enzyme concentration increases from birth to adulthood (Oleinik and Svetchkina, 1992). Trypsinogen-trypsin activity is high in the first week of life and then falls to about one third of this level for the next two weeks. A linear rise occurs to a point at ll weeks with an activity level two thirds that of adult mink (Elnif and Enggaard Hansen, 1988). Chymotrypsinogen-chymotrypsin shows none at birth. It is still low for the next few weeks, then rises slowly until week l0, followed by a significant rise to week 12 (Elnif and Enggaard Hansen, 1988).

Studies of Clausen et al. (2004b) are of interest. Their experimental work indicated that proteinase inhibitors occur in rapeseed, potatoes, legumes -- especially soybeans and egg white proteins.

Proteinases and proteinase inhibitors bind to each other in the ratio of 1:1 (equi-molar ratio) and are excreted as such. It is of interest to note that trypsin inhibitors have approximately 10X higher effect on mink trypsin than on porcine trypsin. Mink trypsin has approximately 10X the proteolytic activity relative to porcine trypsin. There are two isoforms of mink trypsin, one with a low pH_i of 5.3, and the other with a pH_i of > 11.0.

Sucrase

There was no consistent change in activity with advancing age from 2-14 weeks, but the activities remained highest in cortisol treated kits (Sangild and Elnif, 1996).

Digestive Enzymes—Summary

The activity of protein and fat hydrolyzing enzymes in the small intestine of the mink at the age of seven weeks does not substantially differ from that of adult mink, whereas the activity of amylase and invertase were much lower even in 11 week old kits (Oleinik and Svetchkina, 1992).

Dietary Enzyme Supplementation

Studies by Jorgensen et al. (1962) indicate that the addition of protein and carbohydrate cleaving enzymes to practical mink ranch diets did not enhance the growth performance of the mink. These observations are identical to those seen by Leoschke (1962).

Intestinal Transport

Recent experimental work of Buddington et al. (2000) is of interest. Their studies show that the rates of glucose and fructose transport decline after birth for intact tissue but increase for brush border membrane vesicles (BBMV). Roles of absorption for five amino acids that are substrates for the acidic (aspartate), basic (lysine), neutral (leucine and methionine) and imino acids (proline) carriers increase between birth and 24 hours for intact tissue but decrease for BBMV. The proportion of BBMV amino acid uptake that is sodium dependent increases during development but for aspartate is nearly 100% at all times.

Despite the ability of mink to rapidly and efficiently digest high dietary loads of protein, rates of amino acid and peptide absorption are not markedly higher than those of other mammals. Mink provide an interesting comparison with omnivores and companion carnivores (cats and dogs). Mink are born altricial (at an early stage of development), much like laboratory rats and mice. However, in mink the digestive tract develops more slowly and the sensitivity to stimulation by glucocorticoids develops later (Elnif and Sangild, 1996; Sangild and Elnif, 1996).

Food Passage Time

The small length of the intestine of the mink and its physiological structure result in a high speed of passage of the food through the digestive tract that is common with predators. A review of more than eleven reports in the scientific literature indicates an average feed passage time of about 3 hours with a wide range of from 1 to 6 hours (Bernard et.al., 1942; Wood, 1956a; Howell, 1957; Neseni and Piatkokski, 1958; Sibbald, 1962; Leoschke, 1963; Anon., 1977; Hansen, N.E., 1978; Bleavins and Aulerich, 1981; Charlet-Lery et al., 1981; Szymeczko and Skrede, 1990; Enggaard Hansen et al., 1991). It is of interest to note that studies by Leoschke (1963) indicated faster feed passage with raw grains (corn, oat groats, and wheat) than with the identical products cooked for 30 minutes. Studies with both mink and polar foxes indicate that the feed passage was faster when raw fish was replaced with fishmeal. Szymeczko and Skrede, 1990; Szymeczko et al., 1992 N.E. Hansen (1978) reported that when mink rations contained sulfuric acid preserved herring, the feed passage time was increased significantly; this effect could be corrected by supplementing the diet with sodium hydroxide, but calcium hydroxide had no effect.

Studies of Wood (1956b) provide experimental data on the time of feed passage in terms of both time between feeding and first bowel movement and time between initial feeding and complete emptying of the digestive tract. The first study involved a variety of markers (including Sudan III, charcoal, and Chromium Oxide) or no markers with an average time of 1.8 hours from initial feeding to first bowel movement. Here it appeared that employment of the fecal markers extended feed passage time. The second study showed the time of total feed passage as 2.2 hours with continuous feeding, 3.1 hours for 1/2 hour feeding, and 3.3 hours for hourly feeding.

Intestinal Flora

Although mink possess a relatively short intestinal tract and a very short time of food passage from mouth to anus as compared to other monogastric animals, there is evidence of significant bacteria populations in the intestine and colon and of a significant contribution of the intestinal flora to the nutritional well being of the animal.

Williams et al. (1998) found the highest densities of all bacteria groups in the colon, up to 108 c.f.u./g. for total anerobes, densities 2-4 orders of magnitude lower than those of other mammals. In studies of mink kept in the zone of the Chenobyl NPP, Sudenko et al.(1995) found that the bacterial population of the small intestine was 1-2 orders lower than that of mink kept outside of the radioactive zone.

Although there is little microbial digestion of protein by the mink (Skrede, 1979; Szymeczko and Skrede, 1990), studies by Borsting et al. (1995) found that microbial degeneration of cell wall NSP can occur in the mink. Significant fermentation in the intestine/colon was evidenced by the presence of substantial amounts of short-chain fatty acids in the fecal material.

Indirect evidence that there is significant nutritional benefit from intestinal flora synthesis of vitamin K appears in a study by Travis et al. (1961) showing that the employment of sulfaquinoxaline, a coccidiostat, in the dietary regimen of the mink increases blood clotting time.

Studies by Leoschke (1953) and Leoschke et al. (1953) indicated that mink on vitamin B_{12} free purified diets with different carbohydrate resources excreted significantly different levels of vitamin B_{12}. With sucrose as the sole carbohydrate source, fecal levels of vitamin B_{12} were six times that of mink on purified diets containing dextrins (cooked corn starch) as the sole carbohydrate source. Growth performance of the mink kits on vitamin B_{12} free purified diets also evidenced significant intestinal flora synthesis of vitamin B12 which reached the blood circulation of the animal. Mink kits with a vitamin B_{12} deficiency on purified diets with dextrins as the carbohydrate source made surprisingly good weight responses when switched to purified diets with sucrose as the primary carbohydrate resource. An additional observation is of interest: mink kits placed on purified diets containing dextrins developed classic vitamin B_{12} deficiency signs at a shorter time period than mink kits placed on purified diets containing sucrose as the carbohydrate source.

Physiology

Feedstuff Palatability

A thorough discussion of the palatability of dehydrated mink feedstuff is reported in the chapter on Modern Mink Nutrition - Pellets. Thus the material to follow is limited to experimental observations on the palatability of fresh/frozen feedstuffs for the mink.

Experimental studies by Lyngs (1992) indicated that mink in both the early growth period (May-June) and the late growth period (July-August) preferred diets providing poultry offal to high fish diets. Unfortunately, this work did not involve an examination of the taste appeal of individual portions of poultry offal -- heads, feet, or entrails. Observations of mink ranchers throughout North America indicate that mink prefer poultry entrails, especially duck entrails, over almost all other fresh/frozen feedstuffs except liver. Studies by Leoschke (1959) indicated that adult mink males have a strong aversion to experimental diets on protein digestion wherein the sole protein source was poultry feet. Only one-half of the twenty mink protein digestion panel accepted the experimental diet of poultry feet as providing valid data on protein digestibility of this mink feedstuff.

Digestive Physiology

The growth, fur production, and reproduction/lactation performance of mink is dependent upon the availability of the nutrients in the diet to the digestive processes of the animal as well as on the nutritional content of the diet presented to the animal.

Carbohydrate Resources

Multiple studies indicate that the digestive capacity of the mink generally increases with advancing age (Skrede and Herstad, 1978; Elnif and Enggaard Hansen, 1987 and 1988; Oleenck and Svetchina, 1993). Layton (1998) observed that lactating females had a higher capacity for the digestion of carbohydrates than their kits. Russian experimental evidence indicated that the ability of the mink to digest nutrients was similar for mink of different genetic types (Semchenko, 1973). However, Leoschke (1965) provided specific data indicating that the capacity of adult male sapphire mink to digest raw corn and soybean oil meal was significantly less than that of adult pastel males. Studies by Ahlstrom and Skrede (1995) indicated that the mink's capacity for carbohydrate digestion was not affected by fat/carbohydrate ratios.

As early as 1927, Smith commented on the value of cooked cereal grains for fur animals. Since that initial report, multiple experimental studies have appeared in the scientific literature on the digestibility of carbohydrates present in raw and cooked grains and other carbohydrate sources. A few of these reports have provided data which are much lower than those reported by the majority of research workers. This apparent inconsistency in the experimental data on carbohydrate digestibility may be due to multiple factors. One factor may be the actual level of the specific carbohydrate source

28

provided, as relatively low levels may produce less accurate data. Another factor may be the specific analytical methods employed. In some cases carbohydrate digestibility is assessed by not actually measuring the level of carbohydrate provided to the mink, but simply by using N.F.E. (Nitrogen Free Extract) data obtained by the simple mathematics of N.F.E. = Feed Solids - (% Protein + % Fat + % Ash). In this procedure all analytical errors that may occur in the data on protein (Nitrogen content x 6.25), fat, and ash are compounded in the N.F.E. values achieved. In Table 3.15, Carbohydrate Digestibility Coefficients For Mink, I have employed sound judgment or "common sense" in deciding to ignore experimental data found which are simply inconsistent with the bulk of data available on the carbohydrate digestibility of specific raw, processed, or cooked carbohydrate sources.

TABLE 3.15: Carbohydrate Digestibility Coefficients for Mink

Ingredient	% Digestibility	References
Barley, Bran	40	7
Barley, Raw	60	2,5,11
Barley, Cooked	69	2,4,5,11
Corn, Raw	58	3,4,6,8
Corn, Cooked	80	6,8
Corn, Extruded	80	8
Corn Flakes, Toasted	82	82
Corn Gluten meal	60	4
Corn, Hominy By-Product	69	8
Corn Starch, Raw	58	3,6
Corn Starch, Cooked	85	3,4,6,10
Milk	100	2
Milk, Dried	98	2
Oats, Raw w/Hulls	50	4
Oat Groats	68	4,8
Oats, Steam Rolled	81	2,8
Oats, Cooked	84	8
Soybean Oil Meal, 44% Protein with hulls	49	1
Soybean Oil Meal, Cooked	57	1
Soybean Oil Meal, 50% Protein w/o hulls	58	8
Potatoes, Raw	2	6
Potatoes, Cooked	80	1,4,6
Potato Starch, Cooked	77	2
Rye Bran	40	4
Tapioca Meal, Cooked	80	4
Tapioca Starch, Raw	32	6

Tapioca Starch, Cooked	82	6
Wheat, Raw	73	6,8,9
Wheat, Cooked	79	2,4,6,8,9
Wheat Flakes, Toasted	85	8
Wheat Starch, Raw	72	6
Wheat Starch, Cooked	87	6
Wheat Bran	50	4,8
Wheat Germ	68	1,8
Wheat Middlings	67	8
Cereal Mix-Wisconsin*	74	8

1–Smith (1927); 2 –Ahman (1958, 1959); 3–Bernard et. al. (1942); 4 –Glem-Hansen (1977); 5–Jorgensen and Glem-Hansen (1973a); 6–Jorgensen and Glem-Hansen (1975c); 7–Kiiskinen and Makela (1991); 8 – Leoschke (1965); 9 –Ostergard and Mejborn (1989);10 –Roberts (1988); 11–Skrede and Herstad (1978).

* Leoschke & Rimeslatten (1954), Steamed rolled oats, 55; wheat germ meal, 20; brewers yeast, 10; soybean oil meal, 8; wheat bran,5 and alfalfa leaf meal, 2. It is of interest to note that studies by Alden (1988) indicate that the digestibility of commercial cereals in Europe was similar to the Wisconsin cereal mix with carbohydrate digestibility ratings of 70-75%.

The studies by Jorgensen and Glem-Hansen (1975c) indicated that the mink tolerated raw potato starch and raw tapioca starch quite well, while untreated maize (corn) and wheat starch gave them diarrhea. Roberts (1988) reported on studies on the digestibility of simple and complex carbohydrates by adult mink. While lactose had a low digestibility rating of only 50%, the other simple sugars including fructose, glucose, maltose, and sucrose were 100% digested. On the other hand, complex carbohydrates including cellulose, hemicellulose, lignin, pectin-gum, raffinose, and stachyose had 0% digestibility ratings. It is of interest to note that note that the significant level of stachyose in soybean oil meal may be one of the factors responsible for the low carbohydrate digestibility coefficient for soybean oil meal.

Processing of cereal grains by cooking, extruding, and toasting results in a process termed gelatinization wherein microscopic granules of amylose and amylopectin are converted to smaller polymers (dextrins) which possess higher water absorption qualities and enhanced digestibility ratings.

Other procedures which increase the carbohydrate digestibility coefficients of mink feedstuffs include fine grinding, acid addition, and enzyme supplementation.

Fine grinding (micronization) of raw barley, corn, oats, and wheat does enhance carbohydrate digestibility as much as 20% (Jorgensen and Glem-Hansen, 1973; Skrede and Herstad, 1978; Glem-Hansen and Sorensen, 1981). In their most detailed study (1981) grinding maize very finely enhanced carbohydrate digestibility to 48% with the 0.7 mm grinding and a carbohydrate coefficient of digestibility of 68% with a 0.1 mm grind. With mink feed economics in mind, the authors recommended that at least 93% of the milled maize should pass through a 0.5 mm sieve to provide extra digestibility.

Experimental studies with 0.5% glacial acetic acid (80% conc.) indicated that the viscosity of the grain mixture was improved and that the bacteria count in the grain mixture was not only decreased but kept sufficiently low for more than three days (Loftsgard et al., 1972). However, studies by Jorgensen and Glem-Hansen (1973b) indicated that the addition of acetic acid to grains prior to the cooking process did not significantly increase carbohydrate digestibility. That acetic acid lacked the capacity to provide significant hydrolyzation of the grain carbohydrates can be understood on the basis of its relatively low acidity, that is, a free acid rating of only 1% of that of sulfuric acid or 10% of phosphoric acid in 1 Normality solutions.

Treating wheat fractions with amyloetic and cell wall degrading enzymes of a commercial nature including cellulase, hemicellulase, beta-glucanase, pentosanase, pectinase, and xylanase does enhance the digestibility of the carbohydrate sources (Borsting and Damgaard,1991; Borsting et al., 1995), the process being even more effective than boiling of the grain fractions.

Finally, experimental studies have indicated that a good estimation of the digestibility of carbohydrates by the mink can be provided via two analyses of the feed sample—determination of alpha-link glucose with and without previous autoclaving of the feedstuff (Glem-Hansen et al., 1977).

Fat Resources

As stated earlier multiple studies indicate that the digestive capacity of the mink generally increases with advancing age. In terms of a specific fat hydrolase, Layton (1998a, 1998b) noted that gastric lipase activity increased between three and nine weeks. Studies by Hejelsen (2002, 2003) noted that the fat digestibility capacity of mink kits increased linearly between seven and nine weeks, with adult level fat digestive capacity reached by eleven weeks. These observations are somewhat different from those of Skrede (1978c) and Elnif and Hansen (1987) who noted only slightly lower fat digestive capacity of mink kits at seven weeks of age relative to adults. Ahlstrom and Skrede (1995) observed that fat digestibility in mink declined as fat/carbohydrate ratios decreased.

A summary of current experimental data on the digestibility of fats for the mink is presented in Table 3.16.

Multiple factors influence the digestibility of fats by the mink including the proportion of saturated and unsaturated fatty acids present, the relative degree of oxidative rancidity, and the presence of calcium salts which can undermine the capacity of the mink's intestinal tract to absorb specific long chain fatty acids.

Studies by Austreng et al. (1979) confirm certain general facts about the digestibility of fats by animals. The digestibility of individual fatty acids decreased with increasing chain length up to C-18. Unsaturated fatty acids showed higher digestibility ratings than their saturated counterparts. Fatty acid composition, rather than melting point, may be the main factor governing fat digestibility.

Studies by Jorgensen and Glem-Hansen (1973) indicate that the digestibility of fat sources can be estimated on the basis of their content of a specific saturated fatty acid, stearic acid. The higher the level of stearic acid in the dietary fat source, the lower the digestibility of the specific fat. The only exception to this general observation is noted in studies by Austreng et al. (1979), wherein the estimation of total lipid digestibility in mink from the content of stearic acid (C-18-0) in dietary fat resulted in considerable discrepancy with their experimental data. Since this disagreement was particularly severe with hydrogenated capelin oil, it would appear that the stearic acid method should not be used for hydrogenated marine fats.

TABLE 3.16: Fat Digestibility Coefficients for Mink*

Ingredient	Digestibility	References
Fresh/Frozen Feeds		
Beef Meat	81	10
Beef Liver	91	10
Beef Lungs	91	14
Beef Spleen	91	10
Beef Tallow, Raw	68	3,9,12
Beef Tripe (Rumen)	89	2,14,15
Chicken By-Product 50-25-25 Entrails-Feet-Heads	94	10,14
Fish, Racks	94	1,10
Fish, Whole	96	2, 10,16
Hens, Spent	91	21
Horsemeat	93	14
Milk Products	90	10
Pork Offal	85	10
Processed Animal/Fish		
Fish meals	92	2,10
Meat meals	82	2,11
Poultry meal w/Feathers	79	4,10,13
Silkworm Pupa Meal	88	20
Rendered Animal/Fish		
Fish Oils	95	3,6,7,8, 10,14,19
Beef Fat (Tallow) U.S.** Titer42.4C.	93	14
European	73	3,9,12
Pork Fat (Lard) U.S.** Titer 37.2C.	93	14
European	85	3,12,19
Hydrogenated Fish Oil		
Capelin Oil	94	6
Hydrogenated - 21° C.	91	6

Hydrogenated- 33° C.	84	6
Hydrogenated- 41° C.	67	6
Plant Oils		
Lecithin	91	3,12
Rape Seed Oil	95	18
Soybean Oil	95	3,10,17

* Apparent Digestibility values. An estimate of "true" digestibility values taking into account metabolic fat excretion of the mink may be found by adding 2.7% units (Austreng et. al., 1979).

**It is generally acknowledged that beef and pork fat marketed in the United States may have a significantly higher level of polyunsaturated fatty acids relative to similar products available in Europe due to the employment of higher levels of corn in the animal's diet in the U.S. relative to a greater emphasis on barley in Europe. This basic point may be one of the factors responsible for the higher digestibility values noted.

(1) Ahman (1958); (2) Ahman (1959); (3) Ashman (1974); (4) Alden (1987a); (5) Alden (1987b); (6) Austrang, et. al. (1979); (7) Borsting (94); (8) Enggaard Hansen and Glem-Hansen (1980); (9) Glem-Hansen (1970); (10) Glem-Hansen (1977); (11) Glem-Hansen (1979); (12) Glem-Hansen and Jorgensen (1977); (13) Kiiskinen and Makela (1991); (14) Leoschke (1959); (15) Nordfelt et. al. (1954), (16) Roberts (1964); (17) Rouvinen (1990); (18) Rouvinen (1991); (19) Skrede (1984); (20) Steger and Piatkowski (1959); (21) White et al. (2002).

Studies by Borsting et.al. (1994) and Borsting and Engberg (1995) indicated that the level of oxidative rancidity of a mink fat source affects the digestibility significantly. See Table 3.17.

TABLE 3.17. Oxidative Rancidity of Fish Oil and Digestibility Coefficients for the Mink

Fat Resource	Peroxide Number	% Digestibility
	m.e. O_2	
Fresh Oil	0	96
Oxidized	200	92
Oxidized	400	76

It was noted that during the storage of the oxidized fish oils there was a considerable loss of n-3 polyunsaturated fatty acids— especially in the heavily oxidized oil (400 m.e.)—an observation that would explain, in part, the loss of fat digestibility for the mink.

In early studies, Ahman (1962) noted that bone meal and a variety of calcium compounds reduce the fat digestibility capacity of the mink. Later studies confirmed these early experiments and indicated that the level and type of calcium salts in the animal's diet had a profound effect on fat digestibility for the mink. Limestone (calcium carbonate) was the most effective in reducing fat digestibility with a dietary level of 1% reducing fat digestibility by 10%, Ahman (1974) This unique role of limestone in reducing the digestibility of fats by mink was confirmed by Rouvinen and Kiiskinen (1991). The studies by Ahman also indicated that calcium resources for modern mink nutrition such as dicalcium phosphate and tricalcium phosphate and even meat and bone meal were less effective than limestone in reducing fat digestibility.

When sulfuric acid is used for fish preservation, calcium hydroxide is often added to reduce the degree of acidity, that is, to raise the pH level to the point of practical application in mink dietary programs. Studies by Enggaard and Glem-Hansen (1980) indicated that this procedure may reduce the digestibility of crude fat by as much as 6%.

Studies by Rouvinen and Kiiskinen (1990, 1991) show that in addition to limestone, ash provided by fish offal meal and meat and bone meal significantly reduces digestibility. With the ash level rising from 4 to 14% of dry matter, the digestibility of beef tallow was reduced from 82 to 66% and rapeseed oil from 95 to 90%.

The physiological factors responsible for the reduction of fat digestibility for an animal with rising ash levels or the addition of calcium salts is well known. In the intestinal tract fats undergo a process termed saponification wherein the fatty acids are separated from their triglyceride esters to yield free fatty acids which in turn react with the calcium cations present to yield soaps similar to common household soap, a mixture of sodium salts of free fatty acids provided by a variety of sources. One of the most indigestible soaps is calcium stearate formed from stearic acid, an 18 carbon saturated fatty acid. Salts of unsaturated fatty acid, especially the polyunsaturated fatty acids, are more likely to be absorbed from the intestine into the blood circulation of an animal. This fact provides the physiological basis for the fact that in the Rouvinen studies, the high ash diets had a minimal effect on the digestibility of the fatty acids in rapeseed oil.

Protein Resources

Studies by Allen et al. (1964) indicated that by 11 weeks of age, both male and female mink kits have reached full protein digestive capacity. Experiments by Skrede (1978) and Elnif and Enggaard Hansen (1987) indicated that the process of maturing does not result in large differences in digestive capacity of the mink in the interval of 7 to 38 weeks wherein the values of N digestibility increase only slightly with advancing age regardless of the source of the dietary protein. Mink have evolved to provide the kits with maximum capacity to digest protein feedstuffs at an early age. More recent studies by Hejlesen (2002 and 2003) indicate that the protein digestion capacity of mink kits increases linearly from seven to nine weeks with high digestibility protein feedstuffs (85% for adult mink); but with inferior protein resources such as meat-and-bone meal (65% protein digestibility for adult mink), the mink kit's capacity for protein digestion did not increase from seven to nine weeks.

At the same time, it must be admitted that in terms of a relatively low intestine/body length ratio and a very short time of feed passage from mouth to anus, the capacity of the mink to digest protein feedstuffs is relatively limited. This point is well illustrated in Table 3.19: it is obvious that the 5 week old chick has a higher capacity for the protein digestibility challenge of meat-and-bone meal than a 9 month old mink (Skrede et al., 1980). Obviously, all factors considered, a chick assay for the nutritional value of dehydrated protein feedstuffs is not advised since it is not a valid assessment of the biological value of a given dehydrated protein feedstuff resource for modern mink nutrition.

Protein Digestibility

A detailed survey on the digestibility of protein feedstuffs by the mink is provided in Table 3.18: it is of interest to note that studies by Skrede (1978a, 1978c) indicate that the employment of 6.25 as a conversion factor from nitrogen content to protein content is likely to overestimate the protein content of fish by-products containing a high proportion of skin and bones.

TABLE 3.18. Protein Digestibility Coefficients

Ingredients	%Digestibility	References
Fresh/Frozen Feedstuffs		
Beef Gullets	80	2
Beef Intestines	89	9
Beef Liver	89	7
Beef Lungs	80	7, 14
Beef Meat	87	9
Beef Rumen (Tripe)	85	2, 14, 16, 21
Beef Spleen	86	7, 15
Cheese	96	9
Cottage Cheese	93	10
Chicken, Day Old	68	7
Chicken Eggs	90	9, 10
Chicken Entrails	87	7, 14
Chicken Feet	56	7, 14
Chicken Heads	77	7, 14
Chicken Necks	82	7
Chicken Offal w/o Feathers	80	9
Chicken Offal w/ Feathers	50	9
Fish Backs	83	18
Fish Fillets	96	18
Fish Heads	83	9
Fish Intestines	94	9
Fish Racks	80	1, 2, 20
Fish Skin	95	18

Fish, Whole	90	1, 2, 14, 17, 20, 21
Horse Liver	93	15
Horsemeat, Boneless	92	14, 21
Milk	94	1
Pork Backbones	61	9
Pork Heads	56	9
Pork Meat	87	9
Pork Ears	87	7
Pork Feet	81	1
Whale Meat	92	21
Dehydrated Animal/Fish		
Bloodmeal	90	8
Feather meals	18	12
Feather meals, Acid Hydrolyzed	68	12
Fish meals, High Ash	80	9
Fish meals, Whole	83	1, 2, 4, 11, 12, 16, 20
Fish Solubles	77	10
Meat meals - 10% Ash	80	9
Meat meals - 20-25% Ash	71	9
Meat meals - 30% Ash	60	9
Poultry meals w/o Feathers	74	10, 12
Poultry meals w/ Feathers	58	5, 6, 7, 11
Silkworm Pupa Meal	91	19
Skim milk Powder	92	9
Whale meat Meal	91	3
Plant Proteins		
Barley	75	2, 12
Corn Germ Meal	57	8
Corn Gluten Meal	86	9
Corn, Whole	70	7
Oat Groats	77	9
Potato Protein	88	8
Rye Bran	55	7
Soybeans, Raw	62	2
Soybeans, Cooked	67	1,7
Soybean Oil Meal	80	9, 10, 11
Soybean Protein Conc.	92	9

Wheat Bran	65	9, 16	
Wheat Germ	74	1	
Wheat Gluten Meal	92	11, 13	
Wheat Middlings	60	1, 2	
Wheat, Whole	70	2, 7	

* Apparent Digestibility Values without accounting for endogenous nitrogen excretion. Studies by Kiiskinen et al., (1985) indicate that the True Digestibility Values for the mink in terms of dehydrated protein resources could be obtained with an added 6% factor.

(1) Ahman (1958); (2) Ahman (1959, 1962); (4) Ahlstrom et al. (2000); (5) -- Alden (1987a); (6) Berg (1987); (7) Glem-Hansen (1977); (8) Glem-Hansen (1979); (9) Glem-Hansen and Mejborn (1977); (10) Glem-Hansen and Eggum (1974); (11) Kiiskinen et al.(1985); (12) Kiiskinen and Makela (1991); (13) Lau et al. (1997); (14) Leoschke (1959a); (15) Loosli and Maynard; (16) Nordfeldt et al. (1954); (17) Roberts and Kirk (1964); (18) Skrede (1978); (19) Steger and Piatkowski (1959); (220) Szymeczko and Skrede (1990); (21) Wood (1958, 1964).

It should be noted that there is an inverse relationship between the level of ash provided in a mink feedstuff and its degree of protein digestibility, as well illustrated by the data provided on the protein digestibility of poultry entrails, heads, and feet and the lower protein digestibility of the higher ash meat meals. Studies by Skrede (1978) indicated a 0.6% drop in N digestibility for each 1% rise in the level of ash in a mink protein feedstuff. Experimental observations of Makela and Polonen (1978) indicated that limestone grist added for neutralizing acid–preserved slaughterhouse offal actually lowers the protein digestibility capacity of the mink.

It is of interest to note that an in vitro program for the estimation of protein digestibility of mink protein feedstuffs has been developed (Laerke et al., 2003). The process provides a good correlation for most mink feedstuffs but yields a wider variation with specific dehydrated protein feedstuffs including blood meals and wheat gluten.

TABLE 3.19. Protein Digestibility Capacity of Mink vs. Chick or Fox

Amino Acid	Meat and Bonemeal-%		
	Chick	Mink	Fox
	5 Weeks	9 Months	15 Months
Nitrogen	ND	60	80
Essential Amino Acids			
Arginine	93	78	89
Histidine	74	53	88
Isoleucine	76	63	74
Leucine	77	66	75
Lysine	77	65	78
Methionine	68	64	71
Cystine	ND	26	ND
Phenylalanine	76	73	81

Amino Acid			
Tyrosine	74	66	81
Threonine	63	57	78
Tryptophane	53	38	61
Valine	76	63	74
Amino Acid	**Meat and Bonemeal-%**		
	Chick	**Mink**	**Fox**
	5 Weeks	**9 Months**	**15 Months**
Non-Essential Amino Acids			
Alanine	82	67	81
Aspartic Acid	64	25	70
Glutamic Acid	74	55	74
Glycine	77	63	83
HO-Proline	85	70	87
Proline	83	67	83
Serine	77	64	78

Amino Acid Digestibility Coefficients

For the most part, the ability of the mink to acquire amino acids from the multiple proteins present in mink feedstuffs is similar to the digestibility coefficient of that specific product. The primary exception to this general observation would be the limited capacity of the mink to obtain cystine for its blood circulation. This point is well illustrated in Table 3.19 and in studies reported by Glem-Hansen and Mejborn (1985), Skrede (1981), and Ahlstrom et al. (2000). Also, as the ash level rises in a dehydrated protein source, there is a corresponding difference between the protein digestibility of the product and the availability, of cystine to the digestive processes. Borsting (1997) has shown that when crystalline amino acids, in comparison to those in peptide bonds, are employed in mink diets, the digestibility of cystine, threonine, and tryptophan was significantly lower.

Heat Processing—Protein Resources

Heating a protein feedstuff brings about denaturation of the proteins present thereby increasing or decreasing the availability of the amino acids to the digestive processes. A number of early studies –Hodson and Maynard, 1938; Loosli and Maynard, 1939; Bernard and Smith, 1942; Jarl, 1952– noted that the drying and cooking of meat seemed to decrease the protein digestibility coefficients.

Heating With Water or Steam

In terms of certain specific mink feedstuffs, boiling of the protein product enhances the value of the protein source as seen in the following examples.

Whole Eggs

Boiling eggs for at least 5 minutes at 91 degrees Celsius or 196 degrees Fahrenheit has a two fold value for the mink: denaturation of the egg protein avidin which can bind the vitamin biotin in an indigestible linkage and denaturation of the egg protein conalbumin which has the potential to bind iron in a structure unavailable to the digestive processes of the mink (Leoschke, 1978).

Raw Soybean Oil Meal

The heat processing of raw soybean oil meal is essential for the denaturation of a trypsin (a protein digesting enzyme) inhibitor which undercuts the capacity of animal digestive physiology to function in an optimum manner (Skrede and Krogdal, 1985).

Raw Fish Bones

Skrede (1978) found that the cooking of fish backbones altered the structure in such a way as to improve the digestibility of the skeletal proteins.

Aneamogenic Fish

Raw coal fish (*Gadus virens*), Pacific Hake (*Merluccius productus*) and Ocean Whiting (*Gadus merlancus*) contain dimethylamines and trimethylamines and/or formaldehyde. These substances bind and/or convert iron to a structure unavailable to the digestive processes, the result yielding a condition known as "Cotton Mink." Simple cooking of the fish destroys the factors responsible for the anemia (Ender et al.,1972; Costley, 1970).

Thiaminase Fish

Certain species of fish contain a thiaminase enzyme which destroys any thiamine that may be present in the dietary mixture presented to the mink. Cooking denatures the enzyme and thus resolves the problem (Greig and Gnaedinger, 1971).

Heating Without Water—Fishmeal Production

The biological value of a given protein feedstuff for the mink is directly related to two key factors:

1—amino acid pattern as compared to the actual amino acid requirements of the animal for that specific life phase—growth, fur production, and reproduction/lactation and

2—the availability of the constituent amino acids to the digestive processes of the mink.

Without question the dehydration processes required to convert fish products to fish meals may result in a significant loss of biological value for animals via destruction of specific amino acids such as cystine, lysine, methionine, and tryptophan and bonding of other amino acids such as arginine in structures which are unavailable to the digestive processes of animals. The biochemical basis for this potential loss of protein biological value during dehydration procedures has been known since the beginning of the 20th century. In the Maillard Reaction under conditions of heat, and in a semi-arid environment, carbohydrate carbonyl groups react with free amino structures of proteins (Allison, 1949; Varnish and Carpenter, 1975).

Experimental studies involving more than 400 dehydrated protein resources at the National Research Ranch over the period of the last half-century on the biological value of such resources for modern mink nutrition indicated that a large majority of the fishmeal products on the world market are inferior in that they provide only sub-optimum growth, fur production, and reproduction/lactation performance of the mink. In other words, "All Fishmeals Are Not Created Equal. " Obviously, this general statement on the quality of fish meals on the market applies to other dehydrated protein sources for mink nutrition including blood meals, meat-and-bone meals, lamb meals, poultry meals, and poultry by-product meals.

Let me add a personal note here. The reason that I am so sensitive on the topic of the loss of protein biological value with the dehydration procedures required to convert animal, poultry, and fish products to high protein meals is directly related to my graduate school and post-doctoral studies at the University of Wisconsin during the years 1950-1959. In that period I spent a very significant part of six years of my life trying to discover the nature of the liver residue factor required by mink kits placed on purified diets. In the end I found that the liver residue factor deficiency was not that of an undiscovered vitamin but an actual amino acid deficiency brought about, in part, by maltreatment of the protein source casein (dried cottage cheese) via excessive heat treatment to the point of significant loss of protein biological value.

Studies initiated at the University of Wisconsin on the fundamental nutritional requirements of the mink in the 1940s involved placing mink kits on purified diets providing all the known vitamins but no unknown vitamins that might be present in common mink feedstuffs employed on ranches. A "vitamin free" dietary program was achieved by using vitamin-test casein (dried cottage cheese from which all vitamin content had been removed via alcohol extraction).

By 1950 these studies indicated that for the survival of mink on the purified diets, four "unknown" factors were required -- the liver extract factor, the liver residue factor, and unknown factors present in hog mucosa (lining of the intestine) and lard. By 1952 the liver extract factor proved to be vitamin B_{12} and studies on the unknown liver residue factor continued until the nature of this factor was finally resolved by the fall of 1956.

The initial breakthrough in resolving the nature of the liver residue factor occurred in the spring of

40

1955 with a $2,000 shipment of vitamin-test casein. The earlier shipments of the vitamin-test casein had been cream colored while this product was a definite "dark-tan," indicating to me that the product had been "over-heated" in the process of removing the solvent from the alcohol extraction procedures required to yield vitamin-test casein. I immediately knew that something was wrong, that this was an inferior protein resource with the distinct possibility of undercutting the performance of the mink kits in the summer/fall of 1955. I immediately expressed my concerns to the Biochemistry Department secretary, who simply said "Take it or leave it." In other words, the product would not be shipped back to Ohio and I would employ this sub-optimum product in my research studies or cease the operations.

Thank God that my philosophy on mink nutrition research for the past half-century has been simple: "Listen To The Mink." The immediate response of the mink kits placed on the purified diets in early July, 1955, was predictable. Kit losses occurred 3-4 weeks earlier than in the previous 5 years of my experimental studies. Thus I slowly began to realize that the unknown liver residue factor required by the mink on purified diets was not an undiscovered new vitamin but a mink dietary deficiency brought about by sub-optimum delivery of amino acids to the mink kits for the critical fur development phase wherein the animals experienced enhanced requirements for the sulfur amino acids (methionine and cystine) and arginine. The problem with the vitamin-test casein provided in the purified diets as the sole protein resource for the mink was two-fold: relatively low levels of methionine, cystine, and arginine in the casein protein resource for the mink and excessive heat treatment of the casein leading to a loss of sulfur amino acids and the bonding of arginine in a structure unavailable to the digestive processes of the mink.

The problem of the "unknown" liver residue factor was resolved in the fall of 1956 with supplementation of the purified diet with methionine and arginine (Leoschke and Elvehjem, 1959). However, in terms of modern mink nutrition management in years beyond 2000, the basic problem of sub-optimum amino acid nutrition with the employment of second rate protein resources can not be easily resolved with the employment of special amino acid supplementation programs, although DL-methionine is readily available on the market, DL-arginine is simply not available to the world mink industry at a reasonable price. Thus, all factors considered, mink ranchers throughout the world employing dehydrated protein resources such as blood meals, fishmeals, meat-and-bone meals, poultry meals, and poultry by-product meals must be concerned about the biological value of the proteins provided in these dehydrated protein resources.

The reduction of the biological value of proteins present in fish in the dehydration procedures required to yield fish meals is related to multiple factors, including the temperatures employed and the final level of moisture achieved.

41

Temperatures Employed

Studies by Hjertnes et al. (1988) showed that in the preparation of NorSeaMink Herring meal, raising the drying temperature from 60° Celsius to 140° Celsius reduced the true protein digestibility from 92 to 85%. Studies by Ljokjel and Skrede (2000) involved a double screw extruder (Buhler EX-50/134 90 kW) and temperatures of 100°, 125°, and 150° C. True digestibility of total protein and amino acid nitrogen was significantly reduced by extrusion as compared to the unextruded sample. Digestibility of cysteine, asparagine, threonine, serine, proline, glycine, alanine, histidine, lysine, and arginine was significantly decreased at 100°C. The digestibility of all individual amino acids was significantly decreased after extrusion at 125-150°C.

Moisture Content < 6.0%

All factors considered, final moisture levels achieved in the fish meal product may be much more important than temperature levels employed, inasmuch as the Maillard Reaction (which results in the destruction of cystine, lysine, methionine, and tryptophane and the bonding of arginine in an indigestible structure) requires a semi-arid environment. The minimum moisture content of 6.0% standard is seen in Table 3.20.

TABLE 3.20. Fishmeal Standards*

Chemical Analyses	Norway	Denmark	Soviet Union
Crude Protein - min. %	70	73	60
Crude Fat - max. %	10	10	10
Water - %	6-10	6-8	8-12
Ash - max. %	16	13	22
Ammonia Nitrogen - max. %	0.2	—	0.2
Salt, NaCl - max. %	3	—	3
DiMethyl Nitrosamine	Not Detectable		
Total Volatile Nitrogen - max. mg/100 g	—	120	—
Free Fatty Acids - max. % Crude Fat	—	10	—

* Joergensen (1984).

The importance of the concept that higher quality fish meals have minimum and maximum water levels is seen in fish meal specifications included with LT 999 fishmeal produced by Esbjerg Fiskeindustri, Denmark, which states "Moisture, min. 6% and max. 10%." This product is employed in the Scandinavian aquaculture industry wherein a high standard of protein quality in dehydrated protein ingredients is required for top performance. It is of interest that the label also states, "True Protein Digestibility -- Min. 90% -- determined with mink" (Esbjerg, 1998). This use of mink to determine the digestibility of fish meals for employment in the aquaculture industry is related to studies by Skrede et al. (1980) which indicate that the digestive systems of mink and rainbow trout are most sensitive as to protein resources of low digestibility, the mink being a useful tool for determining the protein quality of fish meals in terms of a short digestive tract and short food passage time leading to minimum capacity for the digestion of protein resources.

Extruded Pellets—Protein/Amino Acid Digestibility

Observations on the reproduction/lactation and fur development performance of mink fed commercial extruded pellets in the decades of the 1970s and 1980s in North America indicated that, unfortunately, a high proportion of the pellets marketed provide sub-optimum nutrition of the mink. In all cases of extruded pellet nutrition failure, there was a consistent pattern of pellets with moisture levels < 6.0%, very likely indicating a significant Maillard Reaction. This point is well illustrated in Table 3.21.

TABLE 3.21: All Mink Pellets Are Not Created Equal (Skrede, 1980)
Protein and Amino Acid Digestibility Ratings-%

Dietary Program	Pellet #1	Pellet #2
Amino Acid Composition		
Proteins	80	74
Essential Amino Acids		
Arginine	91	85
Histidine	87	72
Isoleucine	88	80
Leucine	90	80
Lysine	87	71
Methionine	92	84
Cystine*	70	52
Phenylalanine	90	83
Tyrosine*	89	82
Threonine	84	81
Tryptophane	83	67
Valine	88	83
Non-Essential Amino Acids		
Alanine	88	83
Aspartic Acid	75	66
Glutamic Acid	86	73
Glycine	79	73
Hydroxy-Proline	74	70
Proline	84	81
Serine	82	79

* Non-essential amino acids which "spare" the requirement of the adjacent essential amino acid; that is, they reduce the actual requirement for the essential amino acid.

It is obvious that the extruded pellet #2 has undergone significant Maillard Reaction with resultant bonding of amino acids in indigestible structures. Amino acid analyses of both pellets indicate minimal destruction of cystine, lysine, methionine, and tryptophane by the Maillard Reaction but maximal bonding of multiple amino acids in structures not available to the digestive processes of the mink.

Verification of the obviously lower biological value of the proteins provided in pellet #2 is seen in the relatively poor growth performance of mink kits placed on pellet #2 as a nutritional resource.

What is fascinating about the top performance of mink kits placed on pellet #1 and the inferior performance of the mink kits placed on pellet #2 is that the mink nutrition expertise behind both pellet formulations involves the same world recognized mink nutrition biochemist, Dr. Willard Roberts, Ph.D., University of Wisconsin. The sharp difference in the biological value of the proteins present in the two extruded pellets is directly related to the nature of the mink protein feedstuffs employed in the extruder processing. In the case of pellet #1, a high proportion of the protein ingredients were raw poultry offal while in the case of pellet #2, all of the protein resources were dehydrated protein feedstuffs including fish meals and poultry meals. This observation is very important for mink nutrition scientists and mink ranchers in understanding the major significance of the Maillard Reaction in the production of the modern mink nutrition resource pellets.

In the case of pellet #1, the protein feedstuffs were subjected to the destructive effects of the Maillard Reaction which requires the combination of heat and a semi-arid environment only once in the extruder processing. However, in the case of pellet #2, the protein feedstuffs were subjected to the debilitating effects of the Maillard Reaction twice—first in the dehydration processes required to convert fish, animal, and poultry products into high protein meals and then in the extruder processing which again involves heat and a semi-arid environment. The net result of processing of common raw mink protein feedstuffs in two sequences of heat and a semi-arid environment is obvious: a significantly higher proportion of the amino acid present being bonded in structures unavailable to the digestive processes of the mink.

Additional support for the basic concept that the Maillard Reaction provides mink pellets with protein resources of relatively low biological value for the mink is seen in Table 3.22.

TABLE3.22: Amino Acid Composition of Furring Pellets

Dietary Program	Steam Pellet	Extruded Pellets
	#1	#2
% Moisture	10.6	4.6
Amino Acid Composition		
Essential Amino Acids		
Arginine	5.4	5.1
Histidine	2.4	2.4
Isoleucine	2.9	2.9
Leucine	7.8	7.2
Lysine	6.4	6.2
Methionine	2.3	1.9
Cystine*	1.3	1.1
Phenylalanine	4.1	3.9
Tyrosine*	2.6	2.6
Threonine	4.1	3.9
Tryptophane	0.9	0.9
Valine	4.4	4.0
Total Essential Amino Acids	44.6	42.1**
Non-Essential Amino Acids		
Alanine	6.3	6.3
Aspartic Acid	9.1	8.8
Glutamic Acid	13.6	14.1
Glycine	7.2	7.4
Proline	5.7	5.7
Serine	5.2	4.5
Total Non-Essential Amino Acids	47.1	46

* Non-essential amino acids which "spare" the requirement of the adjacent essential-amino acid, that is, reduce the actual requirement for the essential amino acid.

** Level of essential amino acids in this extruded pellet is 94% of that provided by the steamed pellet.

45

In terms of the amino acid composition presented, it is obvious that the amino acid pattern of the two pellet formulations is similar. Too, the quantity of protein provided is more than ample, providing more than 30% of metabolic energy as protein, the recommended level of protein for the critical fur development phase of the mink ranch year. However, the performance of the mink kits placed on the two pellet programs was dramatically different, the #2 pellet providing both inferior growth performance and fur development. It is obvious that this specific extruder engineering provided a combination of heat and an excessively semi-arid environment to the point of significant Maillard Reaction as evidenced by relatively small destruction of cystine, lysine, and methionine but significant bonding of the amino acids present in structures simply indigestible by the mink.

Both Tables 3.21 and 3.22 wonderfully illustrate the basic concept of modern nutrition science that animal protein nutritional deficiencies can be divided into two specific divisions:

1. **Primary Protein Nutritional Deficiency** in which the dietary program creates an amino acid imbalance in protein resources that does not meet the actual amino acid requirements of the animal for top performance, and

2. **Secondary Protein Nutritional Deficiency** in which the dietary program provides ample levels of protein/amino acids, but the dietary environment is inadequate, making the amino acids present unavailable to the digestive processes of the animal.

In summary, the protein requirements of the mink for top performance in all phases of the ranch year are not simple. Ranchers planning their mink dietary programs must consider protein levels not only in terms of quantity but also in the quality of their biological value.

This basic recommendation is especially important with the employment of dehydrated protein resources, a reminder well illustrated by the expression "All Fish meals Are Not Created Equal." This lesson has been learned over a period of 50 years in studies at the National Fur Foods Research Ranch... enough said.

Manure Composition

Daily fecal excretion of the mink varies from 14 to 56 grams/day (AALAS, 1972). Annual fecal excretion per an adult female mink is about 42 kg (Aarstrand and Skrede, 1993; Anthonisen and Loehr, 1974). Data on the composition of fecal dry matter is presented in Table 3.23.

Table 3.23 Composition of Mink Fecal Dry Matter

Parameter	Howell (1976)	Aulerich et al. (1999)
Protein	31	22
Non-Protein Nitrogen	—	5
Fat	4.5	4.5
Fiber	8	12
Carbohydrates (NFE)	32	39
Ash	24	18
Calcium	7.6	4.0
Phosphorus	3.2	2.3

Fecal water content is normally in the range of 60-65% (Ericksson et al., 1983). Obviously, the percentage of ash and fiber in the ranch diet will have a significant effect on fecal composition. A study by Newell (1999) indicated that mink manure was found to contain relatively high levels of nitrogen, 5%; phosphorus, 2.5%; and potassium, 0.5%.

Rouvinen (1996) and Newell (1999) have estimated that 15-20% of all nitrogen excreted by the mink is via the feces. The Newell observations indicated that nitrogen excretion via manure was about 3.3 - 5.1 g/day for a juvenile male and 2.4-4.1 g/day for a juvenile female mink during July-October. According to Newell (2001) the daily nitrogen excretion is estimated at 3.9 g/day for growing mink fed a diet containing 39% protein on a dry matter basis. This figure yields a total nitrogen excretion per ranch year of 1.10 kg/pelt produced, accounting for the nitrogen excretion by both the growing mink and the breeder animals.

For detailed experimental data on characteristics and volume of mink fecal and urine waste in terms of environmental concerns, more specific data my be found in Travis et al. (1978) and Aulerich et al. (1999).

In terms of phosphorus loss via the feces, the Newell data indicated that about 90% of phosphorus is excreted via the manure, representing approximately 0.5 g/day for juvenile males and 0.4 g/day for juvenile females during July-October. These data on phosphorus losses via the feces are not consistent with earlier studies by Leoschke (1960, 1962b) wherein the experimental data indicated that about 65-77% of the phosphorus is excreted via the feces. Obviously, phosphate absorption from skeletal structures or from muscle meats will vary significantly, providing a logical basis for the different experimental observations. Studies by Bursian et al. (2002) indicated that phytase enzyme addition to mink diets did not reduce the phosphate excretion of the mink via the feces.

According to Newell (2001), the daily phosphorus excretion for a growing mink is estimated to be 0.5 g/day with growing mink fed a diet containing 1.2% phosphorus on a dry matter basis. This figure yields an annual excretion of 0.15 kg of phosphorus per pelt produced and accounts for the phosphorus excreted by the growing mink as well as the breeder animals.

Studies on the employment of Yucca Shidigera extracts (containing saponins and urease inhibitors) in mink diets which should reduce the hydrolysis of urea to free ammonia did not significantly reduce ammonia liberation from mink manure (Rouvinen et al., 1996; Korhonen and Niemala, 1999). The Rouvinen study also noted that the employment of Sphagnum peat moss in mink ranch diets did not minimize ammonia liberation from mink fecal material.

Kidney Function

The kidneys are the most critical organs for the excretion of metabolic waste biochemicals including ammonia, primarily as urea, and creatinine (Leoschke, 1984; Wamberg and Tauson, 1998) as well as minerals and toxic substances. With mink ranch diets providing relatively high levels of protein to meet the mink's amino acid requirements, the excess intake of amino acids is catabolized to yield glucose (glucogenic amino acids) or acetoacetate (ketogenic amino acids) and free ammonia which is converted to urea by the liver.

Urine Composition

A poetic approach to this topic is provided by one of my former senior research students at Valparaiso University, Elizabeth Zawadke (1988), a talented and enthusiastic mink kidney physiology student who in 1992 obtained her Ph.D. in biochemistry at the Massachusetts Institute of Technology. She wrote the following poem:

> The wind is very cold
> The smell of mink urine
> penetrates the air
> I am happy once again

A most detailed review of the composition of mink urine is provided in the "Handbook of Biological Data For Mink" (Aulerich et al., 1999). My observations on the composition of mink urine will be limited to those facets of mink urine that have a direct bearing on a mink disease of nutritional origin, urinary calculi or struvite stones composed of magnesium ammonium phosphate hexahydrate, a salt that tends to crystallize in a slightly acidic or alkaline urine but which is highly soluble at a pH of 6.6 or less.

The earliest report in the scientific literature on the pH of mink urine was that of Kubin and Mason (1948) whose observations indicated a pH range of 6.8 to 7.5. Studies by Leoschke and Elvehjem (1954) indicated a mink urinary pH range of 6.4 to 6.8. In another study by Leoschke (1954), the observed mink urinary pH range was from 6.0 to 6.8. In this same study with a dietary program providing commercial mink cereal, 20%; beef liver, 10%; whole ocean whiting, 30%; and boneless horsemeat, 40%, the average urinary pH was 6.1 while the addition of 2% dicalcium phosphate lowered the pH to 6.0 and the addition of 0.5% calcium carbonate or 1.5% calcium carbonate raised the urinary pH to 6.9 and 7.8 respectively.

Experimental data on mink urinary excretion of phosphorus, magnesium, and calcium are presented in Table 3.24.

TABLE 3.24. Urinary Excretion of Phosphorus, Magnesium, and Calcium by the Mink

Mineral	% of Dietary Inake		
	UW*	VU**	
	Average	Average	Range
Phosphorus	23	35	30 - 46
Magnesium	10	13	9 - 18
Calcium	0.9	1.6	1.0 - 2.4

* University of Wisconsin (Leoschke, 1960). Dietary program of fortified cereal, 20; beef liver, 10; whole ocean whiting, 30; and boneless horsemeat, 40%.

** Valparaiso University, Mink Metabolism Laboratory (Leoschke, 1962). Dietary program of fortified cereal, 20; beef liver, 10; and old hen heads and entrails, 70%.

The University of Wisconsin report noted that with a ranch basal diet providing oatmeal, 10; beef liver, 10; whole ocean whiting, 30; and boneless horsemeat, 50, the daily urinary phosphorus was 81 mg/24 hours which was increased to 194 mg/24 hours with the addition of 2% dicalcium phosphate. However, urinary excretion of phosphorus was increased to only 122 mg/24 hours with an equivalent phosphorus intake via raw calf bone. Additional experimental data indicated that as boneless horsemeat was replaced with 10, 20 and 30% raw calf bone, the urinary phosphorus excretion decreased 131 to 113 to 108 mg/24 hours, indicating that the digestive capacity of the mink to digest raw calf bone is relatively limited and implying that the capacity of the mink to digest bone meal would likely be sub-optimum, all factors considered.

The relatively high percentage of phosphorus excretion via the kidneys is in sharp contrast to the studies of Newell (1999) indicating only 10% of phosphorus excretion via the kidneys. Obviously, significant differences in mink feeding regimens were involved in these studies. The experimental work from Nova Scotia would have included a relatively high level of fish bone phosphorus while the University of Wisconsin studies involved a high level of boneless horsemeat and the Val-

49

paraiso University studies involved a high proportion of poultry entrails combined with a small proportion of poultry heads with relatively low ash content.

In terms of environmental pollution via mink ranch operations, it is of interest to note that studies reported by Wamberg and Tauson (1998) indicate a total urinary nitrogen excretion by the mink of about 4 grams/day.

Growth

In terms of the mink nutrition management required for top fur development, it is important to understand the significant difference between body size in terms of skeletal development and body size ascertained by weighing the animals.

Studies by Lohi and Hansen (1990) indicate that mink kits have reached 94% of their body length by mid-August. This observation is of interest in terms of confirming Schaible's 1961 observation. His experimental work indicated that the epiphyses of the long bones of the mink close about 16 weeks of age (that is, August 21 for kits born on May 1), thus actual body length cannot be increased beyond this point although body weight may be enhanced. Hence, extra "size" achieved after mid-August represents fat (adipose) tissue deposition and not actual extended body length.

Considering these factors of growth, it makes "common sense" for ranchers to provide "trim" male kits coming into September, so that extra feed volume and protein intake in the fall period will be supportive of superior fur development without loss of pelt size.

Fur Development

The skin is composed of three distinct layers, as Figure 3.1 shows. The epidermis is the outermost layer wherein keratinocyte cells at the base constantly multiply and differentiate. As these cells mature, they move toward the surface of the skin. Finally these cells keratinize and die, forming the horny layer on the surface of the skin. The epithelium contains two specialized cells: hair follicles which extend deep into the dermis and hypodermis, and melanocytes which exist at the root of the follicles and are the site of melanin synthesis.

The dermis lies below the epidermis and is composed of fibrous connective tissue (elastin and collagen) and ground substance. The elastin fibers are found about the hair follicles and the sebaceous glands and also between the bundles of collagenous fibers. These fibers may be responsible for the suppleness of the skin following the dressing of the pelts. The dermis contains blood vessels, nerves, hair follicles, and three types of glands which open into the epidermis. The sebaceous or oil glands provide a sebum to the hair follicle; it contains cholesterol, waxes, and simple lipids to the hair follicles and acts as a water repellent surface for the hair shafts as well as making the hairs soft and pliable. It also contains the apocrine sweat glands and the exocrine sweat glands which are limited to the foot pads of the mink. The Arrector Pili muscle is a smooth muscle that contracts in response to fright or cold, causing the hairs to stand on end, not only giving the animal a different appearance but also increasing the insulating properties of the fur.

The hypodermis, sometimes called the subcutaneous layer, consists mostly of adipose tissue (adiposities: fat storage cells) along with some fibrocytes and blood vessels and is adjacent to the muscles. This layer provides valuable insulation.

Figure 3.1

Structure of a Hair Follicle

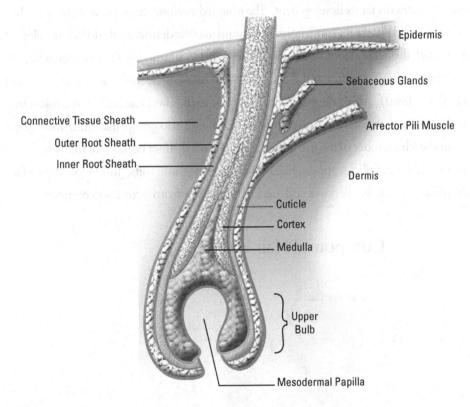

Epidermis

Sebaceous Glands

Arrector Pili Muscle

Connective Tissue Sheath

Outer Root Sheath

Inner Root Sheath

Dermis

Cuticle

Cortex

Medulla

Upper Bulb

Mesodermal Papilla

Hypodermis

Fur Fiber Development

Each hair follicle may yield only one type of hair, that is, either a guard hair or underfur. See Figure 3.2. However, they do share a common epidermal opening from the sebaceous gland which yields a bundle of hair from each skin pore – a single coarse guard hair with multiple softer underfur hairs. The genetic heredity of a particular mink determines how many follicles will be provided a kit at birth. All the hair follicles are formed before the mink fetus is four weeks old (Dolnick, 1959a; 1959b; 1959c). Dietary programs subsequent to birth, however, will determine whether the fur will be of normal color, length, and density.

51

Not all follicles are active and may not yield guard hair or underfur. For those that do, the number of hairs per bundle is directly related to the quality and quantity of protein provided the mink (Dolnick et al., 1960). It is apparent that the supply of nutrients to the hair has a relatively low priority in the animal's metabolism, as witness full size mink kits with second rate fur development because of a secondary biotin deficiency (a gray-banded underfur) or with second rate fur development because of secondary iron deficiency ("cotton" mink) (Wehr et al. (1981).

The hair follicle is a thin sac-like structure first seen in the fetus as a small swelling in the epidermal layer. While the origin of the hair follicle is epidermal in nature, it is from the dermis that the papilla and outer connective tissue sheath is formed. The papilla fits into the bulb of the growing hair and helps to supply nutrients for its development. The rounded swelling that represents the early hair follicle changes into a solid plug of cells and moves downward into the dermis. As the follicle develops, it pushes its way through the deeper layers of the skin and serves to accommodate the newly growing hair.

Distribution is uniform during the early period of hair follicle formation within the skin of the fetus (Dolnik , 1960). As fur development proceeds, each guard hair follicle is flanked by a follicle that will give rise to an intermediate guard hair, forming a trio group. Incoming hairs of the underfur surround each member of the trio group to form three distinct bundles of follicles.

As more and more follicles appear, hairs cluster together into tufts. Just below the surface of the skin, the follicular walls break down and tufts of fur emerge from a common opening.

Compound Hair Follicle

Figure 3.2

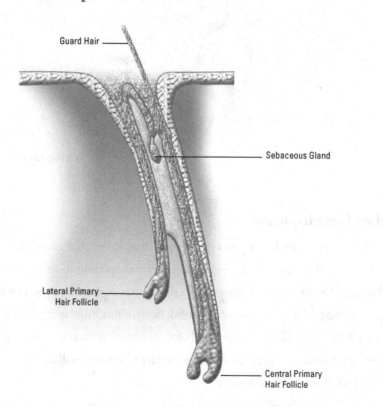

Guard Hair

Sebaceous Gland

Lateral Primary Hair Follicle

Central Primary Hair Follicle

The follicle containing a growing hair lengthens and migrates downward to the lower most limits of the skin and remains there for a period of time. As the hair continues completion of its growth, the follicle with its mature hair will ascend to the level of the sebaceous gland. Thus the depth of the follicle changes with the continuing hair growth cycle. While the hair is growing actively, the base of the follicle lies deep within the skin and the hair pigment which is provided by specialized cells termed melanocytes can be seen on the leather side of the unprimed pelt (Dolnick, 1959b). As the hair growth cycle is completed and the pelt becomes prime, the hair follicle shortens until its base is located just below the sebaceous gland in the upper layer of the skin.

The guard hair/underfur ratio best expresses the density of the follicle population. At birth the guard hair/underfur ratio is about 1:5, about five soft under furs and one coarse guard hair or intermediate guard hair in every skin pore bundle. By July the guard hair/underfur ratio has increased to 1:10, and at pelting the ratio is about 1:25 or more depending on genetic potential and nutrition management (Dolnick et al..1960; 1961). It is very important for ranchers to fully appreciate and understand the basic fact that optimum underfur development of the mink involves the final weeks and days prior to pelting. Hence, the vital importance for ranchers not to pelt too early to minimize labor and prevent loss of sharp color, but to make common sense nutrition management decisions that are supportive of "sharp color" in dark mink and thereby allow maximum underfur development.

Once past the skin pore of the sebaceous gland, the hair shaft is biologically inert and can no longer be modified except through physical influences such as wear and sun light.

Fur Structure

The major structures of the guard hair and fur fiber (underfur) are the same and consist of the following parts as shown in Figure 3.3.

1. **Medulla**—the central core or pith of the hair, composed of very fine cells in which masses of color pigments are found;

2. **Air Vesicles**—air spaces found between the masses of cells composing the medulla;

3. **Cortex**—a layer of cells surrounding the medulla and air vesicles;

4. **Cuticle**—an outer covering of scaly cells which overlap like shingles lying with their free margins pointing towards the hair tip;

5. **The dermal papilla (essential for cyclic hair formation)**—a small plug of connective tissue containing fibroblasts, histiocytes, melanocytes, mast cells, and the blood vessels which provide the growing hair with the nutrients required for development,

6. **Matrix**— the cells of the "generative region" or "matrix plate" just above the papilla, from which all growth occurs.

Fur Color Development

The following commentary is based in part on an excellent article reviewing the studies at Oregon State University on the topic of fur color development in the mink (Stout et al. (1967). Color in the fur of the mink and all other vertebrates is produced in only specialized cells called melanocytes which are formed during active fur growth within the hair bulb. This type of cell is also located in the basal layer of the epidermis and produces skin pigment as well as coloration in the iris and retina of the eyes.

Figure 3.3

Fur Structure

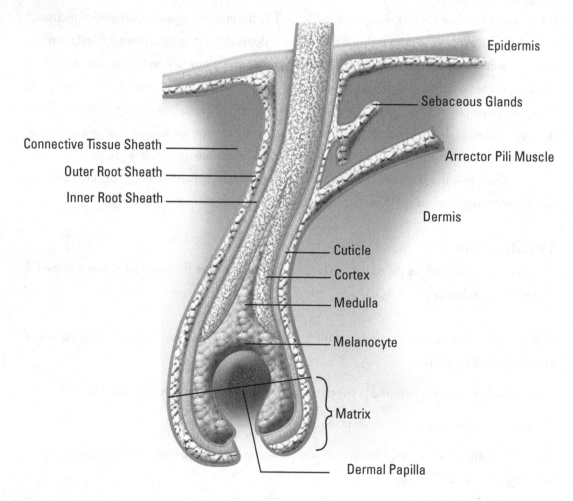

Epidermis

Sebaceous Glands

Connective Tissue Sheath

Arrector Pili Muscle

Outer Root Sheath

Inner Root Sheath

Dermis

Cuticle

Cortex

Medulla

Melanocyte

Matrix

Dermal Papilla

Hypodermis

Fur color originates from the type and amount of pigment that are retained in the hair shaft. Another determining factor is the amount of air in the intercellular spaces of the medulla. Luster is a feature of fur color as well, so that fur color is also affected by the character of its surface scale structure, whether smooth or rough, and the degree of gloss produced by the sebaceous gland secretions. Because guard hair scale edges are folded tightly against the cortex while underfur scale margins are rough, the two types of hair differ in light reflectance and hence luster (Bachrach, 1946).

Fur may appear to have a wide range of color hues, but microscopic examination reveals only black, brown, and yellow pigmented granules. Color diversity depends on various characteristics of the granules including their size, shape, concentration, and spatial arrangement within the fur. The two pigments responsible for all fur color are the melanins, or brown-black pigments originating from the amino acid tyrosine, and the pheomelanin, or red-yellow pigments originating from the amino acid tryptophan. The two pigments are under separate genetic control and are synthesized by distinctly unique biochemical pathways. White hair results from both a lack of pigment and light reflectance off the air spaces.

The primary determinant of fur color in dark mink is the degree of melanin granule deposition within the hair medulla and cortex cells. Under normal physiology, the melanocytes, which are specialized pigment cells in the "generative region" of the hair bulb, continuously produce melanin pigment granules which are then transferred into the cytoplasm of newly formed epithelial cells. These will constitute the medulla and cortex of the growing hair. The melanin pigments are always attached to a protein polymer; therefore, as the cells move upward and die, the pigment is trapped within them.

Photoperiods and Mink Fur Development

Regardless of the species, hair growth is a discontinuous process in all mammals, that is, periods of active growth are followed by variable periods of dormancy. Thus the fur growth of the mink is cyclical occurring at different periods during the year. These cycles consist of periods of active fur growth called anagen followed by periods of inactivity or rest called telogen, separated by periods of reduced activity termed catagen. It is interesting to note that the first fur growth cycle occurs during intrauterine development. Studies by Blomstedt (1987) indicate that the moult pattern of new born kits is distinct from that of adult mink with a whelp coat, a summer coat, and a winter coat. The whelp coat is moulted by the end of July and the summer coat is mature by the end of August. This same study indicated that the female summer coat was moulted by mid-October while in the males, the moult was three weeks later.

The fur growth program of the mink is influenced by the photoperiods of the ranch year which in turn affect the production of the hormone melatonin by the pineal gland. Thus with increasing hours of light each spring, there is a spring moult wherein the new summer fur grows from the head and ends at the tail. With the decreasing photoperiod after June 21 the fall moult of late

August starts at the tip of the tail and ends up at the tip of the nose.

It appears that for the winter fur growth cycle to begin, serum prolactin levels must decline below some critical level. Decreasing day length after June 21 results in an increased production of melatonin, which subsequently inhibits prolactin secretion by the pituitary gland, allowing the winter fur development to begin. The changing day length after June 21, perceived through the mink's eyes, affects the production of melatonin via neural circuitry to the pineal gland. Increasing day length after December 21 results in decreased synthesis of melatonin by the pineal gland and subsequent rise in the blood levels of prolactin, leading to growth of the summer fur coat and subsequent moulting of the winter fur. The moult releases hair for the nesting area and, later, prolactin is supportive of the lactation performance of the nursing mink.

As the fur development phase is completed in early December, the melanocytes cease pigment production with the result that the base of the hair follicle is white, producing the clear white underside of "prime" pelts. The fur growth resting phase between moults lasts until spring when the development of the summer pelt is initiated.

It is of vital importance for mink ranchers to recognize that underfur development may lag behind guard hair production by as much as two weeks. This knowledge leads to the advice, "Don't pelt too early"; wait for full under hair development and a fine cushion prior to pelting.

The follicle containing a growing hair lengthens and migrates downward to the lower most limits of the skin and remains there for a period of time. As the hair continues completion of its growth, the follicle with its mature hair will ascend to the level of the sebaceous gland. Thus the depth of the follicle changes with the continuing hair growth cycle. While the hair is growing actively, the base of the follicle lies deep within the skin, and the hair pigment provided by specialized cells termed melanocytes can be seen on the leather side of the unprimed pelt (Dolnick, 1959b). As the hair growth cycle is completed and the pelt becomes prime, the hair follicle shortens until its base is located just below the sebaceous gland in the upper layer of the skin.

The guard hair/underfur ratio best expresses the density of the follicle population. At birth the guard hair/underfur ratio is about 1:5, i.e., there are about five soft under furs and one coarse guard hair or intermediate guard hair in every skin pore bundle. By July the guard hair/underfur ratio has increased to 1:10 and at pelting the ratio is about 1:25 or more depending on genetic potential and nutrition management (Dolnick et al..1960; 1961). It is very important for ranchers to fully appreciate and understand the basic fact that optimum underfur development of the mink involves the final weeks and days prior to pelting. Hence, the vital importance for ranchers not to pelt too early to minimize labor and prevent loss of sharp color but to make common sense nutrition management decisions that are supportive of "sharp color" in dark mink and thereby allow maximum underfur development.

Once past the skin pore of the sebaceous gland, the hair shaft is biologically inert and can no longer be modified except through physical influences such as wear and sun light.

56

4

Nutrient Requirements

Analogy—Mink and Cat Nutritional Requirements

Without question, the nutritional requirements of the mink, a carnivore, are significantly different when compared to all other domesticated animals which, for the most part, are herbivores and omnivores. Mink ranchers and mink nutrition scientists interested in gaining a deeper understanding of mink nutrition and mink nutritional physiology should give serious thought to following the critical review of the scientific literature on cat nutrition.

Mink (*Mustela vison*) and cats (*Felis catus*) have unique though relatively similar nutritional requirements in terms of high protein requirements, unique amino acid needs, and distinct vitamin requirements. In addition, both mink and cats are subject to similar nutritional diseases such as urinary calculi and steatitis—problems directly related to their common nutritional regimens. It is thus of interest to briefly review the comparative nutritive requirements of mink and cats which set them apart from other domesticated animals.

Inasmuch as the scientific data on cat nutrition and physiology is extensive relative to the limited data provided on the mink, I have included a review of the scientific data on the cat which is the basis for its unique nutritional requirements relative to herbivores and omnivores.

High Protein Requirements

The protein needs of the mink may be as low as 20-25% of metabolic energy (ME) for maintenance and as high as 40-50% ME for the critical reproduction phase of the mink ranch year (Joergensen, 1984). Both the growing kitten and the adult cat have relatively high requirements for protein (Rogers and Morris, 1982, 1983). Kittens require about 1.5 times the protein needed by chicks and pigs. The protein requirement for adult maintenance is about two to three times higher in cats than in adult non-carnivores.

The high protein requirement of cats is not due to a higher than normal requirement for essential amino acids with the sole exception of arginine; in short, essential amino acid requirements are similar to those of other growing mammals (Rogers and Morris, 1979). Cats require a high-protein diet because they have a high requirement for nitrogen, the primary element provided in protein and amino acid resources. Most omnivorous animals given diets low in protein conserve amino acids by reducing the enzymes involved in the first step of amino acid catabolism. However, when cats were given high (700 g/kg) and low (170 g/kg) crude protein diets, there was very little adaptation in the activities of the amino transferase enzymes of the general nitrogen metabolism to dietary protein (Rogers et al., 1977). This lack of enzymatic adaptation in cats is similar to that reported in other carnivorous species such as trout (Cowey et al., 1981). (The very close relationship between trout and mink in terms of digestive capacity of fish meals is of interest (Skrede et al., 1980).)

When omnivorous and herbivorous animals are given diets of varying protein contents, the

59

activities of the urea cycle enzymes are positively correlated with the level of protein in the diet. The increased activity facilitates disposal of the higher nitrogen levels from high protein diets through the urea formation from ammonia. The reduction of urea synthesis activity when a low protein diet is consumed facilitates the conservation and re-utilization of nitrogen for the synthesis of non-essential amino acids. However, for cats given a low protein diet there was no reduction in the activities of the urea cycle (Rogers et al., 1977). Ammonia resulting from the deamination of amino acids by cats is continuously diverted to the urea and lost from the body pool, rather than contributing nitrogen for the synthesis of non-essential amino acids. Thus the high protein requirement of cats is for dispensable nitrogen. The enzymes involved in the first irreversible step of the degradation of the essential amino acids are controlled; otherwise, cats would have a high requirement for essential amino acids as well as for nitrogen (Rogers and Morris, 1980). A benefit to cats (and other true carnivores) arising from the lack of regulation of the enzymes in the catabolism of non-essential amino acids is the immediate capacity to catabolize amino acids as a source of energy and for gluconeogenesis. During starvation, because of their normal high protein diet, carnivores in general are better able to maintain glucose concentration in the blood than omnivores (Kettelhut et al., 1980).

Unique Amino Acid Requirements

For both the mink (Leoschke, 1959) and the cat (Morris and Rogers, 1978), arginine is an essential amino acid while for most species it is considered to be non-essential. For cats the absence of arginine in the diet has a profound effect on physiology. When an essential amino acid is deleted from the diet of cats, the food intake declined by the second day and there was a progressive slow body weight loss, but no other acute clinical signs (Rogers and Morris, 1979). However, when arginine was deleted from the cat's diet, there was a rapid overnight loss of body weight that continued over the next few days (Morris et al., 1979) related to hyperammonanemia. For many mammals, arginine is not an essential amino acid in the diet since the endogenous rate of synthesis is equal to the needs of these animals.

Taurine is an essential amino acid for cats such that a deficiency yields retinal degeneration (Hayes et al., 1975) and dilated cardiomyopathy (Pion et al., 1987). Although taurine has not been shown to be an amino acid requirement for the mink, veterinarian observations with foxes on a pellet program with high meat meal (low taurine content) and minimal fishmeal (higher taurine content) levels have noted multiple cases of of dilated cardiomyopathy as direct cause of deaths (Barkvull, 1987).

In terms of fur production, both the mink and the cat have amino acid requirements which are significantly higher than for growth. In the case of the mink there is an extra requirement for arginine (Leoschke, 1959) and for the sulfur amino acids (cystine and methionine) (Glem-Hansen, 1980) for fur development. In the case of the cat, there is almost a double nutritional requirement for phenylalanine/tyrosine for synthesis of the fur pigment melanin as for kitten growth (Anderson et al., 2002).

It is of interest to note a comment made in an exciting research review (Morris, 2002) on

60

the unique nutrition of the cat: "However, we are unaware of a requirement for a secondary function in any animal being so much greater than the requirement for growth." Obviously, Dr. J. G. Morris knows a little less about the amino acid requirements of the mink as Dr. W. L. Leoschke knows about the amino acid nutrition of the cat.

It should be noted that although cats and mink are similar in their nutritional requirements for high levels of protein and a unique requirement for arginine, there are some significant differences including the fact that even though the mink are strict carnivores, in contrast to the cat, they are able to modulate the rates of intestinal transport of simple sugars and amino acids to match changes in dietary intake of carbohydrates and proteins (Buddington et al., 1991). This unique capacity may be related to the fact that some members of the *Mustelidae* (skunks) are omnivores and that the mink have retained some ancestral traits.

Distinct Vitamin Requirements

Vitamin A

The minimal ability of cats to use carotenoids as a vitamin A resource was one of the first nutritional peculiarities of cats identified (Gershoff et. al., 1957). Studies by Warner and Travis (1966) indicated a minimum ability of the mink to convert carotenoids to vitamin A. For cats, vitamin A has to be present preformed in the diet rather than supplied as carotenoids because of the complete deletion of an enzyme required for the oxidation of carotene to retinal.

Niacin

Molar yield of nicotinic acid from tryptophan varies with species but is of the order of 33 to 60 mg of tryptophan to one mg of niacin synthesis in rats and humans. However, cats have no significant synthesis of nicotinic acid from tryptophan (De Silva et al., 1952; Sudadolnik et al., 1957; Ikeda et al., 1965). Studies indicate that the mink can synthesize only very small quantities of niacin from tryptophane but the synthesis is simply inadequate to meet the requirements for niacin by this species (Warner et al., 1968; Bowman et al., 1968).

Trytophan can be metabolized in either of two pathways: one results in the production of NAD (niacin co-enzyme) and the other yields acetyl Co-A and carbon dioxide. Picolinic carboxylase is the enzyme catalysing the first step of the degradation pathway of trytophan to acetyl Co-A and carbon dioxide; across the species the dietary niacin requirement is inversely related to the hepatic activity of this enzyme (Iketa et al., 1965; Scott, 1986).

Cats possess all the enzymes of the pathway of niacin synthesis from tryptophan, but the activity of picolinic carboxylase is extremely high, the highest of all animals studied (Sudadolnik et al., 1957; Ikeda et al., 1965), precluding any measureable synthesis of nicotinic acid (Da Silva et al., 1952; Leklem et al., 1969). Inasmuch as meat is well supplied with the NAD and NADP coenzymes, and

61

while cats consume a diet of animal tissue, there is no need to produce niacin from tryptophan. The direct production of acetyl Co-A may be energetically more efficient for cats than the oxidation of NAD or the intermediates of the NAD pathway.

Unique Essential Fatty Acid Requirements

There is a consensus that cats have limited capacity to synthesize arachidonate from linoleate, and probably eicopentaenoate and docosahexanoate from alpha linolenate (MacDonald et al., 1983; Bauer, 1997). The limited capacity of the mink to convert linoleic acid to arachidonic acids has been reported by Polonen et al. (2000). In cats this limited synthetic capacity has been attributed to low desaturase activity of cat liver (Rivers et al.,1975; Sinclair et al., 1979)

Urinary Calculi—Struvite Crystals

Cats like mink with similar dietary regimens are subject to urinary calculi, struvite crystals (Fischler, 1955; Leoschke et al., (1952).

Steatitis—Yellow Fat Disease

Cats, like mink with similar dietary regimens, are subject to steatitis, yellow fat disease (Coffint and Holzworth, 1954; Gerschoff and Norkin, 1962; Hartsough, 1951).

Water

Without question, air and water are the most essential nutrients for animals. It has been estimated that a human being can live 4 minutes without air, 4 days without water and 40 days without food (assuming ample adipose (fat) tissue energy reserves). A mink can lose 100% of its body fat and live, 50% of its body protein and survive; however, with the loss of only 10% of its body water, death is assured. Withholding water for as little as three days will kill a mink under average weather conditions (Gorham and Dejong, 1950). Mink with no water for 24 hours refuse to eat, are restless and depressed, exhibit eyes that are shut, and appear sunken. Muscular coordination and convulsions are sometimes observed. Muscles shrink and loss of fluid is such that the pelt can be raised from the body.

Water and salt are not ordinarily thought of as feedstuffs. They undergo no significant changes in the body but are used and excreted in the same form as ingested. However, in passing from infancy to maturity, substantial amounts of these nutrients are incorporated into the anatomy of the body.

Water is supportive of three major functions within the anatomy and physiology of the mink: it gives structure and form to the animal's body, it provides an optimum environment for the digestion of feedstuffs and for cellular metabolism, and it is absolutely necessary for maintenance of optimum body temperature.

Water comprises about 70% of the total weight of an adult animal and is distributed with about 2/3rds intracellular and 1/3rd extracellular compartments. The intracellular fluids exist within tissue cells while the extracellular fluids exist outside of the cells in two portions, an intravascular portion in the blood plasma and the extravascular portion outside of the blood circulation within the interstitial spaces (spaces between the cells in the tissues). There is a basic difference in the composition of the intracellular and the extracellular fluids. The intracellular fluids are high in potassium cations and low in sodium cations with significant levels of magnesium cations, phosphate, and sulfate anions while the principal inorganic constituents of the extracellular fluids including the spinal fluid and the lymphatic fluids as well as the secretions of the digestive system and glands are high in sodium and low in potassium cations with significant levels of bicarbonate anions. Note that the blood has a "salty" taste. The water balance between the intracellular and extracellular compartments is maintained by a process termed osmosis (this movement of water is controlled by concentrations of salts and colloidal particles such as protein molecules). Blood albumin molecules play a key role in maintaining proper water balance between the blood circulation and the adjoining tissues. With liver damage and subsequent sub-optimum synthesis of blood albumins, the net result is an edema as water moves from the blood to the interstitial spaces and ascites (accumulation of fluid in the body cavities). It is of interest to point out that the inorganic cation and anion content of the intracellular and extracellular compartments are all isotonic with one another and therefore the loss of 1,000 mls of intestinal fluid by diarrhea causes the loss of an amount of sodium cations equal to

63

that contained in an equal volume of blood.

Normally the marked difference in the inorganic ionic composition of the fluids present in the extracellular and intracellular compartments is well maintained because they are separated from one another by semi permeable membranes which although permitting the passage of water, glucose, and metabolites, do not readily allow the passage of sodium and potassium cations.

Water Requirements

Introduction

The two resources of water for the mink include first the water content of mink feed and that available from the water cup and second metabolic water or water provided to animal physiology as the direct result of metabolic reactions including (a) the oxidation of simple sugars, fatty acids and glycerol from fats to yield carbon dioxide and water and from amino acids to yield carbon dioxide, ammonia and water and (b) dehydration processes involving polymerization of sugars to yield more complex carbohydrates—the polymeric processes involving polypeptide and protein synthesis as well as the process of creating triglycerides (simple fats) from glycerol free fatty acids. Farrell and Wood (1968) have estimated that water requirements of the female mink are met by:

Feed Resources	66%
Fluid Resources	14%
Metabolic Water	20%

In a later study, with lactating mink, it was noted that metabolic water made up to 10-12% of the total water intake with the higher value during the 4th week of lactation possibly reflecting an increased oxidation of body fat (Tauson et al., 1998). In Tauson's 1999 study with adult mink, metabolic water made up to 14-70% of total water intake.

The fluid balance of the mink is primarily controlled by the ingestion of water and the excretion of urine. Mink urine volume is of immediate interest in an experimental study by Wamberg et al. (1996) with mink under a normal feeding regimen and during fasting. Their study indicated that with fasting, the withdrawal of the dietary water supply caused only a slight and insignificant increase in the daily intake of drinking water. The animals maintained their normal water balance by a dramatic reduction in urine excretion. At the same time, urinary solute excretion declined significantly, due in part to the cessation of dietary electrolyte intake and in part to the reduced formation of urea, whereas urinary osmolality decreased only moderately. The mean 24 hour balances of sodium, potassium, calcium, magnesium, chloride, and phosphate were close to zero and only minor differences between the feeding and the fasting periods were observed. In the fed animals, dietary water intake accounted for about 70% of total daily water intake. In the fasting period with dietary water intake at zero, the animals were capable of a mean water balance of 58 ml/kg/day, which was not significantly different

64

from the corresponding value of 63 ml/kg.day obtained in the feeding period. In the feeding period, the resulting water balance was about 16% lower than the value of 75 ml/kg/day obtained by direct calorimetry in female mink maintained at 18°Celsius (Wamberg, 1994) and the value of 76 ml/kg/day for lactating mink kept at 15° Celsius. These figures are in good agreement with the evaporation losses of 64 ml/kg 0.75/day at 20° Celsius (Tauson, unpublished data).

The water requirements of mink are relatively high compared to those of other domesticated animals. The mink is a carnivore with a high protein requirement which a very significant proportion is not employed for tissue, enzyme, and hormone synthesis but metabolized to yield carbon dioxide, water, and ammonia. The ammonia is converted to urea by the liver and finally excreted via the kidneys and the urinary tract. Kidney excretion of body metabolic products including urea requires water for solvation. A study by Einarsson and Enggard-Hansen (2000) involved dietary programs providing 25, 49, and 60% of ME as protein indicated increased water intake and excretion with the higher levels of protein in the diet reflective of the extra water requirement of the mink for the excretion of urea nitrogen. In earlier experimental work (Berg et al., 1984) involving 53, 35, 19, and 0% of ME as protein, it was noted that total water intake per ingested dry matter correlated positively with the protein content of the dietary regimen. Water excretion in the feces was slightly lower on the zero protein diet than on the other diets, but did not differ between the diets containing varying levels of protein. Instead, urine volume was high on the high protein diet and declined parallel with decreasing dietary protein content. On the zero protein diet, less water was excreted via the urine than with the feces. An even earlier experimental study by Makela and Valtonen (1982) noted that increasing the protein content from 21% to 44% of ME had these results: water intake increased from 133 ml/day to 157 ml/day, urine flow rate rose from 50 ml/day to 70 ml/day, and urine osmolality increased from 1471 mOsm/kg to 2151 mOsm/kg. Important here is that the mink excrete the higher nitrogen load by increasing urine output slightly and increasing urine osmolality by a much greater degree. Due to the ability of the mink to excrete nitrogen load by increasing urine osmolality, increased protein intake has only a slight effect on the demand for water. This observation is in sharp contrast to the salt intake of the mink wherein the response of a higher salt intake is an increased urine flow rate and decreased osmolality (Makela and Valtonen, 1982). Pregnant mink usually produce urine that is more dilute than that of non-pregnant mink (Tauson et al., 1998) as a very significant input of protein is not metabolized to yield energy and urea but provided to the kits for body structure.

The water intake of an animal is obviously related to its metabolic size; thus, a male mink kit weighing 2 kilograms and a female kit weighing one kilogram would have a water requirement of about 190 and 26 grams/day respectively (Jorgensen, 1985).

Water losses from the body include evaporation from the skin's "insensible perspiration." This is very important inasmuch as the mink does not have sweat glands to allow sensible perspiration. In the respired air, urine, feces, and during lactation a very significant amount of water is

65

secreted from the female's body. There are also significant water losses with diarrhea and kidney malfunction. A recent study (Tauson, 1999a) noted that TEWL (Total Evaporated Water Loss) was 3.7 g/kg/hr at 18° C. and 5.5 g/kg/hr at 24° C. They also reported that the mean rate of insensible water loss was 84-90% of TEWL or 3.1 g/kg/hr at 18°C. and 5.0 g/kg/hr at 24° C. A study by Tauson (1999b) indicated that evaporation loss increased from about 50 to 125 g/kg/day metabolic weight as Ta increased. A study by Howell (1976) is of interest relative to the daily water balance of 6 male kits during the month of October wherein a daily water intake of 157 grams was distributed as 33 grams of fecal moisture, 47 grams of urine, and 77 grams retained as tissue growth and respiration. Urine excretion is mostly dependent on the intake of electrolytes, the quantity of protein metabolites (including urea and ammonium cations), and the renal concentration capacity.

Ambient temperature, dry matter, and ME intake also influence the water intake and excretion of the mink.

Ambient Temperatures

Prior to a detailed discussion on ambient temperatures and water requirements, it is of interest to note that studies by Moller (1988) indicate that mink have a clear preference for water at 40° C. rather than at 6° C. There was significantly more wasted water with the colder temperature. This work with adult mink was not confirmed with mink kits given water at 17° and 40° C., but waste of water was higher with the cold water access. It has been confirmed that water intake in the summer months at temperatures of 20°-30°C. is ten fold that of the winter time (Makela and Valtonen, 1982).

A detailed study by Tauson (1999a, 1999b) is of major interest. The experimental plan involved ambient temperatures of 5°, 20°, and 35° C. Water intake was strongly affected by ambient temperature with dietary water being the major source at 5° and 20° degrees C., and even more important at 35° C. Water excretion as urine was highest at the lowest Ta and lowest at the highest Ta. Restriction of access to drinking water resulted in lower total water intake; and excretion, maintained by decreased urinary volume, reflected increased osmolality and increased solute concentration. Cortisol excretion generally tended to increase when water supply was restricted, the increase being significant at 35° C.

Dry Matter Intake

Studies by Makela (1971a, 1971b) indicated that total water inake per gram of dry feed was almost the same in mink fed diets varying from 27 to 39% dry matter. This observation was confirmed by later studies by Kiiskinen and Makela (1981) with mink on a pellet program and by Makela and Valtonen (1982). Studies by Farell and Wood (1968) indicated that the water intake of the mink is directly related to feed intake, a result identical to that observed by Carver and Waterhouse (1962).

Metabolic Energy Intake

Studies by Tauson (1999a, 1999b) indicated that ME intake decreased as Ta increased and ME increased when Ta was decreased to 5° C. This effect was reflected by an increased 24 hour excretion of sodium and potassium as well as an increased osmolality of the urine. Urinary and total water output per kJ ME was not significantly affected by Ta or water supply. Excretion of sodium and potassium cations per kJ ME mainly monitored urinary water excretion, being highest under conditions when urine production was highest.

Variation—Mink Ranch Year
Maintenance

Studies by Farrell and Wood (1968) with female mink under non-stress conditions noted a water consumption of 2.8 g total water/g dry feed or 0.63 g/kcal of apparent digestible energy (ADE). Their water intake data when related to surface area are appreciably higher but when related to ADE of the feed, are in reasonable agreement with values for other species. Water intake/day amounted to 13.3 g water/ 100 g body weight for the average female mink at 780 grams.

Pregnancy

Studies by Tauson et al. (1998) indicated that, compared with females that were not mated, females in the last trimester of gestation had a slightly higher water intake and a lower output, reflecting water retention in the body.

Lactation

Studies by Tauson et al. (1998) indicated that, surprisingly, water volume via the water cup did not increase in lactating mink, but by increasing their feed intake the animals managed to balance the increased water output via the urine and milk production. Lactating mink usually produce a urine that is more dilute than that of non-pregnant animals.

Suckling Kits

Studies by Wamberg and Tauson (1998) with litters of 6-9 kits, 1-4 weeks post partum and employing titrated water (H^3HO) dilution techniques observed that the biological half-life of body turnover in the mink kits increased linearly from 0.9 days in week one (3-5 days post partum) to 1.9 days in week four (22-24 days of kit age).

Daily milk intake for week one was 10.9+/- 4.0 g and for week four, 27.7+/- 1.0. Male kits were 10% heavier than female kits with males having a significantly higher milk intake.

Water Quality

Water quality recommendations are the same for mink as for humans.

Table 4.1. Quality Requirements In Drinking Water For Human Consumption (Danish Ministry for Environment Order No. 6, January 1980*

	Normal Value	Maximum Value
	pmm (mg/liter)	
Nitrate	25.00	50.00
Nitrite	0.00	0.10
Ammonia	0.05	0.50
Phosphorus	0.00	0.15
Chloride	50.00	300.00
Iron	0.05	0.20
Manganese	0.02	0.05
Permanganate	6.00	12.00
Aggressive Carbon Dioxide		0.00
pH	7.0 - 8.0	8.5
Coli bacteria	0 per 100 ml	
Total germ count 37 C.	5 per 20 ml	
Total germ count 21 C.	50 per 200 ml	

* *Mink Production*, Editor: Gunar Jorgensen (1985).

Hardness—the hardness of water should be between 5 degrees and 30 degrees dh (dh – Danish Hardness corresponding to 7 mg calcium/liter)

Nitrates—found in all ground water. Nitrate anion is not toxic per se, but under anaerobic conditions can be reduced to nitrite anion which is very toxic, binding to red blood cells in the same way as carbon monoxide. In young mink kits, nitrite anion can bring about methaemoglobinemia.

Ammonia—in drinking water is an indication that surface water (including that from mink manure piles) is entering the well. An indication that other harmful substances and bacteria may exist in the water.

Energy

Energy is not a nutrient but a property possessed by fats, carbohydrates and proteins. Whereas the major function of proteins is to supply amino acids for the construction of enzymes, muscles, organs, blood, bone, fur, and so on, the primary function of carbohydrates and fats is as energy resources with a secondary role as components of cell structures. A full appreciation of the key role of energy in fur animal nutrition and nutritional physiology requires a clear understanding of multiple definitions directly related to the biology and chemistry of the mink.

1. **Energy—Physical Science** = the capacity to do work

2. **Energy—Chemical Science** = caloric value of a feed product as determined by combustion within a bomb calorimeter, Gross Energy value of a feedstuff

3. **Energy—Biochemical** = caloric value of a feed product as provided to a living animal involving the processes of digestion and nutritional physiology and thus capable of being sub–divided into:

Energy—Digestible = Gross Energy minus fecal energy loss

Energy—Metabolizable = Digestible Energy minus urinary energy and gaseous energy loss. ME = GE – (Fecal Energy + Urinary Energy + Gasous Energy). When specific data on Digestible Energy is not known, an approximation of ME may be estimated as 77% of Gross Energy (Evans, 1967, 1976, 1977) for Canada; 80% of Gross Energy (Chwalibog et al., 1979) for Denmark

Energy—Net = Metabolic Energy minus heat energy loss. The Net Energy represents the fraction of the gross energy that is actually utilized by an animal for productive purposes. While Net Energy is the most precise estimate of a feed's energy value, it is not of practical employment in modern mink nutrition and physiology because of the difficulty of measuring heat losses. Thus Metabolic Energy is employed in calculations considering a mink's feed intake and the requirement of nutrients;

4. **Energy Balance—Negative** = condition wherein an animal mobilizes its energy reserves, primarily fat, to maintain the processes of life as illustrated with mink on a restricted feed intake prior to breeding for proper conditioning for breeding or during the lactation period when extra energy output as energy rich milk may exceed the nursing mother's capacity for adequate energy intake;

5. **Energy Balance—Positive** = condition wherein an animal's energy intake exceeds its daily needs and the excess energy is deposited primarily in the form of fat (adipose tissue) and protein (body structural components and fur)

69

6. **calorie**—the amount of heat energy required to raise the temperature of one gram of water by one degree Celsius measured from 14.5o to 15.5o Celsius (the temperature of the European laboratory in which the original work was conducted)

7. **Calorie**—the kilocalorie = 1,000 calories. Calorie with a capital letter represents an energy value 1,000 times the energy value of a calorie with lower case letter

8. **Joules**—1 calorie = 4.184 joules. Joules in the energy unit to be employed in the international scientific community in the 21st century

9. **Metabolism—Animal**—total summation of all biochemical processes taking place within an animal's physiology

10. **Metabolism—Basal**—the lowest rate of an animal in a state of complete rest(in the case of the mink when they are asleep), in a thermal neutral environment, and in a post–absorptive state. The amount of energy required for the involuntary work of the body including the function of the various organs and the maintenance of body temperature (via oxidative reactions in resting tissue, especially in the maintenance of muscle tone) in a physiological environment without involving feed digestive processing. Thus the Basal Metabolism is the result of the energy exchanges taking place in the cells of the body, i. e., without the involvement of body fat, extracellular fluids, and bone marrow. (It is of interest to note that sodium pump activity involving sodium– potassium ATPase in the muscles and liver represents as much as 10–20% of total basal energy expenditures)

11. **Basal Metabolism Rate—BMR**—Energy requirement of an animal in a physiological environment of minimal energy expenditure

12. **Maintenance Metabolic Rate—MMR**—Energy expenditure of an animal under conditions of no net weight gain or loss under normal conditions of activity and environmental temperature. MMR must be met before any production functions can be accomplished including pregnancy, lactation, kit growth, and fur production

13. **Metabolic Weight = Metabolic Body Size (MBS) = BW$^{0.75}$** (bodyweight kilograms)
All factors considered, an animal's highest priority for nutrient intake is for energy; all other needs such as a requirement for amino acids, simple sugars, and essential fatty acids for cell synthesis become secondary. Thus, for example, if a mink's diet provides an energy level from carbohydrates and fats which is insufficient for its daily needs, the animal will simply convert amino acids into energy as a higher need compared to the synthesis of body structures or fur.

In terms of mink kits in the growth period, experimental studies have shown that the total intake of energy is allotted as follows (Jorgensen, 1985).

70

Energy loss in feces (undigested)	15%
Energy loss in urine (absorbed but not used)*	7%
Energy loss(employed for processes of life and locomotion)	53%
Energy deposited as tissue protein and fat	25%

* as urea, creatinine (Leoschke, 1984), and other energy content chemicals.

Nutrient Resources–Metabolic Energy Values

The metabolic energy (ME) values employed by modern mink nutrition scientists have their origin in the studies of Atwater and Bryant (1899) on the gross energy values for protein, 5.65; fat, 9.4; and carbohydrate, 4.15 Calories/gram. These provided the basis for the commonly used values for ME of 4, 9 and 4 Calories/gram respectively for protein, fat, and carbohydrate in human nutrition science. The gross energy values were modified to yield metabolic energy values by taking into account feedstuff digestibility values and subtracting 1.25 Cal/gram from the gross energy value for protein for the energy lost via urea formation and excretion.

A significant evolution of metabolic energy values for nutrients employed in mink nutrition experimental studies over the last half–century is seen in Table 4.2.

TABLE 4.2. 1945-2000—Nutrient Metabolic Energy Values Mink Nutrition

| Year | Metabolic Energy Value | | |
	Proteins	Fats	Carbohydrates
1945 (1)	4.1	9.3	4.1
1966 (2)	4.5	9.3	4.1
1974 (3)	4.5	9.5	4.1
1977 (4)	4.5	9.5	4.0
2000 (5)	4.5	9.5	4.2

(1) Hodson and Smith (1942, 1945) (2) Ahman (1965, 1966), (3) Ahman (1974), (4) Glem–Hansen (1977a)—ME values employed in the *National Research Council bulletin* Number 7, *Nutrient Requirements of Mink and Foxes*, Second revised edition, 1982, *National Academy Press*, Washington, D.C. 1982, (5). Enggaard–Hansen et al. (1991) and Rouvinen–Watt (2000).

The Metabolic Energy values for digestible carbohydrates and fats are essentially the same as the Gross Energy values obtained with a bomb calorimeter. However, as can be seen from Table 4.3, the Metabolic Energy value for that digestible protein which is metabolized to yield water, carbon dioxide, and energy is about 25% less than the bomb calorimeter value. Significant energy losses occur with the obligate synthesis of urea from the deamination of amino acids, i.e., loss of nitrogen atoms which are converted in the liver to urea, a product with significant heat energy value for a bomb calorimeter but not for animal metabolism (Eggem, 1973; Glem–Hansen, 1974). The urea is subsequently secreted into the blood circulation system and excreted from the body via the kidneys.

TABLE 4.3. Metabolic Energy Value—Mink Nutrient Resources 2000

Nutrient	Kilocalories	Kilojoules
Digestible Fats	9.5	39.7
Digestible Carbohydrates	4.2	17.5
Digestible Proteins as Energy Resource	4.5	18.8
Digestible Proteins as cellular deposition or bomb calorimeter	5.7	23.8

Relative to the ME values of protein as 4.5 kcal/g. as an energy resource for the mink and 5.7 kcal/g. as a cellular synthesis resource, it is of interest to note that an experimental study by Glem–Hansen (1979) designed to study the protein requirements of the mink for the lactation phase of the ranch year noted that as the percentage of ME as protein decreased in the experimental feeding regimens, the energy consumption of the mink also decreased significantly, exhibiting an inverse ratio of feed volume to the level of protein provided the lactating mink. These observations are easily explained by the basic physiological fact that with lower and lower levels of protein in the diet of the lactating mink, a higher and higher proportion of the protein intake is being diverted to the highest priority of milk production where the milk protein has an ME value of 5.7 kcal/g. and less and less as an energy resource with an energy value of only 4.5 kcal/g.

Mink Color Phase and Energy Requirements

It is well recognized that certain of the mutation color phase mink have higher energy requirements than the standard dark mink including blue iris, hopes, mahoganys (demi–buffs), pinks, sapphires, triple pearls, violets, and wilds (Leoschke, 1969). Some blue iris ranchers provide their mink with a higher energy level (>24% fat) with the cooler temperatures of early October to provide optimum health and size at pelting (Fuhrman, 2000; Johnson, 1988; Sandberg, 1990). However, providing blue iris kits with relatively high fat levels in the late growth phase (July–August) may lead to "heavy" males coming into September with resultant "hipper" problems (sub–optimum fur development).

Even within the dark mink color phase, there are specific strains with relatively high energy requirements. This point is well illustrated by the experience of two Canadian mink ranchers who imported bred dark mink from a Wisconsin ranch. By mid–April most of the animals were dead, losses directly related to accidental energy starvation due to a feeding regimen similar to that fed the local dark females with lower energy requirements (Oswald, 1970; Hillard, 1977).

Body Size and Mink Energy Requirements

It is obvious that the energy expenditure of an animal is related to its size (body weight); however, nutrition physiology scientists have noted that larger animal species require less energy per unit of weight than smaller animals. Thus an animal's energy expenditure does not rise in linear relation to its body weight, that is, to an exponential power of 1.0, but more slowly, to the exponential of 0.75. Thus the energy expenditure of an animal is reported as kcal/kg $BW^{0.75}$/day wherein the $BW^{0.75}$ is termed the "Metabolic Body Size" (MBS). Experimental studies on many animals indicate that the energy requirement for maintenance can best be expressed in relationship to an animal's surface area. Metabolic Body Size is a mathematical expression that relates an animal's surface area to its body weight. It is also referred to as an animal's "Metabolic Weight."

The studies of Farrell and Wood (1968) with female mink indicate that the energy expenditures for maintenance purposes tend to vary directly with the body weight and not with the body weight raised to some fractional power. This observation is consistent with interspecies studies on the basal or resting metabolism of horses (Brody, 1945) and of rabbits (Lee, 1939) showing that heat production may not in fact vary with body weight raised to such a fractional power but with body weight raised to unity. This phenomenon is less surprising with mink since the increment of energy expenditure associated with activity is more than twice the basal energy output. There is some evidence that energy expenditure associated with muscular activity may vary directly with body weight (Brody, 1945; Mitchell, 1962). One would therefore expect that the total maintenance energy requirement of the mink would vary with body weight raised to a fractional power closer to 1.0 than to 0.7 because activity is such an unusually large portion of the total maintenance energy requirement of the mink.

Basal Metabolic Rate (BMR) or Resting Metabolic Rate (RMR)

For active animals like *mustelids*, the term Resting Metabolic Rate (RMR) is being used for energy expenditure at rest to replace the conventional term Basal Metabolic Rate (BMR) (Wamberg, 1994). Basal metabolism or Basal Metabolic Rate (BMR) is a function of body size and is generally similar in adult animals when expressed per unit of Metabolic Body Size (MBS) or "Metabolic Weight." Studies by Brody (1964), Kleiber (1961, 1975), and Blaxter (1972) led to the following

73

equation for the estimation of BR for a wide variety of animals: BMR (kcal/day) = 70.5 W $^{0.75}$ (kg). Thus a 10 pound (4.54 kg) animal would have a BMR = 70.5(4.54)$^{0.75}$ = 70.5 x 3.1 = 219 kcal/day, that is, a minimum daily intake of 219 kilocalories energy per day.

Experiments with a variety of animals show that the basal maintenance requirement varies from 51 to 75 kcal/kg body weight 0.^{758}, (Blaxter, 1972). A study by Farrell and Wood (1968) with three sleeping female mink indicated that the BMR, basal heat production, was 84 kcal/kg/ 24 hours or 77 kcal/kg metabolic weight (W$^{0.73}$). The Farrell and Wood data for BMR are supported by a comparative study of the basal metabolism of *mustelids* by Iversen (1972) which indicated that the BMR was about 85 kcal/kg/body weight/day. Experimental data of Perel'dik and Titova (1950) estimated the BMR as 150 kcal/kg/body weight; however, this was considered to be excessive by Harper et al. (1978) due to multiple factors: not only were the mink awake, but their excessive activity was compounded by the fact that they had been fasted for 24 hours.

Maintenance Metabolic Rate–MMR

MMR is also referred to as ME for maintenance, MEm. The Maintenance Energy requirements of an animal include all the physiological processes through which the species maintains its body without any change in animal weight or body composition. These energy requirements can be broken down into special energy requirements for:

1. Basal Metabolism;
2. Thermoregulation;
3. Activity Pattern – Locomotion;
4. Heat Nutrient Metabolism.

Basal Metabolism

The BMR represents essential metabolic processes. It is, in fact, an expression of the size of the body compartment usually called "active cell mass."

Thermoregulation

It is obvious that environmental temperatures influence the MEm requirement. With relatively cold temperatures, mink need an extra energy intake to compensate for the increased heat loss from the body. The temperature at which extra energy is required for the maintenance of an animal's body temperature is called the Lower Critical Temperature (LCT) which varies, in the case of the mink, according to the insulation properties of the mink's fur coat and the nest box environment. Studies by Glem–Hansen and Chwalibog (1980) estimated that for each degree Celsius temperature below the LCT, mink require an extra 3.7 kcal ME/kg MBS per day. Later studies by Wamberg (1994) on

24 hour energy expenditure of adult female mink as THL (Total Heat Loss) was 8.9 W /kg at 18° Celsius and 5.8 W/kg at 24° Celsius under controlled conditions using direct calorimetry, that is, the equivalent of 0.2 W/degrees Celsius/kg. These values agree well with the energy requirement for maintenance reported earlier at 586 kJ/kg/day or 6.8W/kg (NRC 1982). These same studies showed that the contribution of SHL (Sensitive Heat Loss) and EHL (Evaporation Heat Loss) to 24 hour THL in adult female mink were inversely related and markedly dependent on chamber temperature—76 and 24 at 18°Celsius and 41 and 59 at 24° Celsius.

Activity Patterns—Locomotion

It is not surprising that increasing the cage size of mink has been shown to increase the animal's energy output. Farrell and Wood (1968) noted that at a mean temperature of 11° C., the adult mink female requires 258 Calories/kg of digestible energy in large farm cages and 202 Calories/kg in smaller cages. This highly significant decrease in energy needs associated with confinement is expected and is similar in magnitude to that found for the cat by Miller and Allison (1958) and Hodson and Smith's (1945) work on the mink in metabolism cages.

Heat of Nutrient Metabolism

The assessment of the BMR of an animal is achieved with the animal at complete rest and in a post–absorptive state, that is, a physiological environment without the energy enhancement related to feed digestive processing. The rise in heat output following a meal represents energy released during the digestion, absorption, and assimilation of the nutrients in feedstuffs. This is termed the heat increment (HI) of feeding or heat of nutrient metabolism. The amount of heat energy released, HI, represents less than 20% of ME available from fats but as high as 50% of ME available from proteins. The HI directly related to protein intake was formerly termed Specific Dynamic Action, SDA.

Mink Metabolic Energy for Maintenance—MEm

The maintenance energy requirements of an animal must be met before any productive functions can be accomplished. For many animals, the MMR is about twice the BMR (Brody, 1945; Kleiber, 1961). A survey of experimental studies on the MEm of the mink conducted over the period of the last 75 years is presented in Table 4.4.

TABLE 4.4. MEm Maintenance Energy Requirements of the Mink

Year	MEm – Cal/kg/day	Researchers
1927	203–220/BW	Palmer
1945	209/BW	Hodson and Smith
	200/MBS	
1949	174/BW	Oldfield
1950	210/BW	Perel'dik and Titova
1969	202/BW	Farrell and Wood
1978	148/MBS	Harper et al.
1979	126/MBS	Chwalibog et al.
1980	126/MBS	Glem–Hansen and Chwalibog, Chwalibog et al.
1981	156/MBS	Charlet–Lery et al.
1981	126/MBS	Weis
1984	150–170/BW July	Enggaard–Hansen et al.
	170–175/BW August	
1999	128 /MBS (20 Celsius)	Tauson
	155 /MBS (35 Celsius)	

The earliest study on the basic energy requirements of the mink was reported by Palmer. His data on the gross energy values of 264 to 286 Calories/kg/day have been recalculated to approximate digestible energy values by employing the factor of 0.77.

The MEm values of 200/MBS and 209/BW of Hodson and Smith were considered to be excessive by Harper et al. in terms of the fact that the studies involved only 5 days, a period considered too short for this sort of assessment on energy requirements. The relatively high MEm values obtained by Oldfield may be related, in part, to the fact that he reported some difficulties in maintaining a quiescent state in his animals. The relatively high MEm values of Perel'dik and Titova were also considered by Harper et al. to be excessive. These results were probably related to the experimental design wherein animal activity was very likely excessive due to multiple factors. Chief among these was the 24 hours of fasting prior to obtaining the MEm data since it is generally acknowledged that mink react strongly to extended hunger. The Farrell and Wood data on MEm values may also be considered relatively high: These values more than double their own experimental data on BMR of 84 Cal/kg/BW.

It is of interest to note that the Charlet–Lery et al. experimental data of 156 Calories/MBS was found to increase to 192 Calories/MBS for farm conditions.

It is very likely that the relatively lower experimental data found by several researchers of 126–128 Calories/MBS are related to multiple factors including more sophisticated experimental equipment with stricter temperature controls. In particular, Chwalibog et al. employed special respiration units suitable for the mink.

MEm requirements of the mink in the range of 126–128 Calories/ MBS/day are generally higher than those of other domesticated animals: 100–113 Cal/MBS/day for pigs (Kielanowski and Kotarbinska, 1970; Just–Nielsen, 1975; Thorbek, 1975; Close and Mount, 1978), 113 Cal/ MBS/day for chickens, and 103 Cal/MBS/day for bull calves (Thorbek and Henckel, 1976), very likely due to a higher muscular activity of the mink.

MEp—Metabolic Energy–Production

The production functions of the mink include reproduction/lactation, growth, and fur development. Each phase of the mink ranch year requires special consideration relative to any survey of the energy requirements of the mink (MEp), including:

1. **Pre–Breeding;**
2. **Reproduction;**
3. **Lactation;**
4. **Early Growth;**
5. **Late Growth;**
6. **Fur Development.**

MEp—Pre–Breeding Phase

Proper energy management of both male and female mink in the period prior to breeding in late February or early March is very important in the achievement of the top reproduction/lactation performance of the mink in the months to follow. Overweight male kits often lack the vigor required for a successful breeding season. A moderate weight at mating of females appears to increase the appetite of these mink during gestation and improve nutrient supply to the fetuses. Excessively heavy females must undergo proper conditioning (slimming) prior to mating to provide optimum reproduction performance. At the same time, it must be understood that extreme weight losses, greater than 300 grams, have been shown to result in poor reproduction results (Tauson and Alden, 1984). According to Kemp et al. (1993a), the poor reproduction performance of weight loss—stressed female mink is due to depressed ovulation rates rather than reduced embryonic viability.

MEp—Reproduction Phase

Optimum energy management is extremely important throughout the pregnancy period if top reproduction/lactation is to be achieved. Obviously, mink nutrition during pregnancy has a significant impact on the reproduction performance of the mink. Excessive energy provision during this period, that is, 360 kilocalories (1500 kJ)/day, can result in poor reproduction performance

(Ahlstrom and Skrede, 1997). Their experimental data indicated that kit survival rate in mink is dependent on the body weight of the female and the energy intake during the reproduction period. A moderate weight at mating appears to increase the appetite of the female during gestation and improve nutrient supply to the fetuses. On the other hand, excessive feed restriction prior to whelping has been shown to increase preweaning mortality, reduce the number of weaned kits, and result in poor weaning weights for the litters (Kemp et al., 1993b). Since the litter sizes at birth were not significantly different between the experimental groups, it can be assumed that the impaired reproduction/lactation performance was related to poor viability of the kits and reduced nursing capacity of the feed—restricted mink.

The matter of sub–optimum reproduction performance of mink with severe feed restriction is well illustrated by a report from Canada (Friend and Crampton, 1961). In that study, mink nutritional deprivation was achieved not by severe feed restriction but by a relatively low fat diet. The experimental study was initiated to gain an understanding of the low reproduction/lactation performance of mink placed on diets containing as high as 80% codfish (20% whole fish and 60% fish scrap). An insight into the potential nature of the problem was provided with a consideration of the energy content of the ocean fish available to the local mink ranchers. Fish such as salmon, herring, and pilchard have relatively small livers with little oil in them, but they have an oily muscle tissue. On the other hand, cod, haddock, hake and halibut have fairly large livers which contain almost all of the fat in the fish. Thus with a high proportion of codfish scrap in a practical mink ranch diet, the net result would be a relatively lean dietary regimen. The high fish ration provided 390 gross kilocalories/mink/day at 1/2 pound per animal and yielded a 57% whelp. Supplementation of the lean high fish ration with 8% fat provided an 84% whelp, a result significant at the 1% level of probability.

A process termed "flushing" has been shown to enhance the reproduction performance of female mink (Tauson, 1993). Flushing is defined as a period of restricted feeding followed by *ad libitum* feeding just prior to the estrus period or breeding season. The most consistent and positive effects are achieved when the energy intake has been restricted moderately (about 80% MEm) for two weeks. This program is provides a relatively lean diet (14–18% fat) as an asset to "the man with the spoon" in proper conditioning of the mink for breeding. This period is followed by "flushing" which is achieved by feeding *ad libitum* (about 150% MEm) from 3–5 days before the start of the mating season until mating has been achieved for the individual females.

A good insight into the extra energy demands of the mink during pregnancy is provided by Graph 4.1. Protein Deposition in Pregnant Uterus of Mink (Moustgaard and Reis, 1957a;1957b). This graph provides a physiological basis for the observation of Weiss (1981) that the MEp requirement for the mink in gestation is, in the early part, not greater than the MEm requirements; but, in the later part of the pregnancy, it increases and by the conclusion of this period is about 50% greater than MEm requirements.

Investigations of energy consumption of mink at the Agricultural University of Norway by

78

Rimeslatten (1964) showed an increase from 186 to 194 kcal ME/female in average from March to April, indicating that the energy requirement for fetal development is very limited. These observations were confirmed by commentary provided in the Russian Handbook (Penelaik, 1975) stating that the energy requirement for female mink is 185 kcal ME/kg in the period of January–March and 195 kcal ME/kg in April. It is estimated that the average energy requirement during periods of low temperature increases by 1% per degree Celsius below zero (Rimeslatten, 1964).

Of major interest relative to MEp –Reproduction Phase are the studies of Tauson, Elnif and Enggaard–Hansen (1992, 1994). Their experimental data are summarized in Table 4.5. One week energy balance periods included a 24 hour measurement of heat production (HE) by means of indirect calorimetry in open–air circuit respiration chambers. Calorimetry measurements were carried out from mating to parturition.

Heat Production (HE) was calculated from oxygen consumption and carbon dioxide production using the formula by Brouwer (1965):

HE, kJ = 16.18 x Oxygen L. + 5.02 x Carbon Dioxide L. – 5.99 x Nitrogen in urine, g.

ME was calculated from feed intake and the equation: Retained Energy (RE) = ME – HE.

Heat Production (HE) was not significantly affected by stage of gestation. The HE data indicate that the mink, unlike other species, does not have a considerable increase in HE as an effect of gestation. To keep HE and therefore feed requirement low may be a good strategy for a species that in the wild are pregnant where the feed supply is limited. A possible reason for this lack of increase in HE may be that MEm decreases with advancing stage of gestation due to decreased locomotor activity of the pregnant female. Earlier investigations have stressed the difficulty in determining MEm values in adult mink (Chwalibog et al., 1980; Chwalibog et al., 1982) caused by great individual variation in activity. 42% of total HE was via fat oxidation while protein oxidation accounted for 38% of HE. That level of protein oxidation reflects ME intake and tends to decrease in late gestation when protein is needed for retention as Graph 4.1, Protein Deposition in Pregnant Uterus of Mink, shows.

For the unmated females, the ME intake was far below, and for the pregnant females slightly below, the average ME intakes of female mink reported by Enggaard–Hansen et al. (1991). The increases in ME intake during pregnancy are slightly below the levels reported for the cat (Loveridge and Rivers,1989).

Mean RE was low and in some individuals was even negative, indicating that part of the energy requirement for pregnancy may be supplied by mobilization of body reserves. These observations were confirmed by studies of Tauson and Elnif (1994) indicating significant negative energy balance of pregnant mink during late gestation. Additional support for the observations of Enggaard–Hansen et al. (1991) and Tauson and Elnif (1994) is provided in the Ph. D. thesis of B. K.

79

Hansen (1997) where it was noted that in the last trimester of pregnancy, ME intakes were low and were clearly exceeded by HE. It should be noted that the calculated protein deposition in uteri and foetal tissue is in good agreement with that of Moustgaard and Ries (1957a, 1957b). RE deposition in fetal tissue was limited, with the exception of the last few days of pregnancy during which the main fetal growth occurs. See Graph 4.1. It is of interest to note that the above authors calculated that the accumulated energy costs of pregnancy in mink would amount to only about 5 times MEm.

TABLE 4.5. HE, ME, RE, and Energy in Uteri and Foetal Tissue in Relation to Stage of Gestation.

Energy Acronym*	Unmated Female	After Implantation	Close to Parturition
HE, kcal	139	150	139
ME, kcal	129	182	193
RE, kcal	10	32	54
Nutrients & Energy in Uteri & Foetal Tissues			
		Uteri	Foetal Tissue
Protein, g.	0.13	0.39	10.51
Fat, g.	0.05	1.21	<0.001
Carbohydrates, g.	0.01	0.02	0.52
Energy, kcal	1	3	74

* HE, Heat Production; ME, Metabolic Energy; RE, Retained Energy

MEp—Lactation Phase

The requirement of energy in the lactation phase of the mink ranch year is obviously higher than in the gestation period (Crampton, 1964). This point is well illustrated by Graph 4.1. Protein, Fat, Carbohydrate, and Ash Content of Mink Milk (Glem–Hansen and Jorgensen, 1975; Jorgensen, 1984). This graph is obviously a good introduction to the topic of MEp–Lactation Phase. These experimental data are supported by a more recent study on the composition of mink milk by Olesen et al. (1992). During the period from 3 days to 39 days, the level of protein decreased from 38% to 26%, the fat level increased from 38% to 50%, and carbohydrates (Nitrogen Free Extract, NFE) decreased from 11% to less than 1%. The data on mink milk composition indicating an increase in milk fat and dry matter over the entire lactation period coincides with the chemical changes seen in the body composition of kits during the lactation period (Tauson, 1994; Layton, 1998; Layton et al., 2000).

Mink kits are born altricial (very immature physiologically), blind, and nearly hairless. They have very limited locomotor abilities and no thermoregulatory capacity of their own. Moreover, they have almost no modifiable energy reserves inasmuch as the fat content of the kit body at birth is only 1.4% (Tauson, 1994). Their thermoregulation develops only gradually and their survival is dependent on the heat and nourishment provided by the mother (Rouvinen–Watt & Harri, 2000).

They have negligible stores of metabolizable energy at birth with a fat content of only 1% and their glycogen content amounts to about 0.04 kilocalories, inadequate to meet the maintenance energy requirement of a 10 gram newborn kit (4.1 kilocalories/day). It is therefore necessary that suckling be established soon after parturition.

The growth potential of the newborn mink is very high, with a maximum relative growth rate of 23%/day between days 1–2 (Tauson, 1994) and a doubled birth weight within 4 days (Jorgensen, 1960). During the first three weeks of life the growth rate is 12%/day and from birth to weaning, a period of 42 days, an average relative growth rate of 9%/day has been calculated (Tauson, 1994). Another study (Wamberg and Tauson, 1998) noted that kit growth after birth was 2.9 g./day during week one and as high as 5.4 g./day during week 4.

This same experimental report indicated an observed daily milk yield with second year lactation of an average of 87 g./day in week one to an average of 190 g./day in week 4. Calculations of mean intake of mink milk per unit of body weight gain were remarkably stable at 4.0 g./g. during weeks 1–3 postpartum, but increased to 5.6 g./g. in week 4.

The Wamberg and Tauson (1998) report indicated that ME provided the kits by daily milk intake increased from about 110 to 240 kcal/day which corresponds well with calculated daily energy requirements.

TABLE 4.6. Lactation Phase MEp

Lactation Period	Perel'Dik &Tivota(1951)*	Jorgensen et al. 1960	Rimeslatten 1964	Tauson et al. 1998
1st Week	—	—	—	182
1st 10 days	250	190	185	—
2nd Week	—	—	—	236
3rd Week	—	—	—	257
2nd 10 days	350	280	262	—
4th Week	—	—	—	309
3rd 10 days	454	440	312	—
4th 10 days	550	600	510	—
5th 10 days	800	700	800	—

* five kits/liter.

Table 4.6 provides a survey of research on the topic of the energy requirements of lactating mink over the period of the last half century. These data are consistent with studies by other Danish and Russian scientists. In experiments with lactating females having six kits in the litter, Glem–Hansen and Jorgensen (1975) noted an *ad libitum* energy intake of 253 kcal/ME/day during the period from 3–24 days after birth. *The Russian Handbook* (Penalaik, 1975) recommends an energy intake for lactating females from 200–250 kcal ME/day according to body weight, an increase per kit in the litter of 5 kcal ME in the first 10 days after birth, 20 kcal from 10–20 days, 50 kcal from 20–30 days, 70–90 kcal from 30–40 days and 110–150 kcal from 40–50 days after birth.

Danish experiments (Glem–Hansen and Jorgensen, 1975) during the lactation period with diets containing 3,430 to 4,150 kcal ME/ kg dry matter did not show any significant difference in mink production or growth.

Multiple studies (Wamberg et al.,1992; Tauson and Elnif, 1994; Hunter and Lemieux, 1996; Hansen, 1997; Hansen and Berg, 1998) indicate that lactating mink with large litters are likely to be in a negative energy balance in late lactation, employing fat reserves. The studies of Hansen (1997) and Hansen and Berg (1998) indicate that lactating mink with large litters may lose as much as 15–20% of their body weight at parturition, the main part, as high as 10%, during the period from 4–6 weeks of lactation. As this significant loss of weight may lead to nursing sickness, it is of utmost importance to stimulate energy and water intake to improve lactation performance and animal health.

A classic example of a mink ranch dietary program with sub–optimum energy management during the lactation period is that of a Minnesota rancher with a dietary regimen providing 17% ash and 16% fat (Leoschke, 1965) experiencing lactation failure. The problem was resolved within a few days with the employment of higher levels of beef tripe providing an increase in fat content and decrease in ash content.

MEp—Growth Phase

The early growth (June), the late growth (July–August), and the fur development periods of the ranch year will be discussed as a single topic of MEp–Growth inasmuch as most of the experimental data on energy requirements of the mink for growth cover all three of these phases together. Special important considerations for each of these periods will be discussed in sub–sections to follow.

Of significant interest are the observations of Rimeslatten (1964) on the energy consumption of mink over the period of 1950–1962, which yield the data provided in Table 4.7.

TABLE 4.7. MEp—Growth Period–Estimated Energy Consumption

Dates	ME – kcal/day		
	M/F	M	F
May 6–19	2		
May 20 – June 2	23		
June 3–16	67		
June 17–30	149		
July 1–15		229	177
July 16–31		295	222
Aug 1–15		323	243
Aug 16–31		337	253
Sept 1–30		348	259
Oct 1–31		357	268
Nov 1–30		308	223

Obviously, these 1964 figures are for dark mink with genetics significantly different from those of current dark ranch mink of the 21st century in terms of final size, growth rates, and Mep. They do not cover the different energy requirements of the multiple mink mutations existing today within the world mink industry, but they are significant in terms of the simple fact that they are unique.

Additional observations on the energy requirements of the mink for the growth period from July through pelting are of interest. Studies by Evans et al. (1962), Sinclair et al. (1962), and Evans (1963) indicate that the minimum energy requirement of male mink kits for the growth period from July to pelting was in the range of 408–421 kcal/100 grams. The 1963 study indicated that the energy level for male kits could be as high as 462 kcal/100 grams with top performance diets providing high levels of protein in terms of both quantity and quality of the protein resources. The Evans (1963) and Allen et al. (1964) studies indicated that the minimum energy requirement for female mink kits for the growth period from July to pelting was in the range of 346–370 kcal/100 grams.

Glem–Hansen (1980) noted that the female kits had a greater feed intake per kg of body weight than did the male kits. With smaller body size relative to a kilogram weight, the females have a relatively higher maintenance requirement since the energy requirement for maintenance is not directly related to body weight but to the metabolic weight involving the weight in kilograms with an exponential factor of about 0.75.

In addition to these average requirements, there are some unique nutritional factors related to the energy requirements for the early growth, late growth, and fur production phases of the mink ranch year.

Early Growth—June

The importance of a high quality nutritional program for the mink in June with optimum energy levels is seen in Table 4.8:

TABLE 4.8. Eight Week Male Kit Weights and Pelt Length*

Weight – gms	Pelt Length – cms
450	69
500	71
550	72
600	73
670	75

*Jorgensen (1984).

June—Cannibalism

Mink cannibalism, that is, kits attacking, chewing, killing, and eating each other, has been a common problem within the North American mink industry for more than sixty years. Leoschke (1970) pointed out that multiple factors were involved in precipitating the problem including genetics and energy balance. It was noted that certain specific color phases of mink (including blue iris) with high energy requirements were especially sensitive to cannibalism and that raising the basic energy level of mink diets in early June to as high as 28–30% fat would minimize the problem. Later ranch observations indicated that with certain mink population genetics, an alternate program of lower fat and higher protein levels is required to prevent and/or resolve the problem.

Cannibalism Resolution—High Energy Programs

As early as 1956, studies by Leoschke (1960), indicated that the mink, a carnivore, had a relatively high requirement for fat, as high as 26% for top growth performance and minimum pounds of feed per pelt marketed. For instance, in June, 1957, the Riedel Mink Ranch (Riedel, 1957) experienced severe cannibalism problems which were resolved within 24–48 hours by a simple increase in energy content of the ranch diet via raising the fat level to 26–28%.

The experience of North American mink kits on Ross–Wells commercial pellet formulations is of interest. In the 1980s this top pellet program provided top performance of the mink in all phases of the mink ranch year except for the early growth period, June, wherein multiple reports of cannibalism where noted related, in part, to the relatively low fat content of the early growth pellet (20–22% dehydrated basis). The problem was immediately resolved with fat supplementation to yield 26–28% fat levels or a transition to another pellet program emphasizing 26–28–30% fat levels for the June kits. At the National Research Ranch, Oshkosh, Wisconsin, cannibalism problems

84

were kept to a minimum by providing fat levels of 28–30% via supplementation of the early growth pellets with 5–6 pounds of choice white grease or poultry oil per 100 pounds of pellets to yield a high energy mash.

In 1985, a blue iris mink ranch on a redi–mix program providing 23–24% fat experienced a major cannibalism problem which was resolved within a few days with poultry fat supplementation yielding 26–27% fat levels for the June kits (Wrolson, 1985).

Cannibalism Resolution—Lower Energy–Higher Protein Programs

In recent years, a number of North American mink ranches have noted very significant cannibalism kit losses on high energy programs. Among them is the Newman Mink Ranch in Minnesota, which experienced severe losses of kits with cannibalism in June on a high energy dietary regimen. The problem was resolved by a recommendation of Dr. Robert Westlake (1985) to lower the commercial fortified cereal level from 20% to 15% with resultant higher levels of both fat and protein.

The West coast mink rancher Terry Basl (1990) is considered to be the first rancher to employ lower energy diets (21% fat) and high protein levels in combination with feeding the kits four times a day as a resolution to cannibalism kit losses.

Should ranchers use a high energy or a low energy–high protein program for best prophylactic nutrition management in June to prevent cannibalism? The choice is simply made by "Listening To The Mink." Enough said.

Late Growth—July–August

Optimum energy management is critical for the late growth phase of the mink ranch year in terms of both nutrition management and business management, inasmuch as the caloric density of the ranch diet has a major influence on kit growth and feed volume per pelt marketed. Studies by Sinclair et al. (1962) indicated a linear relationship between mink feed volume and the apparent digestible energy content of the mink feed. This relationship simply supports common knowledge that two factors control the feed intake of an animal: metabolic energy content of the feeding regimen and palatability of the diet for the animal. Fat levels in the ranch diet for July and August should be at the highest level in terms of ME kilocalories/ gram consistent with the top performance of the mink.

Proper energy management of male mink kits in the late growth period has a major effect on the quality and market value of pelts. Extra "heavy" male kits coming into the fall fur development in late August can develop (a) the anatomical defect of "weak hips" or "hippers," (b) the physiological problem of "wet–belly" disease, (c) and/or sub–optimum fur development. Each of these potential problems of pelt quality is directly related to sub–optimum mink energy intake management in July and August and deserves special consideration.

"Hipper" Development

The terms "weak hips" or "hippers" have been used to describe male kits in early September or pelts at marketing which possess sub–optimum fur development over the hips. The "hipper" condition is due to incomplete shedding of the summer fur with subsequent inadequate development of the winter fur. It is very likely related to excessive fat (adipose) tissue development in the hip area, resulting in sub–optimum blood circulation to the hair follicles during the moulting period and, thus, sub–optimum fur development.

Field observations on two mink ranches in 1962 and 1988 provide insights into those suboptimum nutrition management programs that develop the condition. The combination of sub–optimum early growth nutrition management and Leoschke's enthusiasm about the value of higher energy dietary programs for the late growth period resulted in an "ideal" plan for hipper development in the mink. As of July 10, 1962, the male kits at the Elcho Mink Ranch in Wisconsin possessed "female" size directly related to nutrition mismanagement via provision of a very lean dietary program in June and early July. The recommendation of Leoschke at that point for a higher energy program (26% fat) was accepted and resulted in the highest incidence of "hippers" (more than 1/3rd of all male pelts) ever achieved in the history of the worldwide fur industry (Langenfeld, 1962).

Another experimental approach to producing "hippers" in male mink provides a higher energy program for the late growth period in blue iris kits. It is well acknowledged that blue iris kits have a higher energy requirement than most mutation mink such as pastels. Thus, the brilliant idea of Leoschke to initiate a higher energy pellet program for the Falcon Mink Ranch in Wisconsin in 1988. With a control pellet formulation providing 20–22% fat and an experimental pellet with 2% choice white grease replacing 2% kibbled corn, the net result was a yield of a high proportion of "hippers" on the higher energy blue iris experimental pellet product (Falcon, 1988).

"Wet Belly" Disease

Intensive studies at Oregon State University by Oldfield and associates have found that the "wet belly" disease of the mink is primarily related to high fat levels in the fall fur development diet and/ or excessive weight of the male kits prior to pelting (Wehr et al., 1983).

Sub–Optimum Fur Development

A number of experimental studies at the National Research Ranch have indicated that the late growth (July–August) performance of male kits can have a profound influence on the following fall fur development. These feeding trials support the basic observation that sub–optimum male kit weight performance in July–August can be very supportive of superior fur development directly due to enhanced feed intake—hence higher protein intake—in the fall months.

The observations at the National Research Ranch have been confirmed by a fascinating Danish study (Hansen et al., 1992), on the relationship between male kit weights at different ages and fur quality of pelts at marketing. See Table 4.9.

TABLE 4.9. Correlation Between Male Mink Body Weights at Different Ages and Fur Quality of Pelts Marketed

Age	Fur Quality Correlation
Birth	+0.16
14 Days	*
28 Days	*
42 Days	−0.16
June	−0.25
July	−0.24
August	−0.24
September	−0.40
October	−0.41
December	−0.34

* No significant correlation

Note: With female pelts, a negative correlation between body weights and fur quality is only found for body weight in December (−0.18).

From the previous commentary, it is obvious that extra "heavy" male kits coming into September can yield a net economic loss to the rancher in terms of sub–optimum fur development and "hipper" and "wet belly" disease problems. The latter two problems can be resolved by sound genetic management, whereby mink with genetics prone to these fur quality problems can be eliminated from the breeder colony over a period of years. The problem of sub–optimum fur development related to "heavy" males coming into September cannot be resolved by genetic management.

This particular problem is obviously a matter of energy management, requiring careful consideration of the actual energy intake of these kits over the period of July and August. It is important to understand that this problem cannot be resolved by such simplistic procedures as lower energy nutrition programs and/or by feed restriction via "The Man With The Spoon." With lower energy diets, the net result is higher feed volume per pelt marketed, a potential for significant infighting between kits, and a minimal effect on final weight of the male kits coming into the fur development period. Restricting the volume presented to the mink is simply impractical: it is labor intensive and too often results in a combination of male kit weights in late August—"skinny," of sub-optimum weight, or extra "heavy." The practical answer to the problem of extra "heavy" male kits coming into September is to give careful thought to the primary two factors that regulate the feed intake of animals: the energy density of the dietary program and the palatability of the dietary program for the mink.

It is well acknowledged that the mink, a carnivore, is very particular about the taste of feed products and that such mink feedstuffs as poultry feet, whole eggs, wheat germ meal, and alfalfa leaf meal have low taste appeal ratings. Even mink nutrition scientists involved in the formulation of commercial fortified cereals and pellets for 21st century mink nutrition acknowledge that these products do not have the highest palatability for mink. This point is well illustrated by Table 4.10.

TABLE 4.10. Male Mink Kit Growth And Fortified Cereal Levels*

Color Phase	July–August grams	September–November	Totals
Darks			
Ranch Mix w/ GnF–20	920	310	1230
Ranch Mix w/ GnF–70	820	450	1270
Pastels			
Ranch Mix w/ GnF–20	970	450	1420
Ranch Mix w/ GnF–70	870	580	1450

* Leoschke (1977)

These experimental data show that the reduction in weight gains of the male kits in the late growth period did not result in "small" mink at pelting. On the other hand, the GnF–70 program provided for extra weight gains in the critical fur production months through a net increase in feed volume and a higher protein intake. with resultant superior fur development.

Ranchers involved in high quality pellet programs in North America have not experienced any significant incidence of "hippers" or "wet–belly" mink but have noted superior fur development in terms of lighter leather, silkier fur, and unique sharp color in pelts marketed.

A good question at this point might be "Why was there superior fur development of male mink kits provided lower palatability dietary programs during the late growth (July–August) phase of the ranch year?" The answer is simple in terms of a careful examination of the dietary environment of commercial ranch mink and that of their distant cousins in the wild. Kits on mink ranches in the late growth period have it "too good," with a limited food hunting area and readily available feed 24 hours a day. In other words, they have nothing to do all day and night long but "eat and sleep." On the other hand, mink in the wild require acres of hunting space and maximum activity to achieve their daily feed requirements—and, in most cases, their prey is relatively lean during the summer months. No wonder ranch raised male mink kits are extra "heavy" coming into September and the critical fur development months.

MEp—Fur Development–September–Pelting

Dietary planning for the critical fur development phase must give serious consideration to that level of energy supportive of top fur development. The National Program of Mink Nutrition recommends fat levels in the range of 20–22% in order to provide maximum protein intake for top fur development and to achieve optimum fall weight gains. Such levels avoid excessive male weights that yield "hipper" and "wet belly" occurrence and result in economic gain at pelt marketing. In mink herds without a history of "wet belly" disease, higher energy levels may be quite satisfactory provided that the protein level is a minimum of 30% ME and the diet contains a minimum of 30% of quality protein feedstuffs—that is, cooked eggs, cheese, whole fish, and muscle products including whole poultry.

For most mink the energy program providing 20–22% fat is quite satisfactory. However, for certain higher energy demanding mink such as blue iris, higher levels of fat may be required for top performance of the mink. The Oregon Sandberg Mink Ranch employs a standard 20–22% fat pellet for the period from July through September and then provides a 22–24% fat pellet from early October to pelting (Sandberg, 1990). It has been noted that with blue iris mink in Minnesota, an 18–20% fat level was unable to maintain body weights in late October.

Protein

Protein polymers consist of as many as 22 different amino acids. Proteins are the main component of the muscles, organs, and endocrine glands. They are the major constituent of the matrix of bones, teeth, skin, nails, and hair as well as of blood components including hemoglobin oxygen carriers, the plasma polymers responsible for the regulation of osmotic pressure and the maintenance of water balance in the body. The blood antibodies as well as enzymes and many hormones are also protein in nature. All of the proteins present in the anatomy and physiology of an animal are in a constant state of flux, that is, a turnover related to degradation and synthesis. Even though some of the constituent amino acids are reused, the recycling metabolic process is not completely efficient. Also, some of the amino acids are used for energy and some proteins are lost from the body. In the case of growing, furring, or pregnant animals, additional body tissue is being synthesized. Since animals cannot synthesize amino acids as plants can, they require an outside resource of protein (amino acids) to survive and reproduce. In terms of nutrition, proteins also serve as an energy resource for fur animals, providing a metabolizable energy (ME) value of 4.5 kilocalories/gram.

In 1838 a Dutch chemist, Gerhardus Johannes Mulder, described certain organic material as "unquestionably the most important of all known substances in the organic kingdom. Without it no life appears possible on our planet, through its means the chief phenomena of life are produced." Jön Jakob Berzelius, a contemporary of Mulder, suggested that this complex nitrogen–bearing substance be called *protein* from the Greek word *proteios* meaning "take the first place" or primary.

Inasmuch as the mink is a carnivore, ranchers must give protein resources primary consideration in planning commercial mink ranch dietary programs. For modern mink nutrition management, without question, protein is the key nutrient since pelt production requires more protein than any other nutrient, and in feed economics, protein is the most expensive nutrient resource, affecting the balance between expenditure per pelt and pelt dollars achieved.

Animals do not require protein of itself, but actually require the individual amino acids present in the feedstuff protein, all of which are essential units for the synthesis of protein polymers required in the anatomy and physiology of animals. However, the animal body can synthesize more than half of the required amino acids through metabolic products of glucose and "extra" nitrogen resources, including amino acids provided in the body's physiology which are in excess of actual requirements of the animal at that specific time period. This process is termed transamination. Thus dietary proteins are considered in terms of their biological value in terms of their capacity to provide an optimum mixture of both essential and non–essential amino acids. The essential amino acids are considered to be indispensable inasmuch as the animal body cannot synthesize them, while the non-essential amino acids (actually required by animal's physiology) are termed dispensable inasmuch as the animal physiology is capable of providing them via metabolic processes.

The amino acid requirements of the mink can be classified as follows:

Essential: Arginine, Histidine, Isoleucine, Leucine, Lysine, Methionine, Phenylalanine, Threonine, Tryptophan, and Valine

Semi–Essential*: Cystine, Tyrosine

Non–Essential: Alanine, Aspartate, Glutamate, Glycine, Hydroxy–Proline, Norleucine, Proline, Serine, Taurine**

*Amino acids without an absolute dietary requirement but the presence of which in the animal's diet can minimize the requirement of a corresponding essential amino acid. This is accomplished by modification of the semi–essential amino acid's structure by metabolic processes, thereby providing a "fill up" capacity. Such action is illustrated by the conversion of tyrosine (non–essential) to phenylalanine (essential). The key role of cystine in fur synthesis is that of minimizing the dietary requirement of methionine (essential) for the metabolic conversion to cystine in the critical fur development phase of the mink ranch year.

**Non–essential for the mink, but an absolute essential amino acid for the cat and the fox.

90

Proteins are very large molecules obtained via the combination of multiple amino acids (monomers) in a linkage termed a peptide bond (secondary amide) to yield structures with hundreds of amino acid units. Smaller chains of amino acids are termed polypeptides, well illustrated by hormones such as human insulin with 66 amino acids and oxytocin with nine amino acids in a cyclic structure.

Physiology

For the most part, protein molecules and large polypeptides cannot be absorbed by the intestinal mucosa to any appreciable degree and hence must be broken down (hydrolyzed) by enzymes in the stomach and small intestine to yield free amino acids and small peptides. The exception is colostrum, the immature milk provided at birth which is unique in terms of high protein levels related in part to immunoglobulins. These immunoglobulins (congenital antibodies) are absorbed intact and provide the newborn with transient protection in the form of a passive immunity, for example to distemper, for up to 8–10 weeks (Hunter and Lemieux, 1996). In specific susceptible individuals, intact proteins may be absorbed which may lead to immunological sensitizations and resultant problems of allergy to specific food proteins.

Protein digestion is initiated in the stomach where hydrochloric acid not only denatures the proteins so that they are more susceptible to enzymatic action but also activates the inactive pepsinogen (secreted into the stomach by the direct physiological reaction to food intake) to the active form of pepsin. Pepsin brings about a breakage (hydrolysis) of specific peptide linkages to begin the process of digestion of the protein. In the small intestine pancreatic secretions include two inactive proteolytic enzymes, trypsinogen and chymotrypsinogen. Trypsinogen is activated to trypsin by enterokinase, an enzyme from the intestinal wall. Chymotrypsinogen is activated by trypsin to chymotrypsin. The active enzymes continue the process of hydrolytic cleavage of the protein by attacking certain peptide linkages. The final hydrolysis of proteins and polypeptides is completed by a group of enzymes termed peptidases which are secreted from the intestinal mucosa. The end products of digestion are amino acids, dipeptides, and tripeptides which are absorbed by the gut muscoa. Hydrolysis in the brush border and also in the cytosol of the mucosal cells insures that most of these small peptides will yield free amino acids for the portal blood circulation leading to the liver. The liver maintains control over the flux of amino acids via catabolic reactions to yield energy and urea as a by–product, synthesis of specific blood proteins, and the release of free amino acids into the general blood circulation.

Requirements—Quantity–Males

Over the period of the last century, the worldwide fur industry has considered the protein needs of the mink in a variety of ways, including:

1. **% Protein—Dehydrated Basis;**
2. **Calorie/Protein or Protein/Calorie Ratios;**
3. **% of Metabolic Energy (ME) as Protein.**

Commentary on each of these perspectives of the protein requirements of the mink follows.

% Protein–Dehydrated Basis

Since the origin of the fur industry in North America, ranchers have planned their mink and fox diets on the basis of percentage of protein–dehydrated basis. In terms of the 21st century understanding of the nutrition of fur animals, this perspective is simplistic because it does not take into account the actual protein intake of the mink, obviously a key factor in both mink nutrition management and mink ranch business management. The feed intake of any animal is related both to the palatability (taste appeal) of the diet and the actual useful (metabolic) energy content of the specific diet. Thus two practical mink ranch diets with identical protein content but different fat content, (for instance, 20% fat versus 28% fat) would provide feeding regimens wherein the mink on the higher energy diet would actually consume 11 % less protein (Leoschke, 2001). Obviously, planning mink ranch diets on the basis of protein content without serious consideration of final metabolic energy content of the diet does not make "common sense."

Calorie/Protein or Protein/Calorie Ratios

The assessment of the protein requirements of the mink on the basis of Calorie/Protein ratios is directly related to experimental studies on protein requirements by Allen et al. (1964). In that study, Energy/Protein ratios (that is, Apparent Digestible Energy/Apparent Digestible Protein, ADE/ADP ratios) were employed under the supervision of the eminent Canadian mink nutrition scientist, E.V. Evans. The inverse of the ADE/ADP ratio is the ADP/ADE ratio, which is essentially identical to the modern day perspective of employing percentage ME as protein in discussing the protein requirements of the mink. The following explanations may be useful:

ADP/ADE = grams digestible protein/digestible energy value of the diet in calories, and

% M.E. as Protein = grams of digestible protein / digestible energy value as calories (expressed as energy value).

92

Thus, all factors considered, it is equally valid to express the protein requirements of the mink as Calorie/Protein ratios or as percentage of protein as ME. Both formulations provide accurate assessments of the protein requirements of fur animals. See Table 4.11.

TABLE 4.11. % M.E. As Protein vs Protein/Calorie Ratios

% M.E. as Protein	25	30	35	40
Protein/Calorie Ratio	5.6	6.7	7.8	8.9

% ME as Protein

Assessment of the protein requirements of the mink on the basis of percentage of Metabolic Energy (ME) as protein (See Table 4.12) has been employed by European mink ranchers for many years and is the basis of all Scandinavian experimental studies. The distinct advantage of this system (as well as Protein/Calorie ratios) is that protein levels are expressed in terms of the energy value of the mink diet. This assessment is very logical since the energy variations of the animal's diet (primarily related to fat content) determine the actual daily protein intake of the animal.

TABLE 4.12. Calculations of Protein as % Metabolic Energy

	Nutrient%		Digestibility		ME*			Total
Protein	36	x	85	=	30.6	x 4.5	=	137.7
Fat	28	x	90	=	25.2	x 9.5	=	239.4
Ash	8							
NFE**	28	x	75	=	21.0	x 4.2	=	88.2
								465.3

* M.E. = Metabolic Energy or actual total energy value as achieved within the digestive and metabolic physiological processes of the mink from each energy resource unit.

** Nitrogen Free Extract = "Carbohydrates" = 100% – (% Protein + %Fat +%Ash)

% Metabolic Energy as Protein = Protein Calories x 100/total ME Calories

=137.7 x 100/465.3 = 29.6% Protein as total ME

To summarize: although both perspectives of Protein/Calorie ratio and percentage of ME as protein are valid, the concept of % ME as protein represents 21st century worldwide concepts of animal nutrition, while the concept of ADE/ADP (or the reverse Protein/Calorie ratio) represents an earlier perspective on animal nutrition.

93

TABLE 4.13. Recommended Protein Levels—% Metabolic Energy

Phase of Ranch Year	North America*	Nordic Association**
Late Growth		
July–August	25	30
Fur Production		
September–Pelting	30	30
Maintenance		
Pelting – Late February	25	35
Reproduction:		
March–April	40	
Whelping/Lactation		
Late April–May	35	40
Early Growth		
June	30	40

* Leoschke (2001), ** Enggaard–Hansen et al. (1991)

Assessment

Multiple procedures have been employed to determine the protein requirements of the mink for each phase of the mink ranch year including nitrogen–balance, kit weight gains, fur quality development, reproduction performance, and lactation results as measured by kit growth. A few words are warranted on nitrogen–balance (N–balance) procedures for the assessment of the protein requirements of the mink.

Of the major energy nutrients required by animals––carbohydrates, fats, and proteins— the protein resources are unique because they contain nitrogen. Thus, for the most part, a qualified measure of the protein content of a given feedstuff can be ascertained by multiplying the nitrogen content by 6.25. It should be noted that the conversion factor of 6.25 of nitrogen to protein is likely to overestimate the protein content of fish by–products containing a high proportion of skin and bones (Skrede, 1977). Thus the percentage of dietary nitrogen retained in the body of an animal can be considered as an assessment of the actual protein requirement of the animal for each specific phase of its life, that is, an indication of the actual utilization of the protein (nitrogen) intake of the animal. This N–balance concept is of interest but has not been given major emphasis in this report inasmuch as for the most part, it often overestimates the actual protein needs of the mink. Table 4.13 illustrates this basic point.

TABLE 4.14. Mink Protein Requirement Assessment

Period	Protein Requirement – Males – % Metabolic Energy	
	N–Balance*	Mink Performance**
Late Growth	41–42	25
July–August		
Fur Development	31–32	30
September–Pelting		

* Glem–Hansen (1980) **See Table 4.13 Recommended Protein Levels – % Metabolic Energy

In another study Chwalibog and Thorbek (1980) concluded that on the basis of N–retention, a crude protein level of 48% (ca. 36% ME) was preferable to a crude protein level of 38% (ca. 27% ME) for the late growth phase, obviously excessive in terms of Table 4:13. A study by Glem–Hansen and Jorgensen (1976) showed that weight gains of mink kits are independent of the nitrogen intake. It is of interest, for instance, that Skrede's 1978a study calls the N–balance assessment of protein requirements of the mink into question. An analysis of nitrogen retention revealed no differences between protein sources of fillet cuttings and filleting scrap. However, the same study indicated that fillet scrap as a protein resource had lower biological value (lower level of essential amino acids) for the mink as illustrated by reduced weight gains of the male kits as well as sub–optimum underfur density.

That the nitrogen retention procedure might provide an inaccurate assessment of actual protein requirements is seen in a report by Sinclair et al. (1962). Their experimental data supported the conclusion that on the basis of N–balance, the protein requirement for fur development was only 20% of ME. However, Allen et al. (1964) showed that this N–balance assessment was invalid. They established that the 1962 data were anomalous, since the animals on the higher protein diet had achieved most of their growth earlier in the late growth period (July–August).

Recommendations—Protein Quantity

For most animal species, the protein requirement is met with as little as 10% of the digestible energy (Crampton, 1964). The mink, as an obligate carnivore, has a demand for a large proportion of protein in its diet, as high as 25–40% of the daily intake of calories and much higher than that of other carnivores such as the dog and the weasel which need is 4–12% of caloric intake (MacDonald and Rogers, 1984).

An animal's requirement for protein depends upon many factors including energy intake, level of activity, physiological status, growth capacity, previous nutrition, individual differences, and genetic differences. An estimated protein requirement always depends on the physiological activity for which the requirement is met since the protein level (quantity and quality) required for top performance in one physiological process may not necessarily be sufficient for another physiological process. Thus the presentation of specific commentary on recommended protein levels in practical mink ranch diets presented in Table 4.14.

95

TABLE 4.14. Recommended Protein Levels—% Metabolic Energy

Period	Protein Requirement – Males – % Metabolic Energy	
	North America*	Nordic Assoc. Agri. Scientists**
Phase of Ranch Year		
Late Growth: July–August	25	30
Fur Production: September–Pelting	30	30
Maintenance: Pelting–Late February	25	35
Reproduction: March–April	40	
Whelping/Lactation: Late April–May	35	40
Early Growth: June	30	40

* Leoschke (2001),**Enggaard–Hansen et al. (1991)

These quantitative recommendations support the top performance of the mink only when a significant proportion of the ranch diet consists of quality protein feedstuffs. This point is well illustrated by the recommendations for levels of quality protein feedstuffs within the National Program of Mink Nutrition for North America: Late Growth, 20%; Fur Development, 30%; Maintenance, 20%; Reproduction/Lactation, 40%; and Early Growth, 30%. The rationale for unique recommendations for North America will be discussed in the paragraphs to follow.

The primary cause of death for mink on sub–optimum levels of protein may be fatty infiltration of the liver with consequent liver dysfunction (Damgaard et al., 1994; Damgaard, 1998). It is likely that the accumulation of fat in liver cells results in their destruction and the subsequent release of alanine transferase (ALAT) since high plasma activities of ALAT are related to a high degree of fatty infiltration of the liver (Damgaard et al., 1998; Damgaard and Clausen, 1999). The enzyme ALAT is an intracellular enzyme that is relatively liver specific in mink (Juokslahti et al., 1980).

Late Growth—July–August

The earliest recommendation of specific levels of protein for the growth period of the ranch year by a qualified fur animal nutrition committee was provided by the National Research Council, National Academy of Science, USA (Harris et al. (1953). The recommendation of 22 to 26% protein for mink kits between the ages of 7 and 16 weeks of age was based on two reports, Bassett (1948) and Basset et al. (1951). It should be noted that this early recommendation of the protein requirement for top performance in the growth period is of questionable value for modern mink nutrition management since the size of the mink kits in these studies was unusually small (Lalor, 1956). It is also noteworthy that the NRC recommendation did not provide a specific relationship between protein levels and energy content of the experimental diets.

The excellent detailed studies on the protein requirements of the mink by Canadian researchers (Sinclair et al., 1962; Evans, 1963; Allen et al., 1964) involved experimental studies over

96

a time period that included early growth through fur development (June–December). Thus the research data are not applicable to the current discussion, which targets only the late growth phase.

In Table 4.14, it is noted that the protein levels recommended in 1991 by the Nordic Association of Agricultural Scientists for both the late growth and the fur production phases of the mink ranch year are identical, 30% of M.E. as protein. Multiple experimental studies indicate that with ranch diets providing significant levels of quality protein resources (minimum 20%), protein levels of 25% ME are ample for the late growth period and even for the fur development phase if the only criterion was pelt length and not fur quality (Skrede, 1978a; Tyopponen et al., 1986, 1987; Kerminen–Hakkio, 2000; Rasmussen and Borsting, 2000; Sandbol et al., 2004).

In summary, it is obvious that the quality of protein in the mink's diet will determine the exact level of percentage of ME as protein required for top performance of the mink for the late growth phase of the mink ranch year. The Nordic Association of Agricultural Scientists took a relatively conservative viewpoint in recommending a protein level of 30% M.E. for the late growth phase of the mink ranch year, as it did not take into account the more recent experimental data.

Fur Development—September–Pelting

The National Research Council (Harris et al., 1953) provided the earliest recommendation of specific levels of protein for the fur production period of the ranch year. The committee recommendations of 22 to 26% for mink kits between the ages of 7 and 16 weeks of age and 16–22% protein from 16 weeks to maturity were based on two reports, Bassett (1948) and Bassett et al. (1951). However, these early observations are of questionable value for modern mink nutrition management because the size of the mink kits was unusually small (Lalor, 1956), no specific data was presented on the caloric density of the experimental diets, and the sole criterion for the requirements was the weight gains of the kits without reference to the quality of fur production on the different protein levels of the research diets. Experimental reports by Canadian researchers (Evans, 1963; Allen et al., 1964) make the same error, assuming that kit weight gains are an adequate criterion for protein requirements of the mink for fur development. This assumption led them to conclude that the protein requirement for fur development is less than that for kit growth:

> "After 16 weeks, the growth ratios were lower and the Apparent Digestible Energy/Apparent Digestible Protein (ADE/ADP) ratio appears to be less critical. Hence, a wider ADE/ADP ratio (lower percentage of ME as protein) could be fed to mink after 16 weeks with satisfactory results."

Their recommendations for the protein needs of the mink were 36% ME as protein for the period of 6–16 weeks (early growth and late growth phases) and only 22–26% ME as protein for the period of fur development, 7–20 weeks.

We now know, without question, that the protein requirements of the mink for top quality fur development are significantly higher than those which provide satisfactory weight gains. The first observation of this basic fact was noted in studies at the University of Wisconsin with mink on purified diets in 1956 (Leoschke and Elvehjem, 1959). Mink kits placed on the experimental diets in early July grew satisfactorily until early September when a nutritional crisis arose as evidenced by anorexia, weight loss, and finally death. Supplementation of the late growth purified diet with two amino acids (arginine and methionine) resolved the nutritional deficiency within 24 hours.

Multiple studies over the last quarter century indicate that although 25% of ME as protein is quite satisfactory for kit growth and optimum pelt length, a protein level of 30% ME is an absolute requirement for top fur development (Stout et al., 1963; Adair et al., 1966; Jorgensen and Glem–Hansen, 1972, 1979; Skrede, 1975, 1978a; Glem–Hansen, 1980a, 1980b, 1982a, 1982b; Lund, 1983; Berg et al., 1983, 1985; Tyopponen et al., 1986, 1987; Lund and Hansen, 1987; Hilleman and Lyngs, 1989; Hansen et al., 1991; Borsting and Olesen, 1993; Damgaard, 1997; Damgaard et al., 1998; Hejlesen et al., 1998; Damgaard and Clausen, 1999; Kerminen–Hakkio, 2000; Rasmussen and Borsting, 2000).

It is of interest to note that the Skrede (1978a) report indicated that top fur development was possible with only 27% ME as protein when that protein resource consisted of fillet cuttings and fish muscle protein with high biological value, on the basis of amino acid pattern and digestibility. The same report indicated reduced underfur density with male skins on diets providing a high concentration of filleting scrap (lower biological value, related to a second rate amino acid pattern and relatively low digestibility for the mink as directly related to high ash content). These observations accord with the thinking of fur animal nutrition scientists throughout the world: that as the biological value of a protein feedstuff decreases, higher and higher quantities of protein as % ME are required for top performance of an animal. Thus with protein resources of high biological value, such as whole egg protein and muscle meats, one can provide for top fur development with a protein level significantly less than 30% ME. The Skrede (1978a) study also indicated that the density of the guard hair was not significantly related to protein levels in terms of protein quantity and protein quality.

Recent experimental studies confirm the work of Skrede (1978a) that protein levels less than 30% ME support top fur development provided that these experimental diets place a major emphasis on the biological value of the constituent protein feedstuffs, or an enhanced ratio of essential to non–essential amino acids, that is, an "ideal pattern" of amino acids. Experimental studies by Hejlesen and Clausen (2001) indicated that dietary protein levels as low as 24% ME, with fats or carbohydrate providing 50% of ME, yielded top fur development. A study by Clausen and Sandbol (2003) indicated top fur development with a dietary plan providing an essential amino acid to total amino acid ratio of 0.52 within a protein level of 25% ME; moreover, with the same dietary plan, raising the protein level to 29% and 33% of ME did not enhance fur development.

More recent studies by Sandbol et al. (2004c) employing the concept of an "ideal protein"

98

pattern of amino acids, noted that while the minimum protein requirement for top pelt length was 24% ME, a protein level of 28% M.E. was required for top fur development. An additional study by Sandbol et al. (2004b) confirmed this conclusion.

Maintenance—December–February

Studies by Hejlesen and Clausen (2002) and Clausen and Hejlesen (2003) indicate that 30% of ME as protein is ample for the period from pelting to late February. The Leoschke recommendation of 25% of ME as protein for this period (See Table 4.14) is based on the logic that the protein requirement for maintenance should not be higher than that for the late growth period.

Reproduction—March–April

All factors considered, without question the reproduction/lactation phases of the mink ranch year are the most critical relative to protein requirements. In work towards developing commercial pellets for modern mink nutrition, the major problem requiring many years of research was to discover dehydrated protein resources with high enough biological value to support top reproduction/lactation performance.

As early as 1957, Moustgaard and Ries (1957) illustrated the importance of protein in the ranch diet for the support of pregnancy in the mink as seen in Table 4.15 and Graph 4.1.

TABLE 4.15 Protein (N) Deposition in the Pregnant Uterus

Days After Mating	Total Deposit In milligrams N*
0	3.9
5	9.1
10	16.2
20	29.4
30	53.7
40	98.6
50	181.4
55	294.0

* Values adjusted to 6 fetuses

GRAPH 4.1. Protein Deposition in Pregnant Uterus of Mink

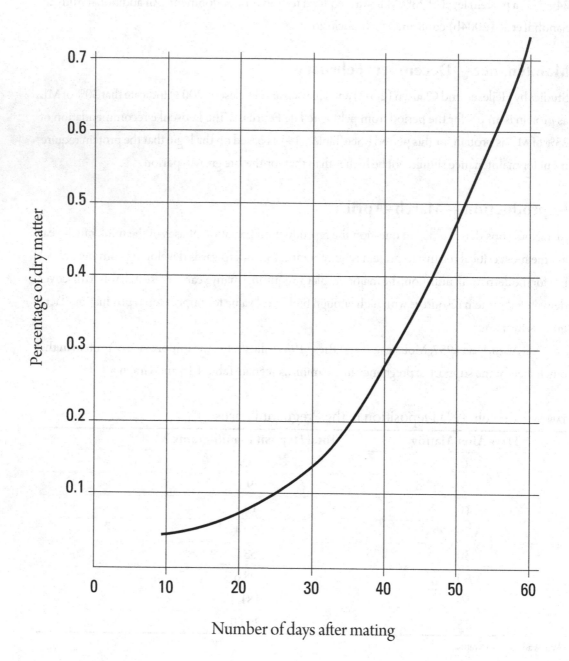

As early as 1967, Ahman reported on experimental studies on the protein requirements of the mink for the combined reproduction and lactation, phases of the mink ranch year. Thus these studies did not differentiate between the protein requirement for pregnancy and that for lactation.

Experimental studies on the protein requirements of mink for reproduction/lactation indicate that protein levels as low as 25–27% ME provided ranch averages at whelping equal to diets with much higher levels of protein; however, mortality from birth to weaning was very high in the low protein group (Jorgensen and Glem–Hansen, 1977; Skrede, 1978b). It would appear to be obvious that the viability of kits at birth and their survival to weaning is of equal importance if not more significant than the whelping average.

It also appears that protein quality (Biological Value) is as important if not more critical than protein quantity for reproduction/lactation. That protein quality is more critical than protein quantity for the mink during reproduction/lactation and fur development has been seen consistently in a half–century of research studies at the National Research Ranch. In early experimental studies (Leoschke, 1972) this distinction was noted. In a series of experimental studies with equal quantity of protein (45% ME), the pregnant mink on a sub–optimum quality of protein diet provided a 6.0 ranch average at whelping but only a 2.0 ranch average at weaning while pregnant mink provided higher quality protein resources yielded a 6.0 ranch average at whelping and a 4.5 ranch average at weaning.

A number of studies indicate that 40% of ME as protein is quite satisfactory for the reproduction/lactation period (Damgaard et al., 2000; Hejlesen and Clausen, 2000, 2001, 2002; Clausen and Hejlesen, 2003; Clausen et al., 2003).

A study by Rasmussen and Borsting (2000) is of interest from another perspective. Their experimental work indicated that 40% of ME as protein was sufficient for mink kits to express their genetic capacity to produce hair follicles.

Although experimental studies indicate that 40% of ME as protein is quite satisfactory for reproduction performance, higher levels of protein in terms of quantity and quality may be required under circumstances wherein significant feed restriction may be necessary ("heavy" pregnant females and/or a warm April) as ranchers try to keep the pregnant mink in top weight condition for whelping. This very point provides the logical basis for the Leoschke recommendation of a relatively high level of protein for the pregnancy phase of the mink ranch year. Obviously, in colder environments, lower protein levels may equally support top reproduction performance.

Whelping/Lactation—Late April and May

Investigations by (Glem–Hansen and Jorgensen, 1973; Glem–Hansen et al., 1973; Glem–Hansen and Jorgensen, 1975) indicated that the quantity and quality of protein provided the mink did not influence the quality of the milk in terms of the composition of protein, amino acids, fat, or carbohydrates. They noted that the composition of fatty acids in the milk clearly reflected the fatty acid composition of the diet presented to the lactating mink. Data on kit weight gains indicated that the quantity and quality of protein provided the lactating mink did influence the quantity of milk provided the young.

In the paragraphs to follow, it is obvious that multiple experimental studies have indicated a wide range of protein requirements for the top lactation performance of the mink (30–40% ME). The key factor is, of course, the quality of protein resources provided the mink, as well illustrated by the work of Glem–Hansen and Jorgensen (1973), Glem–Hansen et al. (1973), and Glem–Hansen and Jorgensen (1975). These studies indicated that 35% ME as protein resources with high biological value (BV) provided top kit weight gains from the 3rd to the 18th day of lactation, while with lower BV protein resources, a protein content as high as 40% ME did not provide top kit weights at 18 days. Simply said, ranchers and mink nutrition scientists cannot afford to underestimate the factor of protein quality (BV) in meeting the protein requirements for top lactation performance.

Too often decisions on the formulation of mink ranch diets are unfortunately related to "bottom line" economics which may prove to be "penny wise and pound foolish," as well illustrated by a commercial pellet program (not National Fur Foods) of a few years ago. The initial pellet product marketed provided top protein nutrition in both quantity and qualify. The net result for a specific mink ranch was a commercial pellet program providing top reproduction/lactation and a 5.7 ranch average at weaning relative to a 4.8 ranch average at weaning for mink on the same ranch with a more conventional feeding program. In the next year, a decision was made by business management to give more thought to feed ingredient economics without consulting the mink nutrition scientist who had developed the pellet program (Leoschke, 1987). Disregarding serious thought about reproduction/lactation performance of the mink, a decision was made to maintain the quantity of protein but to significantly undercut the quality (BV) of protein provided. The net result for the unfortunate mink rancher was that the reproduction/lactation performance of the mink on the commercial pellet was reduced to a 4.5 ranch average at weaning (equal to the mink on the conventional ranch diet) but with complete lactation failure, requiring the transfer of the mink on the commercial pellet to the conventional ranch diet. Obviously, it is short sighted to undercut the protein nutrition of the mink during the critical reproduction/lactation or fur development phases.

Many studies over the period of the last 30 years support the conclusion that a minimum of 40% ME as protein is required for the top lactation performance of the mink (Jorgensen and Glem–Hansen, 1972; Skrede, 1978a, 1978b, 1978c; Glem–Hansen, 1980, 1992; Larsen, 1994; Kerminen–Hakkio et al., 2000; Damgaard et al., 2000; Rasmussen and Borsting, 2000; Fink et al.,

2004; Hejlesen and Clausen, 2002; Clausen and Hejlesen, 2003). A Clausen et al. (2003) study provided experimental data indicating that 30% of M.E. as protein was satisfactory for the lactation period until the kits started to eat, whereas a 40% M.E. as protein level was required for top growth performance of the kits. The Clausen et al. (2003) study was essentially confirmed by Fink et al. (2004) wherein an experimental plan of 60:35:5, 45:40:15 and 30:45:25 resulted in highest kit weights at four weeks with the 30:45:25 program, but highest kit weights at nine weeks with the 45:40:15 program. A report by Clausen et al. (2005) confirms the Clausen et al. (2003) and Fink et al. (2004) studies, indicating that relative to a 52% M.E. as protein program, a 32% M.E. program provided minimum lactating female weight loss and superior kit weights at 4 weeks.

Working with relatively low protein levels of 32:67:1, 32:52:16 and 32:37:31, Fink et al. (2002) showed that the combination of 16 or 31% M.E. as carbohydrate had no adverse effects on glucose homeostasis or glucose metabolism in lactating mink.

At the same time, other experimental work involving protein resources with higher protein quality (BV) supports a 35% ME as protein for the top lactation performance of the mink (Glem–Hansen and Jorgensen, 1973; Glem–Hansen et al., 1973; Glem–Hansen and Jorgensen, 1975; Glem–Hansen, 1979). Other studies by Fink and Tauson (2000, 2004) indicated that the highest milk yield of the mink was with a group of lactating animals fed a diet containing high quality protein resources with a protein quantity of 30% ME in combination with a fat–carbohydrate ratio of 45–25. However, the study also recorded that at nine weeks the highest kit weights were obtained with a medium protein diet providing ratios of 30:40:30 as % ME. It was suggested that that the lower early growth period weight gains for this group may have been due to a sub–optimum capacity of the kits to utilize the high carbohydrate content of the low protein experimental regimen.

It is apparent that Leoschke's recommendation of 35% ME as protein for the lactation period (see Table 4.14) is ample for North American mink ranch dietary programs containing significant levels of liver, cooked/raw eggs, and/or cheddar cheese.

The significant decrease in protein requirements of the mink at the transition from pregnancy to lactation at whelping is supported by Graph 4.2. It is to be noted that the protein level in mink milk as percentage ME is relatively high at whelping when milk volume is minimal and then rapidly decreases in the days following parturition as the level of fat rises.

Studies by Glem–Hansen (1979) stressed the importance of protein quantity and quality for the lactation period. With mink kits on the same diet from weaning at six weeks to pelting, it was noted that the differences in kit size at weaning were not compensated for during the late growth and fur production periods. The pelt length was to a great extent influenced by dietary protein content in the sucking period but there were no statistically differences between groups in pelt quality or color.

GRAPH 4.2. Mink Milk Composition Glem–Hansen and Jorgensen (1957)

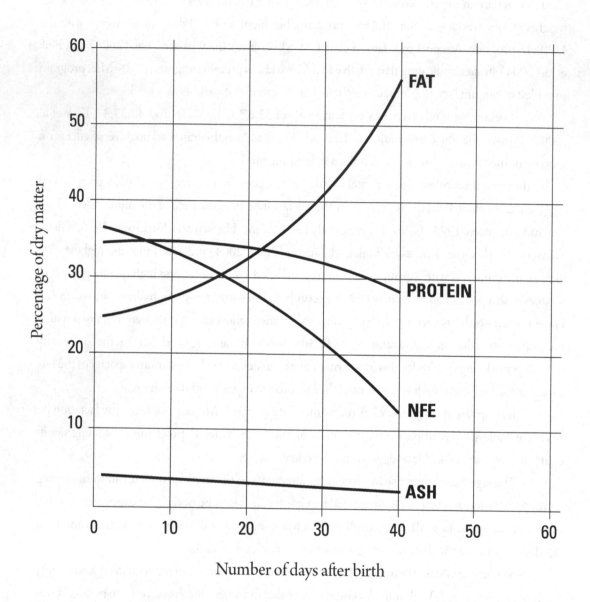

Early Growth—June

Initial studies on the protein requirements of the mink for the early growth period indicated a minimum of 40% of ME as protein (Rimeslatten and Skrede, 1957; Ahman, 1966, 1967; Jorgensen and Glem–Hansen, 1972). One exception was a study by Evans (1963) indicating a protein requirement of 35% ME (a Calorie/Protein ratio of 13). More recently, a report by Kerminen–Hakkio et al. (2000) with a protein mixture containing fish meal 50%, feather meal 20%, corn gluten meal 15%, and soybean oil meal 15% indicated that the protein requirement for very early growth 3–6 weeks was certainly lower than 40% but higher than 32% of ME. A recent study by Clausen et al. (2005b) indicated that from 4 weeks to 7 weeks, the kits need from 44 to 47% M.E. as protein for optimal growth and from 6 to 8 weeks, they need 40% M.E. as protein for optimal growth with the applied nutrient composition. Obviously, the wide variation in the experimental data on the protein requirements of the mink for the early growth period is directly related to the biological value of the various protein resources provided in each of the experimental diets. Very likely, all factors considered, the protein requirement for June would have been met by 30% of ME if the quality of protein had been superior, for instance, a diet providing high quality fish meals. Studies by Leoschke (1976) indicated essentially 90% of normal weight gains with male kits 7–9 weeks old on dry diets providing 27% of ME via high quality fish meals.

In terms of North American June ranch diets providing a minimum of 30 percent of high BV through liver, cooked eggs, and cheddar cheese, whole bird or whole fish, a protein level of 30% ME would be ample for the needs of the mink kits for the early growth phase.

Protein Requirements—Quantity–Females

For the most part, experimental studies on the protein requirements of the mink for the period from early June to pelting have involved only male kits. Thus the recommendations in Table 4.14. Recommended Protein Levels—% Metabolic Energy essentially applies to males for the early growth, late growth, and fur production periods. This is both logical and practical, inasmuch as male and female kits are raised together as litters in June and then as male/female pairs from July to pelting. However, without question, the protein requirements of female kits for growth and fur production are definitely lower than those of male kits as indicated in studies by Allen et al. (1964) and Skrede (1978a). Studies by Leoschke (1976) on the growth of mink kits in June are also of interest. Seven week old male kits placed on a special control diet providing 27% ME as protein from fish meals provided weight gains of 300 grams in the two week experimental period. When brothers of the male kits on the control diet are provided an experimental diet wherein 20% of the fish protein is replaced with powdered egg protein (the very highest protein biological value for animals), the male kits achieved weight gains of 360 grams, an indication that 27% ME as protein is sub–optimum for the early growth period when the sole protein resources are fish meals. When the male kits are paired with their sisters on both the control diet (100% fish meal protein) and the experimental diet (80% fish protein and 20% powdered egg protein) the female kit weight gains were identical. There was no nutritional advantage to the presence of the whole egg protein, which indicates that although 27% ME as protein (fish meal resources) is sub–optimum for male kits in June, the same level of protein is ample for female kits in the early growth period.

On this same topic a later study by Leoschke (1999) is of interest. In an experimental study with a pellet program providing a high level of a second quality fishmeal (moisture content less than 6.0%), the fur development of the male kits was sub–optimum while the fur development of the female kits was optimum. The quantity and quality of protein provided with the second quality fish-meal was inadequate for the top fur development of the male kits but ample for the top fur develop-ment of the female kits because the extra quantity of protein (a quantity above the actual require-ments of the female kits for top fur development) provided a final level of amino acids that met the protein needs of the female kits for top fur development.

Amino Acid Supplementation

Supplementation of most common practical mink ranch diets with amino acids may or may not actu-ally enhance the performance, and in some cases, amino acid supplementation may actually undercut performance. The addition of amino acids to a mink ranch diet is only of value if the specific dietary plan is sub–optimum in either the quantity or the quality of protein provided for that specific phase of the mink ranch year. The need for special amino acid addition to a specific diet should be ascer-tained by "Listening To The Mink," that is, carrying out carefully designed experimental studies.

DL–Methionine

The first report on the value of DL–methionine for modern mink nutrition was that of Watt (1952) in experimental studies with high fish diets. A supplement of 0.05% methionine (about 0.15% dry matter basis) provided extra growth and superior fur development. In Russian experiments (Milova-nov, 1961), the supplementation of the mink's diet with methionine and cystine did not influence the growth rate but significantly improved fur quality. Hooggerbrugge (1968) reported on the supple-mental value of both methionine and lysine to a commercial dry dietary program. However, studies by Jorgensen and Clausen (1964) and Jorgensen and Glem–Hansen (1970) indicated no enhance-ment of mink growth or fur characteristics with methionine supplementation of Danish ranch diets.

DL–Lysine

The first reported beneficial value of lysine supplementation for modern mink nutrition (Hoogger-brugge, 1968) involved a commercial dry dietary program. Ten years later, experimental studies with pellet formulations at the National Research Ranch (Leoschke, 1978), indicated a significant negative effect on growth of mink kits in the late growth period. This observation contrasts with the study by Hooggerbrugge and would seem to indicate a significant potential for superior protein resources in the National Fur Foods pellet formulation. One explanation for sub–optimum growth performance of the mink kits with lysine supplementation is physiological: the intestinal site for absorption into the blood circulation is identical for both arginine and lysine. Hence, excess lysine in the mink's diet is antagonistic to arginine in terms of inhibiting arginine absorption (Damgaard, 1997). It is of parallel interest to note that studies by Dahlman et al. (2002) with foxes indicated that lysine supplementation decreased early growth. Studies by Jorgensen and Clausen (1964), indicated that lysine supplementation of a typical Danish mink ranch diet did not enhance the growth or fur characteristics of the mink.

Protein Requirements—Quality–Biological Value

Introduction

An animal's requirement for protein for each phase of the mink ranch year will be directly dependent on the BV of the protein resources provided the animal. With protein feedstuffs of high biological value or quality, a minimal level of protein will be required. With protein feedstuffs of lower biological value, greater and greater quantities of protein will be required to meet the animal's amino acid needs for top performance.

Thus, all factors considered, without question, the quality (BV) of protein is equal to and more important than the quantity of protein provided the mink during the most critical phases of the ranch year—reproduction/lactation and fur development. The proteins found in different mink feed ingredients, whether of animal or plant origin, are not all the same but show a wide variety in structure and amino acid composition. Hence, a clear understanding of the protein needs of the mink requires an in depth analysis of each protein resource beyond the quantity of protein provided.

The quality of a protein mixture is directly related to the capacity of that protein dietary program to meet the total protein needs of the mink for that specific phase of the ranch year. It must provide a proper complement of amino acids to maintain the protein stores of an animal as represented by the proteins of the tissues and body fluids as well as supply extra amino acids that may be required for tissue synthesis related to growth and fur development or lactation performance. The biological value (BV) of a specific protein feedstuff is related to two factors: amino acid pattern of the protein resource and the digestibility of that protein resource for the mink. This basic point can be well illustrated by two protein feedstuffs commonly found in ranch diets, whole egg protein and poultry feet.

Whole Egg Protein: The world's standard for the highest biological value protein is whole egg protein with an amino acid pattern similar to animals' requirements and a high digestibility rating This observation is self–evident: the protein supports the very life of the embryonic chick.

Poultry Feet Protein: This is a classical example of a protein resource with low biological

107

value for the mink. It has both a low digestibility rating of only 52% (Leoschke, 1959) and a sub–optimum amino acid pattern. The skin and bone proteins provided are primarily collagen in nature, relatively low in critical amino acids including cystine, tyrosine, tryptophane, and valine.

A listing in descending order of the relative quality of protein provided by common mink feedstuffs is seen in tables 4.16, 4.17 and 4.19.

TABLE 4.16: Biological Value of Fresh/Frozen Protein Resources*

Quality Rating	Protein Feedstuffs
Highest Quality	Whole Egg
	Whole Milk Cheeses
	Skim milk Powder
	Liver
	Muscle Meats w/o Bone including Beef and Fish Fillets
	Hearts, Horsemeat, Pork, Poultry, and Whale
	Cottage Cheese
	Kidneys
	Whole Fish, Whole Poultry, and Nutria
	Poultry Necks and Backs
	Beef Tripe (Rumen) and Poultry Entrails
	Beef and Pork Lungs
	Spleens
	Filleted Fish (frames)
	Fillet Scrap w/o heads/viscera
	Fish and Poultry Heads
	Fish Skins
	Poultry Feet
Lowest Quality	Animal and Fish Bones

* A general listing of the quality of a variety of protein feedstuffs for modern mink nutrition as reported by Wood, 1964; Perel'dik et al., 1972; Leoschke (1962, 1984 and 2001, is based on experimental studies with laboratory rats, chicks, dogs, and mink as well as observations of the practices of North American mink ranchers.

Commentary is warranted on a few of these mink feedstuffs as follows.

Cottage Cheese termed casein, as an isolated dehydrated protein resource, has a relatively low level of sulfur amino acids and arginine. "Vitamin Test" casein marketed to research laboratories for employment in purified diets is a product of alcohol extraction followed by dehydration. With excess heat treatment, the final product can be dark tan in color. In this case, the nutritional value for the mink has been undermined inasmuch as a significant portion of the arginine content has been bound in a structure unavailable to the digestive processes of the mink (Leoschke and Elvehjem, 1959).

Kidneys

Perel'dik et al. (1972) have reported that kidneys have a relatively low level of the essential amino acids relative to muscle meats. Beef and Pork Lungs Perel'dik et al. (1972) have reported that lungs contain a high quantity of connective tissue, constituting a product with relatively low levels of the essential amino acids. Pork Lungs as a fresh/frozen product have a specific liability of high bacterial populations directly related to processing procedures. However, properly cooked pork lungs are supportive of the top fur development of the mink as a replacement for horsemeat (Leoschke, 1955).

Spleen

A single report from Denmark has indicated that significant levels of spleen in a mink furring diet can undercut sharp color in dark mink.

Fish By–Products

Higher Ash Studies by Skrede (1978a, 1978b, and 1978e) indicate that these products have relatively low biological value for the mink. They have lower protein digestibility, in that a 1% unit increase in content of ash caused a depression in both AND (Apparent Nitrogen Digestibility) and TND (Total Nitrogen Digestibility) value by 0.6% (Skrede, 1978c). They also contain a relatively high level of collagen protein with known low levels of cystine, tyrosine, trytophane, and valine. Mink ranchers concerned about fur quality will note Skrede's (1978a) findings that while the density of mink guard hair was not significantly related to the employment of higher ash fish by products, the grading of the under fur density of male skins showed that the fillet scrap dietary program provided a lower density of fur relative to fillet cuttings.

Fish Skin

This product contains a large proportion of collagen, resulting in a sub–optimum amino acid pattern even though the protein digestibility is 94% (Skrede, 1978c).

Poultry Feet

These have a very low palatability rating for the mink. In protein digestibility studies at the University of Wisconsin (Leoschke, 1959), only half of the animals would accept the 100% poultry feet dietary regimen as a nutritional program.

Fish Bones These products may impose a severe strain on the digestive system of mink, as noted by Helgebostad (1973).

TABLE 4.17: Biological Value–Plant Protein Resources

Quality Rating	Protein Feedstuffs
Highest Quality	Legumes – Soybean Oil Meal, Peanuts, Peas
	High Protein Wheat Gluten Meal
	High Protein Corn Gluten Meal
	Wheat Germ Meal
	Whole Grains – Barley, Oats, Wheat, and Rice

	Whole Grains–Corn
	Processed Grains–Toasted Products
	Cereal By–Products–Hominy (Corn Milling Product), Wheat Bran
Lowest Quality	Single Cell Proteins from Petroleum

For the most part, plant protein resources have lower digestibility ratings for the mink and provide an amino acid pattern relatively low in lysine and methionine in comparison with animal protein feedstuffs. The exception is soybean oil meal, with a relatively high level of lysine.

Commentary is warranted on a few of these products as follows.

Soybean Oil Meal: Multiple factors must be considered before employing soybean oil meal products for modern mink nutrition including the following: carbohydrate digestibility, palatability, methionine levels, fur characteristics, and the presence of a trypsin inhibitor. The first three factors all are sub–optimum. Low carbohydrate digestibility was noted in four studies—Leoschke (1965), Jorgensen and Glem–Hansen (1975), Johansson et. al. (1977), Jorgensen (1979). Low palatability was noted as early as 1952 by Watt and confirmed by Belzile (1976a, 1976b) and Skrede (1977). The sub–optimum amino acid pattern relative to methionine was readily resolved by the addition of Dl–Methionine as noted by Belzile (1976a).

Studies on fur characteristics by both Skrede (1977) and Johansson et al. (1977) indicated longer and longer guard hairs as the level of soybean oil meal increased in the experimental diets. Jorgensen (1979) has reported the development of dry and less silky pelts with the employment of soybean oil meal, a condition which he attributed to insufficient heat treatment of the soybean oil meal resulting in significant levels of active trypsin inhibitors. Finally, soybean oil meal contains a factor which inhibits trypsin, a pancreatic enzyme involved in protein digestion in the intestinal tracts of all animals. As early as 1917, Osborne and Mendel reported that soybean oil meal should be heat treated in order to support normal growth of animals. Raw soybeans contain a component (antitrypsin factor) that inhibits the action of the trypsin enzyme in protein digestion, but that the problem could be resolved by heating the raw soybean oil meal to denature the enzyme. Heat denaturation of the trypsin inhibitor is especially important in modern mink nutrition as the process also enhances the availability of the sulfur amino acids, cystine and methionine, to the digestive processes of the mink (Hoie, 1953).

Studies by Struthers and MacDonald (1983) and Skrede and Krogdahl (1985) indicated that this process of trypsin inhibition is unique in the mink where there is a linear relationship between % trypsin activity and level of the inhibitor, while all other species have lower activity until 20 micrograms of trypsin inhibitor/ml concentration. The differences in species disappear by concentration of 30 micrograms/ml assay. The experimental work of Skrede and Krogdahl indicated a striking difference in soybean oil meal trypsin inhibitor activity in the intestinal tract of mink and that of chicks. In their studies they noted that the trypsin activity of mink feces was about 20 times higher than that of the chick excreta, increasing with heat treatment. A surplus of fecal trypsin was found in mink fed unheated soybean flakes while chick excreta contained an excess of protein inhibitors. Thus the pancreas of mink and chick appeared to respond differently to dietary proteins and inhibitors.

Multiple studies indicate that the degree of heating of raw soybean oil meal is critical. While processing temperatures of 110° Celsius support soybean oil meal as a valuable protein resource (Table 4.18), temperatures as high as 120°, 130° or 135° Celsius brought about a significant reduction of soybean oil meal protein value in terms of loss of lysine, arginine, and cystine and reduced digestibility of all amino acids (Skrede and Krogdahl, 1985; Ljokjel et al., 2000). Overheating can bring about the Maillard Reaction (destruction of lysine and bonding of arginine in a structure unavailable to the digestive processes of animals); hence a balance must be achieved that denatures proteinase inhibitors and minimizes Maillard Reactions.

TABLE 4.18: Amino Acid Digestibility of Soybean Products–The Value of Heat Processing*

Amino Acid	Soybean Products		
Soya Flakes			
	Raw	Heated 110° Celsius	Commercial Soybean Meal
Arginine	76	95	93
Cystine	27	77	66
Tryptophan	20	89	86

* Skrede and Krogdahl (1985)

There is a wide variety of soybean products on the market including the following.

Soybean flakes—unheated have very low palatability and low amino acid digestibility (Skrede and Krogdahl, 1985).

Soya cake (soybeans pressed to reduce oil content, followed by heating) has low palatability resulting in reduced weight gains (Hoie, 1953).

Soybean meal—solvent extracted (heated to 110° Celsius with short–term toasting) has 44–45% protein content. A study by Seier et al. (1970) indicated that the employment of soybean oil meal as a replacement for 50% of the frozen fish component of the mink ranch diet resulted in a significant reduction in the apparent digestibility values of dry matter, energy, fat, and protein.

Soyflour—solvent extracted (heated to 110° Celsius with short–term toasting) is a dehulled soybean product which contains higher protein, 50%). At higher levels, 18–31%, low palatability resulted in low weight gains but similar fur quality (Belzile, 1976b). In an earlier study, Belzile and Poliquin (1974), a soyflour product with 72% crude protein proved to be satisfactory for the growth and fur development of the mink at levels of 10 and 17% (dry matter basis), while at levels of 24 and 32%, weight gains were sub–optimum, resulting in shorter pelt lengths with similar pelt quality.

Micronized soybean meal (short–term exposure to radiant heat at high temperatures). Even at the low level of 4% micronized soybean meal, kit weight gains were sub–optimum (Johansson et al., 1977). A study by Skrede and Herstad (1978) with a micronized soybean product found that protein digestibility was significantly reduced and that urease activity was very low, indicating overheating. High fat soybean meal (heat treated) produced sub–optimum kit growth performance (Leoschke, 1981).

Soybean meal: pre–hydrolyzed enzymaticly Treatment of dehulled soybeanmeal with pancreatin, pepsin, and papain at specific pHs provided significant solubilization of the nitrogen products present, from 26–36% untreated soybeanmeal to 43–46%. However, the net result of increased solubilization of soybean meal proteins was a negative effect on the growth of the mink (Belzile and Dauphin, 1984).

Soycomil K (Unimills, Holland) achieved 65% protein via the removal of easily hydrolysable carbohydrates. Studies by Hillemann (1981a, 1981b) with 40% of the proteins from fish and slaughter offal replaced with soycomil K found, as a net result, superior fur quality and color although size was non–uniform.

Multiple studies indicate that 20% of the total protein content of the mink ranch diet for growth and fur production can involve soybean protein, (Alden and Johansson, 1975; Hillemann, 1972, 1976; Hoie, 1953; Kiiskinen et al., 1974; Rimeslatten, 1974). In several investigations, a tendency toward undesirable fur quality characteristics in the mink fed 20% of total protein as soybean meal has been noted (Glem–Hansen, 1977). A recent report, Clausen and Hejlesen (2000), recommends a maximum of 4–8% soybean meal for the growing–furring period inasmuch as at the 12% level the pelt length was reduced somewhat.

Corn Germ Meal Produces significant diarrhea problems in both mink and fox when incorporated in pellets. No diarrhea occurs if the pellets are soaked in water (Leoschke, 1987).

Corn Gluten Meal High Protein Danish studies indicate that mink ranch diets should contain corn gluten meal at a maximum of 10% of the total protein provided the mink. The protein resource has a relatively high content of sulfur amino acids but relatively low levels of lysine and arginine. Relative to the amino acid requirements of the mink, corn gluten protein contains an unusual unfavorable leucine/isoleucine ratio.

Wheat Germ Meal Relative to rolled oats as a standard, the palatability of wheat germ meal for the mink is only 55% (Leoschke, 1976)

Whole Grains These contain relatively low levels of lysine, methionine, threonine, and trytophan.

Corn This product uniquely low in tryptophan.

Single Cell Proteins from Petroleum Studies by Aulerich at Michigan State University (1985) and Leoschke at the National Research Ranch (1968) indicate that this protein resource has low biological value for mink kits.

TABLE 4.19: Biological Value—Dehydrated Animal Protein Resources*

Quality Rating	Protein Feedstuffs
Highest Quality	Whole Blood Meals
	Red Blood Cell Meals
	Whole Fish Meals
	Whole Poultry Meals
	Poultry By–Product Meals
	Higher Ash Fish Meals
	Meat Meals
	Meat and Bone Meals
Lowest Quality	Feather Meals

* I have major hesitation about providing a listing of the biological value of dehydrated protein resources for modern mink nutrition inasmuch as the handling and processing of these products is so varied. Experimental studies at the National Research Ranch since over the past fifty years indicate that 80% of the dehydrated animal protein resources available in North America have low biological value for the mink, directly related to sub–optimum handling and processing. This observation is supported by a study of Chilean fish meals (Romero et al., 1994) in which a combination of selected chemical analyses and protein digestibility with rainbow trout indicated only one–third of the fish meals were of premium quality suitable for modern nutrition of the mink. Thus the listing above is applicable only to the highest quality products available to the worldwide fur industry which at a minimum, meet the guidelines for fishmeal quality as established by Scandinavia and Russia. See Table 4.21. Standards For Fish Meal Quality. It is strongly recommended that all dehydrated protein resources for modern mink nutrition be evaluated in research studies wherein the mink have a voice in assessment of biological value. Once again, to put it simply, "Listen To The Mink."

Specific commentary is warranted on a few of the products as follows.

Whole Fish Meals Certain whole fish meals are supportive of "sharp" color in dark mink never achieved with practical ranch diets providing a simple cereal mix in combination with fresh/frozen feedstuffs.

Higher Ash Fish Meals These have lower protein digestibility for the mink: a 1% unit increase in the content of ash results in a decreased protein digestibility value of 0.6% (Skrede, 1978c).

Meat and Bone Meals Over thirty years of experimental feeding trials at the National Research Ranch on the biological value of dehydrated protein resources have indicated that a high quality meat and bone meal is not available in North America. These observations are confirmed by a study of Ahlstrom et al. (2000) in which the meat–and–bone meals studied failed to provide levels of digestible sulfur amino acids required by the mink. A more recent study, Clausen et al. (2002) noted that the employment of more than 3% meat–and–bone meal resulted in sub–optimum kit growth in the summer months and that when the diet contained 12% meat–and–bone meal, the color of the mink pelts was reddish, perhaps due to a relatively low level of phenylalanine and tyrosine in the feed mixture. In two recent experimental studies on the protein requirements of the mink emphasizing this low quality, Dahlman et al. (1996) and Kerminen–Hakkio et al. (2000), fish–bone meal, meat meal, and meat–bone meal were selected to provide an experimental diet with lower protein quality for the mink.

Feather Meals Without question, feather meals have a relatively high content of arginine, a critical amino acid for fur development and reproduction/lactation. However, the basic question remains: What proportion of the arginine in a given feather meal is actually available to the digestive processes of the mink? Only continuing research studies can answer this question.

In many experimental studies, the protein requirements of an animal can be met with protein resources of lower quality by simply increasing the quantity of protein provided. That is, provide an extra quantity of protein to cover for a sub–optimum amino acid pattern or lower protein digestibility rating of a low quality protein resource. The experimental report of Dahlman et al. (1996), however, contradicts this maxim, indicating that increasing the amount of low quality protein would not only be useless but might have some adverse effects on mink production parameters. Higher levels of low quality protein feedstuffs may result in a lower energy value of the diets directly related to high ash levels provided, resulting in lower weight gains of the kits. In the final analysis, the absence of quality protein resources can not be made up by higher levels of protein in the diet presented to the mink. To add high levels of lower quality protein resources decreases the level of other vital nutrients in the diet. Also, in addition to these factors, it strains the metabolism of the mink with energy–wasting reactions.

Still, even considering the data of Tables 4.17, 4.18 and 4.19 and the commentaries that followed, it is important to point out that ranchers should not ignore the lower quality protein resources. All protein feedstuffs have a potential role in the planning of ranch diets that promote top performance and minimum feed costs per pelt marketed. A balance of different protein resources in the mink's diet should provide the animals with an optimum amino acid pattern for each phase of the ranch year. Thus ranch diets should be formulated with the goal of providing an optimum amino acid intake via a proper balance of the highest quality, medium quality, and lower quality protein feedstuffs. Such diets should include optimum levels of the higher quality protein products, as well illustrated in the following table.

TABLE 4.20: Recommendations—Quality Protein Feedstuffs and Phases of the Mink Ranch Year—National Program of Mink Nutrition North America*

	Reproduction Lactation	Early Growth	Late Growth	Fur Production
Ingredients				
Fortified Cereals	15–25	15–25	20–35	20–35
Liver	10	—	—	—
Quality Proteins	30–35	30–35	20–25	30–35

* Leoschke, (2001)

114

Assessment—Protein Quality

The biological value of a specific protein resource for mink nutrition is related to the amino acid pattern provided and the availability of those amino acids to the digestive processes of the mink. In terms of dehydrated protein feedstuffs such as fish meals or poultry meals there is a wide variation in quality due to multiple factors including the nature of original protein source, the freshness of ingredients, (i.e., storage time between obtaining of the feedstuff and processing), and the nature of the dehydration programs.

Chemical Evaluation
Fishmeal Standards

A good starting point for the evaluation of a given dehydrated protein feedstuff is provided by Table 4.21.

TABLE 4.21: Standards for Fishmeal Quality

	Norway(1)	Denmark(2)	Soviet Union(3)
Chemical Analysis			
Crude Protein–%	70	73	60
Crude Fat – %	10	10	10
Water–%	6–10	6–8	8–12
Ash–%	16	13	22
Ammonia Nitrogen–%	0.2	—	0.2
Salt–%	3	—	3
Di–Methyl–Nitrose Amine		not detectable	
TVN, mg/ 100 grams		120	
FFA, % of crude fat		10	

(1) Ugletveit (1975) (2) Konsulenternes (1976) (3) Kuznecow (1976)

Commentary is warranted on the significant value of each of these chemical analyses.

Crude Protein Levels of crude protein in a dehydrated protein feedstuff may be lower than 70% and still be considered a quality protein resource, as illustrated by a poultry meal with 62% protein, 16% fat, and 14% ash. The higher fat level is partially responsible for the lower protein level relative to a leaner herring meal with 70% protein.

Crude Fat Fat levels may be higher than 10% provided that the fat content is well stabilized with anti–oxidants to minimize oxidative rancidity.

Water—6–12% Moisture levels above 10–12% may encourage the growth of molds.

115

Moisture levels below 6% are likely to indicate an over–heating of the protein resource. This action yields a combination of excessive heat and low moisture supporting the Maillard Reaction which significantly reduces the biological value of the product. Studies have shown that specific amino acids including arginine, cystine, lysine, methionine, and tryptophan are heat labile. With sub–optimum dehydration procedures there can be a destruction of cystine, lysine, and tryptophane and the bonding of arginine and methionine in chemical structures which are unavailable to the digestive processes (Allison, 1949; Miller et al., 1965; Varnish and Carpenter, 1975). Reactions which bring about the destruction and/or bonding of amino acids include the involvement of free carbonyl structures from sugars and carbohydrates. In the absence of sugars, heating of proteins may involve cross–linking reactions between the primary amide groups of asparagine and glutamine with the epsilon–amino group of lysine or the guanido tail of arginine. It is of interest to note that even vitamin B–6 can be involved in amino acid bonding in indigestible structures wherein pyridoxine and pyridoxal phosphate may react with certain amino acids during heat treatment to yield biologically inactive pyridoxylamino compounds (Gregory and Kirk, 1981).

It is wonderful to note that a number of fishmeal producers are giving serious thought to setting standards for minimum moisture levels in products marketed to the worldwide fur industry. For instance, Esbjerg Fiskeindustri (1998) was marketing Fishmeal Special A – 999 for early weaned piglets, mink, and aquaculture (animals and fish with high requirements for protein in terms of both quantity and quality). The label states "Moisture–Max 10% Min. 6%." On the other hand, Sopropeche (1991), with a serious concern about cadaverine (evidence of bacterial action on stored fish prior to processing), ignores common sense levels of moisture as its label states "Moisture–Max 5%."

Experimental studies at the National Research Ranch over the period of the last thirty years have indicated that a high proportion of dehydrated protein resources for mink nutrition with moisture levels less than 6.0% have low biological value for the mink (Leoschke, 1985).

Berg (1989) recommends that dehydrated animal feedstuffs be provided dry storage with temperatures no more than 15° Celsius (59° Fahrenheit) for minimum dehydration.

Ash Levels Without question, higher ash levels have a significant effect on the digestibility of protein feedstuffs. Skrede (1978c) found that a 1% unit increase in the content of fishmeal ash results in a decreased protein digestibility of 0.6%.

Amonia Nitrogen and TVN (Total Volatile Nitrogen) Higher levels of ammonia nitrogen or TVN indicate a long storage time between obtaining a fish product and its final processing yields a fishmeal wherein bacterial action on the protein content of the fish results in the destruction of the amino acid content and the release of ammonia and TVN (Bruun de Neergaard, 1972). TVN includes such volatile chemicals as ammonia, trimethylamine(TMA), trimethylamineoxide (TMAO), and biogenic amines. The biogenic amines result from bacterial decarboxylation of amino acids–

cadaverine (lysine), histamine (histidine), putrescine (ornithine, a product of the hydrolysis of arginine), tyramine (tyrosine), and tryptamine (tryptophane). While Sopropeche (1991) has set a limit of a maximum of 250 ppm for cadaverine, Esbjerg Fiskeindustri (1998) has set a maximum of 1,000 ppm for cadaverine and a maximum of 500 ppm histamine for its LT–999 fishmeal. It further notes that typical analyses of its Con–Kix fishmeal has a maximum of 510 ppm for cadaverine, 140 ppm for histamine, and 250 ppm for putrescine. The literature for Con–Kix states, "Con–Kix is a whole-meal made from absolutely fresh fish. This ensures a specifically low level of biogenic amines." The value of histamine determination in stabilized fish meals may be questionable, inasmuch as studies by Romero et al. (1994) have noted that the difference in histamine content between FAQ (Fair Average Quality) and Premium fish meals decreases considerably four months after processing.

Salt The total salt content provided to the mink is the basis of nutritional value of the dietary regimen. Thus, higher salt fish meals may be employed, provided the final level of salt is given serious consideration. With higher salt diets, the key point is the availability of ample water resources.

DiMethyl–Nitrose Amine This toxic chemical brings about a sub–optimum fur development known as "cotton mink," related to sub–optimum iron nutrition.

FFA (Free Fatty Acids) The free fatty acids of themselves are not a negative factor in terms of the requirements of the mink for fat; however, their presence in a fishmeal product is specific evidence of sub–optimum storage of the fish prior to processing. In that case, bacterial actions can bring about hydrolysis of the fish fats to yield FFA.

Lysine Availability

As already noted, dehydration procedures yielding fish meals and other high protein meals for animal nutrition may bring about the Maillard Reaction wherein specific amino acids including lysine and arginine may be bound in structures unavailable to the digestive processes of animals. Consequently, animal nutrition scientists seek an economical procedure involving minimal time to accurately assess via laboratory means the degree of bonding of an amino acid such as lysine. These have the potential to provide a rapid and inexpensive approach compared to biological evaluation of the quality of dehydrated protein resources. However, studies conducted by Leoschke (1991) indicate a relatively poor correlation between laboratory data on percent lysine availability of multiple high protein mink feedstuffs including blood meals, fish meals, powdered eggs, and poultry meals and the actual performance of mink kits on experimental diets planned to evaluate the actual biological value of dehydrated protein resources.

Pepsin Digestibility

Pepsin digestibility data is obtained in a laboratory procedure designed to duplicate an animal's stomach by mixing a sample of a protein feedstuff with pepsin enzyme material in an acidic buffered environment. As seen in Table 4.22. It is obvious that the laboratory program ("in vitro") is misleading. It provides a relatively high assessment of protein value in comparison with biological assessment ("in vivo") which provides a lower value due to multiple factors including the short, 2–3 hours time of feed passage. The "in vitro" program is valued because it is inexpensive and short as a few hours, while the "in vivo" program is relatively expensive and involves a period of two weeks. All factors considered, it is foolish for mink ranchers to "Listen To The Chemist In His Laboratory" instead of "Listening To The Mink In His Cage" in their decision making relative to the choice of dehydrated protein resources for their mink ranch diets. Enough said.

TABLE 4.22: Protein Resources—Pepsin Digestibility Data and Biological Value Assessment via Growth Promotion Ratings

Protein Resources	Pepsin Digestibility(1)	Growth Promotion Rating(2)
Powdered Eggs	98	120
Blood Meals	98	90
Herring Meals	95	90
Soybean Meal	93	84
Meat & Bone Meal	91	60
Poultry meal	90	70

(1) Roberts (1990) (2) National Research Ranch, Leoschke (1976)

Amino Acid Analyses
Chemical Score

It is well acknowledged that the determination of protein quality by feeding tests is time consuming and costly. Thus, calculation of the Chemical Score from the amino acid profile of the protein is considered by some mink nutrition scientists to be acceptable (Roberts, 1990). However, this approach to the amino acid nutrition of the mink is completely unacceptable since it essentially simply ignores the problems of the Maillard Reaction.

Total Amino Acid Value (TAAV)

Studies by Glem–Hansen and Eggum (1973) indicated a correlation of only 0.87 between TAAV and biological value (BV) in experiments with mink comprising 12 feed mixtures. The TAAV proved to be significantly higher than BV, suggesting higher protein utilization in rats than in mink. Additionally, the Maillard Reaction may occur in the production of fish meals making lysine and arginine unavailable to the digestive processes of the mink, but the bond is broken in the acid hydrolysis step preceding amino acid analyses.

Biological Evaluation
Nitrogen Balance

In terms of the energy nutrients, carbohydrates, fats, and proteins (polymers with amino acid monomers), proteins are unique in terms of containing the element nitrogen (N). The most common method of chemical analysis for the protein content of a given feedstuff is to measure the nitrogen content. In most cases the nitrogen content x 6.25 provides a measure of the protein content. This procedure is more rapid and less expensive than assessing the content of a feedstuff for 20 or more different amino acids.

A full understanding of the value of nitrogen balance studies for the assessment of the biological value of protein resources for the mink requires careful attention to a number of specific definitions. Anabolism is a build–up or synthesis of new structures. Catabolism is a breakdown of biochemical structures. Endogenous Nitrogen is nitrogen originating from within the body. For example, the daily creatinine loss from the body is in a quantity directly related to the muscle mass of the body and not its protein intake. Flux implies a continuous moving or passing by as of a stream. Metabolism is constituted by the total complex of biochemical reactions taking place in a living organism. Metabolic Nitrogen is nitrogen originating from the body such as bile, digest residues, and epithelial cells abraded from the alimentary tract.

Nitrogen Balance refers to the long–term relationship between the intake of food nitrogen and the loss of nitrogen compounds from an animal's body. Food nitrogen is primarily protein nitrogen with smaller quantities of nitrogen provided by nucleic acids, phospholipids, and vitamins. Animal nitrogen losses include those exiting the body via the urine and feces. Nitrogen equilibrium exists in an animal when the nitrogen intake equals the waste nitrogen output, that is, zero N–balance.

Keeping these definitions in mind aids in understanding certain critical bodily processes. Animal tissues exist in a dynamic state in which there is a constant flux with a steady, simultaneous anabolism and catabolism of constituent protein structures. Thus, a given amino acid may exist at one point within a muscle cell and later, after catabolism and release from the muscle protein, undergo anabolism to be part of a pancreatic trypsin enzyme secreted into the intestine. From there, it may be released after hydrolysis and re–absorbed into the blood circulation and be employed in the synthesis of a liver cell, or undergo catabolism to yield energy and carbon dioxide, water and free ammonia which is converted by the liver to urea and excreted via the kidneys in the urine. It has been estimated that in terms of the human animal, the dynamic change in the proteins and other structures of the body results, within a 20 year period, in almost every cell of the body being replaced except for part of the skeletal structure.

The total of all the simultaneous gains and losses of nitrogen in the different components of the body is reflected in the overall nitrogen balance and is "zero", neither positive nor negative in a mature animal at constant weight. With tissue synthesis related to kit growth, fur development, pregnancy or milk synthesis in the lactation period, the mink would be in positive N–balance, while in the case of a diseased animal "off feed" with weight loss, a mink would be in negative nitrogen balance as total nitrogen losses exceeded nitrogen intake. Another example of negative N–balance for the mink would be ranch diets with low levels of carbohydrates and fats (low caloric density) wherein the mink would be forced to metabolize protein (amino acids) for their energy value.

As early as 1940, Cornell University (Loosli and Smith, 1940) provided preliminary data on nitrogen balance studies with mink. With a dietary regimen containing low levels of protein, urinary losses of nitrogen products exceeded the nitrogen intake of the animals. The losses resulted from the breakdown of protein structures such as the muscles, inasmuch as the dietary intake of protein was insufficient to meet the amino acid requirements of the mink.

A later study on urinary nitrogen excretion of the mink is of interest. Experimental data of Taranov (1977) found that mink kits on fishmeal diets excreted more total nitrogen in their urine than mink kits fed a fresh fish diet, indicating that the biological value of fish proteins was being undercut through the Maillard Reaction in the processes required to yield fish meals.

Reports by Glem–Hansen and Jorgensen (1973a, 1973b, 1976) provided detailed experimental data on the metabolic fecal nitrogen and endogenous urinary nitrogen of the mink required to calculate the biological value of protein feedstuffs. Their study indicated that the average amount of metabolic N in the 113 balances to be 0.085 grams N per day per mink with an average weight of 1.2 kg. The amount of endogenous N in the study was 0.245 grams N per day per mink with an average body weight of 1.2 kg.

The biological value (BV) of a given protein resource or mixture of protein feedstuffs for the mink can be assessed via the following equation:

$$\mathbf{BV} = \frac{N \text{ intake} - (\text{fecal } N - \text{metabolic } N) - (\text{urinary } N - \text{endogenous } N) \times 100}{N \text{ intake} - (\text{fecal } N - \text{metabolic } N)}$$

The biological value computed is thus the percentage of actually digested nitrogen from protein (amino acids) that is actually utilized by the body for tissue repair. (It should be noted that BV is similar, but not identical to NPN (Net Protein Utilization.) This analysis of the protein requirements of animals is no longer employed within the world animal nutrition science and rarely in mink nutrition science. The metabolic N appears in this equation in both the numerator and the denominator. This results in only small differences in the BV caused by changing the metabolic N within certain limits. On the other hand, the amount of endogenous N influences the BV a great deal.

Cages specifically designed for N–balance and metabolism studies with mink have been designed by Sinclair and Evans (1962) and Jorgensen and Glem–Hansen (1973).

It is of value to acknowledge the experimental work of Glem–Hansen and Eggum (1974) on the nutritional correlation between mink and laboratory rats relative to the biological value of protein resources via N–balance studies. (Note: Net Protein Utilization (NPU) = Biological Value (BV) x digestibility coefficient for the particular protein.) Their study showed correlation coeffi-

120

cients between the two animal species for Total Digestibility (TD), BV, and NPN of 0.90, 0.88, and 0.92 respectively. The regression equations for calculation of TD, BV, and NPU for mink on the basis of experiments with rats are:

TD (mink) = 1.325 x TD (rats)–36.25

BV (mink) = 0.77 x BV (rats) + 5.38

NPU (mink) = 0.795 x NPU (rats)–0.23.

The utilization of protein is about 20% lower in mink than in rats. This inferior NPU is a result of a lower BV as well as a lower TD. The difference in TD is most probably due to the extremely rapid rate of passage of feed through the digestive tract in mink (Wood, 1956). The lower BV might be caused by a higher requirement in mink than in rats for the sulfur containing amino acids (SAA) cystine and methionine. SAA have been shown to be the first limiting factor for protein utilization in typical diets for the mink, (Glem–Hansen, 1974, 1977). An analysis of the proportion between protein from the body to protein from hair in mink as compared to other animal species leads one to expect that the requirement of SAA is considerably higher for mink than most other animal species (Jorgensen and Eggum, 1971).

Animal Evaluation

In addition to using rats and mink in N–balance studies, scientists have studied the growth of mink and a variety of other animals to assess the biological value of protein feedstuffs. This procedure is as valid as N–balance studies and more practical in terms of time and labor.

Laboratory Rats

Over a period of many years, the laboratory rat has been used to assess the biological value of protein resources for animals. But, all factors considered, using rat data to determine mink nutritional requirements is certainly questionable, as there are significant differences between them. These include the rate of growth of the young animals, the time of feed passage, and the proportion of protein from the body and from hair and fur in the two animals. Has already been noted in an earlier discussion (Glem–Hansen and Eggum, 1974) that there is poor correlation between NPU values of protein feedstuffs for rats and mink. In fact, as early as 1951, the noted Cornell University professor L. A. Maynard (1951) commented "...interpretation of data on biological value from one species to another must be with reserve."

Chicks

The sharp differences in the capacity of chicks (5–8 weeks) and mink (8–10 months) to digest protein feedstuffs essentially makes using chicks to assess biological value of protein resources for modern mink nutrition invalid. Skrede, Krogdahl, and Austreng (1980) noted that in terms of meat and bone meal, the digestibility values of the mink for all of the amino acids was lower than that of the chicks. Especially low values were seen for arginine, histidine, and tryptophan. In a later study, Skrede and Krogdahl (1985) noted that in the case of soybean meal, the digestive capacity of the mink relative to chicks was significantly lower for cystine, threonine, and phenylalanine.

Ferrets

Studies by Leoschke (1976b) indicated that a valid assessment of the relative quality (biological value) of dehydrated protein feedstuffs for the mink could be achieved with the measurement of weight gains of 6–7 week old male kits over a two week period. The experimental diet provided 28% ME as protein. Later studies with ferret kits (Leoschke, 1992) indicated that the same experimental plan could employ ferret kits provided that the level of protein was reduced to 20% ME as protein.

Mink—Short Term

For the past decade, National Fur Foods has been able to employ as many as 2,000 animals each year in mink nutrition experimental studies. In the early growth phase (June) a minimum of 400 male kits were used. In the late growth and fur development period each control and experimental group consisted of 100 genetically matched brother kits and 100 genetically matched sister kits. For the reproduction/lactation and weaning phase of the mink ranch year, the control and experimental groups consisted of 200 matched breeders. In terms of research planning, labor, and available mink, the experimental studies were limited to two control and four experimental groups each summer and fall and to a control and a single experimental group in the winter and spring.

With multiple new dehydrated protein resources for modern mink nutrition coming on to the worldwide market each year, a decision was made in 1970 to initiate screening studies on the quality (biological value) of a larger number and wider variety of dehydrated protein feedstuffs for the mink each June (Leoschke, 1976). These studies involved as many as 8 control groups and 16 experimental groups of 20 genetically matched male kits paired in pens for psychological reasons. Identification was maintained by clipping a toe nail of one member of the pair.

If daily weight gain is the only criterion for assessing the quality (biological value) of mink protein resources, it is obvious that the younger the animal the better, inasmuch as the net weight gain of an animal is the highest in the immediate period after birth (A mink kit doubles his birth weight in 4 days while a human infant may take 3–4 months to do so.) However, it also must be noted that to employ genetically matched brothers in experimental feeding trials prior to weaning

122

is simply impractical. Finally, only male kits were involved because of their greater growth potential.

Historically, the most important criterion in these experiments is to provide a protein level (% ME as protein) which is below the actual requirement for that specific phase of the animal's life. These relatively lower levels of protein allow the quality to be expressed in the growth performance of the animals. At higher levels of protein, the quantity may be at such a level as to actually "cover up" or mask the deficiency of amino acid pattern or digestibility that may exist in the protein resource. Multiple studies indicate that for the June early growth phase, a ranch diet with 7 grams of digestible protein per 100 Calories, or with 30% of ME, is optimum in quantity, provided that careful consideration has been given to the proportion of quality protein feedstuffs. In the experimental diet for assessing protein quality of dehydrated protein feedstuffs, the level of protein was planned to be about 6 grams of digestible protein per 100 Calories equivalent to 28% of ME as protein. At first glance, this reduction of about 10% of the protein requirements of the mink in the experimental diets appears to be a relatively shallow approach. However, if you give serious consideration to the key nutritional factors of palatability of the diet and unknown digestibility values, the decision to employ this particular protein level may have made "common sense." The experimental data provided in the thirty years of the experimental trials support this decision.

The initial protein quality assay diet contained fishmeal, 50; potato flour, 15; beet pulp, 3; vitamins, 1; trace mineral salt, 1; and fat (pork choice white grease), 30%. Careful dietary planning resulted in both control and experimental dietary regimens providing 32% protein, 33% fat and about 8% ash.

With one set of genetically matched brothers on a high quality ranch diet containing fortified cereal at 15% with 85% fresh/frozen feedstuffs (control diet) and the other set of animals on the experimental diet outlined in the previous paragraph provided as a wet mash, kit growth on the experimental formulation was 87% of the weigh gains of their brothers on the control ranch diet. Sub–optimum growth performance on the experimental dietary formulation was very likely related to multiple factors including taste appeal of the dietary mix and the relatively low level of protein provided. The first factor is minimized in the experimental work of the next thirty years inasmuch as all the control and experimental diets were formulated with dehydrated ingredients.

Initially, the control diet animal protein was provided as high quality NorSeaMink herring meal. Later the fishmeal resource was changed to provide multiple sources of animal protein wherein the protein mixture contained 50 parts of NorSeaMink herring meal, 30 parts of an anchovy meal and 20 parts of a high quality blood meal. This decision was made on the common sense fact that in practical ranch diets, a variety of protein resources are employed. In the experimental diets the NorSeaMink herring meal was replaced with an equal quantity of protein from the test protein resource. Thus half of the total protein resource originated with the new test protein. The validity of the experimental dietary design for assessing the quality (biological value) of dehydrated protein resources for modern mink nutrition is supported by the experimental data provided in Table 4.23.

TABLE 4.23. Powdered Eggs and Mink Kit Growth Performance

Protein Resources–Research Diets	Male Kit Growth Grams 2 Weeks
Control Diet	
100% protein as herring meal	300
Experimental Diet	
70% protein as herring meal and 30% protein as egg powder	360

The kit growth data shows that the 28% ME quantity of protein provided by dehydrated protein resources is sub–optimum for the June early growth phase unless the protein quality is enhanced by including whole egg protein, the world's standard of protein quality for animal nutrition. In these experiments, the performance of genetically matched kit sisters was strikingly different from that of their kit brothers. Their growth on both the control and experimental diets was essentially identical, that is, they showed no enhanced weight gains with the rise in biological value of the protein resources by the inclusion of powdered eggs indicating that the level of 28% ME from dehydrated protein resources was beyond their actual protein requirements for the early growth period.

At the present time at the National Research Ranch, the preliminary screening of dehydrated protein resources for protein quality is reported as a numerical value for a biological rating calculated as follows:

Biological Rating = Growth Promotion–% relative to Control Diet =
GP–% x 100/Vol Feed Volume % relative to Control Diet.

Palatability is a major factor in determining feed volume on experimental diets. A dehydrated protein resource with low palatability could produce low growth because of lower feed volume, not directly related to the biological value of the protein feedstuff. With feed volume data on both the control and experimental feeding regimens, scientists can make a more accurate estimate of the protein quality of a less palatable feed product. Normally, at National Fur Foods, a 95% biological rating is the minimum requirement for a specific dehydrated protein resource to be considered for scheduling in long term feeding trials involving the late growth, fur production, and reproduction/lactation phases. Thus a dehydrated protein resource with lower palatability for the mink might achieve only a GP rating of 86% and automatically be rejected for long term experiments on biological value. However, with feed volume intake data of 90% (relative to the control group), the product would be given serious thought for long term feeding trials on the basis of the following calculations:

Biological Rating = GP–%/ x 100/Volume–% = 86 x 100/90 = 96

Two interesting but unrelated notes on protein biological rating research studies also deserve mention here. Over many years, studies have shown a constant pattern of dark kits not providing as high biological ratings as pastel kits on identical growth promotion diets, since dark mink are more sensitive to protein quality. Hence, in more recent years, only dark kits have been employed in the short term feeding trials conducted in June. Poultry meals with positive salmonella EIA tests generally yielded higher biological ratings than poultry meals with negative salmonella EIA tests. These results indicate that in the primary goal of killing salmonella bacteria, the rendering processes are also enhancing the Maillard Reaction to the point of yielding a protein resource with a lower biological value for the mink.

Mink—Long Term

"All Fish Meals Are Not Created Equal" is a mink nutrition statement that also applies to blood meals, egg powders, and poultry meals. Thus experimental studies on the biological value of dehydrated protein resources must be conducted throughout the mink ranch year. This point is well illustrated by research data from the National Research Ranch facilities as provided by Tables 4.24 and 4.25.

TABLE 4.24. Fishmeal Quality and Fur Development—Mink

Fishmeal Resource	Males	Breeder Selection	Females
Quality Fishmeal	27		3
Commercial Fishmeal	14		16

This experimental data emphasizes the importance of protein quality in diets supportive of the top fur development, inasmuch as the quantity of protein provided in both diets was identical. In the case of the male kits, the biological value of the commercial fishmeal was sub–optimum as evidenced by the relatively poor quality fur production. In the case of the female kits which require a lower quantity of protein, the "extra" quantity of protein was able to support top fur development, to "mask" the inferior quality of protein provided by the commercial fishmeal.

TABLE 4.25. Fish Meal Quality and Reproduction/Lactation—Mink

Fishmeal Resource	Kit	Counts	Days	Post–Whelping
	0	10	21	42
Quality Fishmeal	6.2	5.0	4.9	4.7
Commercial Fishmeal	5.5	4.0	3.7	3.5

The data show that the commercial fishmeal with protein content of relatively low biological value supports neither the top reproduction nor top lactation performance. Some progress was made in two later studies with two new commercial fish meals: kit counts at whelping were excellent but lactation failure followed in both cases to the point that ranch management had to "farm out" more that 20% of the kits.

The importance of year round experimental studies on the biological value of dehydrated protein

125

resources is seen in the interesting observation that fish meals that may support the top fur development may prove inferior in supporting top reproduction/lactation performance. Obviously, the amino acid requirements for quality fur development are distinct from those required for the best lactation performance.

Amino Acids

A minimum of 20 amino acids are required to synthesize protein structures of animal and plant cells. A good question might be, "Why so many different amino acid structures?" The answer is simple: in order to provide living structures (muscles, skin, bone, fur) and physiological function (hormones, enzymes). In the human animal, for example, as many as a million different protein structures are required. Furthermore, each of these amino acid polymers must be unique to perform its specific function in support of a living system. Thus multiple varieties of amino acid structures must be available for the processes yielding these thousands of different protein structures. A short review of the processes yielding protein molecules would be worthwhile for understanding the requirement for such amino acid structures as cysteine (monomer) and cysteine (dimer) as well as that for hydrophilic and hydrophobic amino acid structures.

Protein Structures

Primary Structure—Polypeptides

The initial or primary structure of a protein polymer requires the linkage of individual amino acids in a chain–like structure. Within that structure a carboxylic acid functional grouping from one amino acid combines with the alpha amino functional grouping of a second amino acid to yield a secondary amide functional grouping, also termed a peptide bond, via a dehydration process.

Secondary Structure

Coiling of the chain–like polypeptides via hydrogen bonds between peptide structures provides a carbonyl dipole and an alpha amino dipole (similar to the magnetic attraction between the north and south poles of magnets) as well as hydrogen bonding between amino acid tails or residues, such as the hydroxy group of serine and the phenolic residues of tyrosine.

Tertiary Structure

Finally, a functional protein polymer such as muscles, digestive enzymes, or the insulin polypeptide results from the interaction of polypeptide tail or residue structures via varieties of chemical bonding such as the following.

126

Secondary Amide or Peptide Linkages

These are dehydration linkages between the carboxylic acid residues of aspartic acid or glutamic acid structures and the epsilon amino group of a lysine or the guanido tail of an arginine residue.

Salt Linkages

These involve ionic or salt bonding via the carboxylate anionic residues of aspartate or glutamate and the cationic residues of lysine or arginine.

Disulfide Linkages

Arising from a combination of two sulfhydryl groups of cystine residues is a disulfide bond (cystine linkage). This biochemical structure is very important in fur development as well as the unique structure of human insulin. Although human insulin and whale insulin have identical amino acid sequences, the polypeptides differ in their three dimensional structures via different disulfide bondings between the available cystine hydrosulfhydryl structures.

Hydrophobic Bonding

This "water exclusion" bonding occurs between hydrophobic amino acid residues of multiple amino acids including isoleucine, leucine, methionine, phenylalanine, tryptophan, and valine. These water–free regions of certain enzymes are critical for specific biochemical reactions to take place.

Amino Acid Roles

As noted earlier, animals do not have a specific requirement for proteins per se, but for the constituent amino acids in the protein resources.

In terms of the protein nutrition of the mink, amino acids are required for the synthesis of those proteins involved in growth, cell maintenance, and cell repair (including the synthesis of enzymes, hormones, and hemoglobin). Some of these amino acids that must be provided by the daily feed intake are termed essential (indispensable) inasmuch as the animal's own physiological processes can not synthesize them. Other amino acids, termed non–essential (dispensible), may be provided in the ingestion of food and/or the physiology of the body wherein they are synthesized via the carbon skeletons from carbohydrate metabolism, as illustrated by the synthesis of alanine from private. Intermediates of the citric acid cycle also may yield non–essential amino acids, as seen in the synthesis of aspartate from oxaloacetate and glutamate from alpha–keto–glutarate. In addition the alpha amino group may be acquired from other amino acids by transamination or from the ammonia available from the deamination of dietary amino acids as their carbon skeletons are employed as energy resources. Other non–essential amino acids may be synthesized from essential amino acids, as illustrated by the synthesis of cysteine from methionine.

127

The critical nature of the essential amino acids for the fur nutrition of the mink is seen in the observations of Borsting and Clausen (1995a) that 40% of ME from protein with a dietary program low in essential amino acid concentration was required to obtain optimal fur quality, whereas a similar fur quality was obtained with a diet containing only 30% ME from protein resources with high levels of essential amino acids.

Some amino acids are termed anaplerotic ("fill–up") amino acids inasmuch as they can replace some but not all the animal's need for a specific essential amino acid. Thus, the value of cysteine in reducing the mink's requirement for methionine, and of tyrosine for reducing the mink's requirement for phenylalanine.

Another critical classification of amino acids, hydrophobic or hydrophilic, is based on the nature of the amino acid structure beyond the first two carbons. In terms of the initial two carbons, all amino acids are essentially the same (with the exception of proline and hydroxyproline), wherein the number one carbon is a carboxylic acid structure and the second carbon has a primary amine structure. The nature of the structure of each amino acid remaining, termed the "tail" or "amino acid residue," may or may not have an affinity for the water surrounding all protein structures in living cells. If there is a strong attachment for water (a polar structure), the amino acid is said to be hydrophilic (water loving); if there is a repelling action toward water (non–polar structure), the amino acid is termed hydrophobic (water hating).

Essential Amino Acids
Arginine: Glucogenic, Hydrophilic, Basic

Without question, carnivores including the mink, cat (Morris and Rogers (1978), and ferret (Deshmukh and Rusk, 1989; Deshmukh et al., 1991) have uniquely high nutritional requirements for arginine.

Arginine is a key amino acid for animal physiology in multiple ways involving both catabolic (metabolic breakdown) and anabolic (metabolic synthesis) reactions within the body. Arginine is a key component of the urea cycle, a biochemical pathway absolutely essential for animal life in terms of its role of converting toxic ammonium cations from amino acid catabolism to the less toxic urea molecule. Arginine is also involved in many anabolic reactions of the body, including yielding the muscle component creatine required for the synthesis of creatine–phosphate, the major energy storage resource of the body and also a precursor of the polyamines required for cell synthesis.

In terms of mink physiology, arginine is critical for all phases of the mink ranch year and especially for the reproduction, lactation, and fur development periods.

Pregnancy: In fetal development, as the rate of cellular replication increases, more and more ornithine (formed from arginine via hydrolysis) is needed for the synthesis of the spermine and spermidine precursors of the polyamines required for cell synthesis (*Nutrition Review*, 1979). The diversion of arginine to polyamine synthesis results in reduced urea cycle activity with a resul-

128

tant increase in free ammonium cations in the urine of pregnant mink. This condition increases the potential for struvite (magnesium ammonium phosphate hexahydrate) bladder stone (urinary calculi) formation, especially with mink with a large number of fetuses.

Lactation: The milk of the mink, like that of the cat, has a relatively high level of arginine. A comparison of the amino acid composition of mink milk and cow's milk indicates that, for the most part, the amino acid composition is similar, with one exception: the level of arginine in mink milk is 1.6 times the level present in cow's milk (NRC cat bulletin, 1986).

Late Growth: Studies at the University of Wisconsin (Leoschke and Elvehjem, 1956; 1959) indicated good growth of male mink kits on a low protein (19%) and low fat (11%) diet wherein the single protein resource was vitamin–test casein. This study is of major interest as it makes the point that mink have a relatively low arginine requirement in the late growth period, but a much larger one during fur production. This point is underscored by two key observations. According to Block and Bolling (1945), casein, dehydrated cottage cheese, has relatively low levels of arginine, yielding a level of only 0.8% arginine in the purified diet provided the experimental mink. Arnold et al. (1936), working with baby chicks, noted that the arginine present in vitamin–test casein had a relatively low biological value, that is, limited availability to the digestive capacity of the baby chick.

In terms of this experimental study with baby chicks and later studies by Leoschke and Elvehjem (1959) with mink, it appears that in the conversion of dehydrated cottage cheese (casein) to vitamin–test casein via alcohol extraction there is a significant loss in the availability of the arginine content to the digestive processes of baby chicks and mink. Very likely, all factors considered, the procedure responsible for this minimal biological value of vitamin–test casein for arginine is the final heat/dehyration processing step wherein there is a potential for the Maillard Reaction (Allison, 1949).

Fur development: In terms of the essential amino acid composition of animal tissues, for the most part, leucine is consistently present at the highest percent. In mink fur, the arginine composition is equal to that of leucine (Moustgard and Riis, 1957; Disse, 1957).

The major importance of arginine for the fur development of mink was first noted in studies with purified diets at the University of Wisconsin (Leoschke and Elvehjem, 1956; 1959). Mink kits placed on the experimental dietary program in early July exhibited satisfactory growth until early September, concurrent with the stress of fur development. At that time they became anorexic and died. Some mink kits exhibited only slight growth depression in the fall and survived this sub–optimum nutritional regimen in September; however, these exhibited the same signs in May of the following year with the initiation of summer pelt development. In both the fall and spring, the animals responded dramatically when the purified diet was supplemented with arginine and methionine. The resolution of the nutritional deficiency began with Leoschke's acute observation as a post–doctorate student: Noting that a new shipment of vitamin–test casein was dark tan instead of the usual cream color, he suspected a significant Maillard reaction. His complaints, however, were ignored; the net result is that the mink on the experimental diets became anorexic and died in early

129

August instead of early September as in past years. This observation led to the recognition that there was inadequate bioavailability of the arginine in the over heated vitamin–test casein product.

A further important understanding of the arginine nutrition of animals is that an excess of dietary lysine antagonizes arginine nutrition (Damgaard, 1997; 1998), that is, a relatively high ratio of lysine/arginine leads to a sub–optimum absorption of arginine.

One accurate assessment of the nutritional status of arginine of an animal on a specific diet is the examination of the urinary orotic acid excretion. With sub–optimum available dietary arginine, the level of orotic acid excretion rises significantly (Damgaard, 1998). Thus the level of urinary orotic acid excretion can assess the degree of arginine bonding as the direct result of the Maillard Reaction. A study by Leoschke (1988) indicated that mink fed a dietary program with a low moisture (3.4%) fish meal excreted more orotic acid than those provided a higher moisture (9.6%) fish meal. Summary: With the development of mink pellets providing top performance of the mink in all phases of the mink ranch year and the increasing employment of dehydrated protein resources including fish meals, blood meals, and poultry meals in standard ranch diets, it is obviously important that mink nutrition consultants fully acknowledge certain basic facts about arginine. Arginine is a critical amino acid for the mink throughout the ranch year, and the biological availability of arginine in dehydrated protein resources will vary considerably in different products on the market, a variation directly related to the quality of processing and the degree of the Maillard Reaction resulting from the dehydration procedures.

The statement that "All Fishmeals Are Not Created Equal" applies to all dehydrated protein resources for modern nutrition of the mink. At the very least, mink nutrition consultants should maintain the simple goal of a minimum 6.0% moisture in all dehydrated protein products considered for mink dietary programs as advised by Scandinavian and Russian mink nutrition scientists (Jorgensen, 1985).

One final interesting note: Most mink ranchers are familiar with the putrid odor of old mink feed that has been "on the wire" for too long during a hot summer day. A significant component of that odor is putrescine. In the process of putrefaction of meat by bacterial action, arginine is hydrolyzed to ornithine, which in turn is decarboxylated to yield the volatile putrescine.

Histidine—Glucogenic, Hydrophilic, Basic

Histidine is a component of multiple protein structures in animal bodies including hemoglobin, the oxygen carrying pigment of red blood cells, and myoglobin, the oxygen storage pigment of the muscles. In both cases histidine is chelated with iron. The pKa of histidine, 6.0, provides a critical cellular acidity for an environment necessary to multiple biochemical reactions in cells.

In terms of the histidine nutrition of the mink, ranchers should be concerned about long term freezer storage of mink feedstuffs such as horsemeat and fish products that contain high levels of polyunsaturated lipids. Oxidation products of these lipids can react with histidine, reducing its bioavailability for the mink (NRC cat bulletin, 1986). Ranch diets containing high levels of acid–preserved feedstuffs are especially prone to the bacterial decarboxylation of histidine to yield histamine, a toxicant for mink. Bacteria responsible include *clostridium, proteus, salmonella,* and *escherichia coli* within a pH range of 5.0 to 8.0 (Woller, 1977).

Isoleucine—Ketogenic, Glucogenic, Hydrophobic

The three–branched chain amino acids required in animal nutrition include isoleucine, leucine, and valine. In some animal species, there is a significant level of antagonism between these branched chain amino acids as illustrated by the fact that high levels of leucine in the diet of the cat decreases plasma levels of isoleucine and valine (Hargrove et. al., 1984). Among all animal protein feedstuffs for the modern nutrition of the mink, bloodmeals are unique for significantly low levels of isoleucine (Block and Bolling, 1945).

Leucine—Ketogenic, Hydrophobic

Leucine is one of the key branched chain amino acids (of animal nutrition), along with isoleucine and valine. In terms of animal feedstuffs, the amino acid pattern of proteins present indicates that leucine and lysine are the essential amino acids present in highest proportion (Block and Bolling, 1945). Blood meals are unique in unusually high leucine/isoleucine ratio.

Leucine is often used as an indicator amino acid in breath test studies undertaken to examine the effects of dietary composition on animal physiology. Studies by Borsting and Riis (2000) indicated that the oxidation of leucine was lower during rapid growth (August) compared to the period of fur development, whereas the opposite was the case for methyl C^{14}–L–methionine. Another study indicated about equal oxidation earlier in November, due to a higher requirement for fur development in early November than in later November.

Lysine—Ketogenic, Hydrophilic, Basic

Lysine supports an optimum basicity required for cellular function. Relative to animal protein resources for mink nutrition, plant proteins, for the most part, provide sub–optimum levels of lysine and methionine, with the exception of soy proteins which have satisfactory levels of methionine and cysteine.

For mink nutrition involving dehydrated protein resources, lysine is an amino acid of critical concern inasmuch as it may be destroyed via the Maillard Reaction in processing fish and other animal protein feedstuffs to yield high protein fish meals, blood meals, and meat meals. There is also significant potential for the Maillard Reaction to take place in the extrusion process yielding mink pellets. In past years, too many companies in North America marketed pellets that provided sub–optimum protein nutrition due to excessive heat treatment.

These companies had a specific guideline for maximum moisture (to minimize mold and bacterial growth) but ignored the critical importance of low moisture (less than 6.0%) wherein significant amino acid destruction may take place. The Maillard Reaction involves an interaction between the epsilon amino group of lysine and the carbonyl grouping of reducing sugars. In the absence of sugars, heating and dehydration of proteins may produce cross–linking reactions particularly between lysine and glutamine or lysine and asparagine, yielding structures which are not available to the digestive processes of animals.

The earliest report on lysine supplementation of a mink research diet is that of Travis (1956) wherein the control purified diet contained only 15% protein (6% vitamin–test casein and 9% zein). Zein is a corn protein constituent which, like gelatin, is a classic example of protein feedstuffs with a sub–optimum amino acid pattern. Zein protein is simply an amino acid polymer without significant levels of lysine and tryptophan. In a simple 12 day experiment, with the control diet containing 0.48% lysine, raising the lysine level to a total of 1.56% provided the highest growth response.

One of the first reports on the employment of lysine supplementation of pellet formulations was by Dr. Hoogerbrugge of the University of Utrecht, Netherlands (Hoogerbrugge, 1968a, 1968b).

Mink nutrition scientists must acknowledge the basic antagonism between lysine and arginine in animal nutrition: high levels of lysine in a ration may limit arginine absorption. Experimental studies at the National Research Ranch with lysine supplementation of pellet formulations during both reproduction/lactation and late growth/fur development have yielded only sub–optimum performance of the mink.

It is of interest to note that with cadavers, bacterial decarboxylation of lysine leads to the production of a volatile amine, cadaverine.

Methionine—Cysteine–Cystine

Without question, the sulfur amino acids, methionine, cysteine, and cystine (SAA), are the limiting amino acids for fur (Glem–Hansen, 1976, 1977). The 1977 Neils Glem–Hansen thesis represents the most thorough study ever conducted on the requirement of the mink for the SAA. In terms of body composition, the adult mink contains higher levels of SAA than any other domestic animal, largely because 16% of the protein content is in the pelt.

The contrast in the SAA content of the growing pig and that of the growing mink is of interest. The SAA content of the growing pig is relatively constant towards maturity; however, young mink show an increase in the content of SAA from 3.6 g. to 5.6 g./16 g. of nitrogen (100 grams of protein) as growth progresses to adulthood. In a detailed study on the deposition of amino acids in the mink, Glem–Hansen (1974) has shown that the requirement for SAA increases rapidly in relation to the other amino acids as growth progresses through the fall fur development period. Complementary investigations by Jorgensen and Eggum (1971) have shown that mink fur contains 17% SAA.

Methionine—Glucogenic, Hydrophobic

The first report on the value of supplemental DL–methionine for mink on high fish diets was that of Watt (1952), in which the addition of DL–methionine at the level of 0.05% of the dry matter yielded superior growth of the kits and enhanced fur quality. A later report (Watt, 1953) confirmed the earlier observation that DL–methionine was more effective in male kits with their relatively higher protein requirements, than in female kits. Still later Leoschke's 1956 study with mink kits on a research diet containing 19% vitamin–test casein as the sole protein resource showed that while the experimental diet was satisfactory for the late growth period (July–August), as the fur development phase began in September, the mink kits lost weight and died unless the dietary program was supplemented with DL–methionine. A study by Chavez (1981) indicated enhanced fur development when the control ranch diet was supplemented with 0.5% DL–methionine. Belzile (1976) has shown the value of DL–methionine for mink fur development with experimental diets employing relatively high levels of soybean oil meal.

The methionine in mink feedstuffs can be destroyed in the production of high protein dehydrated fish meals, blood meals, or mink pellets if excessive heat treatment is involved or by oxidative damage. This last process occurs when high fat mink feedstuffs are stored in the frozen state for a long period of time and products of the oxidation of polyunsaturated fatty acids (oxidative rancidity) may react with methionine (NRC cat bulletin, 1986).

Both rats (Wretlind and Rose, 1950) and cats (Teeter et al., 1978a) can utilize D–methionine but the mink cannot convert D–methionine to L–methionine to meet its methionine requirements (Glem–Hansen, 1982). Experimental studies by Elnif and Hansen (2005) involving the employment of C–13 and C–12 isotopes in expired air from D– and L–methionine indicated that less than 30% of the D–form was oxidized by adult female mink.

Studies by Borsting and Riis (2000) find a high rate of methionine decarboxylation during fur development, indicating a high rate of cysteine synthesis as methionine contributes sulfur groups for cysteine. Cysteine can replace part of the methionine requirement of the mink in an anaplerotic reaction ("fill–up"). In the cat (Teeter et. al., 1978b), 50% of the methionine requirement can be provided by cysteine. A recent study by Clausen et al. (2005c) indicated that the optimal proportion of met/cys for skin length and fur quality (NS) in the fur production period was 0.16/0.06. However, skin length was the same if the met/cys ratio was 0.12/0.06 with the level of cysteine the same and the level of methionine reduced. Experimental studies on the utilization of Methionine–Hydroxy–Analog (MHA) by Sandbol et al. (2004a) indicate that mink are only partially able to utilize MHA in the growing period, even though an earlier study by Sandbol et al. (2003a) had indicated MHA as a good source of methyl– groups.

Both choline (a B–vitamin) and methionine are important methyl– group donors in animal physiology. With relatively low levels of choline in an animal's diet, the requirement of methionine increases significantly (Borsting and Riis, 2000). Betaine, the anhydride or inner salt of

133

carboxymethyltrimethyl ammonium hydroxide, is also a potential resource of methyl groups for mink nutrition (Borsting and Riis, 2002; Sandbol et al., 2003a.; Clausen et al., 2004). In a Sandbol et al. (2003a) study it was noted that when 20% of the methionine was replaced with betaine, the mink had longer pelts (NS) and a better fur quality (p. = 0.0007).

The combination of DL–methionine and ammonium chloride has been employed for the acidification of the urine of mink to prevent bladder stones (Westlake and Newman, 1985). With the catabolism of methionine, one of the products is sulfuric acid. As to whether ranchers should use DL–methionine as a supplement, it is important to recognize that the product is not palatable for the mink, Leoschke (1967), an observation confirmed by Sandbol and Clausen (2004). Ranchers still wishing to do so should initiate the program in early July when the animals are eager and not in late August when relatively high levels of methionine can depress feed volume intake.

Cysteine—Non–essential, Glucogenic, Hydrophobic

The percentage of cysteine is second highest in the amino acid composition of mink fur with proline being present in highest quantity. Cysteine can replace part of the methionine requirement in an anaplerotic ("fill–up") reaction. In terms of cat nutrition, cysteine can replace up to half of the methionine requirement (Teeter et. al., 1978b). Because cysteine also has the capacity to form several complexes with ferrous ions, cysteine supplementation has proven to enhance iron absorption (Skrede, 1986).

Cysteine, with its free sulfhydryl grouping, is especially sensitive to destruction in heating procedures required to convert feedstuffs into high protein dehydrated feed products (Ahlstrom et. al., 2000; Tyopponen et al., 1988). A study by Skrede (1979b) revealed a tendency towards reduced cysteine content after heat drying of filleted scrap, thus confirming earlier studies of heat treatment in the absence of carbohydrates in which cysteine was found to be more heat–labile than other amino acids. It is of interest to note that in one study by Szymeczko and Skrede (1990), cysteine digestibility in whole fish was 85% but only 62% in fish meal. Cysteine is also sensitive to destruction in extrusion processes for mink pellet production if excessive heat treatment is employed (Ljokjel and Skrede, 2000).

Cystine—Non–essential, Glucogenic, Hydrophobic

Cystine is a dimer of cysteine present in critical R–S–S–R cross–linkages in protein structures and especially important in hair and fur synthesis. Studies by Chavez (1981) indicated that supplementation of a mink ranch diet with 0.5% DL–methionine yielded enhanced fur development while the supplementation of the ranch diet with 0.8% L–cystine (equivalent basis with methionine) provided no significant advantage on size or pelt quality. Borsting and Clausen (1996) found that the digestibility of added crystalline cystine was only about 60–70%, while that of crystalline methionine was 95–100%.

Phenylalanine/Tyrosine

An animal's requirement for phenylalanine and tyrosine are always discussed together inasmuch as the actual requirement for these amino acids is the combined total of the essential amino acid, phenylalanine, and the non–essential amino acid tyrosine. Tyrosine is an anaplerotic amino acid replacing a significant portion of the animal's requirement for phenylalanine.

Phenyalanine—Glycogenic, Ketogenic, Hydrophobic

Phenylalanine is an important amino acid in the synthesis of multiple proteins of animal physiology and the critical melanin pigment of mink fur. Phenylalanine via hydroxylation to yield tyrosine and an additional hydroxylation step to yield dihydroxyphenylalanine (DOPA) are the initial steps along the biochemical pathways leading to the synthesis of melanin, eumelanin (brown– black), and pheomelanin (reddish–brown) (Ozeki et al., 1995).

Phenylalanine is the only high calorigenic essential amino acid. That is, it provides high specific dynamic action, the increase of heat production as the direct result of the ingestion of food, and the metabolic processes required for the catabolism of these nutrients.

It is well known that the non–essential amino acid tyrosine can replace part of an animal's requirement for phenylalanine. At the present time, the proportion of required phenylalanine that can be provided by tyrosine is unknown. In the case of the cat, half of the phenylalanine requirement can be met by tyrosine (Anderson et al., 1980).

Tyrosine—Non–Essential, Glucogenic, Ketogenic, Hydrophilic

Tyrosine is an important amino acid for the synthesis of multiple proteins in animal physiology and the melanin pigment of mink fur.

In the case of the cat, the animal's requirement for phenylalanine/tyrosine for proper melanin pigment formation is over twice the requirement for maximal growth. This experimental observation is directly related to the Michaelis Constant, K_m, for the enzymes involved in the tyrosine anabolic pathways within the physiology of the cat. The K_m value for tyrosine aminotransferase, leading to the synthesis of the melanin pigment is nearly 50 times that of acyl transferase, leading to the synthesis of proteins (Morris et al., 2002). Their studies indicated that the "red" hair of black cats raised on a low phenylalanine/tyrosine dietary regimen could be converted to a new coat of "black" hair over a period of time with simple supplementation of the research diet with tyrosine. Comparably, studies by Clausen et al. (2003) indicated that dark mink raised on a ranch diet providing 12% meat–and–bone meal yielded pelts with a reddish tone, an observation directly related to the relatively low level of tyrosine and phenylalanine in the meat–and–bone meal.

Tyrosinemia Type II

Hereditary tyrosine type II in the mink was first reported by Christensen et al. (1979). The disease was initially discovered on Danish farms when a rather high mortality among standard (dark) mink was observed. Actually, evidence of the disease had been reported earlier by Schwartz and Shackelford (1973) whose clinical observations of pseudodistemper were identical to the initial observations of Christensen et al. (1979). Initially, the disorder was considered as a simple Mendelian autonomic recessive character. However, studies by Marsh (1995) indicated that tyrosinemia is actually a double recessive gene. Marsh felt that a careful examination of ranch genetic records could actually eliminate tyrosinemia from the mink colony in one year via pelting of all carriers.

The biochemical features of the disease correspond to tyrosinemia type II and clinical signs correspond to tyrosinemia type I (Richner–Hanhart syndrome) in humans. An interesting facet of the disease is the wonderful "black" fur exhibited by the mink, the very reason these unique strains were promoted in the genetic management of the ranch.

Studies by Christensen et al. (1980) indicated that the affected kits had 20–100 times higher plasma tyrosine concentration than that found in normal kits. In another study, Henriksen (1980), the serum tyrosine was 20–50 times the normal serum values of 0.01–0.03 mM.

In all forms of hereditary tyrosinemia type II, the defect is with hepatic tyrosine aminotransferase (EC 2.6.1.5). This specific tyrosine aminotransferase enzyme has been isolated from livers of mink with tyrosinemia II. Enzyme kinetics indicated that insufficient binding of the co–factor pyridoxalphosphate (PLP) to the hepatic tyrosine aminotransferase is the likely cause of the disease. In this study by Christensen et al. (1989), dietary treatments of the affected mink with PLP were successful in eliminating the disease.

Taurine

Although cats are unable to decarboxylate appreciable amounts of cysteine sulfinic acid to taurine, a key component of bile salts (Hayes and Carey, 1975), there is no reported evidence in the scientific literature to date that mink are unable to obtain taurine from cysteine sulfinic acid. That is, they may not have a specific dietary requirement for taurine.

Threonine—Essential, Glucogenic, Hydrophilic

Threonine is unique in terms of the biochemistry of E. coli wherein it is an essential amino acid which can be converted into another essential amino acid, isoleucine.

Tryptophan—Glucogenic, Ketogenic, Hydrophobic

In animal nutrition and physiology, tryptophan is unique in terms of being a precursor of the B–vitamin, niacin, as well as a precursor for the neurotransmitter, serotonin. The latter point is of special interest to fur ranchers in terms of studies by Rouvinen et al. (1999) that indicate that tryptophan has psychopharmacological properties for fox. Dietary supplementation of silver fox with tryptophan resulted in reduced fear and enhanced exploratory behavior in the female fox.

In most animals, the niacin requirement can be met by synthesis of the B–vitamin from tryptophan. However, the cat (Carvalho da Silva et al., 1952; Suhadolnick et al., 1957; Ikeda et al., 1965) and the mink (Warner et al., 1968) have a very limited capacity for the conversion of tryptophan to niacin.

It is of interest to note that tryptophan is one of the amino acid precursors of pheomelanin, the red–yellow pigment of mink fur.

Tryptophan is sensitive to significant loss in the processing of mink feedstuffs. This destruction can stem from dehydration procedures and possible Maillard Reaction, oxidative damage (extensive freezer storage) wherein its bioavailability is reduced (NRC cat bulletin, 1986), and the use of acids as silage preservation (as high as 50% destruction) (Skrede, 1981).

Valine—Glucogenic, Hydrophobic

Even though valine (5–C) is a branched chain amino acid like leucine (6–C), it is not ketogenic like leucine, but glycogenic.

In studies reported by Borsting and Riis (2000), it was noted that the oxidation rate of valine in fasted mink was almost equal to the oxidation rate of leucine. However, in the fed state, the oxidation of valine within 4 hours rose to 32% compared to the 23% level found for leucine. One explanation for this difference might be the fact that valine is a glycogenic amino acid which can be readily utilized to cover the glucose demand of the mink, whereas leucine is strictly a ketogenic amino acid. This finding indicates a high utilization of glucogenic amino acids for glucose synthesis in the post–prandial period in the mink. This basic point is also indicated by the rise of glucagon (a hormone involved in glucose utilization) levels after feeding with high protein diets (Borsting and Damgaard, 1995; Borsting and Gade, 2000).

Non–Essential Amino Acids
Alanine—Glucogenic, Hydrophobic

Alanine is a component of multiple proteins within animal anatomy and physiology. Alanine, like glycine and glutamic acid, is one of only three non–essential amino acids which is highly calorigenic, a significant Specific Dynamic Action (SDA) increased heat production as the direct result of the ingestion of food and the metabolic processes required for the catabolism of these nutrients.

Asparagine—Glucogenic, Hydrophobic

Asparagine is a key component of multiple protein structures within animal anatomy and physiology.

Excessive heating and dehydration of mink feedstuffs to yield high protein meals may result in an interaction between the available asparagine primary amide groups and the epsilon amino

137

group of lysine and/or the guanido group of arginine to yield a secondary amide bonding which is unavailable to the digestive processes of animals.

It is of interest to note that asparagine, a non–essential amino acid for most animals, is essential for early post–weaning kittens (Kamikawa et al., 1984) but not required after the first few weeks.

Aspartic Acid—Glucogenic, Acidic, Hydrophilic

Aspartic acid, like glutamic acid, is a key component of multiple proteins in animal anatomy and physiology which are required for an optimum protein pH environment for many protein functions.

Cysteine/Cystine—Glucogenic, Hydrophobic

See Methionine/Cysteine/Cystine

Glutamic Acid—Glucogenic, Hydrophobic, Acidic

Glutamic acid, like aspartic acid, is a key component of multiple proteins in animal anatomy and physiology required for an optimum pH needed for protein function.

Glutamic acid, like alanine and glycine, is one of only three non–essential amino acids which is highly calorigenic, that is, it produces significant Specific Dynamic Action (SDA). Cats are less tolerant of an excess of glutamic acid than rats, chickens, and most other species (Deady et al., 1981). High levels of glutamic acid in kitten diets yields a classic thiamine deficiency syndrome (NRC cat bulletin, 1978), likely related, in part, to the extra requirement for thiamine as a co–factor for the enzyme involved in the oxidative decarboxylation of glutamic acid.

Glutamine—Glucogenic, Hydrophobic

Excessive heating and dehydration of mink feedstuffs to yield high protein meals may result in an interaction between available glutamine primary amide groupings and the epsilon amino group of lysine and/or the guanido group of arginine yielding a secondary amide bonding which is unavailable to the digestive processes of animals.

Glycine—Glucogenic, Hydrophobic

Glycine is a key amino acid in multiple protein structures in animal anatomy and physiology including the connective tissue, collagen, the most abundant protein of mammals and the main fibrous component of skin, bone, tendon, cartilage and teeth. In collagen every third amino acid residue is glycine. Inasmuch as collagen is the key structural protein of bone, mink feedstuffs such as fish scrap (filleted fish) with a high bone content have a high ash/protein ratio and relatively low glycine digestibility rating for the mink (Skrede VI, 1981; Ahlstrom et al., 2000).

138

Glycine, like alanine and glutamic acid, is one of only three non–essential amino acids which is highly calorigenic.

Hydroxy–Proline—Glucogenic, Hydrophilic

Hydroxy–proline is a key component of the skeletal protein structures within animal anatomy. Hence, it is not surprising that mink feed products such as fish scrap (filleted fish) with high bone content have a relatively low hydroxy–proline digestibility rating for mink (Skrede VI, 1981). The examination of the hydroxy–proline content of a poultry by–product mixture or poultry meal would provide an insight into the level of collagen protein provided via skins and bone, inasmuch as hydroxy–proline is present at a low percentage in muscle meats and a high percentage in skin and bone.

Proline—Glycogenic, Hydrophobic

Proline is a key component of the skeletal protein structure within animal anatomy. Hence, it is not surprising that mink feed products such as fish scrap (filleted fish) with a high bone content have a relatively low proline digestibility rating for the mink (Skrede VI, 1981). In the rat, proline in an anaplerotic reaction may "spare" at least one–half of the arginine requirement (Rogers et al., 1970; Rogers and Harper, 1970). However, in the cat, proline does not "spare" the arginine requirement (MacDonald and Rogers, 1984).

Serine—Glucogenic, Hydrophilic

Serine, although classified as a non–essential amino acid, is very important in terms of the total anatomy and physiology of animals. It is a precursor of other non–essential amino acids such as cysteine and glycine. Serine is also a key component of phosphoproteins such as casein in milk wherein serine provides a convenient hydroxy–group for esterification with phosphoric acid. Serine is also a constituent of cephalin, a phospholipid.

Tyrosine—Glucogenic and Ketogenic, Hydrophilic

See Phenylalanine/Tyrosine.

Amino Acid Summary

Mink are distinct from most mammals in terms of amino acid requirements beyond those needed for optimum growth. For top fur development, they need extra arginine and sulfur amino acids and very likely, similar to the cat, extra phenylalanine and tyrosine for top dark (melanin pigment) fur color development.

Amino Acid Requirements

The amino acid requirements of the mink as a carnivore are unique relative to all other domesticated animals because they have a relatively high proportion of body protein as fur (see Table 4.26), their body amino acid composition is unique, and the amino acid composition of mink milk is also unique (see Table 4.27).

TABLE 4.26. **Amino Acid Composition of Mink Fur, Mature mink, and Adult Pig**

Amino Acid	Mink Fur*	Mature Mink**	Adult Pig***
Alanine	1.5	6.0	6.5
Arginine	7.5	6.4	6.1
Aspartate	2.2	8.0	8.0
Cystine	12.6	2.8	1.0
Glutamate	11.5	14.0	14.0
Glycine	5.6	8.6	9.3
Histidine	1.0	2.4	2.8
Isoleucine	2.6	3.2	3.9
Leucine	7.5	7.2	7.1
Lysine	3.5	6.6	6.9
Methionine	1.0	2.6	2.0
Phenylalanine	2.6	3.6	3.7
Proline	17.9	7.2	—
Serine	8.0	5.0	3.1
Threonine	5.6	4.3	3.5
Tryptophan	0.4	1.0	1.2
Tyrosine	5.2	2.8	1.9
Valine	5.4	4.8	4.8

* Jorgensen & Eggum (1971) **Moustgaard & Riis (1957), Disse (1957) ***Buraczwski (1973).

It is readily seen that the content of the sulfur amino acids (SAA) cystine and methionine in the mature mink is almost double that of the adult pig. The obvious reason for this high level of SAA in the mature mink is related to the fact that the fur of the mature mink contributes 16% of the total amount of body protein but as much as 49% of the SAA (Jorgensen and Eggum, 1971). At birth, mink kits have very little fur; in the next four months they develop a summer coat and moult in late August and early September. This phase is followed by development of the winter coat, a process completed by early December. Thus the mink has a unique requirement for SAA as well as a requirement for two coats of fur, each with a high level of SAA.

An N–balance study by Glem–Hansen (1974) indicated that the level of SAA was the first limiting factor for protein utilization in twelve traditional Danish mink ranch diets, that is, he found a direct correlation between the level of SAA in the protein content and the biological value of the dietary regimen for the mink. This observation was confirmed by other N–balance experiments (Glem–Hansen, 1976) wherein N–retention was used as a parameter for protein utilization, as shown in a linear correlation between percent of protein utilization and the SAA content of the protein resources.

TABLE 4.27. Amino Acid Composition of the Milk of Cows, Mink, and Cats

Amino Acid	Cows*	Mink**	Cats***
Arginine	3.6	6.1	6.4@
Cystine	1.6	1.0	1.2
Histidine	2.8	1.7	2.7
Isoleucine	5.2	3.7	4.3
Leucine	10.0	11.3	11.8
Lysine	8.8	5.5	5.7
Methionine	2.4	1.3	3.2
Phenylalanine	5.2	3.4	3.0
Threonine	4.0	3.9	4.6
Tryptophan	1.6	1.2	—
Tyrosine	5.2	2.1	4.5
Valine	6.8	5.6	4.7

* Rogers et. al. (1986), **Moustgaard and Riis (1957), ***Davis et. al. (1994) @ highest level of 14 species.

This data shows that the milk provided by mink and cats, carnivores with many nutrient requirements similar to the mink, to their young is exceptionally high in arginine relative to the level of this amino acid provided by cows and other animals, including humans. It is of interest to note that North American mink ranchers for many years have emphasized the employment of liver or cooked eggs (rich arginine resources) in their ranch diets for the critical reproduction/lactation period.

Requirement Data

The most detailed experimental report ever published on the protein and amino acid requirements of the mink was provided by Glem–Hansen (1992). The summary of the amino acid requirements for the periods of weaning to mid–August (early growth and late growth phases of the mink ranch year) and from mid–August to pelting indicated that the amino acid requirements of the mink were identical for both periods except for the sulfur amino acids (cystine and methionine) which showed a 50% higher requirement during the furring period.

The studies on the amino acid requirements of the mink by Borsting and Clausen (1996) (provided in Tables 4.28 and 4.29) led to an improved understanding of the amino acid requirements of the mink.

TABLE 4.29. Amino Acid Requirements of Mink and Pigs—Expressed as Grams Apparent Digestibility/Dietary Energy Level

Essential Amino Acid	Mink (1)		Pig (2)
	Mcalories ME	/MJ ME	/MJ ME
Histidine	< 1.60	< 0.38	0.20
Isoleucine	< 2.60	< 0.62	0.32
Leucine	< 5.00	< 1.20	0.64
Lysine	2.70	0.65	0.57
Methionine	1.60	0.38	0.14
Cystine*	0.70	0.14	0.14
Phenylalanine	< 2.90	< 0.69	0.32
Tyrosine*	< 1.80	< 0.43	0.32
Threonine	1.70	0.40	0.36
Tryptophan	0.50	0.12	0.10
Valine	< 3.50	< 0.84	0.42
Crude Protein	30% ME		

(1) Borsting and Clausen (1996), (2) Boisen (1996) *Non–essential amino acids which can replace part of an animal's requirement for the corresponding essential amino acid in an anaplerotic reaction.

TABLE 4.30. Essential Amino Acid Requirements of Mink (1)

Essential Amino Acid	%Protein
Arginine**	6.4 – 8.0 (2)
Histidine	< 2.4*
Isoleucine	< 3.9*
Leucine	< 7.5*
Lysine	4.1
Methionine	2.4
Cystine***	0.9
Phenylalanine	< 4.3*
Tyrosine***	< 2.7*
Threonine	2.5
Tryptophan	0.8
Valine	< 5.3
Crude Protein	30% ME

(1) Borsting and Clausen (1996), (2) Damgaard (1997).

142

*The requirements of these amino acids were not fulfilled when levels were reduced by 50% of that in modern mink feed with 30% of ME from protein, mainly due to negative effects on growth, health, and mortality. However, the requirements for all of these amino acids are probably lower by about half the values shown, and thus close to the amino acid requirements of the pig (see Table 4.29), inasmuch as the effect of this 50% reduction was not very severe.

**The experimental data on the arginine requirement of the mink deserves special attention. In the Damgaard experiment conducted in the late growth period (July–August), criteria for the argininine requirement was the urinary orotic acid excretion and plasma concentrations of ammonia. 8.0 grams of arginine/100 grams of protein (16 grams of nitrogen), resulted in minimum urinary excretion of orotic acid while a level of 6.4 grams of arginine/100 –234–grams of protein (16 grams of nitrogen) supported growth performance and prevented increased plasma concentrations of ammonia.

Arginine is a key component of the urea cycle wherein excess ammonia in animal metabolism is converted to urea for excretion from the body by the kidneys. With sub–optimum levels of arginine in animal's diet, there is a distinct potential for sub–optimum urea cycle function and subsequent rise in plasma concentrations of ammonia.

Research studies with a great variety of animals including dogs and cats indicate that the urinary excretion of orotate is elevated with arginine deficient diets (Visek, 1979). This inverse relationship between the orotate excretion of an animal and the available arginine in its diet is related to the fact that mammalian cells have two carbamyl phosphate synthetases: CP–1 in the liver mitochondria involved in the synthesis of urea and CP–2 in the cytoplasm involved in pyrimidine synthesis. With sub–optimum physiological levels of arginine or any other amino acid within the urea cycle, carbamyl phosphate synthesized within the mitochondria enters the cytoplasm and the pyrimidine pool, resulting in increased production of orotate and subsequent enhanced urinary orotate excretion. Thus the measure of urinary orotate can be used to assess the dietary arginine levels required for the optimum physiology of the animal. Urinary orotate can also be used to assess the arginine value of dehydrated protein resources for modern mink nutrition, with 24 hour urinary orotate excretion as a measure of the available arginine content to the digestive (Leoschke, 1988).

*** Non–essential amino acids which can replace part of an animal's requirement for the corresponding essential amino acid in an anaploerotic reaction.

A comparison of the essential amino acid requirements of the mink with that of growing pigs is of interest. Apart from methionine, the quantity of specific essential amino acids (lysine, cystine, threonine and tryptophan) needed to support normal growth and health of the mink is similar to that of the pig. However, the mink still seems to require a total protein level about two times higher than that of the pig, since the mink's protein requirement, 30% ME in a feed mixture of standard composition, corresponds to 15.9 grams of digestible protein/MJ compared to only 8.0 grams digestible protein/MJ for pigs at 50 kg (Boisen, 1996). This sharp contrast in the protein requirements of the pig and the mink is obviously related, in part, to the key role of gluconeogenic amino acids in the nutrition and physiology of the mink.

As to methionine, the mink requirement is 2.7 fold that of the pig. For cystine the requirement of the mink was equivalent to that of the pig, indicating that the high requirement of sulfur–containing amino acids for the additional protein synthesis in the hair of the mink is almost exclusively covered by methionine.

The data provided in Tables 4.29 and 4.30 apply only to males. Borsting and Clausen (1996) found that in general the effects of protein and amino acid supply were smaller for females than for males, therefore the female data were omitted from their report.

In Table 4.30, the experimental data are not detailed on the actual minimum requirement for six amino acids—histidine, isoleucine, leucine, phenylalanine, tyrosine, and valine. Fortunately, more recent experimental data from Sandbol et al. (2004) provide a clearer insight into these actual amino acid requirements for the period from early July to pelting.

The concept of an "ideal protein" for animal nutrition has been considered for many years.

143

Sanbol and his associates patterned their "Ideal Protein" amino acid profile for modern mink nutrition on the basis of the current norm, the amino acid profile of the mink body plus pelt, and the present cat norm (Table 4.31).

TABLE 4.31. Essential Amino Acid Profile (Relative to Lysine) for Mink. Norm (N), Mink Body and Pelt (M), Cat Norm and "Ideal Protein" (IP)

Amino Acid	N(1)	M(2)	C(3)	IP
Arginine	115	119	125	115
Histidine	56	38	38	40
Isoleucine	96	54	63	65
Leucine	185	122	150	150
Lysine	100	100	100	100
Methionine	59	32	50	47
Cystine	22	67	44	39
Met + Cys	81	99	94	86
Phenylalanine	107	62	50	65
Tyrosine	67	56	56	65
Phe + Tyr	174	118	106	130
Threonine	70	74	88	70
Tryptophan	22	22	19	22
Valine	130	78	75	80

(1) Borsting and Clausen (1996); Borsting (1998), (2) Chavez (1986), (3) NRC (1986); (4) Sandbol et al.(2004)

On the basis of the Sandbol et al. (2004) experimental data, it is possible to provide an estimate of the amino acid requirements of the mink for the growth and fur development period which is closer to mink protein nutrition reality than the Borsting and Clausen data of 1996 and 1998. See Table 4.32.

TABLE 4.32. Essential Amino Acid Requirements of the Mink for Late Growth and Fur Development

Amino Acid	Borsting & Clausen (1996) Borsting(1998)	Sandbol et al. (2004) "Ideal Protein"*
Arginine	4.7**	4.7
Histidine	<2.3	1.6*
Isoleucine	<3.9	2.7*
Leucine	<7.6	6.2*
Lysine	4.1	4.1
Methionine	2.4	1.9
Cystine	0.9	1.6
Met + Cys	3.3	3.5
Phenyalanine	<4.4	2.7*
Tyrosine	<2.7	2.7*
Phe + Tyr	<7.1	5.4*
Threonine	2.9	2.9
Tryptophan	0.9	0.9
Valine	<5.3	3.3*
Protein	30	30

* Not actual minimum amino acid requirements, but levels of essential amino acids supportive of the sound performance of the mink for growing and fur development within a 30% protein regimen. **Lower than Damgaard (1997) (see Table 4.30), inasmuch as the data presented are based on mink kit performance while the Damgaard data were based on urinary orotic acid excretion and plasma ammonia concentrations.

Additional insights into the amino acid requirements of the mink relative to the "Ideal Protein" have been provided by Nielsen et al. (2004). Their study focused on the late growth period from July 1 through July 26 and involved decreasing the level of each single essential amino acid by 20%. Only the reduction in the level of valine resulted in significant changes in the ratio of weight gain/initial weight. There was a significant difference in the growth rate, with reduced levels of lysine and threonine but this result was most likely due a lack of *ad libitum* feeding as initially planned. There was no significant difference with the reduced levels of the other essential amino acids indicating that, for the "ideal protein," the level of estimated amino acid requirement could be lowered.

Obviously, the amino acid profile or pattern of an "Ideal Protein" for mink nutrition will vary and be specific for each phase of the mink ranch year. This point is well illustrated by experimental data reported by Perel'dik and associates for tryptophane. Perel'dik (1974) reported that, for the growth and fur development stage, the requirement was 0.53 grams/Mcalories ME, while that for reproduction/lactation was more than 50% higher, at 0.86 grams/Mcalories ME (Perel'dik, 1970).

Relative to the essential amino acid requirements of the mink, the following commentary is warranted.

145

Methionine—Cystine

The initial studies on the unique requirement of the mink for sulfur amino acids (methionine and cystine) by Glem–Hansen (1974, 1976) have now been up-graded with new insights into the critical role of sulfur amino acids in the nutrition of the mink with the studies of Borsting and Clausen (1996). Their studies indicate that the unique requirement for the mink for sulfur amino acids is not a specific requirement for a combination of these two amino acids, but a specific requirement for relatively high levels of methionine. Obviously, this was a complete surprise in terms of the relatively high level of cystine in mink fur compared to methionine. However, the experimental data is very clear: there were no significant interactions between cystine and methionine, nor can either replace the other in the mink's diet. In summary, the high sulfur amino acid requirements of the mink are directly related to methionine and not cystine. Specific requirements of the mink for methionine and cystine found by Borsting and Clausen (1996) were confirmed in a later study by Clausen, Therkildsen and Borsting (1998).

At the same time, it is of major value to the world mink industry to examine more exactly the requirement of the mink for the sulfur amino acids in the late growth and fur production phases of the mink ranch year. The requirement of sulfur amino acids for fur production reported by Glem–Hansen (1972) was relatively high, 0.72 g. methionine+cystine/MJ compared to the Borsting and Clausen (1996) report of 0.52 grams of methionine+cystine/MJ. Studies by Sandbol et al. (2003a) indicated that the requirement for methionine per se for the furring period was fulfilled at 0.31 g. of digestible methionine/MJ. The experimental data also indicated that a balance of 60:40 between methionine and cystine should be sufficient. The Glem–Hansen suggested level of 0.48 grams of methionine+cystine/MJ for the late growth phase of the mink ranch year is more than adequate, consistent with the report of Borsting and Clausen (1996) that special methionine supplementation of commercial mink ranch diets in July–August was not warranted.

New insights into the mink's requirements for methionine during the late growth phase of the mink ranch year are provided by Sandbol and Clausen (2004). Considering the present norm of a total of 0.22 grams of sulfur amino acids/100 kcal ME including 0.16 grams of methionine and 0.6 grams of cystine, the researchers provided an experimental plan with 0.16, 0.14, and 0.12 grams of methionine (maintaining 0.22 grams total sulfur amino acids). 0.16 grams/ 100 kcal ME of methionine provided the best weight gains for the whole period extending into pelting. Although there was no significant difference in kit growth with reduced levels of methionine in the furring period, there was a tendency for more dead kits during the furring period with decreasing methionine content. The 0.12 grams 100 kcal M.E. level of methionine provided the shortest pelts and poorest fur quality.

Table 4.33 provides some interesting insights into the variation in the mink's requirements for methionine and cystine during different periods of the ranch year.

146

TABLE 4.33. Sulfur Amino Acid Requirements of the Mink for Late Growth and Fur Development–% Protein

	Late		Fur	Fur Production Period(3)	
Growth(1)	ProductionTotal(2)		EarlySept	MidSept–October	November
< 3.0	3.3		2.6–2.7	3.7–4.1	3.0–3.1

(1) Glem–Hansen (1972), (2) Borsting and Clausen(1996), (3)Glem–Hansen(1982).

Obviously, the requirement for sulfur amino acids reaches its peak in the period of mink kit ages 20–24 weeks (mid–September through October) with a significant slowing of the furring process through November. It should be noted that an earlier report by Glem–Hansen (1980) indicated a requirement for higher levels of the sulfur amino acids than reported in 1982, but these experimental data were in error, inasmuch as the calculations included D–methionine which is now acknowledged to have no nutritional value for the mink (Borsting and Clausen, 1996). The need of the mink for extra sulfur amino acids in the period from 20–24 weeks is supported by Glem–Hansen and Hansen (1981) wherein they assess the deposition of cystine in bodies. In the period from October 12th to December, the level of cystine as percent of total amino acids present rose from 3.1 to 4.0 in male and from 4.1 to 4.5 in female kits. These observations of a higher SAA requirement for fur production relative to growth is supported by experimental data from Russia (Milovanov, 1961) which indicated that supplementation of practical mink ranch diets with methionine and cystine did not influence the growth rate of the mink but did enhance the fur quality significantly.

It is of interest to note that mink have higher requirements for the sulfur amino acids than fox, a difference directly related to the higher ratio of fur to body weight (Pereľdik, 1972).

Phenylalanine–Tyrosine

The levels of phenylalanine and tyrosine must be at least 0.53 grams/100 kcal to obtain dark color in mink pelts. Experimental studies conducted by Clausen and Hejlesen (2004) indicated that cracklings (non–fat residue after removal of fat from adipose tissue) do not provide optimum levels of phenylalanine and tyrosine for top sharp color in dark mink.

Gluconeogenesis

In the previous commentary on the amino acid requirements of the mink, the only experimental data provided was that for the ten essential amino acids plus the two anaplerotic amino acids, cystine and tyrosine. Without question, the non–essential amino acids have a special role in the nutrition and nutritional physiology of the mink in terms of a process termed gluconeogenesis, the synthesis of glucose from non–sugar precursors.

The mink is an obligate carnivore with a demand for a high level of protein in its diet, as high as 20–40% of ME. This protein requirement is relatively high compared to that of other carnivores;

147

for instance, the dog and weasel have protein needs of 4–12% of ME (MacDonald and Rogers, 1984). Although the essential amino acid requirements of the mink and pig are similar, with the exception of methionine, the actual protein requirement of the mink is about twice that of the pig. The major factor in this "extra" protein requirement relative to the pig is the process of gluconeogenesis. Thus the mink, like the cat, has a demand for dietary protein that exceeds the animal's essential amino acid requirements. Support for this conclusion is provoided in studies by Damgaard, Clauson, and Borstring (1998) and Borsting and Gade (2000). The mink's high demand for non–essential amino acids is related to a high rate of catabolism (breakdown) of these amino acids. Studies by Sorensen, Petersen, and Sand (1995) indicate that the activities of the key enzymes of gluconeogenesis from the liver of the mink were higher than those of rats or cats, This high rate of gluconeogenesis in the mink stems from the fact that in the wild its intake of carbohydrates is relatively limited.

All factors considered, in terms of the protein nutrition of the mink and its amino acid physiology, the concept of an "Ideal Protein" may not be applicable. The "Ideal Protein" concept refers to a specific pattern of dietary amino acids which would conform exactly to the pattern of the animal's amino acid requirements and be expressed in terms of a single quantity. Thus it could provide a reference against which an actual protein feedstuff can be evaluated. The concept of an "Ideal Protein," may not be applicable, however, inasmuch as the key physiological factor of required gluconeogenesis and consequent obligate nitrogen loss is not given serious consideration (Borsting and Gade, 2000). Initial studies by Rouvinen et al. (1995) relative to the concept of the "Ideal Protein" involved mink dietary supplementation with lysine, threonine, or methionine or their combinations to achieve an ideal protein composition. The experimental data indicated an increased utilization of the amino acids, but did not improve total nitrogen digestibility or nitrogen–balance. Contrary to the expectations, none of the experimental treatments resulted in reduced fecal or urinary nitrogen excretion.

Protein Feedstuff Processing and Amino Acid Loss

Every modification of a mink protein feedstuff —including dehydration procedures, acid preservation, and alkali preservation—has a distinct potential to undercut the biological value of the proteins present for the mink by amino acid destruction and/or bonding of the amino acids in structures which are unavailable to the digestive processes of the mink.

Dehydration Programs

It is well known that sub–optimum dehydration procedures can bring about the Maillard Reaction and other chemical processes wherein arginine, cystine, lysine, and tryptophan are especially heat labile with resultant destruction of cystine, lysine, and tryptophan and the bonding of arginine and methionine in structures indigestible for the mink. During the critical reproduction/lactation and fur development phases it is especially important that the dehydrated protein resources be of high biological value. Special concern must be given to ample levels of available arginine for the lactation

148

period, inasmuch as mink milk, like the milk of cats, has an unusually high level of arginine, highest of the 14 species examined by Davis et al. (1994). During the fur development period, ample levels of available arginine and the sulfur amino acids must be provided.

Acid Preservation

In the formic acid or sulfuric acid preservation of fish viscera as silage, the studies of Johnsen and Skrede (1981) and Skrede (1981) stress that a 50% destruction of tryptophan is possible

Alkali Preservation

Cysteine amino acid residues are readily oxidized to cystine disulfide structures in neutral or slightly alkaline aqueous solutions, especially in the presence of trace minerals including copper and iron. In a more alkaline environment there is a distinct potential for the loss of both cystine and cysteine via air oxidation to yield cysteic acid structures. These structures occur normally in the outer part of the sheep's fleece, where the wool is exposed to light and weather (Martin, 1945).

Fats and Essential Fatty Acids

Fats, fatty acid tri–esters of glycerol (triglycerides), belong to a class of biochemicals known as lipids which also includes monoglycerides, diglycerides, phospholipids (cephalin and lecithin), sphingolipids, glycolipids, cholesterol, and phytosterols. The sterols can exist as free alcohol structures or esterified with long–chain fatty acids. The triglycerides are usually called fats if solid and oils if liquid at room temperature.

Inasmuch as the mink is a carnivore, fats play a much greater role in meeting its daily energy requirements than they do in omnivores such as the pig or herbivores such as the rabbit. In terms of metabolic energy value, see Table 4.34.

TABLE 4.34. Nutrient Metabolizable Energy Values for the Mink

Nutrient	Metabolizable Energy Value/Kilocalories/Gram
Fat	9.5
Protein	4.5
Carbohydrates	4.2

Fats represent the major energy resource for the mink both in the wild and on commercial mink ranches. In terms of mink feed economics, fats represent by far the most economical energy resource for modern mink nutrition, a pound of lard providing 3 times as much metabolic energy (ME) for the mink as a pound of a quality commercial mink cereal or 10 pounds of lean fish.

Special attention must be given to the basic fact that the fatty acids present in triglycerides can be divided into two very significant classifications—saturated fatty acids with only single bonds

149

between the carbon atoms and unsaturated fatty acids with single or double bonds between the carbon atoms as illustrated in Table 4.35.

TABLE 4.35. Common Fatty Acids in Mink Nutrition

Fatty Acid	Designation*	Major Resources
Myristic	14:0	Animals and Plants
Palmitic	16:0	Palm Oil, Animals and Plants
Palmitoleic	16:1	Animals and Plants
Stearic	18:0	Animals, especially Cattle Fat
Oleic	18:1	Olive Oil, Animals and Plants
Linoleic	18:2	Animals and Plants
Linolenic	18:3	Linseed Oil and Horsemeat
Cetoleic	22:1	Fish Products
Arachidonic	20:4	Animal Fats only
Eicopentaenoic	20:5	Fish Products
Docosahexanoic	22:6	Fish Products

* 14:0 refers to 14 carbons without any double (alkene) bonds.

Fatty acids without alkene bonds have no room for additional hydrogen atoms and are thus termed "saturated" fatty acids while fatty acids with one or more alkene bonds have the capacity to add hydrogen or other atoms and are therefore termed "unsaturated." Fats with relatively low melting points, termed oils, generally have a higher content of short–chain fatty acids as found in butter and palm oil or polyunsaturated fatty acids as found in vegetable oils. Softer animal fats such as poultry fat contain relatively high levels of oleic (18:1) and linoleic (18:2) acids as Table 4.36 illustrates.

TABLE 4.36. Composition of Animal, Fish, and Plant Fats

Fat	Pork	Chicken	Horse	Corn	Soya	Linseed
References	(1)	(1)	(1)	(2)	(3)	(3)
Fatty Acid						
14:0	1.7	1.0	5.8	0.0	—	—
16:0	28.6	23.0	30.2	10.6	10.7	5.0
16:1	2.6	5.6	6.6	0.1	0.2	—
18:0	17.8	7.3	5.6	1.8	3.7	4.9
18:1	35.7	40.1	27.0	27.3	24.8	20.2
18:2	8.4	20.9	10.4	53.2	53.7	16.0
18:3	0.7	1.1	10.2	1.2	7.2	52.5
18:4	—	—	—	—	—	—
20:1	1.9	1.0	.05	0.1	—	—
20:4		0.3	1.4	—	—	—

Fat	Capelin	Herringmeal	Salmonmeal
References	(3)	(4)	(4)
Fatty Acid			
14:0	7.1	6.0	5.5
16:0	11.9	14.4	16.4
16:1	10.2	5.7	7.6
18:0	1.3	2.2	3.7
18:1	13.4	14.7	17.9
18:2	1.4	1.8	3.3
18:3	0.7	0.7	0.9
18:4	3.9	1.2	1.7
20:1	12.8	12.7	7.2
20:4	0.5	0.9	2.0
20:5	9.9	6.0	7.5
22:1	3.4	—	—
22:5	0.8	0.9	2.6
22:6	9.4	8.0	9.9

(1) Yu and Sinnhuber (1967), (2) USDA (2007), (3) Kekela et al. (2001), (4) Fleming (1999).

In a chemical process termed catalytic hydrogenation, industrial plants can convert vegetable oils like corn oil or soybean oil containing high levels of linoleic acid (18:2) to stearic acid (18:0) and thereby yield solid fats similar to margarine. This same process of hydrogenation can also occur naturally in the rumen of beef cattle and dairy cows and thus yield the highly saturated nature of beef fat and cow's milk. An exception to the observation that most animal fats are highly saturated is horsemeat with a high content of linolenic acid (18:3).

151

In addition to the major source of energy for mink nutrition, fats act as physiological carriers for feed flavors and the fat–soluble vitamins, A,D, E, and K. This process includes absorption from the intestines into the blood circulation and via the lymphatic system and final passage to the liver and body cells. Fats are also the only source of the essential fatty acids required for the synthesis of prostaglandins, important for the regulation of body metabolism. Metabolism is a word which simply represents the sum total of all biochemical reactions within a living organism.

Fats provide a major energy storage nutrient in the adipose tissue for the mink during the winter months when feed resources may be limited. Fats also provide thermal insulation and physical protection for the internal organs. Like carbohydrates, they have a "sparing" effect on dietary protein requirements, that is, optimum levels of both fats and carbohydrates in a mink's diet can lead to minimal protein levels required for top performance of the mink in all phases of the ranch year.

Physiology

Digestion and absorption of fat into the mink's lymphatic pathway and finally to blood circulation is facilitated by the combined actions of the enzyme pancreatic lipase, bile salts, and phospholipids such as lecithin, and the peristalic action of the small intestine. The action of pancreatic lipase is to bring about hydrolysis of the ester linkages of triglycerides to yield free fatty acids (FFA), diglycerides, and 2–monoacylglycerols. The endoplasmic reticulum of the intestinal cell is the site of resynthesis of triglycerides from the absorbed products of digestion. The combination of lipoprotein apopeptides (protein polymers lipid addition) with these triglycerides yields chylomicrons which are sent into the blood circulation via the lymphatic pathway and the thoracic duct. Once within the blood circulation, the lipoprotein lyase releases free fatty acids (FFA) for uptake by the liver, adipose (fat) tissue, and other organs.

Requirement—Quantity

As a carnivore, the mink has unique dietary requirements for fat as a major energy resource relative to other domesticated animals, omnivores or herbivores. Multiple experimental studies indicate that fat is the energy resource of primary consideration. It enhances performance in all phases of the mink ranch year. It returns the best investment value in terms of feed volume per pelt marketed. In terms of kilocalories for the ranch feed dollar, common fat resources both natural and rendered are the farmers' "Best Buy." Mink also have a more efficient digestive program for fats than for carbohydrates and protein resources. The fat content in mink ranch diets is the key factor in the total energy content of the diet and has a major effect on the daily feed intake of the mink. Thus, all calculations relative to protein, vitamin, mineral, and essential fatty acid levels should be calculated on the basis of percentage ME instead of on the historical perspective of percent of nutrient on a dehydrated basis.

Specific guidelines from Scandinavia for optimum levels of fat throughout the ranch are provided in Table 4.37.

TABLE 4.37. Recommended Distribution of Metabolic Energy (%) From Protein, Fat, and Carbohydrates in Mink Diets–Nordic Association of Agricultural Scientists*

Period	Protein**	Fat	Carbohydrate
December–Whelping	35	20–50	25
Whelping–June	40	40–50	20
July–August	30	35–55	30
September–Pelting	30	30–55	25

* Enggaard–Hansen et al. (1991). ** Obviously minimum levels must be maintained. Thus higher percentage of Metabolic Energy as fat must be achieved via corresponding changes with higher protein and/or lower levels of carbohydrates.

Since fat resources provide mink ranchers with the most economical energy resource in terms of kilocalories of ME for the ranchers' feed dollar, effective nutrition management must provide the highest levels of fat consistent with the top performance. Specific recommendations in terms of practical mink nutrition, North America, are presented in Table 4.38.

TABLE 4.38. Recommended Fat Levels—North America*

Phase	Period	Fat Levels
Maintenance	December–February	14–20
Reproduction	March–April 20	14–18
Lactation	April 20 – May 10	18–20
	May 10 – May 20	20–22
	May 20 – June 1	22–24
Early Growth	June 1 – June 10	24–26**
	June 10 – July 1	26–28**
Late Growth	July 1 – Aug 10	26–28**
	Aug 10 – Sept 1	16–18***
Fur Production	Sept 1 – Pelting	20–22****

* Leoschke (1985, 1987) **Higher levels of fat may be required for specific color phases of mink such as sapphire, blue iris, pink, demi–buff, and wilds.

** Recommendations are related to singled mink and may not be applicable to doubled mink wherein the combination of lower fat levels and restricted feed intake via "The Man With The Spoon" may not be applicable. A more practical approach to providing "trim" male kits in September would be to provide optimum mink nutrition management as of July first with a program such as GnF–35 fortified cereal, which has significantly lower palatability ratings.

*** Higher levels of fat yielding optimum percent of ME as protein may be applicable for superior feed economics provided that the net result is not an increase in "Wet Belly Disease" (Leoschke, 1959a). In terms of this specific disease, the most practical approach would be the specific elimination of mink sensitive to "Wet Belly Disease" and their close relatives by genetic selection.

The percent fat recommendations in Table 4.38 for mink ranches in North America are supported for the most part by the scientific literature. One exception is the reproduction phase wherein the lower range of fat as percent ME is higher for North America. This difference recognizes that higher quality protein resources such as liver, eggs, and whole bird are more available in North America. Once again, the compensation principle holds: keeping in mind top performance, higher protein quality can result in significantly lower protein levels required for top performance of the mink. Thus, the following reports from the scientific literature for each phase of the mink ranch year.

Reproduction

Studies at Michigan State University by Travis and Schaible (1961) on fat levels in the range of 23–44% dry matter indicated no differences in the performance at breeding and reproduction, although it was difficult to maintain optimum body weights on the higher fat levels. In the Nordic Handbook, Ahman (1966), a fat level of 25–35% ME is recommended for the reproduction period. Perel'dik et al. (1972) reported Soviet Union fat levels of 24–35% ME for the reproduction period. Glem–Hansen (1977) reported observations in Denmark over the years 1974–1976 wherein fat levels of 17–33% ME were employed that the reproduction performance improved with increasing fat content. Skrede (1981) has reported on a two year study of carbohydrate ratios (% ME) of 32:24, 38:18, 44:12, and 50:6 wherein the protein level was constant at 44% ME. The two highest fat/carbohydrate ratios promoted better reproduction and lower kit mortality.

Ranch diets containing high levels of lean fish can provide fat levels so low as to undercut reproductive performance. Studies by Friend and Crampton (1961) indicated that a ranch diet with 83% lean fish (codfish) providing only 390 gross Calories/mink/day (with 0.5 pounds of feed) provided only a 57% whelp while the same diet wherein 8% of fish waste product was replaced with hydrogenated animal fat yielded a dietary program with 560 Calories/mink/day and an 84% whelp.

Lactation

In the Travis and Schaible (1961) studies with mink on fat levels ranging from 23–44% dry matter, there was no difference in lactation performance as measured by kit weights. Jorgensen et al. (1963) designed an experimental feeding trial wherein fat replaced carbohydrates in the lactating diet. The change in the dietary regimen from fat at 34% ME and carbohydrates at 19% ME to a program with fat at 44% ME and carbohydrates at 9% ME did not have a significant effect on kit weights at seven weeks. The Nordic Handbook (Ahman, 1966) gives a fat recommendation of 35–50% ME for the lactation period. Relative to the Soviet Union, Perel'dik et al. (1972) recommended a fat level of 35–47% ME for lactation. Danish experiments during the lactation period (Glem–Hansen and Jorgensen, 1974) involving diets containing from 3430 to 4150 Calories ME/kg dry matter indicated no significant difference in mink reproduction or kit growth. Studies by Skrede (1981) with fat: carbohydrate ratios (% ME) of 32:24, 38:18, 44:12, and 59:6 indicated that the high fat: carbohydrate ratios supported the best preweaning growth of kits as shown by kit weight data at three weeks and at six weeks.

The National Program of Mink Nutrition recommends increasing levels of a special fat supplementation of ranch diets during the lactation period as follows:

Dietary Fat Levels—Lactation Period

May	1st	10th	20th	30th
Rendered Fat	1	2	3	4
Fat Level	20–22	22–24	24–26	26–28

These recommended higher caloric dietary programs are consistent with experimental data on the changes in the fat content of mink milk during the lactation period and the proximate analyses of mink kits as noted in Table 4.39.

TABLE 4.39. Correlation Between Mink Milk Energy Concentration and Mink Kit Composition

Lactation Week	1	2	3	4
Milk Composition*				
Dry Matter– %	19	23	23	26
GE, kJ/kg	4560	5540	5910	6440
Kit Composition**				
Crude Protein–%	66	56	54	51
Fat, %	22	31	34	37
Kits/Litter:	5.6	5.3	5.1	5.1
Litter Weight. g.	97	242	498	697

* Olesen et al. (1992) **first week of lactation, Tauson (1994), 4th week of lactation, Glem–Hansen (unpublished data) and 2nd and 3rd weeks of lactation linear estimates based on cited data.

Growth—Fur Development

As early as 1950–1953, Hoie (1954) conducted studies on optimum levels of fat in mink ranch diets. In these experiments the fat level ranged from 7 to 33% of dry matter with a carbohydrate range from 6 to 38% of dry matter. This wide variation in fat and carbohydrate levels provided for satisfactory performance of the mink. The only exception was with an experimental diet high in protein and low in both fat and carbohydrates wherein minimal weight gains were obtained, related in part to a poor appetite of the mink for the very lean feeding program. Mink nutrition managers should note the key role of fats in contributing to the palatability of mink rations as they carry feed flavors.

Studies by Skrede (1983) involved fat/carbohydrate ratios with a wide range of protein content expressed as % M.E.

Protein	Fat & Carbohydrates
40	25:35 to 55:5
35	30–35 to 51:14
28	30:42 to 66:6
24	31:35 to 56:10

Increasing fat: carbohydrate ratios caused more rapid initial growth in all experiments with resultant similar body weights, body lengths, and pelt lengths at harvesting, although carbohydrate levels above 30% of ME created, in general, lower final weights and smaller skin lengths. These studies showed that total consumption of ME tended to increase with increasing fat: carbohydrate ratios. An observation confirmed in later studies by Ahlstrom (1994) and Ahlstrom and Skrede (1995) wherein fat:carbohydrate ratios as % M.E. ranged from 65:5 to 40:30. The highest fat: carbohydrate ratio promoted an increase in ME consumption per body weight gain.

A recent report on the value of relatively high fat: carbohydrate ratios for the furring period is of interest. The experimental plan of Clausen et al. (2005d) involved ME from protein in the range of 18–30%, ME from fat varied from 40–58%, and carbohydrate varied from 18–36% of ME The best weight gains from September to pelting was with dietary programs providing 24–27% M.E. from protein, 52–58% M.E. from fat, and 18–24% ME from carbohydrates. Fur quality was good with the same amounts of ME from fat and carbohydrates and within the range of 24–30% of ME from protein. Even higher recommended levels of fat are those of Ahlstrom (1995), a recommendation of as much as 60% ME originating from fats.

All factors considered, it is important to emphasize the basic fact that very lean dietary programs (very low fat level programs) can undercut the growth performance of the mink. Studies by Allen et al. (1964) indicated that diets containing gross energy levels below 4.9 Cal/g. (15% fat) did not support optimum growth of males, inasmuch as the male kits were not able to consume enough of the low energy diet to satisfy their energy requirements. Gross energy levels as low as 4.5 Cal/g. (8% fat) appeared to be satisfactory for the growth of female kits. Belcher et al. (1959) indicated that mink kits on a lean fish ration exhibited poor growth relative to mink kits on a higher energy horsemeat diet. They also noted that the replacement of ground corn (low carbohydrate digestibility rating for mink) with lard at 10% level resulted in a marked improvement in kit weight gains – superior, in fact, to that of the kits on the horsemeat diets.

Requirement—Quality

The quality of fat in a mink's diet has a profound effect on the performance of the animal. Qualitative factors include levels of polyunsaturated fatty acids (PUFA), Peroxide Values, and Free Fatty Acid levels.

Polyunsaturated Fatty Acids (PUFA)

The ingestion of large quantities of PUFA via fatty fish, rancid horsemeat, or fish oils in combination with sub–optimum levels of vitamin E can have adverse effects on the performance of the mink including "yellow fat" disease (Leekley and Cabell, 1959; Lalor et al., 1952), muscle dystrophy (Brandt, 1984), iron deficiency anemia (Tauson and Neil, 1991), and sub–optimum reproduction (Frindt et al., 1996). In addition, unstabilized PUFA in fresh/frozen feedstuffs for mink nutrition can undercut "sharp color" in dark mink at pelting since PUFA–containing fats on the surface of the fur can undergo oxidative rancidity with subsequent loss of color quality. This point is well illustrated by Table 4.40 (Stout et al., 1967).

TABLE 4.40. Mink Fur Color As Influenced by Ration Composition

Diets	Fur Color*
100% Purified Diet**	3.00
87% Purified Diet with 13% Rockfish (dehydrated basis)	2.25
87% Purified Diet with 13% Chicken Offal (dehydrated basis)	2.25

 * Fur grader evaluation: 3–dark, 2–medium and 1–brown
** Fats provided include vegetable oils stabilized by natural anti–oxidants (tocopherols) and lard (rendered pork fat) stabilized by synthetic anti–oxidants.

This negative effect of poultry offal on fur color was still noted a quarter of a century later by Tauson and Neil (1991).

All factors considered, all fresh/frozen mink feedstuffs without anti–oxidant stabilization have the potential to undercut "sharp color" in dark mink at pelting as noted in Table 4.41. This potential is noted in increasing order.

TABLE 4.41: Fat Resources in Modern Mink Nutrition

Fat Resource	Stability
	Highest
Anti–oxidant Stabilized Rendered Beef Tallow	/
Anti–oxidant Stabilized Rendered Pork Fat	/
Anti–oxidant Stabilized Rendered Poultry Fat	/
Anti–oxidant Stabilized Dehydrated Protein Feedstuffs	/
Beef By–Products	/
Pork By–Products	/
Chicken By–Products	/
Turkey By–Products	/
Duck By–Products	/
Fish–Lean	/
Fish–Fatty	/
Horsemeat	/
Fish Oils	**Lowest**

Studies at Oregon State University (Stout et al., 1960) indicated that dark kits on experimental diets containing rancid sardine oil or herring oil showed a "brownish" coloration. Even as early as 1937, fox ranchers noted that feeding cod liver oil to silver fox resulted in fox fur that was brown in color (Martin, 1987). This early observation of a lack of sharp color in foxes fed fish oil was confirmed 50 years later by Rouvinen et al. (1991).

Continued observations of mink ranchers over a period of many years indicate that high levels of PUFA in practical mink and fox ranch diets containing poultry products, fish products, and horse-meat yield dark pelts with sub–optimum color. Years ago the feeding of rancid horsemeat (stored for long periods with sub–optimum freezer temperatures) provided ranchers with the "off color" darks referred to as "fire engine red" dark pelts.

In terms of Table 4.40, it should be noted that the purified diets with 100% stabilized fats provided the sharpest color in dark mink. The specific correlation between the use of commercial pellets (with 100% stabilized fat content) for fur animals and resultant "sharp black color" has been noted since as early as Smith (1932) with observations of silver fox raised on Purina fox pellets: "The outstanding feature was the color of the pelts which was far superior to that of the other pelts on the same sale. The lustrous black of the pelts and the absence of brown rusty color were the interesting features of the pelts."

This observation of "Sharp Dark" color in silver foxes raised on pellets was still current sixty to seventy years later with mink raised on commercial pellets providing 100% ant–oxidant stabilized fats. In Scotland, Jim Baxter (1990), an experienced fur grader for Hudson's Bay Company, commented that in his twenty years of grading mink pelts, he had never seen "black" mink pelts until he graded dark pelts raised on commercial mink pellets wherein at least 10% of the pelts were "black". A decade later, Joe Poquette (2002), an experienced mink fur grader, evaluating dark mink on a pellet program in Wisconsin remarked, "My God, what dark mink."

A high degree of "sharp color" in dark mink can also be achieved on non–pellet programs wherein a high percentage of stabilized fats can be achieved with a commercial cereal program involving GnF–70 at 55%, lard (stabilized pork fat) at 15%, and 30% fresh/frozen mink feedstuffs. This GnF–70 feeding program employed in Iowa provided the sharpest dark color in mink ever seen by Harold Smith (1972) who had a reputation of noting an "off color" mink a mile away. A 1990 program initiated by Jack Brennan, National Fur Foods, adds mink pellets crumbled "on top" of conventional mink ranch or redi–mix programs, thereby achieving higher levels of stabilized fats in the mink dietary regimen and gaining "sharper color" in dark mink pelts marketed. A regular ranch diet with 20% fortified cereal and 80% fresh/frozen feedstuffs supplemented with 20% Gro–Fur crumlets "on top" can almost double the proportion of stabilized fats from 12% to 21% of total fats.

Peroxide Number

The quality of fat for mink nutrition may be ascertained via analyses for the Peroxide Number and the percentage of Free Fatty Acids (FFA) (Marcuse, 1972). Peroxides and FFA arise in mink feedstuffs during storage in processes termed oxidative rancidity and hydrolytic rancidity. In the process of oxidative rancidity, the unsaturated fatty acids, especially PUFA, are subjected to air oxidation yielding hydroperoxide, peroxide, ketone, and aldehyde structures. In the process of hydrolytic rancidity, simple lipids termed triglycerides undergo breakdown to yield FFA, diglyceride, monoglyceride, and free glycerol structures. For many years in human nutrition, peroxide numbers less than 20 have been considered satisfactory. However, more recent studies from Denmark (Berg, 1986) and Danish Fur Breeders Association Voluntary Feed Control Program (2003) indicate that mink may be more sensitive than human animals to peroxide levels in their feed components. Accordingly, these researchers make specific recommendations of maximum peroxide values of 6.0. The key problem with peroxide numbers is that over a period of time, the peroxides yield aldehyde and ketones (which can be determined by the anisidine number) and thus a mink feed product which has undergone extensive oxidative rancidity may still show a relatively low peroxide number despite the odor and appearance of the product.

With ample vitamin E supplementation of the ranch diet, feed products with high peroxide numbers may not be a problem, although an odor factor may undercut feed consumption. An experimental study conducted at the Elcho, Wisconsin, mink ranch of Associated Fur Farms (Leoschke, 1962) used a choice white grease product (lard–rendered pork fat) with a peroxide number of 60 (obtained by bubbling pure oxygen through the product for 24 hours without the use of anti–oxidants) at a 2% level in the ranch diet from early March through late June. The rancid lard provided about 20% of the total fat in the diet, yielding a peroxide number of 12 for the fat without consideration of peroxides present in the other 98% of dietary ingredients. This experimental diet yielded excellent reproduction/lactation with only one anomaly, bigger kits at weaning. This experiment was conducted at a time in mink nutrition history when the value of higher fat levels for lactation and early growth was unknown.

Peroxide poisoning may result from the consumption of peroxides, aldehydes, and ketones which are formed during the oxidation of polyunsaturated fatty acids present in ingredients such as poultry fat and fish oils. The net result is a loss of appetite, weight loss, general unthriftiness and diarrhea (Rouvinen–Watt et al., 2005).

Free Fatty Acids

In terms of fat quality for human nutrition, FFA values less than 10% are considered acceptable. This standard is consistent with European and Soviet Union standards for fish meals (Ugletveit, 1975; Konsulenternes, 1976; Kuznecow, 1976). NorSalmOil, a fishoil used extensively in aquaculture, has a quality standard of less than 4.5% FFA (Skrede, 1984). The presence of FFA in a fish meal may not have a significant adverse effect on the nutrition of the mink directly related to the FFA content itself. For, in the process of fat digestion, a combination of FFA, monoglycerides, and diglycerides are presented to the absorption processes at the intestinal cell wall.

Peroxide numbers and percentage FFA data on mink feedstuffs such as commercial fish meals are important to mink nutrition. Fish meals with high peroxide numbers and/or FFA provide inferior protein nutrition since they indicate poor processing, particularly a long storage time prior to production wherein bacterial action can destroy amino acids present in the fish protein structures. Thus it is vital that mink ranchers insist on pertinent quality data: peroxide number, percentage FFA, percentage moisture, and percentage of ammonia or TVN (Total Volatile Nitrogen). If the percentage of moisture is less than 6%, it may indicate excessive heat treatment, and a significant percentage of TVN may indicate significant bacterial action on the fish prior to processing.

Requirements–Essential Fatty Acids

In terms of animal nutrition, the essential fatty acids (EFA) are PUFA with the following designations: linoleic (18:2), alpha–linolenic (18:3) found in highest concentrations in plant oils, and arachidonic acid (20:4) found only in animal fats. Arachidonic acid is not considered to be essential for humans and other animals, with few exceptions including the cat and the mink, inasmuch as most animals can convert linoleic acid to arachidonic acid. The essential fatty acids are important in animal anatomy and physiology as key constituents of cell membranes and as precursors of a class of hormones termed prostaglandins. EFA as well as other PUFA have a major role in providing optimum insulation qualities to the sub–cutaneous fat of the mink. Studies by Rouvinen and Kiiskinen (1989) indicated that the amount of unsaturated fatty acids in the skin and subcutaneous fat increased towards winter. This increase was more notable in the mink than the fox, since the mink's thermophysiological properties are less efficient than those of the blue fox.

160

Linoleic Acid

Glem–Hansen (1978) has studied the mink's need for linoleic acid during lactation. He estimated the requirement with kit growth as the parameter and indicated that the content of linoleic acid should be 5% of ME. Perel'dik (1972) found that pregnant and lactating mink and growing kits require 1.5% EFA in the feed. Studies by Garcia–Mata (1980) on the composition of mink milk indicated 13.7% linoleic acid and 1.2% alpha linolenic acid. Studies by Jorgensen and Glem–Hansen (1974) indicated a correlation coefficient of – 0.90 between the mink milk contents of linoleic acid and the loss of weight of lactating mink from 3–18 days after whelping. With mink milk content of 4.5% linoleic acid, the weight loss was 86 grams relative to a weight loss of 45 grams when the linoleic acid content of the milk was 12.2%.

Hillemann (1992) and Hillemann and Mejborn (1983) found that fur quality was improved when at least 12–20% of the dietary fat consisted of linoleic resources. In their experimental study, this was achieved via replacing tallow and lard with poultry fat and vegetable oils.

Relative to the growth and fur development period of the ranch year, data provided by Juokslahti and Tanhuanpaa (1984) on Finnish mink ranch diets is of interest. They recommended a content of EFA of 17–18% of all fatty acids provided in the mink ranch diets. A study by Hilleman and Mejboen(1983) involved diets containing lard, poultry fat, tallow, soya oil, rapeseed oil, and palm oil. The content of linoleic acid varied from 6–43% of total fat. Fur quality was superior with the highest levels of linoleic acid, very likely related, in part, to the proportion of naturally stabilized fats presented to the mink. For fur development, their study supported a recommendation of a minimum of 20% of linoleic acid in the fat of mink feed. This recommended level may not be practical, all factors considered. A study by Walker and Lishchenko (1966) of a Canadian mink ranch diet indicated the following fatty acid composition; myristic, 3.3; palmitic, 25.7; stearic, 16.6; palmitoleic, 5.4; oleic, 35.3; linoleic, 10.7 and linolenic, 2.1.

Alpha Linolenic Acid

There is relatively limited experimental data on a specific requirement of animals for alpha–linolenic acid in addition to linoleic acid and arachidonic acid. However, the physiological importance of alpha–linolenic acid cannot be ignored. Alpha–linolenic acid is a precursor of docosa–4,7,10,13,16,19 hexaenoic acid for most animals, but not for the mink. It is present in large concentrations in the lipids of the brain, retina, and male reproductive tissue of animals. There is also a specific requirement for alpha–linolenic acid or its metabolic derivatives in fish, particularly trout (Anon, 1979). Studies by Walker and Lishchenko (1966) indicate significant levels of alpha–linolenic acid in the adrenals, brain, red blood cells, heart, kidney, liver, spleen, and blood plasma of mink. The wide variation in alpha–linolenic acid content apparently indicates specific selection of this EFA. Relative to the biosynthesis of eicopentaenoic acid (EPA) and docosaheaenoic acid (DHA) from alpha–linolenic acid (trienoic), studies by Polonen et al. (2000) and Kakela et al. (2001) indicated that the use of linseed oil, a rich resource of alpha–linolenic acid (n–3), failed to increase the proportion of longer chain n–3 PUFA in liver membranes of the mink. It is apparent that there were ample quantities of EPA and DHA in the dietary program of the mink in the wild, and thus no requirement of a bio-synthesis program for the synthesis of these longer chain n–3 PUFA existed. This same inability of the mink to synthesize arachidonic acid (n–6 tetraenoic) from linoleic acid (n–6 dienoic) was also noted.

Arachidonic Acid

There is relatively limited experimental data on a specific requirement of the mink for arachidonic acid. Studies by Walker and Lishchenko (1966) indicate that the content of arachidonic acid in the adrenal, brain, red blood cells, and spleen is higher than the content of alpha–linolenic acid. The most specific experimental data to support an absolute requirement of the mink for arachidonic acid is provided in studies at the University of Wisconsin (Tove, 1950) which indicated that mink kits placed on purified diets providing only vegetable oils as a fat resource developed a nutritional deficiency that was only resolved by the addition of animal fat (lard) to the dietary regimen. The study indicates that the mink may not have the capacity to convert linoleic acid (n–6 dienoic) to arachidonic acid (n–6–tetraenoic). More recent studies by Polonen et al. (2000) find that the mink, similar to the cat, cannot convert linoleic acid (n–6 dienoic) to arachidonic acid (n–6–tetraenoic).

According to Wamberg et al. (1992), the milk glands of North American mink do not possess desaturation or chain–lengthening enzyme systems. This phenomenon appears to be a striking feature of the phylogenetic adaptation to their carnivorous living habits, which they share with marine aquatic mammals such as whales, seals, and polar bears.

Resources

Obviously, multiple sources of fat have been employed in practical mink ranch diets from the beginning of the fur industry. Fat resources for modern mink nutrition management can be conveniently classified as follows.

Animal Fats
Lard–Pork Fat

This product provides high digestibility and palatability ratings for the mink and serves as a key resource for arachidonic acid. Lard that has been subjected to excessive heat processing in the preparation of fast food chain French fries yields a product termed "spent" oil which has an inferior nutritional value for the mink in terms of peroxides, FFA, and low palatability. Studies by Rietvelt (1976) indicated only an 80% growth performance of mink kits relative to brothers on a quality lard product.

Tallow–Beef Fat

Relative to lard, tallow has a lower digestibility rating and lower palatability for the mink.

Poultry Fat

This fat has an excellent digestibility and palatability rating for the mink and a higher content of linoleic acid than pork or beef fats.

Fish Oils

The value of fish oils for mink nutrition is directly related to its quality. Fish oils are highly digestible for the mink and provide significant levels of the longer chain PUFA as DHA and EPA. A number of experimental studies indicate that quality fish oils have a useful role in modern mink nutrition including the work of Ahlstrom, O. and Skrede (1995) Borsting et al. (1998), Brandt et al. (1990), Polonen et al. (2000), Rouvinen and Kiiskinen (1989), Rouvinen et al. (1989), Skrede (1984) and Tauson and Neil (1991).

Some special dietary conditions must be considered with the use of fish oils in mink diets. Studies by Brandt et al. (1990) indicated that supplementation of C. sprattus oil with both vitamin E and sodium selenite was required for top performance. The Borsting et al.(1998) study also indicated that high levels of vitamin E were required with ethoxyquin–stabilized oil of good quality for top performance of the mink for growth and fur development. They noted that the feed intake, body weight, and skin length as well as plasma alpha–tocopherol decreased with a decrease in fish oil quality. With high levels of fish oil, Tauson and Neil (1991a) noted significant iron deficiency anemia and white underfur as well as sub–optimum weight gains of the kits.

163

The Borsting study indicated that a reproduction/lactation diet containing 3% fish oil (33% of ME) as the only fat source and 50 mg of alpha–tocopherol/kg was sub–optimum, producing smaller litter size and high kit mortality. Thus, even fish oil of good quality cannot be used as the only fat resource for breeding mink, and even moderately oxidized fish oil caused depletion of vitamin E stores and anemia in the females at the end of the nursing period.

For fresh/frozen animal feedstuffs in mink nutrition, concern about the quality of processing and freezer storage conditions is warranted. The adverse effect of fresh/frozen fish and poultry offal products on sharp color in dark mink (Stout et al., 1967) has already been discussed in this report.

Plant Fats
Corn Oil

This oil is an excellent fat resource for mink nutrition and especially rich in linoleic acid.

Lecithin – Rape Seed

A study by Clausen and Hejlesen (2001) indicated that rape seed lecithin at both 4 and 8 percent of the ration resulted in lowered mink kit weight gains and smaller pelt size. The low palatability of lecithin products for the mink has been noted again and again in past years by commercial ranchers mixing their own mink feed. At a low level of 0.5%, lecithin products provide excellent emulsifying action for cleaner mixers and potentially higher digestibility ratings for fats such as beef tallow, raw or rendered, usually less digestible.

Linseed Oil

This fat resource has a relatively high content of alpha–linolenic acid (>10%). Studies by Polonen et al. (2000) indicated relatively low palatability of linseed oil for the mink kits as evidenced by sub–optimum kit growth performance. This study also noted the minimum incorporation of alpha linolenic acid, n–3, in the phospholipids of the mink livers, in sharp contrast to the effect of fish oil n–3 fatty acids, DHA and EPA.

Palm Oil

Studies by Hillemann and Mejborn (1983) using palm oil noted that mink on this specific dietary plan had sub–optimum color. This poor performance may have been due to a minimum level of linoleic acid, inasmuch as the study showed a direct relationship between the linoleic content of six fat resources and the fur quality of the mink.

Rapeseed Oil

Multiple studies on the value of rapeseed oil, termed Canola oil in North America, have indicated the value of this fat resource for modern mink nutrition (Hillemann and Mejborn, 1983; Rouvenen and Kiiskinen, 1989; Rouvinen et al., 1989; Tauson and Neil, 1991). The Tauson and Neil study indicated that the mink fed rapeseed oil had significantly higher T–4 values and elevated ME intake compared with the control group. The fur quality characteristics were also superior on the rapeseed oil based diet, very likely related to a relatively high proportion of stabilized fat in the final diet.

Soybean Oil

Soybean oil is a rich source of linoleic acid, with significant quantities of alpha–linolenic acid. Studies by Hillemann and Mejborn (1983), Rouvinen and Kiiskien (1989) and Rouvinen et al.(1989) indicate top performance in terms of growth and fur development. The Hilleman and Mejborn study indicated top pelt prices for the mink kits provided soybean oil.

Toxicity

For the most part, over the period of the last fifty years within the world fur industry, dietary programs with a fat deficiency have been more common than dietary programs containing fat toxicity. The most likely case of fat toxicity in practical mink nutrition would be that of a dietary program with excessive energy (excessive fat content) wherein limited feed intake of the mink would yield sub–optimum protein nutrition to the point of sub–optimum performance of the animals. A good example of excessive fat (excessive energy) levels in a practical mink ranch diet would be relatively high fat levels in the reproduction period in combination with a relatively warm spring, resulting in sub–optimum nutrient intake and/or pregnant animals with excessive weights for top whelping performance.

Carbohydrates

Carbohydrates are named on the basis of their empirical (ratio of atoms) formula, CH_2O. The formula for glucose illustrates the ratio—$C_6H_{12}O_6$ or $C_6(H_2O)_6$—implying that glucose is a hexahydrate of carbon and giving rise to the name carbohydrate, "carbo" from the Latin and "hydrate" from the Greek. Animals can synthesize the simple sugar, glucose, from glucogenic amino acids, or the glycerol component of fats (triglycerides) in a process termed gluconeogenesis; however, unlike plants, animals cannot synthesize simple sugars from the very basic resources of carbon dioxide and water.

Carbohydrates are represented by monosaccharides (simple sugars) including glucose (blood sugar), fructose (fruit sugar), mannose and galactose (a component of milk sugar); the disaccharides such as lactose (milk sugar), sucrose (table sugar); and maltose and the larger polymers including the digestible polymers, starch and glycogen (animal starch). As well, there are dextrins, smaller polysaccharides obtained via the hydrolysis (breakdown) of starches and the indigestible fiber resources cellulose, hemicelluloses, protopectins, pentosans, and lignins which are useful as bulk for proper consistency of the animal's feces.

The primary role of carbohydrates in animal nutrition is an energy resource. They are unique in that simple sugars and highly digestible carbohydrates provide an almost immediate energy resource as compared with fats and proteins, which require delayed metabolic processes to yield ATP (adenosine tri–phosphate) energy resources.

Like fats, carbohydrates have a "sparing" effect on dietary protein: optimum levels of both carbohydrates and fats in a mink's diet can lead to minimal protein levels required for top performance. Carbohydrate hydrolysis products such as simple sugars can also reduce the mink's requirement for non–essential amino acids in reverse gluconeogenesis metabolism.

Mink in the wild have a relatively low intake of carbohydrates inasmuch as the animals are carnivores whose carbohydrate intake is limited to the starches and sugars that may be present in the gastro–intestinal contents of its prey and the glycogen content of the liver and muscle tissue of its kill.

Although carbohydrates may play a minor role in the diet of the mink in the wild, these sugars, starches, and fiber resources play a major role in farmed mink nutrition. With proper nutrition management, a pelt may be produced with as little as 110 pounds of ready–mix or ranch feed "on the wire." This quantity of mink feed represents about 36 pounds of feed solids distributed as shown in Table 4.42. Carbohydrates are so significant in modern mink nutrition because they represent a full one–third of the total nutrient intake required to yield a pelt for marketing.

For North American mink feed economics (and very likely applicable throughout the world) carbohydrates represent one of the most economical energy resources available. It is second only to fat in terms of kilocalories of ME (metabolic energy) for the mink rancher's feed dollar (Leoschke, 1987a; 1987b). Thus, to minimize feed costs per pelt marketed, ranchers are advised to use the highest levels of fat and quality fortified cereals throughout the ranch year, consistent with the very top performance of their mink.

166

TABLE 4.42: Nutrient Requirements per Pelt

Nutrient	Pounds/Pelt
Protein	13
Carbohydrate	12
Fat	8
Minerals	3
Vitamins	+
Total	36

Physiology

The primary role of carbohydrates in mink nutrition is that of an economical energy resource. At the same time, the key role of the fiber content of carbohydrate resources cannot be ignored. Fiber (indigestible carbohydrate polymers) represents a plant's structural components (cellulose and lignins) and cellular adhesives (protopectins and hemicelluloses). A comical note: "Cellulose for the mink is like giving a man a bottle of champagne without a corkscrew". Fiber is useful for proper fecal formation (one gram of fiber provides about 6 grams of feces); it also has a physiological value in encouraging intestinal tone and mobility. At the same time, excessive levels of fiber in mink ranch diets can undercut sodium absorption significantly, leading to a possible secondary deficiency of sodium if the initial dietary plan has been relatively low in sodium (Enggard–Hansen et al., 1985; Moller, 1986).

For the most part, with the exception of lactose (milk sugar containing glucose and galactose) and sucrose (table sugar containing glucose and fructose) as well as molasses (containing both glucose and fructose), the final monosaccharide provided to the blood circulation of the mink is glucose. Thus, the following commentary on the key role of glucose in the physiology of the mink.

Glucose is an absolute essential energy resource for human red blood cells and brain; it also provides oxaloacetate (via pyruvate) for the function of the metabolic energy program known as the citric acid cycle. Sub–optimum glucose and starvation can lead to ketosis (high levels of ketone bodies in the blood) and fatty degeneration of the liver as seen with mink with anorexia.

In contrast to the cat and leopard frog, carnivores like the mink, which have no transporter activity in the intestinal brush border to regulate transportation of glucose in response to changing levels of dietary carbohydrate, are able to enhance transport activity with higher levels of carbohydrate intake (Buddington et. al., 1991). This gives the species more flexibility in its nutrient metabolism. The domestic cat, on the other hand, is unable to increase sugar uptake from the gut lumen and will develop diarrhea when fed diets high in carbohydrates (Maskell and Johnson, 1996).

As a result of evolutionary adaptation to diets which are low in carbohydrates, carnivores have higher activities of gluconeogenic enzymes than omnivores. The activity of liver enzymes involved in amino acid catabolism is high and unaffected by dietary protein level in carnivores. Fur-

167

thermore, the activity of these enzymes is the same in the fed and fasted states, because the carnivorous species cats and mink can synthesize sufficient glucose from glycogenic amino acids (Stryer, 1988). Mink can also synthesize glucose from glycerol, a component of fats (triglycerides). Thus mink are capable of glucose utilization in both the fed and fasted states.

Mink are able to maintain glucose homeostasis when fed almost carbohydrate–free diets during pregnancy and lactation, for example, dietary programs of % of ME from protein, fat, and carbohydrates at 61:38:1 or 47:52:1. However, this was only the case when the proportion of ME from protein was above 33%, that is, a ME program of 33:66:1 was unable to support lactation at 3 weeks postpartum (Borsting and Gade, 2000; Fink and Borsting, 2002; Damgaard, Borsting and Ingvartsen, 2003). Experimental work of Borsting and Damgaard (1995) indicated that females nursing six kits had to cover about 73% of their glucose needs at peak lactation via gluconeogenesis when fed only 12% of ME as carbohydrates. The studies of Fink and Borsting (2002) indicated that the lactating mink are capable of utilizing digestible carbohydrates up to 32% of ME without critical elevated plasma glucose concentration. This finding demonstrates that mink have a high activity of the gluconeogenic enzymes but also a large glycolytic capacity whereby they are able to adapt to a wide variation in dietary protein, fat, and carbohydrate supply. At the same time, it should be emphasized that the risk of hypoglycemia is highest during late pregnancy and lactation when glucose requirements are the highest.

It is of interest to note that mink are able to maintain glucose homeostasis when glycogen stores are almost exhausted (Petersen et al. 1995). Furthermore, mink are able to store excess glucose in the form of glycogen, which indicates that the risk of overloading mink with glucose under farmed conditions is low.

Fink and Borsting (2002) have provided some interesting insights into the carbohydrate physiology of the mink. Their experimental work employing four week postpartum lactating mink with litters of 6–7 kits indicated a glucose turnover rate of 4–5%/ minute and an approximate daily glucose flux of 12–17 grams/ day. The study indicated that dietary carbohydrate levels from 1 to 32% of M.E. did not significantly affect these values. A later report (Fink et al., 2002) involved a postprandial examination of the blood profile of lactating mink provided with a wide range of carbohydrate, fat, and protein levels—including one dietary plan with essentially zero carbohydrates. Researchers observed that there was no postprandial rise in blood glucose levels with those animals on carbohydrate free diets. Relative to changes in blood concentrations of hormones directly related to carbohydrate metabolism, it was noted that postprandial lactating mink on all dietary programs exhibited a rise in insulin levels. Glucagon levels were relatively higher with mink on the low protein dietary regimens relative to animals on higher protein programs. Glucagon/insulin ratios decreased with mink on diets providing normal carbohydrate levels but increased with animals on carbohydrate free diets.

Multiple field observations indicate that mink kits as early as July are intolerant of lactose. As noted by Atkinson (1996), at whelping the level of lactose in mink milk is relatively high but the level of lactose in the milk declines as lactation progresses. The level of activity of the intestinal enzyme lactose, which digests it, also declines, so adults have little remaining capacity to digest milk sugar.

Requirements

No critical studies have shown that mink actually have an absolute requirement for carbohydrates per se as long as gluconeogenesis is supported by ample levels of glucogenic amino acids and glycerol. As noted earlier in this chapter, Borsting and Gade (2000) reported that mink are able to maintain glucose homeostasis when fed carbohydrate–free diets during pregnancy and lactation as long as the proportion of ME from protein was above 45%. This study on the excellent performance of mink on low carbohydrate dietary regimens is confirmed by a report of Damgaard et al. (2000) wherein experimental diets providing carbohydrate levels from 1% to 25% ME were supportive of top reproduction and lactation and even excellent early growth. A study by Lebengartz (1968) on carbohydrate–free diets noted normal growth for two months, which was followed by weight loss. The final weights of the animals on the carbohydrate free diets were less than for animals receiving 15% ME as carbohydratres. They showed poor pelt value as well. However, the sub–optimum performance of the mink kits could also have been related to protein levels inadequate in both protein quantity and quality.

Perel'dik et al. (1972), citing the Scandinavian work of Rimeslatten (1959), Ahman (1961) and Joergensen (1967), recommended that carbohydrates supply not less than 10% and not more than 30% of ME; the best results will be obtained, it was suggested, when 15 to 25% of the ME is supplied by carbohdyrates. Leoschke's recommendations (1980) were more specific, with 15–30% ME for growth and fur development and 10–20% of ME for pregnancy and lactation. As noted by Lebengartz (1968), excessive carbohydrate levels –– as high as 48% of ME –– with moderate levels of fat, 20% of ME, produce smaller mink and lower pelt quality relative to mink fed a diet containing 16% ME as carbohydrate and 48% ME as fat. Essentially, mink kits on the 48% ME carbohydrate diet were undergoing an energy starvation program.

The recommendations of Perel'dik (1972) and Leoschke (1980) support the carbohydrate recommendations of the Nordic Association of Agricultural Scientists (Enggard–Hansen et al., 1991) shown in Table 4.43.

169

TABLE 4.43: Recommended Distribution of ME (%) From Protein, Fat, and Carbohydrates in Mink Diets*

Period	Protein	Fat	Carbohydrates
December – Whelping	35	20–50	25
Whelping – June	40	40–50	20
July–August	30	35–55	30
September–Pelting	30	30–55	30

* It is obvious that minimum protein levels must be maintained and thus changes in protein levels must be achieved via corresponding changes in % ME as fat or carbohydrates.

Fat and carbohydrates are the most economical energy resources for modern mink nutrition; therefore, it makes common economic sense for mink ranchers to plan dietary programs with the highest levels of fat and carbohydrates consistent with the top performance of the mink. Quality fortified cereals are not only an economical resource of energy for the mink, but, more importantly, they are a significant nutrition source that enhances the performance of the mink.

As of the early 1960s, National Fur Foods Co., New Holstein, Wisconsin, had a fortified cereal program of 15% XX–15 for the period from January through June and 20% GnF–20 for the remaining months of the ranch year. In the last four decades, extensive research studies at the National Research Ranch facilities and field observations from throughout North America have provided new insights about how quality fortified cereals enhance the performance of the mink in all phases of the mink ranch year.

Pelting–February

The two periods of the mink ranch year in which proper weight conditioning of the mink is absolutely required for top performance in the weeks to follow are the pre–breeding phase from pelting through February and the pre–fur development months of July–August.

Nutritionally, this observation is based on the difference in the nutritional environment of the mink in the wild and and their distant relatives on the mink ranch. In the wild, mink experience a constant battle for nutritional survival in terms of limited access to prey, an effort requiring major energy expenditure for the physical activity needed to obtain nutrients. In sharp contrast is the nutitional situation of mink on commercial mink ranches: here mink have a very restrictive exploratory environment and limited physical activity, along with unlimited access to nutrients. In short, their daily routine is "eat and sleep." All factors considered, there are only two practical programs for controlling the weight condition of mink in the period prior to breeding or fur development.

1. Mink Body Weight Management simply involves restriction of mink feed intake via "The Man With The Spoon." This daily feeding can, however, be labor intensive and impractical as some mink are underfed and others overfed. It can also be counterproductive: in Wisconsin in 1970, excessive feed restriction led to "protein starvation" and thus sub–optimum reproduction/lactation of the mink;

2. Mink Nutrition Management can be an asset to "The Man With The Spoon" in optimum conditioning of the mink for breeding and fur development through unique mink nutrition management programs which employ lower caloric density diets and/or dietary programs with reduced palatability.

The caloric restriction program requires special attention to several features: genetically, for instance, some unique strains of dark mink require higher energy programs for top reproduction/lactation performance, weight conditions of the animals, and weather factors. Warmer temperatures require leaner diets prior to breeding and/or whelping.

The only practical way to reduce the caloric density while maintaining optimum protein levels as % of ME is to replace digestible fat (9.5 kilocalories/gram) with digestible carbohydrates (4.2 kilocalories/gram.).

For the pre–breeding (maintenance) period, lower caloric density diets can provide leaner diets and thus higher feed volume with equal energy intake. It follows that the National Program of Mink Nutrition recommends GnF–20 (15% protein) at the enhanced level of 25% or, preferably, the combination of GnF–35 (30% protein) at 35% in combination with GnF–20 at 5%. These two fortified creals constitute a total level of 40% which provides ample protein levels for the maintenance period and decreased fat levels, as low as 14–16%.

March–April

Leaner diets may be required during the gestation period, especially with a relatively warm spring. It is well known that "heavy" pregnant females in late April and early May are likely to have problems whelping and, what is more common, provide sub–optimum lactation; thus the recommendation of 25% XX–25 (30% protein) fortified cereal in combination with a special low energy, high protein reproduction "trimmer" pellet (crumbilized) containing 48% protein and 10% fat (dehydrated basis) recommended to be fed at 5–10–15–20% "on top" of the ranch mix or ready–mix, that is, at the level required to achieve a desired lower energy feeding regimen.

May–June

For this period of lactation and early growth, the emphasis in dietary planning must be on higher energy content programs achieved via increasing levels of special fat supplementation to maintain the weight of the lactating mink and the high energy demands of the growing kits. See the preceding chapter on Fats and Essential Fatty Acids. This emphasis on high fat:carbohydrate ratios is supported by a study of Skrede (1981). At the same time, it is important to recognize the key role of

171

carbohydrates in the lactation phase. A study by Polonen et al. (1993) demonstrated that a dietary corn syrup supplement helped maintain body weight of lactating mink. Borsting and Gade (2000) found that gluconeogenesis accounts for over 70% of the glucose requirements in lactating mink when provided a relatively low carbohydrate dietary program providing 12% ME from carbohydrates.

A study by Borsting and Damgaard (1995) showed that large litter size increases glucose production rates in the mink.

July–August

Feeding higher levels of quality fortified cereals during the summer months has the potential to enhance the fur development in the fall months. A comparison of the physical activity and nutritional environment of mink in the wild and on commercial mink ranches has already been made in the sub–chapter 1. Pelting–February. Thus male kits on commercial mink ranches could well have a "wonderful life" during the summer months with resultant "heavy" male kits coming into September as fur development begins. However, this constitutes poor nutrition management, for the high feed volume is second rate and resultant protein intake in the Fall is second rate, leading in the Fall months to sub–optimum fur development. This basic observation is reported in Danish studies by Hansen et al. (1992) and confirmed in later studies, Lohi (2000). Experimental data from the National Research Ranch (Leoschke, 1980), predated and predicted the Danish results. Basically, the question is, "What program of mink ranch management will provide for "trim" male kits coming into September?" Consider the following options.

Mink Body Weight Management

Restricted feed intake by "The Man With The Spoon" is obviously impossible. It is both labor intensive and chaotic, as mink in litters or paired may fight over rations.

Mink Nutrition Management With Lower Energy Diets

Lower caloric diets achieved by providing the mink with lower fat levels would not result in "trim" male kits coming into September but "heavy" male kits. This results from higher feed volume for each of the 62 days of the late growth period and higher feed costs per pelt marketed.

Mink Nutrition Management With Higher Levels Of Quality Fortified Cereals

The dietary programs GnF–35 (30% protein), GnF–50 (34% protein) and GnF–70 (42% protein) are supportive of "trim" male kits coming into the fall. It is important to emphasize the basic nutrition management point that all of these programs provide a carbohydrate intake similar to that provided by the GnF–20 (15% protein) with 80% fresh/frozen feedstuffs.

The value of these higher fortified cereal programs for providing "trimmer" male kits coming into September is well illustrated by Table 4.44.

172

TABLE 4.44. Dietary Management for "Trimmer" Mink in September
National Research Data—1980* Male Kit Growth—Grams

Program	July–August	SeptemberNovember	Totals
Darks			
GnF–20	920	310	1230
GnF–70	820	450	1270
Gain GnF–70/GnF–20	−100	+140	+40
Pastels			
GnF–20	970	450	1420
GnF–70	870	560	1450
Gain GnF–70/GnF–20	−100	+130	+30

* Leoschke (1980)

This experimental data shows that common commercial mink ranch dietary programs emphasizing high levels of fresh/frozen feedstuffs for the summer months undermine the weight gains of the male kits in the fall fur production period. This has the effect of reduced feed volume, including sub–optimum protein intake.

Final Note: It is important to emphasize the fact that the nutritional basis for the optimum performance of the kits on the GnF–70 program is not a lower energy program but a modern mink nutrition program providing a total feed mixture with slightly lower palatability for the mink.It is common knowledge that fresh/frozen feedstuffs have higher taste appeal than a high protein fortified cereal wherein higher protein levels are achieved through dehydrated protein feedstuffs – fish meals, blood meals, and poultry meals with high biological value for the mink. Mink ranchers should note that provision of "trim" male kits as of September did not imply or cause lower kit weights as of November. The GnF–35, GnF–50, and GnF–70 programs simply illustrate the value of mink nutrition research studies for enhancing performance in all phases of the mink ranch year.

September–Pelting—Fall Fur Development

Rancher observations over the years and experimental data support the basic fact that the protein requirements of the mink for top fur development reaches a peak in September and early October, with a gradual decline in amino acid needs in the weeks to follow. Thus, the months of October and November provide an opportunity for mink nutrition advisors to sharpen the color quality of pelts through enhancing carbohydrate levels that support superior fur development. The net result of higher carbohydrate levels in October, November, and December is lower levels of fat and protein with resultant advantages to the rancher of lower feed costs per pelt marketed and, what is more important, sharper color in mink in the final weeks prior to pelting.

The mink nutritional science basis for superior color in pelts marketed from ranches em-

173

ploying higher carbohydrate levels in the late fall is simple: Lower levels of polyunsaturated fatty acids (PUFA) are provided by the fresh/frozen portion of the ranch diet and higher levels of saturated fatty acids are provided. The mink's higher energy intake from carbohydrates results in a glucose intake beyond the immediate energy needs of the animal. The simple sugars are converted to saturated fatty acids. The mink's metabolic program in itself is simply unable to synthesize the PUFA responsible for "off color" mink fur in the final weeks prior to pelting.

There is, then, a solid physiological basis for the recommendation from National Fur Foods to raise the carbohydrate levels in the mink's diet in October, specifically the recommendation of raising the level of GnF–20 at l% per week for the five weeks of October (on October 1, 8, 15, 22 and 29), replacing fresh/frozen products with high PUFA levels such as poultry by–products, and maintaining the 25% level of GnF–20 through November until pelting in December. A superior alternative program that will provide a higher level of stabilized fats in the ranch diet is to employ GnF–35 in early July and then continue the GnF–35 program into October in combination with 1–2–3–4–5% increments of GnF–20 fortified cereal. The resultant November dietary regimen containing GnF–35 and GnF–20 at a total of 40% fortified cereals will produce even superior "sharp color" in pelts marketed. Table 4.45 shows support for these recommendations within the National Progrsm of Mink Nutrition

TABLE 4.45. National Research Ranch—Experimental Data—Pellet Program—2006— "Sharp Color"—Dark Pelts—Males

Diets Control	Experimental	
% Metabolic Energy	30–50–20*	30–40–30*– High Carbohydrate
Breeders	0	6
Black	10	7
Black–XXD	34	44
XXD	2	6
XXD–XD	—	1
XD	5	2
XD–DK	11	—
DK	—	—
???	4	—
$$$	$53.33	$57.76

*% ME as protein, fat and carbohydrates

174

Resources

In terms of higher carbohydrate digestibility ratings for the mink and greater water absorption (hydration) of the carbohydrate resources for better feed consistency "on the wire," ranchers cannot underestimate the value of processing cereal grains by boiling, extruding, "popping," and toasting or similar heating. In many cases, the starch granules of raw grains and vegetables resist digestion unless they are ruptured by heating, which in addition achieves the breakdown (hydrolysis) of the starch polymers in a process termed gelatinization or dextrinization.

A brief additional commentary is warranted. Sucrose (table sugar) has been used in purified dietary studies at the University of Wisconsin and Cornell University at levels as high as 60% (Tove et al., 1949). Molasses is an excellent carbohydrate resource with high palatability for the mink although higher levels may bring about osmotic diarrhea.

Special commentary is warranted on the nutritional value of lactose (milk sugar) in modern mink nutrition. All factors considered, mink kits are very sensitive to high levels of lactose in their dietary program. In recent years, Rouvinen et al. (1996) reported very significant diarrhea in a ranch diet with 6% whey powder. Earlier studies at the National Research Ranch (Leoschke, 1973) indicated that Gro–Fur pellet formulations with 9% skim milk powder was quite satisfactory for pastel mink but dark kits experienced significant diarrhea.

Vitamins

In terms of animal nutrition, vitamins are distinct from carbohydrates, fats, and proteins inasmuch as they are:

    1. not energy resources;

    2. required in very small amounts, that is, in terms of milligrams and micrograms per day;

    3. co–factors for the multiple enzymes required for animal physiology.

Vitamins are broadly classified as either fat soluble or water soluble. This critical difference of classification is related to the fact that the fat soluble vitamins (A, D, E, and K) are stored in the liver and in the fat depots of the body with minimal opportunity for excretion mechanisms after excessive accumulation. On the other hand, the water soluble vitamins (B and C) have minimal liver storage opportunity within the animal body. Mink have a maximum of two weeks storage capacity on a vitamin B_1 (thiamine) free diet wherein there is a great opportunity for daily excretion via the kidneys and urine.

Animals may have (1) an absolute requirement for a specific vitamin such as vitamin B_{12} wherein there is simply sub–optimum synthesis of the vitamin by the intestinal flora from dietary

175

cobalt, (2) no requirement of a specific vitamin as in the case of vitamin C, ascorbic acid, or (3) a nutritional environment as in the case of choline, wherein although there is significant synthesis of choline from the amino acid methionine, the synthetic processes are simply unable to meet the animal's requirement for choline.

Within mink nutrition management programs, there is the potential for both primary and secondary deficiencies of a vitamin. In a primary vitamin deficiency nutritional regimen, there are simply sub–optimum levels of the required vitamin. A secondary vitamin deficiency occurs when initially ample quantities of the specific vitamin exist but certain mink feedstuffs in the ranch dietary program bring about sub–optimum levels of the required vitamin. Two specific cases illustrate the secondary deficiency of a vitamin.

1.A Biotin deficiency can occur in mink on a mink ranch diet providing ample biotin levels. However, the presence of raw eggs in the dietary regimen can bring about a biotin deficiency as the avidin protein present in raw egg whites combines with the biotin in a structure unavailable to the digestive processes of the mink;

2. A thiamine deficiency can occur in mink on a mink ranch diet providing ample levels of thiamine. Adding fish species such as fresh water carp, smelt, or ocean herring yields a thiaminase enzyme with the capacity to destroy (hydrolyze) all the thiamine present in the ranch diet within a few hours.

Quantity Assessment

Purified Diets

Purified diets are an ideal program for the study of the vitamin requirements of an animal, as they can be formulated to provide a nutritional regimen with exact knowledge of the vitamin content. Most mink ranchers are familiar with the advertisement "Ivory Soap— 99.44% Pure", which asserts that the soap chemists knew almost the exact composition of the product. Likewise with purified diets, the scientists know essentially 100% of the composition they employ in animal nutrition experimental studies.

The high "purity" in these diets is well illustrated in Table 4.46.

TABLE 4.46. Purified Diet Composition—University of Wisconsin–1945

Nutrient Content	%
Carbohydrate Resource	
Sucrose – Table Sugar*	66
Protein Resource	
Vitamin Test Casein**	19
Fat Resource	
Cottonseed Oil***	8
Cod Liver Oil****	3
Mineral Resource	
Salts IV*****	4
Vitamin Resources	
B Vitamins as pure crystalline products.	
Fat–Soluble Vitamins via cod liver oil fortified with vitamin E	
(alpha–tocopherol acetate) and vitamin D$_3$	

   * 100% pure carbohydrate.
  ** Vitamin test casein is prepared by subjecting casein (dried cottage cheese) to alcohol extraction to remove all the B–Vitamins. This procedure is followed by heat treatment to remove the alcohol.
 *** Contains vitamin E; thus, this purified diet is not appropriate for studying the vitamin E requirements of an animal that requires specially distilled lard to remove any vitamin E content.
**** a rich source of the fat–soluble vitamins A and D.
***** crystalline salts – 100% pure (Philips and Hart, 1935).

With the research tool of purified diets, scientists can determine whether a given vitamin is essential for health and top performance, the exact quantity of the vitamin required for each phase of life, and the signs and symptoms indicating dietary deficiency of a specific vitamin.

The use of purified diets for the study of mink nutritional requirements was initiated at the University of Wisconsin in 1945 (Schaefer et al., 1947), under the leadership of Dr. C. A. Elvehjem. The Biochemistry Department is the oldest such department in the world and has been at the forefront of vitamin studies. In 1913, the first vitamin—vitamin A—was discovered there. The department later provided key research on vitamin D, and in 1935 Dr. Elvehjem solved the disease of pellegra with the discovery of niacin. At that time Wisconsin produced more mink pelts than any other state or province in North America. Thus, it was natural that researchers would be interested in employing mink as a unique animal tool for studying vitamin requirements.

Much information in the field of nutrition had already been gained through the use of different species in experimentation. Nutrition scientists wondered whether a carnivorous animal such as the mink would have the same nutritional requirements as the common laboratory animals or whether a need for new vitamins might be established.

The original purified diet employed in research studies with mink kits was modified over the years with new insights into the nutritional requirements of the mink to a final version formu-

177

lated by nutrition scientists at Cornell University with added amino acid supplementation and a more detailed trace mineral fortification (McCarthy et al., 1966). See Table 4.47 for this research diet.

TABLE 4.47. Purified Diet for Mink Research Studies—Cornell University

Nurient Content	%
Carbohydrate Resource	
Sucrose – Table Sugar	39
Protein & Amino Acid Resources	
Vitamin Test Casein	30
L–Arginine Hcl	0.5
L–Cystine	0.25
DL–Methionine	0.25
Fat Resources	
Cottonseed Oil	9
Lard	10
Fiber Resource	
Powdered Cellulose	5
Mineral Resource	
Mineral Mixture*	5
Vitamin Resources	
Vitamin Mixture**	1
Choline Dihydrogen Citrate	0.66
Anti–Oxidants	
Ethoxyquin	125 mg/kg

* Mineral fortification includes: calcium, 0.56; phosphorus, 0.53 (additional P in casein); sodium, 0.26; potassium, 0.58; chloride, 0.40; iodine, 0.00041 (additional I in casein); copper, 0.002; iron, 0.015; cobalt, 0.0002; zinc, 0.004; manganese, 0.0061 (additional Mn in casein); magnesium, 0.0625; sulfur, 0.101 (additional S in casein) and the following as ppm: selenium, 0.25 and molybdenum, 1.50.

** Vitamin fortification included (mg/ kg diet): thiamine HCl, 2; pyridoxine, 2; riboflavin, 4; calcium D–pantothenate, 15; niacin, 40; i–inositol, 250; p–amino–benzoic acid, 500; menadione (synthetic vitamin K), 5; folic acid, 1; biotin, 0.25; vitamin B–12, 0.04. Also fat–soluble vitamins as follows (IU/kg diet): dl–alpha–tocopherol acetate, 40; vitamin D–3, 1,200 and vitamin A, 12,000.

It is of interest to note that additional studies conducted at Cornell University by McCarthy (1964) employing this purified diet with five times the concentration of thiamine Hcl, pyridoxine Hcl, Riboflavin, Menadione; double the levels of folic acid biotin; and a special supplementation of vitamin C (L–ascorbic acid) at 99 mg/kg diet did not enhance the growth of mink kits.

Fat Soluble Vitamins
Vitamin A

Vitamin A is a nutrient termed all–trans retinol, with specific roles in the anatomy and physiology of animals. It is required for the integrity of the epithelial tissues and optimum structures of connective tissue and nerve tissue. The epithelial cells include the mucous membranes, the cornea, trachea, lungs, kidneys, bladder, urinary tract, testicles, vagina, and gastro–intestinal tract. Thus the following observations on a vitamin A deficiency in animals.

Maintenance of Epithelial Tissues

In a vitamin A deficiency, symptoms arise in these cells inasmuch as the vitamin is required for the formation of glycoproteins and/or mucopolysaccharides which are critical constituents of collagen, elastin, cartilage, and other connective tissues. Thus with the defective glycoprotein structures of a vitamin A deficiency, the epithelial cells become keratinized and dysfunctional keratinization of the mucous membranes of the eyes may lead to "night blindness" and finally permanent blindness (exophthalmia). Loss of epithelial integrity in the intestinal tract can lead to a greater susceptibility to bacterial infections. The loss of epithelial quality in the bladder and urinary tract can lead to bladder stone formation wherein cellular debris can act as a nucleus for the formation of urinary calculi.

Vision

Because vitamin A is a precursor of a component of the visual pigment, rhodopsin, vitamin A deficiency has effects on the vision of an animal and can lead to permanent blindness.

Bone Growth

Inasmuch as the formation of collagen and cartilage requires vitamin A, bone growth is defective. Nerve function is compromised as well and leads to a lack of coordination of the animal's legs.

Obviously, animals provided dietary regimens with sub–optimum vitamin A fortification will have relatively poor reproduction, lactation, growth, and fur development.

Physiology

In most animals, vitamin A can be derived from multiple provitamin A carotenoids present in the dietary program. However, in the mink these caronoids are not converted to vitamin A in the intestinal mucosa (Warner et al., 1963; Warner and Travis, 1966).

Vitamin A is absorbed into the mucosal cells of the intestine after emulsification with fats and bile salts in the intestine, forming micelle structures which facilitate absorption. In the muscosal cells, vitamin A is esterified with palmitic acid which enters the blood circulation via the lymphatic system, is removed from the blood by the liver, and stored. In terms of this physiological sequence,

179

it is obvious that optimum absorption of vitamin A and other fat–soluble vitamins (D, E, and K) is facilitated by an emulsification process involving dietary fats. Hence, with mink dietary programs very low in fat, the absorption process for vitamin A may be minimal.

It is generally agreed that the blood levels of vitamin A are homeostatically regulated to a relatively constant level, while the liver level reflects the dietary vitamin A level provided to the animal. Hence, plasma vitamin A levels do not reflect an animal's vitamin A status except in severe deficiencies or toxicity.

Mink appear to have a greater ability to build up reserves of vitamin A in the liver than do foxes (Rimeslatten, 1968). Those stored in mink livers appear to have a relationship to the sex of the mink(Belcic, 1981 and Rimeslatten, 1968).

Requirements

Studies by both Bassett (1961) and Rimeslatten (1966) found that 400 IU/kg mink body weight was required for maximum growth of mink kits. This level of vitamin A intake of the mink is equivalent to about 6,000 IU/kg dietary dry matter with a ranch diet providing 4,080 Calories ME (Travis et al., 1982). Experimental data provided by Abernathy (1960) and Warner and Travis (1966) indicated that at 100 IU vitamin A/kg body weight/day there is no significant liver storage of vitamin A; at 400 IU vitamin A/kg body weight/day, the storage was still extremely low. At the 1,000 IU level, liver storage of vitamin A increased two–to threefold.

Studies by Rimeslatten (1968) indicated no significant difference in the reproduction performance of mink provided 6,000 IU or 48,000 IU vitamin A/kg dry matter. However, analyses for vitamin A in the liver of the kits showed detectable amounts only at the higher level of vitamin intake.

Resources

Commercial fortified mink cereals generally contain synthetic vitamin A in the form of relatively stable retinol esters including retinol acetate and retinol palmitate. These ester structures are more stable than free retinol which is sensitive to oxidation by peroxides and other structures commonly present in mink fresh/frozen feedstuffs. The stability of these vitamin A esters in commercial cereal mixtures is seen in a study conducted by National Fur Foods Co. (Leoschke, 1957). A fortified cereal containing 13,200 IU vitamin A/kg was placed in a warehouse in January and a sample obtained six months later for a vitamin A analysis which provided 13,600 IU vitamin A/kg.

Without question, anti–oxidants including vitamin E and vitamin C (ascorbic acid) are useful for minimizing the loss of vitamin A content in mink feed. Moreover, as early as 1948, Bassett et al. provided experimental data indicating that vitamin C supplementation of mink diets (containing relatively low vitamin A levels) actually enhanced vitamin A levels in the blood plasma and the livers of mink kits.

When ranchers employ acid preserved fish at a pH of 4.5 or lower, partial isomerization of vitamin A from the all–trans structure to the cis form occurs during storage. This structural change reduces the utilization of the vitamin to about 75% (De Ritter, 1976 and Fog, 1974).

Nutritional Deficiency

Studies on vitamin A deficiency in mink have been conducted by Helgebostad (1955), Stowe et al. (1959), Abernathy (1960), and Bassett (1961). With dietary programs deficient in vitamin A, mink develop diarrhea, inferior fur development, night blindness, and poor coordination in the rear quarters. Their eyes are also affected—the lenses become opaque and the conjunctivas encrusted, and metaplasia of epithelial tissues and fatty infiltration of the liver occurs. The skull does not enlarge normally; as a result, the cerebellum is compressed and herinates into the foramen magnum. Damage to the cerebellum results in muscle incoordination.

Toxicity

Multiple studies have been conducted on vitamin A toxicity in the mink. Experimental work by Helgebostad (1955) indicated that mink kits can tolerate 40,000 IU of vitamin A/kg body weight/day, the equivalent of a one kilogram kit consuming 100 grams/day of a diet providing 400,000 IU vitamin A/ kg dry matter. However, at a level of 2,000,000 IU vitamin A/ kg solids, significant vitamin A toxicity symptoms were noted including anorexia, cautious gait, tenderness on moving about, decalcification of the skeletal structure, spontaneous fractures, skin lesions, loss of hair, exophthalmia, and hyperesthesia of the skin. These observations were confirmed by Rimeslatten (1966).

In terms of the reproduction/lactation performance, the animals are less tolerant of vitamin A. Ranch diets containing 300,000 IU/kg dry matter bring about significant loss of reproduction/lactation performance as evidenced by reproduction failure, smaller litter size, and poor kit survival (Adair et al., 1977 and Travis, 1977). This level of vitamin A toxicity for the reproduction/lactation phase is consistent with observations of Friend and Crampton (1961b) wherein mink were fed a diet containing 10% whale liver. Calculations indicated that the mink on this feeding regimen consumed about 280,000 IU vitamin A/day provided by the whale liver alone, not counting other vitamin A resources in the diet.

Vitamin D

There are two precursors of vitamin D: the plant sterol ergosterol, which can be converted to vitamin D_2 (ergocalciferol) by ultraviolet light present in sunshine, and the animal sterol present in the skin, 7–dehydrocholesterol, which upon exposure to sunlight is converted to vitamin D_3 (cholecalciferol). It is of interest to note that Bassett et al.(1951) suggested that a diet of natural feedstuffs without a vitamin D supplement is probably adequate for growing mink exposed to sunlight.

Vitamin D is required for optimum skeletal development in the young, and a deficiency termed rickets can lead to a deformed skeleton.

Physiology

Vitamin D is not physiologically functional until modified in the liver with the addition of a hydroxy group at carbon–25 and later by the kidney wherein another hydroxy group is added to yield 1,25–dihydroxycolecalciferol (1,25–OHD_3), the metabolically active form of vitamin D. From the kidney via the blood circulation, 1,25–OHD_3 reaches the intestinal mucosa cells wherein it regulates the synthesis of a calcium–binding protein. This protein in turn transports calcium throughout the body via the blood circulation. In addition to 1,25–OHD_3, the parathyroid hormone (PTH) plays a key role in calcium physiology. The formation of 1,25–OHD_3 is regulated according to the body's requirement for calcium by the blood serum level, which in turn regulates the release of PTH. A low serum calcium level initiates the release of PTH, which yields an increased formation of 1,25–OHD_3, increasing the formation of calcium–binding protein, enhancing calcium absorption from the intestine.

Requirements

Studies by Bassett et al. (1951) indicated that the mink requirement of vitamin D is less than 820 IU/ kg feed solids. At the same time, the experimental data indicated significant skeletal healing with a vitamin D supplementation program of 40 IU/mink/day, equivalent to a male kit consuming 100 grams of a diet (dehydrated basis) providing 400 IU vitamin D/kg. Field observations of Leoschke (1958), support a recommendation of 400 IU/kg mink feed solids as adequate to prevent rickets in mink.

Resources

Commercial mink fortified cereals generally provide ample levels of synthetic vitamin D for the requirements of the mink for top skeletal development.

Nutritional Deficiency

As early as 1941, Smith and Barnes had produced rickets in newly weaned kits by providing a diet deficient in vitamin D and containing 0.06% calcium and 0.54% phosphorus. Symptoms of vitamin D deficiency in mink includes lameness and bent bones but not the more severe symptoms such as recurrent spasms, severely enlarged joints, and cranial enlargement as seen in foxes on the same diet (Harris et al., 1951).

Toxicity

Experimental studies conducted in Denmark and Norway indicate fox and mink are relatively resistant to large doses of vitamin D_3. Work by Hilleman (1978) indicated that mink fed levels of 10,000, 25,000, and 40,000 IU vitamin D/kg of dry matter from July to pelting did not show any significant differences in pelt characteristics. Studies by Helgebostad and Norstoga (1978) indicated no toxic effects of vitamin D at a level of 50,00 IU/ kg feed solids. These same studies indicated that toxic effects were noted within a short period when the level of vitamin D supplementation was raised to 100,000 IU/kg dry matter. This same level of vitamin D toxicity had been reported earlier by Pereldik et al (1972).

Clinical signs of vitamin D toxicity in mink include loss of appetite, nausea, loss of weight, difficulty in moving, and production off dark colored feces. Analysis of blood serum showed markedly elevated levels of calcium. Calcium deposits were demonstrated in the kidneys and in some cases in the muscles, gastric mucosa, bronchi, and larger blood vessels (Helgebostad and Norstoga, 1978). It is important for mink nutrition advisors to acknowledge the real potential for vitamin D toxicity in mink, inasmuch as many ranch diets contain high levels of ocean fish (Jorgensen, 1977).

Vitamin E

Vitamin E, alpha–D–tocopherol, was first discovered in 1922 as a fat–soluble substance in vegetable oils required to maintain pregnancy in rats. Later research studies indicated that vitamin E had significant additional roles in animal nutrition and physiology including as a factor in the development of muscle dystrophy, and as a very important natural anti–oxidant, that is, a free radical scavenger. As a free radical scavenger vitamin E is part of the cellular defense system protecting the lipid moiety in membrane structures from deleterious peroxidation reactions.

Of the eight naturally occurring vitamin E isomers, alpha–tocopherol is the most potent in preventing clinical signs of vitamin E deficiency. Tyopponen et al.(1984) have conducted an interesting experiment on the mink's physiological deployment of 4 of the vitamin E isomers provided in the dietary regimen as seen in the table below:

Deployment Tocopherol Isomer*	alpha	beta	gamma	delta
Diet	1.0	0.07	0.55	0.10
Liver	1.0	0.04	0.12	0.00
Plasma	1.0	0.00	0.13	0.00
Adipose Tissue	1.0	0.00	0.19	0.00

* Ratio to alpha–Tocopherol as 1.0.

Natural alpha–D–tocopherol consists of a single RRR configuration while synthetic vitamin E, all–rac–alpha–tocopherol acetate, consists of an equal quantity of 8 stereoisomers. Relative to the nutritional value of natural and synthetic vitamin E for the mink, a study conducted by Jensen (2004) is of interest. The experimental diet supplemented with synthetic all–rac–alpha–tocopherol acetate provided only about 16% of the natural vitamin E isomer and yet the natural vitamin E isomer was dominating in the plasma and the milk of the lactating mink as well as in the plasma and tissues of their kits. In kit heart muscle the natural isomer made up to 60% of the total alpha–tocopherols.

It is important to note that the trace mineral selenium has a positive interrelationship with vitamin E in animal nutrition and physiology, with the capacity to minimize vitamin E deficiency symptoms and vitamin E requirements. It is also important to note that other trace minerals including copper, iron, and manganese have a negative effect on vitamin E nutrition, with a pro–oxidative effect on the vitamin E content of feedstuffs, that is, they have the capacity to accelerate vitamin E loss via oxidation. The multiple roles of vitamin E in mink nutrition and physiology have been well documented over the period of the last half–century. An excellent review of the topic was presented by Helgebostad (1976) at the first International Scientific Congress in Fur Animal Production, Helsinki.

Vitamin E—Essential For Reproduction/Lactation

Helgebostad's 1974 study of a primary nutritional deficiency of vitamin E in the mink deserves marked attention, especially in one detail. The dietary program involved beef tallow as the primary fat resource. Of all animal fats, beef tallow has the lowest level of polyunsaturated fatty acids (PUFA) which through oxidative rancidity yield peroxide structures that destroy vitamin E and support "yellow fat" disease in the mink. This disease is a classic example of secondary vitamin E deficiency; that is, a dietary program is planned with ample levels of vitamin E, yet yields a vitamin E deficient state because the vitamin is destroyed by specific components of the diet. Thus by using tallow as the primary fat resource, Helgebostad was able to achieve a primary deficiency of vitamin E in the mink. Experimental work in mink with a primary vitamin E deficiency indicated low fertility in female mink, anemia, and decreased growth of mink kits prior to weaning.

Vitamin E and Muscle Dystrophy

A number of studies indicate that mink ranch diets providing both vitamin E and antioxidants may yield a pathology in the mink which is distinct from "yellow fat" disease but involves multiple muscles of the mink anatomy (Knox, 1977; Brandt, 1983; Brandt, 1984; Henricksen, 1984). The prevailing pathological findings are severe epicardial, pleural, abdominal, and muscular transudation; cardiac muscle degeneration; and suppressed serum alpha tocopherol levels.

An inherited progressive muscular dystrophy of the mink, which resembles the amyotonic form of human muscular dystrophy can also occur (Hegreberg et el., 1974 and 1976). Clinically, the earliest sign is a progressive muscle weakness and atrophy. Muscle enzyme activities in serum including creatine phosphokinase, aldolase, and glutamic oxaloacetate transaminase are unusually elevated to pathological levels. The urinary creatine/creatinine ratio is elevated, usually greater than 2/1 for the affected mink compared to less than 1 for normal healthy mink. Genetic studies indicated an autosomal recessive mode of inheritance.

Vitamin E—Natural Antioxidant–Free Radical Scavenger

Multiple experimental studies have shown that vitamin E, alpha tocopherol, functions not only as an essential vitamin in animal physiology but also as a natural biological antioxidant reacting with peroxides and free radical structures as a free radical scavenger. Peroxides and free radical structures are formed in animal physiology by normal metabolic processes which may be key factors in the aging process. Vitamin E in its role as a natural antioxidant reacts with these peroxide and free radical structures to inactivate them; thus in the process, vitamin E is destroyed (Ender and Helgebosted, 1975). The enzyme glutathione peroxidase, containing the trace element selenium, also functions to detoxify peroxide structures.

It is very important for every mink rancher to recognize that a process of oxidative rancidity involving the formation of peroxide structures from polyunsaturated fatty acids (PUFA) in animal and fish products begins as soon as they are processed. As a result, the vitamin E content of these mink feedstuffs is lost (Dam, 1962). This point is well illustrated by the following experimental data from Friend and Crampton (1961).

Vitamin E Destruction in Mink Feed Products Room at Temperature—11 Days

Product	Vitamin E Loss%
Cod Fish Scrap	77
Whole Codfish	72
Whale Meat	70
Whale Liver	44*

* 25% loss of vitamin E after only 36 hours at room temperature.

Obviously, the process of oxidative rancidity and subsequent destruction of vitamin E is reduced by frozen storage of mink feedstuffs, but freezing does not entirely eliminate the process. Several factors play into the loss, including reatively low freezer temperatures (below –5° F.) and /or extremely long periods of freezer storage (human milk stored at –70° F.) The net result of sub–optimum freezer environments is too many cases of sub–optimum reproduction/lactation and kit growth performance by mink on diets containing a high percentage of frozen mink feedstuffs such as whole bird (Sturdy, 1980), fish(Belcher et al., 1959), and rabbit offal (Zimmerman, 1992).

The function of vitamin E as an antioxidant is corroborated by the similar actions of various synthetic antioxidants such as BHT (Butylated Hydroxy Toluene) (Leekley and Cabell, 1959).

In more recent years, Santoquin (6–ethoxy–1,2–dihyroxy–2,2,4 trimethyl–quinoline), also termed EMG (Ethoxyquin), has been used in Scandinavian mink ranch diets that contain high levels of fatty fish. Ethoxyquin is also commonly employed in the production of fish meals for aquaculture, early weaned piglets, and fur animals. The Danish firm Esbjerg Fiskeindustri, Esbjerg, Denmark, adds 100–200 ppm ethoxyquin, before and after dehydration procedures.

Antioxidants in mink ranch diets and commercial fish meal production are highly effective in preventing and delaying oxidative processes related to PUFA and thereby protecting naturally occurring vitamin E resources. This basic point is well illustrated by the experimental data of Travis and Pilbean (1978) wherein fish were frozen and stored from 2/11 to 6/30 with (w/) and without (w/o) antioxidant additions.

Stored Fish w/ and w/o Ethoxyquin

Westcoast Sole	Vitamin E Loss–%
w/o Santoquin	32
w/ Santoquin	2

Vitamin E and Selenium

The exact interrelationship between vitamin E and the trace element selenium is not fully understood. It is known that selenium is a key co–factor for glutathione peroxidase (Se–GSH–Px) enzyme which destroys peroxide structures and thereby minimizes the chemical work required by vitamin E in peroxide removal in animal physiology. Studies by Stowe and Whitehair (1963) showed that a level of 0.1 ppm selenium as sodium selenite added to a vitamin E deficient diet prevented all lesions except for minor accumulations of amorphous non acid fat material at adipose interstices. Brandt (1984) noted that supplementation of an experimental mink diet with selenium enhanced both the levels of plasma vitamin E and plasma concentration of Se–GSH–Px enzyme.

Physiology

In carnivores, vitamin E crosses the placenta with difficulty, and little tocopherol will appear in the milk (Gallo–Torres, 1972). It is therefore reasonable to assume that the stores of vitamin E in mink pups at weaning are very low.

It is well established that absorption of alpha–tocopherol from the small intestine of monogastric animals is dependent on multiple factors including pancreatic enzymes, bile salts, and dietary ingredients (Gallo–Torres et al., 1971). For modern mink nutrition, the nature of the dietary fat resource in the mink's diet is especially important in terms of final levels of vitamin E achieved in the blood plasma. Experimental studies by Eskeland and Rimeslatten (1979) and Brandt (1984) indicated that adsorption/recovery of vitamin E was significantly lowered by the dietary presence of PUFA. Brandt's work here is especially interesting inasmuch as he employed S. Sprattus fish oil with a combination of high PUFA and vitamin E as the fat resource.

Earlier studies of Muralidhara and Hollander (1977) with rats prepared for the experimental work of Eskeland, Rimeslatten, and Brandt on losses of vitamin E in the gastro–intestinal tract prior to absorption. These authors also emphasized the basic fact that increased deposition of PUFA in the tissues increased the extent of PUFA oxidation in cellular membranes to yield peroxide structures which destroy vitamin E structures.

All factors considered, it is obvious that the reduced levels of plasma vitamin E in these studies is not directly related to a lack of vitamin absorption into the blood circulation, but much more likely directly related to the factor of vitamin E loss in a series of destructive sequences. First, vitamin E may be lost through mixing with other feedstuffs. Even if a fish oil product is relatively stable, in the new environment it is vulnerable to peroxide structures present in other feedstuffs in the feed mixture. Second, it may be destroyed by the relatively high body temperatures within the gastro–intestinal tract. Support for the concept that lower blood levels of alpha tocopherol in these experimental studies are related to loss of vitamin E via oxidative processes more significantly than through the absorption process is seen in the experimental data provided by Eskerland and Rimeslatten (1979). Lower vitamin E absorption was observed with mink on high fat diets; within these groups, the lowest absorption was noted in the mink on diets containing high levels of PUFA. They found that mink on dry diets containing stabilized fat resources had twice the vitamin absorption of mink on conventional mink ranch diets with fresh/frozen feedstuffs.

Requirements

Multiple structures in the tocopherol class of vitamin E are widely distributed in nature including alpha, beta, gamma, and delta tocopherol. Vitamin E, alpha tocopherol, is the organic structure with the highest vitamin E activity for animals in general. Tyopponen et al. (1984) recommend that vitamin E analyses of the plasma could be used as a routine method for observing the vitamin E status of the mink. Their studies indicated that plasma alpha tocopherol had a linear relationship to the log of dietary doses, with an apparent half–saturation of the vitamin E binding capacity at 13 mg of vitamin E/ kg of diet. Their studies also showed that at the given levels of dietary vitamin E supplementation, 25 to 150 mg/kg of alpha tocopherol acetate, the liver and adipose tissue continued to accumulate alpha–tocopherol. These sites were more responsive and thus more reliable organs through which to establish the vitamin E status of the mink.

The vitamin E requirement of the mink depends on multiple factors, including the caloric density of the diet and the composition of the diet in terms of sulfur amino acids (cysteine and methionine), selenium, and especially, fat content in terms of both quantity and quality (Stowe and Whitehair, 1963; Brandt, 1980). In terms of fat quality, the key factors include the proportion of saturated, monounsaturated, and polyunsaturated fatty acids (PUFA) present as well as the degree of oxidative rancidity of the fats as measured by peroxide number.

187

Growth

Studies by Stowe et al. (1960) and Stowe and Whitehair (1963) indicated that the vitamin E requirement of the mink for growth was less than 25 mg/kg feed solids. Other studies by Petersen (1957), Ahman (1966), and Ender and Helgebostad (1975) indicated that the actual requirement of vitamin E for mink kits was met by 5 mg vitamin E/kg dry feed, provided the diet contained only minimal quantities of PUFA and the fat contained resources that had not undergone oxidative rancidity.

Reproduction/Lactation

In terms of the reproduction/lactation performance of the mink, Travis and Pilbean (1978) found that increasing the vitamin E level from 31 mg/kg to 138 mg/kg feed solids did not enhance the performance of the mink.

PUFA and Vitamin E Requirements

Multiple studies indicate that with the inclusion of relatively high levels of PUFA in mink ranch diets, higher levels of vitamin E may be required to provide top reproduction/lactation performance of the mink (Tauson, 1993; Rouvinen, 1991) or to prevent a vitamin E deficiency known as "yellow fat" disease (Helgebostad and Ender, 1955; Ender and Helgebostad, 1975; Brandt et al.,1990; Rouvinen, 1991; Engberg et. al., 1993; Tauson, 1993; Engberg and Borsting, 1994). When large amounts of fish are fed to the mink, it is recommended that the level of vitamin E supplementation be as high as 200–300 mg/kg dry matter (Treuthardt, 1992).

As a guideline for vitamin E fortification of the mink ranch diet with a wide variety of PUFA content, for those with relatively low levels, Harris and Embree (1963) recommended 0.6 mg of vitamin E per gram of PUFA to supplement the basic dietary vitamin E levels.

For ranch diets with high levels of fatty fish, in addition to vitamin E supplements, ranchers are to use synthetic antioxidants such as ethoxyquin (Santoquin) or BHA. These will minimize oxidative rancidity of fats during the period of storage as well as during the period of feed processing and while "On The Wire" (Leekley and Cabell, 1961).

Resources

In choosing vitamin E resources to fortify the mink's diet ranchers must take into account the stability of the vitamin E content through storage, mink feed processing, time "on the wire" in the mink yard, and passage through the gastro–intestinal tract of the animal. Some fish oils with relatively high levels of vitamin E, when added to mink diets, yield "yellow fat" disease, a secondary vitamin E deficiency directly related to the destruction of vitamin E content through oxidative rancidity. Obviously, all factors considered, the vitamin E resource of choice for modern mink nutrition is stabilized vitamin E. This product is achieved via esterification of alpha tocopherol's alcohol functional grouping with organic acids (for example, acetic acid to yield alpha tocopherol acetate). Studies supported by National Fur Foods (Leoschke, 1982) indicated that the loss of vitamin E in their pellet formulations was only 20% after six months storage.

Nutritional Deficiency
Primary Vitamin E Deficiency

Studies by Stowe et al.(1960) and Stowe and Whitehair (1963, 1965) achieved a primary vitamin E deficiency in the mink using lard prepared by molecular distillation to remove traces of vitamin E content. The use of rendered pork fat also yielded an experimental diet containing relatively low levels of PUFA to minimize the potential for secondary vitamin E deficiency. A vitamin E deficiency is yielded in the presence of sub–optimum levels of vitamin E in the mink's diet whereby vitamin E loss is created by PUFA in the mink dietary fat resources. These PUFA undergo oxidative rancidity yielding peroxide and free radical structures destructive of vitamin E. The basic nature of this experimental plan was repeated years later in the research work of Helgebostad (1974) with the employment of beef tallow (very low levels of PUFA) in his studies on a primary deficiency of vitamin E during the critical reproductive/lactation phase of the mink ranch year.

These experimental studies on primary vitamin E deficiency in the mink indicated major gross lesions consisting of intercostal adductor and cardiac myopathy, hepatic peripherolubular fatty infiltration. Urinary incontinence was common, with no anemia or "yellow fat." Serum tocopherol values below 50 micro–grams/100 mls serum were common. Interestingly, skeletal and cardiac myopathy in the absence of any form of steatitis ("yellow fat" disease) was produced with kits weaned on tocopherol deficient and PUFA low basal ration. Hematologic alterations with the vitamin E deficiency included erythrocyte fragility leading to anemia (Hartsough and Gorham, 1949; Leader, R.W., 1955). In addition, extensive hemorrhage with hemoglobinuria may be present as observed in field studies (Hartsough and Gorham, 1949; Gorham, 1962). Prolonged alpha–tocopherol deficiency tended to decrease serum albumin and produced an increase in beta–globulin fractions. Elevated serum glutamic–oxaloacetate transaminase values were associated with tocopherol values below 50 gamma/100 mls of blood serum. Anti–body production was not impaired by tocopherol deficiency.

Even with a vitamin E deficiency mink kits' weight gains were not affected by the alpha–tocopherol deficiency. However, studies by Helgebostad (1971) indicated loss of fur pigment in dark mink with vitamin E deficiency.

For their experimental study on primary vitamin E deficiency Stowe et al. (1963) were able to produce a secondary deficiency of vitamin E via the addition of cod liver oil with high PUFA content. A significant development of "yellow fat" disease followed; that is, mink exhibited yellow, acid–fast pigmentation deposition in the adipose interstices.

Secondary Vitamin E Deficiency—"Yellow Fat" Disease–Steatitis

History

"Yellow fat" disease (steatitis), recognized in Wisconsin as early as 1942, was associated with the feeding of a rather high level of animal or fish products that had been stored for more than six months (Hartsought and Gorham, 1949). Disastrous outbreaks of steatitis occurred in the summers of 1947 and 1948 throughout North America (Gorham, 1951). Ender and Helgebostad (1944, 1953) were the first fur animal scientists to describe the clinical and pathological signs of "yellow fat" disease as observed in experimental feeding trials involving high levels of herring and round coal fish and resulting in high mortality of male mink kits. Based on the micro– and histopathology picture in the mink, they concluded that the disease had been caused by a high level of unsaturated fat provided by the experimental rations. This means to resolve "yellow fat" disease with vitamin E supplementation (Hartsought and Mason, 1951) might have been discovered at an earlier date had American veterinarians been aware of the original paper of Ender and Helgebostad published in 1944 in the Norwegian trade journal, *Norsk Pelsdyrblad*.

The name "yellow fat" disease for this disorder peculiar to the mink was first coined by McDermid and Ott (1947).

Etiology

Without question, all factors considered, the development of "yellow fat" disease or steatitis in fur animals requires a combination of an unusually high dietary intake of polyunsaturated fatty acids (PUFA), a relatively low level of vitamin E in the ranch diet, and young eager male kits with aggressive behavior relative to feed intake.

In terms of the physiology involved, the primary factor in the development of steatitis is a high consumption of PUFA including linolenic acid, trienoic (three double or alkene bonds) found in horsemeat (Brooker and Shorland, 1950) and linseed oil, and eicosapentaenoic and docosahexenoic, found in fish and fish oil. It is of interest to note, in an analysis of Angola fish oil (Danse and Steenbergen–Botterweg, 1976), PUFA content was trienoic, 0.7; tetraenoic, 0.6; pentaenoic, 11.6; and hexanoic, 7.2%. A relatively low level of vitamin E in the mink ration is also necessary. This situation is exacerbated by the use of mink feedstuffs with high PUFA levels which have been in freezer storage for a long time (Gorham and Nielsen, 1953; Mason and Hartsought, 1951) and/or stored under sub–optimum freezer conditions (Belcher, Evans and Budd, 1959). These environmental conditions promote the development of "oxidative rancidity" of the fats present with resultant high peroxide values and subsequent destruction of vitamin E. This reaction affects not only frozen feedstuffs but also later destroys the vitamin E present in fresh feedstuffs and the fortified cereal mixed with the rancid frozen feedstuffs to yield, finally, a mink feed "on the wire" with sub–optimum vitamin E content.

One last requirement for the development of "yellow fat" disease is a relatively high feed intake by the male kits. This situation occurs when the larger mink (usually the finest kit in the litter) consumes more than his share of the feed; other mink kits, shoved away from the feedboard, have a relatively low intake of PUFA and so are not affected with "yellow fat" disease (Mason and Hartsought, 1951). This specific point is well illustrated by commentary of McDermid and Ott (1947): "One consistent factor found in analyzing the diets fed by ranchers is the high level of animal protein. The old axioms 'A growing kit cannot eat too much' and 'You can't feed too much horsemeat' are apparently the cause of most of the trouble."

Typical "yellow fat" disease involves adipose cell degeneration (steatose), inflammation and fibrosis of adipose tissue (steatites), and an accumulation of a lipofuscin pigment (brownish–yellow). Microscopically the most significant observation is a non–suppurative inflammation of the subcutaneous and viscera fat. In chronic cases of the disease, anemia is often found with a pronounced fragility of the red blood cells and an increase in the numbers of leucocytes and thrombocytes. In acute cases, a hemoglobinuria is noted, the urine being dark brown in color. Ender and Helgebostad (1953) noted that in addition to the regular observations on "yellow fat" disease of the mink, there was a prolonged clotting time of the blood; by treatment with vitamin K it was reduced to normal. A hyaline muscular degeneration is commonly seen.

Studies by Danse and Steenbergen–Botterweg (1976) provide some very interesting observations of the early stages of the morphogenesis of "yellow fat" disease. Their studies indicated that the initial change in the development of "yellow fat" was the appearance of lipofuscin–laden macrophages in the interstices of adipose tissue without microscopical enzyme, histochemical, or electron microscopical changes of the fat cells.

In addition, increased lipofuscin accumulation was seen throughout the reticuloendothelial system. Since the accumulation of lipofuscin in macrophages is coupled with an overloading and a decreased digestive function of these cells, the activity of the reticuloendothelial system may be depressed during early stages of "yellow fat" disease. The low acid phosphatase activity of macrophages in mink adipose tissue supports this suggestion (Danse, 1976). Microscopic lesions in the adipose tissue are indistinguishable from that of Weber–Christian disease in humans (Quotrup, Gorham and Davis, 1948).

It is important to emphasize the basic point that "yellow fat" disease of the mink is distinct from a simple vitamin E deficiency. A vitamin E deficiency in the mink does not result in nonsuppurative inflammation of the subcutaneous tissue and visceral fat, edema, and pigmentation of the affected adipose tissue. Vitamin E deficient diets must be supplemented with a significant percentage of PUFA (trienoic and higher degrees of unsaturation) in the fat presented to the mink to evidence steatitis disease (Stowe, Whitehair and Travis, 1960). It is quite apparent that the production of the yellow pigment in adipose tissue as noted in steatitis requires the presence of oxidative rancidity in the mink's dietary fat resources to the point of the creation of PUFA with free radical structures capable of polymerization to yield conjugated alkene bonds producing the typical yellow coloration of the adipose tissue.

Experimental Production

The mink disease of "yellow fat" has been experimentally produced over a period of years by a wide variety of dietary management including:

Linseed Oil (Lalor, Leoschke and Elvehjem, 1951);

Cod Liver Oil (Howell, 1955; Stowe, Whitehair and Travis, 1960; Stowe and Whitehair, 1963);

Marine Oils (Dalgaard–Mikkelsen, Kvorning, Momberg–Jorgensen, Hagen–Petersen, and Schambye, 1953; Helgebostad and Enger, 1955; Dalgaard–Mikkelsen, Momberg–Jorgensen, and Peterson, 1958; Danse, 1976).

Oxidized Fish Oils (Ender and Helgebostad, 1971 and Danse and Steenbergen–Botterweg, 1976). Both papers were excellent examples of experimental design in terms of providing specific peroxide values of the oxidized fish oils employed (500 in the 1971 study and 350–450 in the 1976 study) and the peroxide values in the actual fat regimen provided the mink (100–150 in the 1976 study) (Danse, 1976, 1978).

Salmon Waste (Leekley and Cabell, 1961).

Tuna Waste (Adair, Stout, and Oldfield, 1958).

Prophylactic Nutrition Management
Dietary Change

Immediate change in diet to a large amount of fresh horsemeat and liver (Hartsought and Gorham, 1949) will bring about cessation of kit losses.

Synthetic Antioxidants

The use of phenolic–type natural antioxidants (tocopherols) and/or synthetic antioxidants such as BHT and Ethoxyquin is able to minimize oxidative rancidity of fats by reacting with the free radical structures that arise.

BHT

Studies by Leekley and Cabell (1959), Cabell and Leekley (1960b), Leekley and Cabell (1961) and Leekley, Cabell, and Damon (1962) indicated that BHT (Butylated Hydroxy Toluene) at a level of 112 grams/ton of wet feed was very effective in preventing "yellow fat" disease in male kits on diets containing high levels of salmon waste with no residual effect over a three year period. However, the steatitis was not prevented by the addition of BHT directly to the salmon waste prior to freezing and storage. It was necessary that the anti–oxidant present in the fortified cereal be added to the final diet presented to the mink each day.

Moreover, Travis and Schaible (1961) conducted studies with BHT at 10X the levels used in these experimental studies with no detrimental effects on reproduction/lactation performance.

DPPD

Studies by Cabell and Leekley (1960) and Leekley and Cable (1961) indicated that this powerful anti–oxidant, at one time used in poultry diets, at a level of 112 grams/ton of wet feed was the most effective anti–oxidant for the prevention of "yellow fat" disease in the mink. In these experiments. the discontinuation of DPPD supplementation on November 20th immediately resulted in deaths from steatitis within a few days. Although there were no negative effects on reproduction in the first year of using of DPPD, there was apparent build–up of the anti–oxidant in the tissues, with very significant toxic effects on reproduction in the experimental mink in the subsequent two years.

Methylene Blue

Studies by Dalgaard et al. (1958) indicated that the addition of 50–100 mg of methylene blue/kg of mink fodder had the same protective effect as about 50 mg of vitamin E (alpha–tocopherol acetate) in decreasing mortality from steatitis and in reducing the level of peroxides in the fatty tissues.

Ethoxyquin

Santoquin (di–hydro–ethoxy–tri–methyl–quinoline) at 112 gms/ton of wet feed was very effective in the prevention of "yellow fat" disease in the mink (Leekley et al., 1962).

THBP

THBP (2,4,5 tri–hydroxy–butyro phenone) at 112 grams/ton of wet feed has also been shown to be very effective in preventing steatitis in mink (Leekley et. al., 1962).

Natural Anti–oxidants
Wheat Germ Meal

For many years wheat germ meal has been a component of simple fortified cereals for mink nutrition in the belief that it uniquely supported reproduction/lactation performance, with its natural vitamin E content. Not until 1951, with reports by Gorham (1951) and Gorham, Baker, and Boe (1951) was it known that wheat germ meal was valuable in preventing "yellow fat" disease. At levels of 5 and 10%, wheat germ meal providing 3 to 6 mg/mink/day of vitamin E was effective in preventing steatitis, the higher level of wheat germ meal supporting the greatest reduction in the level of yellow pigmentation in the adipose tissue.

Even prior to 1950 and the discovery of the role of vitamin E in the prevention of "yellow fat" disease in mink, Gorham (2001) had been told by an Oregon mink farmer that wheat germ meal was effective in its prevention. His observation supports a basic concept of mink nutrition research, "Listen To The Mink" and/or, as a secondary resource, the mink rancher before planning and initiating scientific studies on the nutritional requirements of this animal.

193

Vitamin E Toxicity

A recent research report by Tauson (1993) is of interest on this point. The experimental plan involved mink diets containing 95% fatty fish products and fish oil supplemented with vitamin E at levels of 27, 108, and 324 mg/kg DM. The reproduction/lactation performance was satisfactory at the lower levels of vitamin E supplementation, but inferior at the 324 mg level.

An earlier experimental study involving relatively high levels of vitamin E supplementation, Brandt (1984) and a field observation of Wilson (1983) with mink kits on a high vitamin E level commercial mink pellet, indicated toxic effects of vitamin E yielding deaths due to severe hemorrhaging. The Brandt observations involved mink dietary programs with 30 and 100 mg vitamin E/Mcal equivalent to about 120 and 410 mg/kg DM within a mink dietary program providing 4,080 kcal ME/kg DM. The Wilson observation involved a mink pellet program providing about 200 mg/kg DM.

With the present knowledge of animal physiology, it is easy to understand how high levels of vitamin E can bring about hemorrhaging, an obvious manifestation of a secondary deficiency of vitamin K. An animal dietary program with an ample level of vitamin K yields a vitamin K deficiency directly related to other factors present in the diet, in this case a vitamin E metabolic product antagonistic to vitamin K.

One of the metabolic by–products of vitamin E catabolism is alpha–tocopherol–paraquinone, a structural analog of vitamin K hydroquinone. The vitamin E paraquinone structure is a competitive inhibitor of a vitamin K dependent enzyme involved in the carboxylation of glutamyl residues of key protein structures required for blood coagulation (March et al., 1973). The mink ranch observed by Wilson resolved the vitamin E toxicity problem through adding vitamin K to the water initially at the rate of 10 mg/mink for five days and then every second day. The initial outbreak of kit losses with haemorrhage was explosive in September and recurred in October after the vitamin K therapy had ceased. High levels of vitamin E in the liver and adipose tissue of the mink had brought about a second series of kit losses related to vitamin E toxicity.

The Wilson observations are especially interesting in terms of two factors. First, reproduction/lactation performance of the mink on the commercial pellet with 200 mg vitamin E/kg DM was optimum. Second, even though mink in North America were providing excellent growth and fur development, mink kits in Scotland on the "identical" pellet formulation were hemmorrhaging. The key point is that the two commercial pellet formulations were actualy not "identical." Mink in Scotland had been provided a pellet formulation engineered in Denmark and employing Danish pork fat, but the "identical" pellet formulation in North America involved American pork fat.

Why should there be a significant difference between Danish pork fat (lard) and American pork fat (lard)? Danish pork fat is unique in its levels of polyunsaturated fatty acids (PUFA) content, a difference directly related to the swine dietary programs in the two geographical areas. In Denmark, swine diets have a high level of barley and other cereal grains and minimal levels of corn, while in North America the primary carbohydrate resources for swine nutrition are corn products provid-

194

ing relatively high levels of PUFA as linoleic acid. With the lower level of PUFA in the Danish swine diets, the potential for the fats to undergo oxidative rancidity was also significantly lower. There was, in turn, a lower potential for the vitamin E content to be destroyed by the free radical and peroxide structures formed in the process of oxidative rancidity.

Vitamin K

Vitamin K was named by the Danish scientist, Henrik Dam, who noted that baby chicks on vitamin K free diets had an impaired blood coagulation (*koagulation* in Danish). Thus with a vitamin K deficiency, an animal may bleed to death with a minor injury.

Physiology

Vitamin K is known to be an essential co–factor for a carboxylase enzyme that converts glutamyl residues of a number of precursor proteins to Gla (gamma carboxyl–glutamyl) residues required by a variety of coagulation proenzymes for proper function.

Inasmuch as vitamin K is a fat–soluble vitamin, it is absorbed along with fats into the body's lymphatic system and finally into the blood circulation. Hence, the absorption of vitamin K requires the presence of bile salts and a healthy functioning intestinal tract.

In the case of many animals, vitamin K is not a dietary requirement inasmuch as their intestinal flora synthesize ample quantities of vitamin K. However, in the case of the mink, synthesis of vitamin K by its intestinal flora is very limited, a circumstance directly related to the mink's relatively short gastro–intestinal tract, the absence of a caecum, and a relatively undifferentiated large intestine which is only one third of body length (Kainer, 1954a).

Requirements

Traditional mink ranch dietary programs, make it almost impossible to achieve a vitamin K free feeding regimen. Studies by Travis et al. (1961) employing a semi–purified diet determined that the vitamin K requirement of the mink was less than 13 mg of menadione sodium bisulfite (synthetic vitamin K) per ton, equivalent to less than 0.037 micrograms per 100 kcal of metabolic energy (ME). This level of synthetic vitamin K yielded no change in prothrombin time.

Very likely there may be significant synthesis of vitamin K by the mink's intestinal flora (Travis et al., 1961). With the addition of sulfaquinoxaline, a coccidiostat, clotting time significantly increased.

Resources

Synthetic vitamin K (menadione sodium bisulfite) is the most common resource of vitamin K in North America. One of the commercial products common in both North America and Europe is Heterozyne K. In Russia the commercial vitamin K source is called Vicasol.

195

Nutritional Deficiency
Primary Vitamin K Deficiency

Although a vitamin K deficiency in mink raised on practical ranch diets is rarely seen, an exception may be seen with pink and/or blue iris mink. Klevesahl (1970) noted significant losses of pink mink while blood sampling for the detection of Aleutian disease of the mink. The addition of 12 grams of Heterozyne K/ton of ranch mix resolved the problem. Zimbal (2001) reported the loss of 1% of his blue iris mink while taking blood samples for Aleutian disease. Vitamin K supplementation of the ranch diet solved the problem.

Secondary Vitamin K Deficiency

Mink on dietary programs containing high levels of vitamin E may exhibit symptoms of a vitamin K deficiency followed by hemorrhage and death. Brandt (1984) reported severe hemorrhages in mink fed vitamin E at levels of 30 and 100 mg/Mcal ME. Wilson (1983) noted multiple deaths of dark mink kits fed a commercial mink pellet containing 220 mg of vitamin E/kg even though the pellet contained 1.2 grams of synthetic vitamin K/ton. Very likely high levels of vitamin E in the dietary programs resulted in the formation of alpha–tocopheryl–para–quinone a known antagonist to vitamin K (March et al., 1973).

It is of interest to note that a secondary deficiency of vitamin K may also be found in the "yellow fat" disease of mink related to the employment of fatty feedstuffs containing high levels of polyunsaturated fatty acids (PUFA). These PUFA are subject to oxidative rancidity leading to the formation of peroxide and free radical structures which can destroy vitamin E via oxidation of vitamin E to para–quinone structures. It is possible that within an oxidative rancidity environment that vitamin K as well as vitamin E may be destroyed. Studies by Ender and Helgebostad (1953) noted that in addition to the regular observations of "yellow fat" disease of the mink there was a prolonged blood clotting time. The percent of prothrombin and reconvertin in the blood was low in advanced cases of the disease. Treatment of the mink with vitamin K resolved the problem of mink kit losses due to hemorrhage.

Toxicity

Synthetic vitamin K provided mink at a high dosage can be toxic to mink. Pereľdik (1972) reported that the employment of a commercial vitamin K product, Vicasol, at a level of 6 mg/kg body weight produced symptoms of dyspepsia, nausea, and intensified saliva production. The addition of 10 mg of Vicasol daily by pregnant mink caused intoxication within 7 days and kits incapable of survival.

Water Soluble Vitamins

Vitamin C—Ascorbic Acid

Vitamin C, also termed ascorbic acid, is a relatively simple compound synthesized from the common sugars glucose and galactose in most animal physiology. Exceptions include primates (humans and monkeys), guinea pigs, fish, and a number of exotic species. The inability of these species to synthesize ascorbic acid is due to a common defect, the absence of the microsomal enzyme L–gulonolactone oxidase.

Scurvy, a vitamin C deficiency in humans, has been known since the period of the Crusades. A deficiency of ascorbic acid impairs a number of physiological activities including the biosynthesis of collagen. Scurvy reduces the activity of dopamine–beta–hydroxylase and tyrosine hydroxylase (leading to melanin pigment synthesis), as well as limiting the hydroxylation of lysine and proline.

Requirement

No scientific data supports the concept that mink require vitamin C in their diets, as they have the physiological capacity to synthesize ascorbic acid (Bassett et al., 1948 and Petersen, 1957). Whereas vitamin C has significant anti–oxidant activity and is thus of value in mink physiology, in terms of practical mink nutrition, it must be noted that vitamin C is unstable in feed mixtures with a pH above 7.0. Studies at Cornell University by McCarthy (1964) employed a purified diet without vitamin C content but with a special supplementation of L–ascorbic acid at 99 mg/kg. This diet, however, did not enhance the growth of mink kits.

Thiamine—Vitamin B_1

Initial studies on the nutritional factors required by animals led to the discovery of a fat–soluble A and a water–soluble B. Further research led to the discovery of multiple members of the B–complex group. The initial discovery of the water–soluble B group was vitamin B_1 or thiamine. The name thiamine was chosen for its unique double heterocyclic structure containing a sulfur atom (thio) and a simple primary amine structure.

In fact the term vitamin in modern animal nutrition is a derivative of "vital amines," unspecified nutritional factors "vital" for the existence of life and containing a simple primary amine structure. Later studies, however, indicated that this definition did not apply to all vitamin structures.

Physiology

In animal physiology, thiamine is activated by adenosine triphosphate (ATP) to yield thiamine pyrophosphate (TPP). TPP is required both for the oxidative decarboxylation of alpha–keto carboxylic acids which include pyruvate, alpha–keto–glutarate, and keto–analogs of leucine, isoleucine, and valine and for the action of the transketolase enzyme of the pentose phosphate pathway. Inasmuch

197

as pyruvate is the end product of simple sugar catabolism, the higher the level of carbohydrate (resources for the simple sugars) the higher the requirement for thiamine in an animal's diet. Thiamine supplementation has been used as an appetite stimulator on experiments involving low palatability soybean oil meal (Watt, 1953). Thiamine is water soluble and hence excess intake is mainly excreted via the urine (Jorgensen et. al., 1975).

Requirement

The mink's requirement of thiamine depends on body size, the environmental temperature, and the composition of the diet inasmuch as the thiamine requirement of an animal increases when the carbohydrate content of the diet increases. In general, requirements of the mink are considered to be 2.5–3 mg/kg feed solids for the reproduction and lactation periods and 1.5–2 mg/kg in the growth period (Unter, 1974). This recommendation of 1.5–2 mg/kg in the experimental study is consistent with earlier studies of Leoschke and Elvehjem (1959) which indicated a mink kit growth requirement of 1.2 mg/kg purified diet equivalent to 33 micrograms per 100 kcal ME. Deposition of thiamine in intestinal organs and muscles, as well as urinary excretion, has been studied by Jorgensen et al. (1975); based on their work, it may be concluded that urinary excretion of thiamine is a relevant parameter by which to estimate the thiamine status in mink.

Resources

Cereal grains and brewers yeast, as well as commercial fortified cereals, are excellent sources of thiamine for mink nutrition.

Nutritional Deficiency
Primary Thiamine Deficiency

Studies by Leoschke and Elvehjem (1959) indicated that mink placed on thiamine–free diets exhibited a reduction of feed volume within a single week. Within three weeks, the mink had developed typical symptoms of a thiamine deficiency including anorexia, loss of weight, lack of muscle coordination, extreme weakness, and finally paralysis and death. In the gestation period, a thiamine deficiency leads to embryonic death (Zimmerman, 1981). The physiological basis for these symptoms is directly related to the fact that thiamine pyrophosphate (TPP) is required for the oxidative decarboxylation of pyruvate to yield acetyl Co–A and carbon dioxide. Thus, with a thiamine deficiency in the mink, pyruvic acid and its reduction product, lactic acid, accumulate in the blood and tissues, resulting in neurological symptoms of lack of appetite. Animals displaying these symptoms due to a thiamine deficiency may recover following a single subcutaneous injection of a thiamine hydrochloride solution, 100 mg/ ml, 1.0 cc and 0.5 cc for males and females, respectively (Rouvinen et al., 1997).

198

Secondary Thiamine Deficiency

A number of nutritional factors can bring about a secondary deficiency of thiamine on ranch diets initially containing ample levels of thiamine—high dietary blood levels, sodium bisulfite, and fish containing the thiaminase enzyme. Stomach hydrochloric acid can convert hemoglobin present in blood to hemin which in turn can inactivate thiamine (Helgebostad and Dishington, 1977). In a laboratory study, they noted that 23% of the thiamine content of mink feed with 30% blood was destroyed within two hours when stored at a temperature of 37° Celsius. A number of studies indicate that sodium bisulfite is a useful chemical for the preservation of mink feedstuffs and/or mink rations (Quist, 1964; Moller–Jensen and Jorgensen, 1975; Wehr et al., 1977). The Moller–Jensen and Jorgensen study demonstrated that almost all of the thiamine in a ranch diet was destroyed within 24 hours when sodium bisulfite is present. Fish–related secondary thiamine deficiency in mink directly related to a thiaminase enzyme in Columbia River smelt was first reported by Long and Shaw (1943). As early as 1932, a unique spastic paralysis in foxes had been noted on the Chastek Fox Farm, Glencoe, Minnesota, related to the feeding of fresh carp to foxes (Green, 1938). The thiaminase enzyme has been shown to be present in the major visceral tissues of the carp with the spleen, liver, pancreas, gastro–intestine, and gills containing the highest concentration. It is also relatively high in kidneys and blood, but absent in the somatic muscle (Sealock et al., 1943). Spitzer et al. (1941) were the first researchers to suggest that the vitamin B_1 destructive agent in raw fish might be enzymatic in nature. It is of interest to note that the thiaminase enzyme destroys both thiamine and the physiological active form of the vitamin, thiamine pyrophosphate (Deutsch and Ott, 1942).The thiaminase enzyme destroys thiamine by a hydrolysis reaction yielding free pyrimidine and thiazole heterocyclic structures (Ceh et al., 1942). Observations of the presence of the thiaminase enzyme in fish, mollusks, and crustaceans are provided in Table 4.48. Also of interest is Table 4.49.

TABLE 4.48. Occurrence of Thiaminase in Fish, Mollusks, and Crustaceans

Species	Habitat	Source	References
Alewife (*Pomolobos pseudoharengus*)	F*	Lake Michigan	Gnaedinger (1965)
Alewife (*Alosu pseudoharengus*)	F	Lake Michigan	Neilands (1947)
Anchovies, striped (*Anchoa hepsetus*)	S	Gulf of Mexico	Jones (1960)
Anchovies (*Engraulis mordax*)	S	Pacific	Stout et al. (1963)
Bass (white) (*Lepibema chrysops*)	F	Great Lakes	Deutsch & Hasler (1943)
Bass (white) (*Morone chrysops*)	F	Great Lakes	Deutsch & Hasler (1943)
Black quahog (*Artica islandica*)	S	Atlantic	Lee (1948) Greig & Gnaedinger (1971)
Bjorken (*Abramis blicca*)	F	Lake Malaren	Lieck & Agren (1944)
Bowfin (*dogfish*)(*Amia calva*)	F	Arkansas	Gnaedinger (1965)
Bream (*Abramis brama*)	F		Kuusi (1963)
Buckeye shiner (*Notropus atherionoides*)	S		Lee (1948) Deutsch & Hasler (1943)
Buffalofish (*Ictiobus prinellus*)	F	Arkansas	Borgstrom (1961)
Bullhead (*Ameirurus m. melas*)	F	Great Lakes	Deutsch & Hasler (1943)
Bullhead (*Ictalurus spp*)	F	Arkansas	Greig & Gnaedinger (1971)
Burbot (*Lota lota maculosa*)	F	Great Lakes	Deutsch & Hasler (1943) Gnaedinger (1965)
Burbot (*Lota lota*)	F	Lake Erie	Deutsch & Hasler (1943) Gnaedinger (1965) Greig & Gnaedinger (1971)
Butterfish (*Poronotus triacanthus*)	S	Gulf of Mexico	Lee et al. (1955)
Capelin (*Mallotus villous*)	F	Arctic	Ceh et al. (1964).
Carp (*Cyprinus carpio*)	F	Great Lakes	Deutsch & Hasler (1943) Gnaedinger (1965)
Catfish (channel) (*Ictalurus lacustris punctatus*)	F	Great Lakes	Deutsch & Hasler (1943)
Catfish (*Ictalurus nebulosus*)	F	Nova Scotia	Neilands (1947)
Chub, creek (*Semotilus a.atromaculatus*)	F	Great Lakes	Deutsch & Hasler (1943)
Clams (*chowder, steamer, cherrystone*)	F		Melnick et al. (1945)
Clara (*Mya arenaria*)	S	Atlantic	Neilands (1947)
Crucian (*Cyprinus casarassius*)	F	Sweden	Lieck & Agren (1944)
Fathead minnow (*Primephales p. promelas*)	F	Great Lakes	Deutsch & Hasler (1943)
Garfish (*garpike*)	S		Borgstrom (1961)
Garfish (*Belone acus*)	F	Sweden	Lieck & Agren (1944)
Goldfish (*Carassius auratus*)	F	Great Lakes	Deutsch and Hasler (1943) Gnaedinger (1965)
Herring (Baltic) (*Clupea harengus var.*)	S	Baltic	Kuusi (1963)

Herring (*Clupea harengus*)	S	Atlantic	Deutsch & Hasler (1943)
			Nielands (1947)
Ide (*Leuciscus idus*)	F	Lake Malaren	Lieck &Agren (1944)
Lamprey eel (adult) (*Petromyzon marinus*)	F	Great Lakes	Borgstrom (1961)
Lobster (*Homarus americanus*)	S	Atlantic	Nielands (1947)
Mackerel , Pacific (*Scomber japonicus*)	S	Pacific	Borgstrom (1961)
Menhaden (*Brevoortia tyrannus*)	S	Chesapeake Bay	Greig & Gnaedinger (1971)
Menhaden, large scale (*Brevoortia patronus*)	S	Gulf of Mexico	Jones (1960)
Menomonee whitefish (*Prosopium cylindracem quadrilaterale*)	F		Deutsch&Hasler (1943)
Moray eel (*Gymnothorax ocellatus*)	S	Gulf of Mexico	Lee et al. (1955)
Minnow, mud (*Umbra limi*)	F	Great Lakes	Deutsch & Hasler (1943)
Minnow (*Fundulus heteroclitus*)	F	Nova Scotia	Neilands (1947)
Minnow, fathead (*Pimephales p. promelas*)	F	Great Lakes	Deutsch & Hasler (1943)
Minnow (*Fundalus diaphanus*)	F	Nova Scotia	Neilands (1947)
Mussel (*Elliptio complanatus*)	F	Nova Scotia	Nielands (1947)
Mussel, bigtoe (*Pleurobema cordatum*)	F	Tennessee River	Gnaedinger (1965)
Mussel (*Mytilus edulis*)	S	Atlantic	Nielands (1947)
Quahogs (*Venos mercenaria*)	S		Soc Exp Bio 1942 60:268–269
Razor belly, scaled sardine (*Harengula pensacolae*)	S	Gulf of Mexico	Lee et al. (1955)
Rudd (*Leuciscus erythrphtalmus*)	F	Lake Malaren	Lieck & Argen (1944)
Sauger pike (*Stizostedion c. canadense*)	F	Great Lakes	Deutsch & Hasler(1943)
Scallup (*Placopecten grandis*)	S	Atlantic	Nielands (1947)
Sculpin (*Myoxcephalus quadricornis thompsonii*)	F	Lake Michigan	Gnaedinger (1965)
Shad, gizzard (*Dorosoma cepedianum*)	F	Lake Erie	Gnaedinger (1965)
Shiner, spottail (*Notropis hudsonius*)	F	Lake Michigan	Deutsch & Hasler(1943)
			Gnaedinger (1965)
Shiner, buckeye (*Notropus atherinoides*)	F	Great Lakes	Wolf (1942)
Smelt, freshwater (*Osmerus mordax*)	F	Great Lakes	Deutsch & Hasler (1943)
			Gnaedinger (1965)
	F	Nova Scotia	Neilands (1947)
Spratt (*Clupea sprattus*)	S	Atlantic	Deutsch & Hasler (1943)
Stoneroller, central (*Campostoma anomalum pullum*)	F	Lake Michigan	Gnaedinger (1965)
Sucker,common white (*Catostomus c. commersonii*)	F	Great Lakes	Deutsch & Hasler (1943)
			Gnaedinger (1965)
Sucker, longnose (*Catostomus colostomies*)	F	Great Lakes	Bell & Thompson (1951)
Sugar pike (*Stizostedion canadense*)	F	Great Lakes	Deutsch & Hasler (1943)

Trench *(Tinca vulgaris)*	F	Lake Dreuagen	Lieck & Agren (1944)
Vae *(Ahiamas vinabra)*	F	Lake Malaren	Lieck & Agren (1944)
White bass *(Lepimbema chrysops)*	F	Great Lakes	Deutsch & Hasler (1943)
Whitefish *(Prosopium cylindraceum quadriaterale)*	F	Great Lakes	Deutsch & Hasler (1943)
Whitefish *(Coregonus clupeaformis)*	F	Great Lakes	Deutsch & Hasler (1943)
			Bell & Thompson (1951)

* F, freshwater, S, salt water

TABLE 4.49. Fish, Mollusks, and Crusteans Habitat Without Thiaminse Enzyme

Species	Habitat	Source	References
Ayu *(Plecoglossus altivelis)*		F	Borgstrom (1961)
Bass, smallmouth *(Micropterus d. dolomieu)*		F	Great Lakes	Deutsch & Hasler (1943)
Bass, large mouth *(Huro salmoides)*		F	Great Lakes	Deutsch & Hasler (1943)
Bass, rock *(Ambloplites r. rupestris)*			Deutsch & Hasler (1943)
Black backs *(Pseudopleuronectes americanus)*		S	Atlantic	Neilands (1947)
Bluegill *(Lepomis m. macrochirus)*		F	Great Lakes	Deutsch & Hasler (1943)
				Gnaedinger (1965)
Chub, bloater *(Coregonus hoyi)*		F	Lake Michigan	Deutsch & Hasler (1943)
				Gnaedinger (1965)
Cod *(Gadus morrhua)*		S	Atlantic	Deutsch & Hasler (1943)
				Neilands (1947)
Crappie *(Pomoxis nigro–maculatus)*		F	Great Lakes	Deutsch & Hasler (1943)
Croaker *(Micropogon undulates)*		S	Gulf of Mexico	Lee et al. (1955)
				Gnaedinger (1965)
Cunner *(Tautogolabrus adsperus)*		S	Long Island Sound	Lee (1948)
				Soc Biol 1945 53:63
Cusk *(Bromse bromse)*		S	Atlantic	Neilands (1947)
Cutlassfish, silver eel *(Trichiurus lepturus)*		S	Gulf of Mexico	Lee et al. (1955)
				Gnaedinger (1965)
Dogfish *(Squalus acanthias)*		S	Atlantic	Neilands (1947)
Dogfish *(Amia calva)*				Deutsch & Hasler (1943)
Eel *(Anguilla rostrata)*		F		Neilands (1947)
Garpike, Northern longnose *(Lepisosteus osseus oxyurus)*		F		Deutsch & Hasler (1943)
Goosefish *(Lophins piscatorius)*		S		Nielands (1947)
Haddock *(Melanogrammus aeglefinus)*		S	Atlantic	Deutsch & Hasler (1943)
Halibut *(Hippoglossus hippoglossus)*		S	Atlantic	Neilands (1947)
Hake, Pacific *(Merluccius productis)*		S	Pacific	Stout et al. (1963)

Hake, silver (*Merluccius bilinearis*)	S	Atlantic	Rouvinen et al. (1997)
Hake (*Urophycis spp.*)	S	Pacific	Stout et al.(1963)
Hake (*Urophycis spp.*)	S	Gulf of Mexico	Lee et al. (1955)
Herring (*Leucichthys artedi areturus*)	F	Lake Superior	Deutsch & Hasler (1943)
Jackfish (*Esox lucius*)	F		Bell & Thompson (1951)
King whiting (ground mullet) (*Menticirrhus americanus*)	S	Gulf of Mexico	Gnaedinger (1965)
Lemon sole (*Pseudopleuronectes americanus dignabilis*)	S		Deutsch & Hasler (1943)
Ling (*Lota lota masculosa*)	F		Bell & Thompson (1951)
Lizard fish (*Synodus foetens*)	S	Gulf of Mexico	Lee et al. (1955)
Lobster (*Homarus americanus*)	S	Atlantic	Nielands (1947)
Lumpfish (*Cyclopterus lumpus*)	S	Atlantic	Greig and Gnaedinger (1971)
Mackerel (*Scomber scombrus*)	S	Atlantic	Deutsch & Hasler (1943)
Mullet (*Mugil spp.*)	S	Gulf of Mexico	Gnaedinger (1965)
Oyster (*Ostrea edulis*)	S	Atlantic	Greig and Gnaedinger (1971)
Perch, white (*Morone americana*)	F		Nielands (1947)
Perch, yellow (*Perca flavescens*)	F		Deutsch & Hasler (1943)
Periwinkle (*Littorina litorea*)	S	Atlantic	Greig & Gnaedinger (1971)
Pikerel, Northern (*Esox lucius*)	F	Great Lakes	Deutsch & Hasler (1943)
Pike, walleye (*Stizostedion citreum*)	F	Great Lakes	Deutsch & Hasler (1943)
Plaice, Canadian (*Hippoglossoides platessoides*)	S	Atlantic	Nielands (1947)
Pollock (*Pollachius virens*)	S	Atlantic	Nielands (1947)
Porgy, scup (*Stenotomus aculeatus*)	S	Gulf of Mexico	Lee et al. (1955)
Porgy, scup (*Stenotomus chrysops*)	S	Chesapeake Bay	Greig & Gnaedinger (1975)
Pumpkinseed (*Lepomis gibbosus*)	F	Great Lakes	Deutsch & Hasler (1943)
Redfish (*Sebastes marinus*)	S		Deutsch & Hasler (1943)
Sculpin (*Myxocephalus octodecemspinosus*)	S	Atlantic	Greig & Gnaedinger (1971)
Salmon, Atlantic (*Salmo salar*)	F	Great Lakes	Nielands (1947)
Salmon, coho (*Oncorhynchus kisutch*)	F	Lake Michigan	Borgstrom (1961)
Seabass (*Centropristis striatas*)	S	Chesapeake Bay	Borgstrom (1961)
Sea catfish (*Arius felis*)	S	Gulf of Mexico	Lee et al. (1955)
Sea catfish (*Galeichthys felis*)	S	Gulf of Mexico	Lee et al. (1955)
Sea raven (*Hemitripterus americanus*)	S	Atlantic	Grieg & Gnaedinger (1971)
Sea robin (*Prionotus spp.*)	S	Gulf of Mexico	Lee et al. (1955)
Sea robin (*Prionotus carolinus*)	S		Gnaedinger (1965)
			Soc Exp. Biol (1945) 60:268
Sheepshead, freshwater drum (*Aplodinotus grunniens*)	F	Lake Erie	Grieg & Gnaedinger (1971)
Shrimp, brine (*Artemia salina*)	S	Lab grown	Grieg & Gnaedinger (1971b)

Skate (*Raja senta*)	S	Atlantic	Grieg&Gnaedinger (1971)
Smelt, pond (*Hypomesus olidus*)	F		Borgstrom (1961)
Spot (*Leiostomus xanthurus*)	S	Gulf of Mexico	Lee et al. (1955)
			Gnaedinger (1965)
Squid (*Loligo brevis*)	S	Gulf of Mexico	Lee et al. (1955)
Starfish (*Asterias vulgaris*)	S	Atlantic	Grieg & Gnaedinger (1971)
Tautog, blackfish (*Tautoga onitis*)	S	Long Island Sound	Lee (1948)
Trout, brown (*Salmo trutta fario*)	F	Great Lakes	Deutsch & Hasler (1943)
Trout, lake (*Christivomer n. namaycush*)	F	Great Lakes	Deutsch & Hasler (1943)
Trout, rainbow (*Salmo gairdnerii irideus*)	F	Great Lakes	Deutsch & Hasler (1943)
Tullibee (*Leucichthys tullibee*)	F		Bell & Thompson (1951)
White trout (*Cynoscion nothus*)	S	Gulf of Mexico	Gnaedinger (1965)
White trout (*Cynoscion avenarius*)	S	Gulf of Mexico	Lee et al. (1955)
Whiting (*Merluccius bilinearis*)	S	Atlantic	Deutsch & Hasler (1943)
Witch flounder (*Glyptocephalus cynogossus*)	S	Atlantic	Greig & Gnaedinger (1971)
Yellow tails (*Lamanda ferruginea*)	S	Atlantic	Deutsch & Hasler (1943)

Using non–thiaminase fish in a mink's diet can still lead to a thiamine deficiency if the fish had consumed significant quantities of thiaminase fish. For instance, as Pacific hake consume anchovies (Stout et al., 1963).and chubs may eat small alewives and freshwater smelt (Gnaedinger, 1963).

Practical steps should be taken when using thiaminase fish in mink diets. Ranchers may cook the thiaminase fish for 15 minutes at 180–200° F. (Gnaedinger, 1963) or at 90° C. for 10 minutes (Alden and Tauson, 1981). Another approach is to prepare perch, and bream thiaminase fish as a silage with 4% sulfuric acid and 0.5 – 1.0% formic acid to yield a pH of 3.4. This process results in a product with low thiaminase activity (Alden and Tauson, 1977; Makela, 1979). Helgebostad (1968) found that storage temperatures near freezing results in a significant reduction in thiaminase activity. A good approach to the thiaminase fish problem with minimum labor is that of feeding of thiaminase and non–thiaminase fish on an every other day program. The effectiveness of alternate day feeding programs or one day out of three with the thiaminase fish is dependent upon the level of employment of the thiaminase fish and on the specific period of the ranch year. An alternative day feeding program that may be functional from July to pelting may not be useful to meet the high thiamine requirement of kits in the month of June. Another nutrition management program is that of extra supplementation of the ranch diet with thamine. Rouvinen (1997) has recommended adding 25 mg thiamine HCl/kg of diet, that is, 50 times the mink's normal thiamine requirement. Another approach related to special vitamin B_1 supplementation is that of Mullen (1992) recommending 0.6# of thiamine HCl/ton of ocean herring.

Reproduction/Lactation: Studies by Kirk (1962a and 1962b) indicated that thiaminase fish twice a week in combination with non–thiaminase fish for 5 days a week from January to mid–June yielded equal reproduction/lactation performance. However, when the program of thiaminase and non–thiaminase fish was intensified to an alternate day feeding program, the net result was absolute reproduction/lactation failure.

Late Growth/Fur Development: As early as 1942, Green et al. provided experimental data that supported an alternative day feeding program. These observations were supported, in turn, by later studies (Long and Shaw, 1943; Ender and Helgebostad, 1945; Gorham and Nielsen, 1953; Kirk, 1962a; and Leoschke, 1955).

Toxicity

Perel'dik et al. (1972) has reported that thiamine dosages exceeding the requirement of thiamine do not produce a toxic effect.

Riboflavin—Vitamin B₂

Riboflavin is also termed vitamin B₂ inasmuch as it was the second water–soluble vitamin to be discovered and identified. The structure of riboflavin consists of an isoalloxazine ring attached to a ribityl side chain. The ribityl structure is similar to the simple pentose sugar ribose. The name riboflavin is derived from its ribityl side chain and the fact that it is key component of flavin mononucleotide (FMN) and flavin adenine dinucleotide (FAD) co–enzymes involved in the metabolism of amino acids, fatty acids, and the energy–yielding citric acid cycle.

Physiology

In animal physiology, riboflavin is activated by ATP to yield FMN. In a second biosynthetic step, FMN is combined with a second molecule of ATP to form FAD. Within the citric acid cycle, FAD is involved in the conversion of succinct to trans– fumarate and in fatty acid, and amino acid catabolism; FMN is involved in the dehydrogenation of alkane structures to yield trans–alkene functional groupings. Inasmuch as FMN is involved in multiple steps of fatty acid catabolism, it is understandable that the riboflavin requirement of an animal increases with the employment of higher levels of fat in dietary programs. Riboflavin is water soluble and hence excess intake is mainly excreted via the urine (Jorgensen et al., 1975).

Requirements

Employing purified diets, Leoschke (1960) determined that the riboflavin requirement of mink kits was met at a level of 1.5 mg/kg or 40 micrograms per 100 kilocalories ME. Studies by Rimeslatten (1968) indicated that with a purified feed mixture, a reduction in riboflavin content from 2.5 to 1.5 mg/kg did not yield any change in performance of the mink. The deposition of riboflavin in intestinal organs and muscles has been studied by Jorgensen et al. (1975) who found that the riboflavin content was not influenced by riboflavin contents in the diet varying from 4.5 to 26 mg/kg. Based on these studies it may be concluded that the urinary excretion of riboflavin is a relevant parameter for the estimation of the riboflavin status in mink. Earlier studies by Eichel (1965) support this concept. It is generally acknowledged that an animal's requirement for a specific vitamin during pregnancy and lactation is higher than that for the growth period. Perel'dik (1972) has recommended not less than 0.25 mg of riboflavin per 100 kilocalories of ME.

Resources

Brewers yeast, liver, and kidneys are excellent sources of riboflavin while cereal grains are good sources. Quality commercial fortified cereals provide ample levels of riboflavin.

Nutritional Deficiency

Mink kits fed purified diets unsupplemented with riboflavin exhibited the effects of a riboflavin deficiency as early as two weeks after being placed on the experimental diet (Leoschke, 1960). Signs of riboflavin deficiency included loss of appetite, weight loss, and extreme weakness. Akimova (1969) stated that poor reproduction/lactation performance of the mink was noted in animals fed diets deficient in riboflavin during growth and fur development, even though adequate quantities of riboflavin were provided from mid–December to late June. Studies by Helgebostad (1980) noted embryonic death via a riboflavin deficiency induced by feeding 10–20 mg galactoflavin (an anti–riboflavin chemical) per animal daily during pregnancy. Feeding 10–20 mg galactoflavin in combination with 50–150 mg riboflavin provided normal reproduction/lactation. However, it is of interest to note that males fed 30 mg galactoflavin/animal/day from mid–December to mid–March had normal fertility.

Perel'dik (1972) has reported inborn deformities including cleft palate, shortening of the bones of the extremities, and abnormal skeletal development in some litters of mink placed on riboflavin deficient diets. In the case of dark kits, he noted a lack of fur color in the new born kits.

Toxicity

Perel'dik (1972) has noted that riboflavin is not very toxic to mink in doses many times exceeding the recommendations.

Niacin

Niacin as a specific chemical was known long before its animal nutritional role was discovered by Elvehjem at the University of Wisconsin in 1935. It was first isolated as an oxidation product of the

206

natural alkaloid, nicotine, from which its name is derived, niacin being preferable to nicotinic acid in terms of public health concern about the negative value of the nicotine content of tobacco.

Physiology

Niacin is a component of two enzymes of metabolism, namely NAD (Nicotinamide Adenine Dinucleotide) and NADP (Nicotinamide Adenine Dinucleotide Phosphate). NAD is the co–enzyme for a number of dehydrogenations involved in the catabolism of fatty acids, simple sugars, and amino acids. NADP also participates in dehydrogenation reactions particularly in the hexose monophosphate shunt of glucose catabolism. Reduced NADP, or NADPH, has an important role in the synthesis of fatty acids and steroids.

Niacin is unique among all the vitamins inasmuch as most animals, but neither the mink (Warner et al., 1968) nor the cat (De Silva et al., 1952), can convert the essential amino acid, tryptophan to niacin as NMN (Nicotinate Mononucleotide) and in another step to NAD. It should be noted that studies by Bowman et al. (1968) have indicated a slight increase in urinary N–1–methyl–nicotinamide by mink following the ingestion of 2.5 mmoles of L–tryptophan/kg body weight. However, this level of conversion of L–tryptophan to niacin by the mink is very limited.

Requirement

Studies by Warner et al. (1968) involving purified diets indicated that the requirement of the mink for niacin is a minimum of 0.5 mg per 100 kilocaloriesof ME or 10 mg/kg of diet. Mink milk is unusually high in niacin content: Jorgensen (1960) found 16 mg of niacin in 100 grams of mink milk, about 20 times the level found in cow's milk. Kon and Cowie (1961) noted that the level was twice that found in the milk of the sow. It is unlikely that supplementation of the typical ranch diet with niacin is required. Studies by Rimeslatten (1966a) and Utne (1974) indicated that common mink ranch diets contain 50 to 75 mg/kg or 1.25 to 1.87 mg of niacin per 100 kilocalories ME.

Resources

Brewers yeast, baking yeast, and liver are excellent sources of niacin while cereal grains are good sources. Quality commercial fortified cereals provide ample levels of niacin.

Nutritional Deficiency

In studies with mink kits on purified diets w/o niacin, Warner et al. (1968) noted that more than half of the animals died within 6 days after being placed on the experimental diet. Symptoms of a niacin deficiency in the mink are non–specific and include anorexia, loss of weight, a bloody stool, coma, and death. It is of interest to note that unlike a niacin deficiency in the dog with the observation of "black tongue," there was no discoloration of the buccal mucosa.

207

Pyridoxine, Pyridoxamine, Pyridoxal—B$_6$

Vitamin B$_6$ exists in three interconvertible forms: pyridoxine, pyridoxal, and pyridoxamine. Of these, pyridoxine is the most commonly used as a supplement in animal feeding programs. Although vitamin B$_6$ is relatively heat stable, various thermal reactions reduce the content of biologically available B$_6$ in processed foods. Thus, considerable amounts of B$_6$, especially pyridoxal and pyridoxal phosphate with free aldehyde functional groupings may react with free amino functional groups of amino acids and proteins during heat treatment to yield less biologically active pyridoxylamino structures (Gregory and Kirk, 1981).

Nutrition researchers, veterinarians working with mink, and mink ranchers may be interested to know that the modern genetic deficit disease of tyrosinemia involves a B$_6$ co–enzyme. Enzyme kinetics with livers from mink with tyrosinemia type II initiated by Christiansen et al. (1989) indicated that an insufficient binding of the co–factor pyridoxal phosphate (PLP) to hepatic tyrosine transferase is the physiological factor in the disease. A nutritional cure of the disease was achieved with high levels of PLP.

Physiology

The physiologically active form of B$_6$ is the phosphorylated form brought about by ATP and pyridoxal kinase.

Requirement

In studies with mink kits, Bowman et al. (1968) found that the lowest level of vitamin B$_6$ which supported growth and normal trytophan metabolism and prevented deficiency symptoms was 1.6 mg/ kg purified diet or 40 micrograms per 100 kilocalories of ME. As noted in multiple animal studies, higher levels of vitamin B$_6$ may be required for the reproduction/ lactation phase of the mink ranch year. A number of studies support a level of 9 mg/kg diet or 225 micrograms per 100 kilocalories ME for pregnancy and lactation (Rimeslatten and Aam, 1962; Jorgensen, 1965; Baalsrud and Kveseth, 1965)

Diets with high levels of protein or specific amino acids such as methionine or tryptophan increase the need for vitamin B$_6$.

Studies by Jorgensen et al. (1975), indicate that the pyridoxine content in the blood is a good parameter for estimating the B$_6$ status in mink. A study by Eichel (1965) on the relationship between B$_6$ intake of the mink and urinary excretion is of interest; however, the experimental data overestimated the B$_6$ requirement of the mink by a factor of 3 to 4.

Resources

Brewers yeast and baking yeast are excellent sources of vitamin B$_6$. Quality commercial fortified cereals provide ample levels of vitamin B$_6$.

Nutritional Deficiency

Studies by Bowman et al. (1968) with mink kits on purified diets w/o vitamin B_6 observed deficiency symptoms of reduced feed intake, reduced weight gains, ataxia, acrodynia, muscle incoordination, convulsions, irritability, and apathy terminating in death. Additional symptoms included diarrhea, a brown exudate around the nose, excessive lacrimation, difficulty in opening the eyes, and swelling and puffiness around the nose and facial region. Similar obnservations were noted by Helgebostad (1961) who also noted that the mink fur may lack normal pigmentation as well as show dusty fur with scales and singe.

Studies by Akimova cited in *Feeding Fur Bearing Animals* (Perel'dik, 1972) indicated that a vitamin B_6 deficiency in the mink during growth and fur production had a negative effect on the reproduction/lactation performance of the mink in the following spring. A deficiency of vitamin B_6 during the reproductive cycle reduced the number of females conceiving and lowered the number of kits per litter according to both Rimeslatten and Aam (1962) and Helgebostad (1963). There is a high mortality of the kits, a skewed sex ratio (low male:female ratio), and low kit birth weights (Rimeslatten and Aam, 1962 and Rimeslatten, 1964).

Studies by Helgebostad et al. (1963) and Helgebostad (1963) employing the B_6 anti–vitamin (desoxypyridoxine) resulted in B_6 deficiency with resorption of embryos in females and sterility in males likely related to degeneration of testes. Normal function of males and females in terms of reproduction was achieved when the desoxypyridoxine experimental diet was supplemented with high levels of vitamin B_6.

Toxicity

Inasmuch as vitamin B_6 is water soluble with excessive levels being excreted in the urine, the possibility of vitamin B_6 toxicity is remote.

Pantothenic Acid

Pantothenic acid is essential for all living organisms and thus is widely distributed in nature, which is the basis of its name from the Greek *panthos* "universal." In terms of initial observations of experimental studies with laboratory rats placed on purified diets without pantothenic acid, the vitamin for many years has been referred to as the "grey hair" vitamin.

Physiology

Pantothenic acid is a key component of co–enzyme A (Co–A) which functions as an acetyl group transfer co–factor for many enzymatic reactions in the metabolism of animals and plants. Co–A is involved in the acetylating of amines, the oxidative decarboxylation of pyruvate (the end point of the glycolysis pathway of the catabolism of simple sugars), the synthesis of citrate from acetyl Co–A

209

and oxaloacetate, the oxidative decarboxylation of alpha–keto–glutarate later in the citric acid cycle, and with the beta–oxidation scheme of fatty acid catabolism.

Requirement

The minimal pantothenic acid requirement for kit growth has been found to be 5 mg/kg of diet equivalent to 0.20 mg per 100 kcal ME, whereas 8 mg/kg of diet was required for optimal performance in terms of maximal feed efficiency and normal serum cholesterol values (McCarthy et al., l966). Multiple studies by Skrivan et al. (1979a, 1979b) indicated that 5 to 6 mg/kg body weight daily is required for optimal reproduction/lactation of the mink.

Resources

Yeasts, liver, and skimmed milk powder are excellent sources of pantothenic acid for mink nutrition while cereal grains and by–products are good resources for this vitamin. Quality commercial fortified cereals provide ample levels of pantothenic acid.

Nutritional Deficiency

Studies by McCarthy et al. (1966) indicated that signs of a pantothenic acid deficiency in mink included reduced feed intake with concomitant weight loss, anorexia, and reduced serum cholesterol. The gross pathological findings included diarrhea, cachexia, emaciation, and dehyration with hemorrhagic gastric ulcers and petechial hemorrhages observed in the intestinal tract. Vomiting followed food or water intake during later stages of pantothenic hypovitaminosis, as well as loose mucoid like feces yielding melena. A report by Glem–Hansen (1977) indicated that mink fur color turned to grey on a pantothenic acid deficient diet with, in the final stages, a loss of fur.

Toxicity

Pereľdik (1972) has reported that pantothenic acid at high levels is non–toxic to the mink, likely related to water solubility and high level of urinary excretion.

Biotin

As is the case with most animals, biotin is not required as a dietary component inasmuch as intestinal flora synthesis of biotin is sufficient to meet the animal's biotin requirements. However, with mink on experimental purified diets without biotin, a biotin deficiency does result, apparently due to sub–optimum intestinal flora synthesis (Travis et al., 1968). Very likely this limited production by the intestinal bacteria is related to the mink's relatively short gastro–intestinal tract and extremely rapid rate of feed passage from mouth to anus.

Physiology

Inasmuch as ferrets, a species closely related to the mink, lack an active transport system for biotin (Spencer and Brody, 1964), it is likely that the mink may also rely on passive diffusion for biotin absorption.

In terms of animal metabolism, biotin is a co–enzyme for a number of acyl Co–A carboxylases and the alpha–ketocarboxylase involved in the conversion of pyruvate to oxaloacetate (initiating the citric acid cycle), and the decarboxylase involved in the conversion of oxaloacetate to pyruvate. Biotin requiring pyruvate carboxylase is also involved in a regulatory pathyway in gluconeogenesis. Biotin requiring acetyl Co–A carboxylase catalyses an essential step in fatty acid biosynthesis. Biotin is also a co–enzyme for a number of enzymes involved in the catabolism of amino acids.

Requirement

Schimelman et al. (1969) estimated that the biotin requirement of the mink for growth was less than 0.125 mg/kg of purified diet equivalent to 0.003 mg 100 kilocalories ME. This level of biotin was the lowest employed in the experimental studies and is still considered to be relatively excessive (Glem–Hansen, 1977).

Conventional dietary programs for the mink apparently provide ample levels of biotin, inasmuch as a biotin deficiency has never been observed in the worldwide mink industry unless raw eggs containing biotin–binding avidin were present (Tauson and Neil, 1991). However, at the same time, ranchers must be concerned about the biotin nutrition of their mink when employing high levels of fish and/or fatty poultry by–products which have been in frozen storage for a long period of time. During this period the polyunsaturated fatty acids (PUFA) present may undergo oxidative rancidity and thereby produce a chemical environment for the destruction of biotin, vitamin E, and other vitamins (Schimelman et al., 1969; Helgebostad and Enders, 1958; Perel'dik, 1972).

Resources

Excellent sources of biotin include brewers and torula yeast, as well as beef and pork liver. Good resources include cereal grains and their by–products. Quality commercial fortified cereals provide ample levels of biotin for normal mink ranch diets without the employment of raw poultry eggs.

Nutritional Deficiency
Primary Biotin Deficiency

All factors considered, it is practically impossible to experience a ranch diet capable of yielding a biotin deficiency in the mink. Thus, researchers must employ a purified diet without biotin to allow an examination of biotin deficiency signs in mink. Classical biotin deficiency in the mink was indicated by spectacle eyes caused by a bloody exudate, crusty feet from a yellow exudate, and a dermatitis of the foot pads. The underfur was grayish–white, similar to the cotton fur of mink with secondary iron defiency (Travis et al., 1968).

Secondary Biotin Deficiency

Mink ranch diets with ample levels of biotin can yield a secondary biotin deficiency if raw eggs or raw egg whites are included in the dietary regimen. Both the egg yolk and egg whites as well as oviduct tissues contain a glycoprotein termed avidin which binds biotin in a structure that is unavailable to the digestive processes of animals (Fraps et al., 1943). Avidin is an antimicrobial protein useful for minimizing bacterial destruction within eggs during the incubation period. Even though the amount of avidin in egg yolk is more than offset by the amount of biotin present, there is still a considerable excess amount of avidin in egg white. Turkey eggs contain levels of biotin similar to chicken eggs but more than twice as much avidin (Stout and Adair, 1970; Wher et al., 1980). More to the point, fish eggs from the species employed in mink ranch diets do not contain avidin (Adair et al., 1974).

The first study on experimental biotin deficiency obtained by using 30% of the protein in raw egg white was reported by Helgebostad et al.(1952). Similar studies with raw egg white were reported by Ender & Helgebostad (1958, 1959). Later studies of secondary biotin deficiency in the mink involved spray–dried egg powder (Wehr et al., 1980; Aulerich et al., 1981) and turkey by-products (feet, heads, and entrails) from old breeder hen turkeys which contained significant levels of egg sacs and under developed eggs (Stout et al., 1966; Stout and Adair, 1969).

Although the fur quality of mink on biotin deficient diets was affected dramatically, growth was not affected indicating that mink fur development, including underfur pigmentation, has a relatively low priority in the life–sustaining functions involved in body growth. This nutritional physiology point is supported by observations of mink on Ross–Wells extruded pellets. Performance on these pellets was excellent in terms of reproduction/lactation, but numerous litters on a number of farms were "hairless" at birth and even a few days after whelping (Kilgore, 1974). When Dr. Gaylord Hartsought, (1974), one of the top fur animal veterinarians in the world, saw these litters of hairless kits he commented, "I guess that we will have to cook our eggs a little more." Apparently the free biotin content of the Ross–Wells extruded pellet was ample for reproduction/lactation of the mink but sub–optimum for fur development.

Studies on biotin needs of the mink that involved poultry egg resources extended over a longer period of time than did studies involving purified diets without biotin, making possible additional observations relative to fur characteristics. Studies at Oregon State University indicated "cotton–like" underfur developed in mink on both secondary iron deficiency (Costley, 1970) and secondary biotin deficiency (Stout et al., 1966; Stout and Adair, 1969). However, there were distinct differences in the characteristics of the underfur in the two nutritional deficiency syndromes. In the case of the iron deficiency anemia, the underfur was totally white while with biotin deficiency, the underfur was grey or grey–banded. Bands of fur result when two periods of high nutrient requirements by the mink overlap in time. These periods probably lie between mid–August and early October when rapid growth coincides with rapid fur development, creating a stress in termsof nu-

212

tritional physiology. At this time any marginal deficiency of a nutrient manifests itself, resulting in a lapse in the body's color formation processes and showing up as a white band in the fur. Such is the case with the turkey waste graying syndrome. Interestingly, however, when the use of turkey by-products created a secondary biotin deficiency, sapphire mink were not affected whereas dark mink grayed completely. This effect ocurred because pigment concentration in the fur of sapphires is low as compared to dark mink. Male kits were much more affected and more extremely affected by the secondary biotin deficiency, an effect directly related to their greater demand for biotin for growth and fur development.

Also observerved in mink kits on biotin deficient diets was major chewing of the fur on each other's bodies and significant tail chewing (Helgebostad et al., 1952; Stout and Adair, 1969; Wehr et al., 1980; Aulerich et al. 1981). This syndrome can be seen within two weeks after placing 6–7 week old kits on experimental diets containing 20% protein content as powdered whole eggs (Michels, 1998). An English mink rancher, Cobbledick (1997), noted that with their "wilds" mink, extra biotin supplementation of the ranch diet resulted in calmer mink and less fur chewing. Reproduction/lactation performance of mink on biotin deficient ranch diets containing 5–10% spray-dried eggs was a complete failure (Aulerich et al.,1981).

It should be noted that raw eggs and powdered eggs can be employed in practical mink diets provided one of the following two programs is applied: an alternate day feeding program (Pereľdik et al., 1972), or extra biotin supplementation at the level of 250 mg of d–biotin/100# of powdered eggs. Wehr et al. (1980) found that this level was ample for providing top performance of the mink.

Choline

Choline was first isolated from hog bile and thus the origin of its name from the Greek word for bile, *chole*. The term lipotropic applied to choline refers to its action as a dietary substance that decreases the rate of deposition of abnormal amounts of lipid in the liver and that accelerates the removal of excess fat from it. The term neurotransmitter is applied to the acetyl derivative of choline to designate its action as the chemical mediator of synapses of the autonomic nervous system. Fibers utilizing acetylcholine as a mediator are referred to as cholinergic. In pigs, rats, and mice, choline is required for the prevention of fatty liver and kidney hemorrhage and in poultry for the prevention of perosis.

Choline is not a vitamin but a nutrient that is more or less essential in specific animals depending upon the degree of biosynthesis from the amino acids methionine and betaine (methyl group donors) and serine (ethanolamine resource via decarboxylation).

Physiology

Lecithin (phosphatidyl choline) amd choline containing sphingomyelin are key components of biological membranes as well as lipoproteins. The neurotransmitter acetylcholine is synthesized from choline and acetyl Co–A via the enzyme termed choline acetylase. The breakdown (hydrolysis) of acetylcholine is via acetylcholine esterase. Choline is activated via cystosine triphosphate (CTP) to yield CDP–choline, the precursor of lecithin and sphingomyelin. Choline is part of a labile methyl pool capable of contributing methyl groups for the synthesis of the amino acid methionine and other methylated compounds including purine and pyrimidine bases necessary for proper growth and cell function. Choline, in terms of its role as a methyl group donor, is important in certain detoxification processes of the body; one illustration is the curative effect of dietary choline supplements for mink suffering with fatty livers after being fed cottonseed oil cakes infected with the toxic fungus *Aspergillus niger* (Perel' dik et al., 1972).

Requirement

According to Perel' dik et al. (1972), mink and fox are not capable of meeting their choline requirements by biosynthesis alone. Since the choline requirement of mink and fox has not been established, Perel' dik et al. (1972) based their recommendation on established dog nutritional requirements—20–40 mg of choline/ kg of live weight for prophylactic dosage and 50–70 mg/ kg or 1% of feed solids as a medicinal dosage. The Russian recommendation of 40 mg of choline/kg body weight was confirmed by Juokslahti et al. (1978) who found that the 40 mg level prevented fatty liver and enhanced hepatic function. For the breeding season, the Juokslahti group recommended the standard mink ration + 60 mg/mink/day.

Inasmuch as the amino acid methionine is a potential source of methyl groups, choline can spare the methionine requirement of the mink and vice versa, that is, higher levels of methionine in the ranch diet can reduce the mink's requirement for choline. Thus, the quantity and quality of protein in an animal's diet can affect the choline requirement, and, conversely, a choline deficiency is usually coincident with some degree of protein deficiency.

Inasmuch as vitamin B_{12} is involved in the biosynthesis of the amino acid methionine, the level of this vitamin would have a significant affect on the choline requirement of an animal.

Resources

The richest resource for choline is egg yolk. Other excellent sources of choline in practical mink nutrition include animal liver, brain, and yeasts. Meat, fish, and cereal grains are satisfactory sources of choline. Quality commercial fortified cereals provide ample levels of choline for modern mink nutrition.

214

Nutritional Deficiency

Without question, the classic sign of choline deficiency in animals is fatty livers or hepatic fat infiltration. However, it very important for mink veterinarians and mink ranchers to acknowledge the basic fact that the existence of a fatty liver in a mink autopsy is, in many cases, not a sign of a choline deficiency. It has been well known for years that any mink "off feed" for a period of time will exhibit a "fatty liver." Studies on the vitamin B_{12} requirement of the mink (Leoschke et al, 1953) indicated excessive fat infiltration of the liver in all B_{12} deficient mink. Such an effect is not directly related to a choline deficiency, but is related to the fact that mink in a state of vitamin B_{12} deficiency itself typically show anorexia, refusal to eat for a period as long as two weeks before dying of malnutrition.

Studies on the value of choline for the treatment of fatty liver in mink have been conducted by Juokslahti et al. (1978). Their work indicated that on specific dietary programs choline supplementation did enhance liver function, as evidenced by reduced aspartate transferase and lactate dehydrogenase levels in the blood, markers of a diminished damage to liver cells. In addition, blood bilirubin values were reduced, an indication of improved liver function.

Normally only 2–4% of total liver weight is fat; however, with fatty livers, the fat content may exceed 50%. Studies at Michigan State University by Aulerich et al. (1984) indicated that on an experimental diet specifically designed for minimum percentage of metabolic energy (ME) as protein (i.e., 50% fat and 36% protein), mink nutrition stress did enhance hepatic fat infiltration. However, these same studies also indicated that adding lecithin (phosphatidyl choline) or adding choline at 0.3% level did not decrease fatty infiltration of the livers but actually increased liver weights and elevated fat content. Of interest are more recent experimental studies by Clausen and Sandbol (2005) showing that fatty infiltration of the liver in mink could be achieved by feeding a low protein diet and fasting for 48 hours.

Toxicity

Choline is highly soluble in water and thus excessive levels are excreted by the body via kidney function.

Folic Acid

Folic acid and vitamin B_{12}, as well as the trace minerals iron and copper, are critical micronutrients required for the prevention of anemia. A deficiency of folic acid in the mink yields a characteristic macrocytic, hypochromic anemia called megablastosis (Whitehair et al, 1949).

Physiology

Folic acid is a key vitamin for metabolic function as a carrier of one–carbon units as derivatives of formate, formaldehyde, or methanol. These one–carbon units are generated primarily during amino acid metabolism and are used in the interconversions of amino acids and in the biosynthesis of the purine and pyrimidine components of nucleic acids required for cell division.

Requirement

Although folic acid is produced by intestinal flora, the level of microbial folate is not sufficient to meet the requirements of the mink (Aulerich et. al., 1995). Considering the observations of Schaefer et al. (1946), a level of 0.5 mg/kg of purified diet or 0.135 mg per 100 kcal of ME is more than adequate for growth and health of mink kits. The Nordic Handbook (Ahman, 1966) recommends 0.8 mg/kg feed solids. Jorgensen (1985) recommends a level of 0.6 to 0.9/kg of ready–mix solids. All factors considered, it is apparent that the folic acid requirements of the mink for top reproduction performance is significantly higher than that required for top kit growth (Aulerich et al., 1997). In these studies, a folic acid level of 1.2 mg/kg feed solids provided sub–optimum nutrition for reproduction, while levels of 5 and 12 mg/kg feed solids yielded superior litter size, related in part to reduced kit mortality at birth. No significant differences in body weight or hematocrit of the adult females was seen at the termination of the experiment. In subsequent experimental feeding trials (Aulerich and Bursian, 1997) enhancing the dietary level of folic acid from 5 mg to 15 mg/kg dry matter did not provide for increased reproduction performance of dark or pastel mink. Nor did the higher levels of folic acid improve the lactation performance, as evidenced by kit weights and survival at three and six weeks of age.

Field reports from American mink ranchers following these Michigan State University studies indicated that supplemental folic acid with practical diets produced larger litters, larger kits, and reduced kit mortality at birth; in addition, it lessened deformity in kits (Disse and Disse, 1995).

Another facet of the folic requirements of the mink deserves special attention: the additional supplement of folic acid required in ranch diets that use formic acid as a feed preservative in silage programs. The amount may go as high as 10 mg/ kg dry matter (Polonen et al., 1997; Polonen et al., 1999). These studies showed that mink on unsupplemented feed oxidized formate into carbon dioxide at a rate 37% less than those on supplemented diets. The oxidation rate increased linearly with the level of supplementation to a maximum at 10 mg/kg dry matter.

These observations also indicated that mink oxidize formate readily, but at a slightly slower rate than rats do. However, if extra folic acid is not added to the feed during the period of intense growth (4 weeks forward), hepatic H–4–folate levels may decline to levels susceptible to formate accumulation. Inasmuch as folic acid is involved in one–carbon metabolism, it is obvious that extra folic acid levels in the mink's diet would be required as the concentration of formic acid was increased in the animal's diet.

It is of interest to note that folic acid conjugates, including the heptaglutamate of folic acid and pteroyl–tri–glutamic acid, were not effective against folic acid deficiency in the fox, an observation which may or may not be applicable to the mink (Tove et al., 1949).

The folic acid nutrition of the mink is critical, as mink can store only a minimal amount of this least stable vitamin. In addition, the metabolic activity of folic acid is dependent upon the presence of vitamin B_{12}.

216

Resources

The richest sources of folic acid are liver and yeast products. Quality commercial fortified cereals provide ample levels of folic acid.

Nutritional Deficiency

Very likely, all factors considered, the anemia in mink observed by Kennedy (1946) was the very first reported cases of a folic acid deficiency in mink on practical ranch diets. The nutritional problem was resolved with dietary supplementation of liver, a rich source of folic acid. Studies by Schaefer et al. (1946) with mink on purified diets without folic acid indicated that the characteristic symptoms of folic acid deficiency included anorexia, severe weight loss, extreme weakness, and irritability. Ulcerative hemorrhagic gastritis led to a bloody, watery stool. Advanced folic acid deficiency is always shown in changes in the blood picture, a characteristic macrocytic, hypochromic anemia called megablastosis. Leucopenia also occurs as well as lesions of the mucous membranes and depigmentation (Whitehair et al., 1949).

Toxicity

Perel'dik et al. (1972) has reported the maximal single dose is equal to 2 mg/ live weight of mink or fox.

Vitamin B$_{12}$ Cobalamine

Vitamin B$_{12}$ and folic acid, as well as the trace minerals iron and copper are critical micronutrients required for the prevention of anemia. A deficiency of vitamin B$_{12}$ in humans yields pernicious anemia. Vitamin B$_{12}$ has a pink color related to the presence of the trace mineral cobalt, hence the extra nomenclature of "cobalamin." Feces of both humans and mink contain significant quantities of vitamin B$_{12}$, its presence directly related to microorganisms in the colon which synthesize the vitamin. However, in the human the vitamin is not absorbed from the colon, and in the mink sub–optimum absorption and/or minimal intestinal flora synthesis results in an absolute dietary requirement for vitamin B$_{12}$ (Leoschke et al., 1953).

Physiology

Vitamin B$_{12}$ is an absolute requirement for the function of folic acid in animal physiology, inasmuch as vitamin B$_{12}$ is required for folate uptake by cells and for regenerating tetrahydrofolate (THF) from which is made 5,10 methylene THF required for thymidylate synthesis, a step of metabolism absolutely required for cell synthesis. Vitamin B$_{12}$ is also critical for methyl group transfer including the methyl group of methionine. Studies by Tauson and Neil (1993a, 1993b) indicate that oral vitamin B$_{12}$, but not B$_{12}$ injection, had some effect in preventing the anemia associated with feeding high levels of fish of the cod (Gaddae) family, indicating an influence on intestinal iron absorption.

217

Requirement

Studies by Leoschke et al. (1953) indicate a maximum of 30 microgram per kilogram of purified diet or 0.8 microgram per 100 kcal of ME as a growth requirement. Perel' dik et al. (1972) have recommended 80 micrograms/kg feed solids.

Resources

The richest sources of vitamin B_{12}, once known as Animal Protein Factor (APF), are liver, egg yolk, and whole fish. Milk is a good resource, but all plant products lack this specific vitamin.

Nutritional Deficiency

Mink on purified diets without vitamin B_{12} exhibit anorexia, severe weight loss, and fatty degeneration of the liver very likely directly related to the anorexia. In terms of reproduction/lactation performance, Perel' dik et al. (1972) have reported sub–optimum reproductive capacity—still–births and high kit mortality.

Toxicity

As a water–soluble vitamin it is expected that vitamin B_{12} is relatively non–toxic and does not cause hypervitaminosis in excessive intake (Perel'dik et al., 1972).

Inositol and PABA

No specific studies have been reported on the requirements of the mink for inositol or para–amino–benzoic acid (PABA). Experimental studies with mink on purified diets at the University of Wisconsin and Cornell University indicated that 250 mg inositol and 500 mg of PABA provided ample levels of these two nutrients (McCarthy et al, 1966).

Unidentified Factors

When introductory studies on the nutritional requirements of the mink at the University of Wisconsin (1945) began using purified diets (Schaefer et. al., 1947), it was noted that the mink kits were unable to survive on this regimen without the addition of raw whole milk or fresh raw whey (Schaefer et. al., 1948). By June 1950, it had been determined that mink may actually require four unknown factors as follows:

1. **Liver Extract (methanol) Factor** (Tove et. al., 1950a);
2. **Liver Residue Factor from methanol extraction** (Schaefer et al., 1947);
3. **Hog Mucosa Factor** (Tove et al., 1950b); and
4. **Lard Factor** (Tove, 1950).

218

Let me depart from my research format for a moment to present the followimg short memoir of my academic experience at the University of Wisconsin. I do this with the desire to provide non–research readers an insight into the excitement of making discoveries, which, I have found, are achieved partly through diligent effort and partly through unforseen and accidental events.

Liver Extract Factor

On June 10th, 1950, I arrived at the Biochemistry Department, University of Wisconsin, with a $100/month research assistantship to work on mink and fox nutrition experiments. I had never seen a mink before in my life and I had no interest in spending the next half century studying the nutritional requirements of this carnivore; in fact, one hour after obtaining my Ph.D. degree, I intended to leave town. In actuality, I ended up with nine years of mink nutrition research studies at the University of Wisconsin as a graduate student and post–doctoratal student. The graduate studies were required for the M.S. and Ph.D. degrees, something like a union card or a plumber's license for a potential college teaching career.

The mink research project involving purified diets was under the direction of Dr. C.A. Elvehjem, a scientist who deserves special praise for his patience with a project which lasted more than a decade before his research objectives were fully realized. During the first year I worked with Bob Lalor, an excellent compatriot in mink nutrition research who taught me much about working with mink as a laboratory animal. By the start of my second year as a graduate student, I knew that the liver extract factor was vitamin B_{12}, but I had no experimental evidence to support my conclusion. I maintained this perspective even though Dr. Elvehjem had just published a paper stating that the unknown liver extract factor was not vitamin B_{12} (Tove, 1950a). In addition Sam Tove had conducted studies involving the vitamin B_{12} content of mink feces from animals receiving purified diets containing sucrose or dextrins (partially hydrolyzed corn starch) as the primary carbohydrate resource. His work indicated no significant difference in the fecal content of vitamin B_{12}, even though mink kits placed on purified diets with sucrose as the carbohydrate resource lived longer before developing the liver extract factor deficiency than mink kits placed on purified diets with dextrins as the carbohydrate resource. He reasoned that if the mink kits lived longer on purified diets with sucrose as the carbohydrate resource and the liver extract factor was vitamin B_{12}, extra intestinal bacterial flora synthesis of vitamin B_{12} would have explained the difference in mink kit survival. Thus he concluded that the unknown liver extract factor was not vitamin B_{12}.

However, with diligent effort over a period of a year, I was able to prove that Sam Tove was in error in his judgment on the nature of the unknown liver extract factor required by the mink.

Mink kits with a liver extract (vitamin B_{12}) deficiency develop anorexia (refusal to eat food for a week to ten days), lose weight, and finally die. Autopsies reveal an enlarged, orange colored liver. Later observations would indicate that the enlarged fatty liver was directly related to starvation, as depot fats are mobilized and transported to the liver. With mink in an anorexic state, Tove

219

would inject them with vitamin B_{12}. When he found no response in livability, he concluded that the unknown liver extract factor was not vitamin B_{12}. Both Bob Lalor and I felt that a lack of response to the vitamin B_{12} injections indicated that an anorexic animal could not live on vitamin B_{12} alone and required energy and protein resources as well as vitamin B_{12}. Bob and I tried to assist the mink in recovery by providing both vitamin B_{12} and the purified diet containing vitamin B_{12} as a slurry (mixed with water) delivered via a stomach tube. The anorexic mink did not appreciate our concern and feeding efforts, simply throwing up the slurry upon being returned to their cages.

Obviously, to resolve of the nature of the liver extract factor required a combination of mink nutrition management and mink psychology. The three pound anorexic mink was "scared to death" in the presence of a 150 pound human placing a stomach tube down his throat. Thus, I decided to change tactics and try a little kindness—as well as a purified diet slurry containing vitamin B_{12}. Using an enlarged tip Mohr pipette as a delivery system, in combination with love and concern, I placed the starving mink kit on my lap and held him with my left hand while attempting to place the tip of the Mohr pipette in his mouth. The immediate reaction of the kit was clenched teeth. I said to the mink, "I know how you feel, but just try to relax." After 5 minutes the starving kit started licking the purified diet slurry off the tip of the Mohr pipette. Within 3–5 days of this kind treatment, I was able to get the starving kits back on solid feed—but only as long as the slurry contained vitamin B_{12}.

I was—almost by chance–to resolve the problem of inconclusive data on intestinal flora synthesis of vitamin B_{12} while conducting a literature research for a required term paper in Dr. Bauman's course on vitamins. I was more interested in conducting mink nutrition research than in completing another term paper, but I had no choice in the matter. In the course of my library work, I came across an article on the riboflavin content of poultry feces at defecation and hours later as the feces were exposed to the environment. The experimental data indicated enhanced synthesis of riboflavin by the bacteria even after defecation. I immediately thought of Tove's experiments on the vitamin B_{12} content of mink feces from animals on a sucrose–based purified diet and on a dextrin–based purified diet. Sam had simply collected the feces from the mink pens and analyzed them for vitamin B_{12}. I wondered if the bacterial synthesis of vitamin B_{12} after defecation at different rates of synthesis accounted for the fact of no significant difference in fecal content of vitamin B_{12}. Thus over a period of two weekends, I set up a program of collecting feces from mink fed purified diets with different carbohydrate resources every 30 minutes for 24 hours and immediately dropping the feces into glass jars containing 100% ethanol to stop any continued intestinal flora synthesis of vitamin B_{12}. I must admit that the first weekend was particularly exasperating inasmuch as I missed Homecoming at Valparaiso University, and more importantly, my former girlfriend, Jean, was getting married in Western New York—to another sailor, not me. However, the extra physical effort paid off in wonderful data—the feces from the mink fed the purified diet with sucrose (previous mink kit history of longer liveability) provided 5 times the vitamin B_{12} content as the feces from mink fed a purified diet with dextrins as the carbohydrate resource (previous mink kit history of shorter life span prior to development of the vitamin B_{12} deficiency) (Leoschke, 1952 and Leoschke et al., 1953).

220

Liver Residue Factor

It took me almost two years to resolve the nature of the Liver Extract Factor, but it took almost six years to resolve the nature of the Liver Residue factor. So much for the brilliance of W. L. Leoschke! Mink kits placed on purified diets (supplemented with vitamin B_{12}) in early July developed another deficiency syndrome during the fall fur development phase in September and October: their under-fur was depigmented and of poor quality. Those mink that did survive the fall, winter, and spring on the purified diets without the liver residue died of the liver residue deficiency in May as the summer furring period progressed. Neither Tove, Lalor, nor Leoschke had considered the logical point that the relationship between fur synthesis and the liver residue factor deficiency might be a clear cut amino acid deficiency.

The initial breakthrough in resolving the nature of the liver residue factor occurred in the spring of 1955 with a $2,000 shipment of vitamin–test casein. The earlier shipments of the product had been cream colored, but this product was a definite "dark tan" indicating to me that the product had been "over–heated" in the process of removing the solvent from the alcohol extraction procedures required to yield vitamin–test casein. I immediately knew that something was wrong, that this was an inferior protein resource with the distinct possibility of undercutting performance of the mink kits in the summer/fall of 1955. My concerns were confirmed that summer/fall when mink kit losses on the purified diets occurred 3–4 weeks earlier than in the previous five years of my experimental studies. Thus, I slowly began to realize that the unknown liver residue factor required by mink kits on purified diets was not an unknown vitamin but a mink dietary deficiency brought about by an actual sub–optimum delivery of amino acids to the mink during the critical fur development phase wherein the animals experienced enhanced requirements for the sulfur amino acids (cystine and methionine) and arginine. The problem with the vitamin–test casein provided in the purified diets as the sole protein resource for the mink was two fold: there were already relatively low levels of methionine, cystine, and arginine in the casein protein resource, and the excessive heat treatment used to remove alcohol from the alcohol–extracted vitamin test casein product in the spring of 1955 simply exasperated the problem by a significant degree through destruction of and/or bonding of cystine, methionine, and arginine in structures unavailable to the digestive processes of the mink kits. See Table 3.21 for a review of these data.

It is of interest to note that previous but somewhat parallel studies by Arnold et al. (1936) that included data on chick feather development found that the arginine in casein was only partially available to the chick. In that study, which involved liver as a survival ingredient for chicks on purified diets, it was Elvehjem who suggested that the factor essential for the chick might be arginine inasmuch as liver is a rich source of arginine. Now, almost two decades later, Leoschke arrives at the related insight that arginine might be as critical for mink fur development as for chick feather development.

221

The problem of the "unknown" liver residue factor was resolved in the fall of 1956 with supplementation of the purified diet with methionine and arginine (Leoschke and Elvehjem, 1959). An amusing footnote to this story underscores the occassional stubbornness of researchers. Once I discovered that the liver residue factor involved amino acids, I asked Dr. Harper, an authority on the amino acid nutrition of rats, to recommend an amino acid profile that would most likely be critical for the mink. When his list included arginine, my immediate response was, "obviously not arginine," since most nutrition textbooks list it as a non–essential amino acid. Obviously, today all mink nutrition researchers and mink dietitians—including W. L.Leoschke—have learned to give serious consideration to the role of arginine in the nutrition of the mink.

Hog Mucosa Factor

Related limited studies by Tove et al. (1950b) indicated that with the purified diet supplemented with 20% liver, the mink still lost weight and experienced anorexia, matted fur, and death. When dessicated hog mucosa (intestinal lining) was introduced into the diet, the mink lived for more than two years. Yet, the exact nature of the hog mucosa factor remains unknown and there is even some question whether the "unknown factor" actually exists, since mink kits placed on 100% raw and pasteurized milk supplemented with trace minerals for two years did not develop this specific dietary deficiency (Leoschke, 1960).

Lard Factor

Studies by Tove (1950) indicated that mink placed on purified diets containing cottonseed oil as the sole fat resource after a period of many months developed a nutritional deficiency that could be resolved with the introduction of pork fat (lard) into the dietary regimen. With the completion of my post–doctoral studies at the University of Wisconsin, the search for this unknown lard factor ceased. It is likely that the factor could be arachidonic acid, 20:4 omega 6, a polyunsaturated fatty acid present only in animal fats. This possibility is supported by the basic fact that cats, carnivores with many nutritional requirements similar to the mink, are unable to convert the essential fatty acid, linoleic acid, 18:2 omega 6, to arachidonic acid, 20:4 omega 6. Most other mammals, in fact, are capable of this process. The enzyme defect in cats—and perhaps in the mink—is the lack of delta–6 desaturase, a critical step in the conversion of linoleate to archidonate (Hassam et. al., 1977). Support for the hypothesis that the mink cannot convert linoelic acid to arachidonic acid is seen in a recent report by Polonen et. al. (2000) concerning mink kits on a dietary regimen with a high level of soybean oil. The fatty acid composition of the total lipids of the livers revealed that the high proportion of linoleic acid characterstic of plant oils was not effectively metabolized to the longer chain and more unsaturated fatty acids such as arachidonic.

Minerals

Minerals have multiple functions in the anatomy and physiology of mink, including the following:

1. Structural role in the skeleton, teeth, and soft tissues; blood composition and the blood clotting mechanisms, as well as nervous system components;

2. Regulatory role in maintenance of osmotic pressure and acid–base balance;

3. Essential role as co–factors for the activity of many enzymes;

4. Role in vitamin structure, illustrated by cobalt in vitamin B_{12} and sulfur in biotin;

5. Key role in fur pigmentation, as an iron deficiency results in "cotton" mink.

In terms of animal nutrition, the mineral requirements can be divided into two groups. The first group consists of the macro minerals calcium, chloride, magnesium, phosphorous as phosphate anions, potassium, sodium, and sulfur both as part of the structure of vitamins and amino acids and as sulfate anions, which are required in relatively large quantities. The second group consists of the micro minerals, or trace elements, including copper, iron, iodine and iodide, manganese, molybdenum, selenium, and zinc. These are needed in relatively small quantities, often expressed in parts per million.

Absorption of minerals from the gastro–intestinal tract of animals involves multiple factors including (a) relative solubility in the specific environmental pH; (b) length of time of feed passage; (c) presence of factors that interfere with absorption such as coalbumin in egg white and DMNA or formaldehyde presence in certain fish products which reduces iron absorption; and (d) the presence of specific vitamins such as vitamin D required for the absorption of calcium and phosphate ions or vitamin B_{12} and ascorbic acid which enhance iron absorption.

Studies by Jensen and Lohi (1988) indicate that a significant positive correlation exists between the levels of calcium, magnesium, sodium, and selenium in mink feed and the concentrations of these minerals in the hair of the mink. A high level of potassium in the feed was only slightly reflected in the fur. Levels of phosphate or copper in mink fur were unaffected by feed concentration. The amount of zinc in the hair did not correspond to the variation in concentration of zinc in the feed but was negatively correlated relative to the concentration of most other minerals in the feed. The levels of iron in the mink fur was directly related to interaction with other minerals in the ranch diet.

In terms of practical mink nutrition, ranchers must be aware of the ash content of the diets they choose. Ranch diets with 35–40% of bone–in feed products contain 7–8% ash (dehydrated basis) provide ample levels of calcium and phosphate. Excessive ash levels can undermine the nutrition of the mink by creating excessive bulk, especially harmful during lactation and early growth when the mink have maximal nutrient requirements and minimal feed intake capacity (that is, small stomach volume relative to nutrient needs) as ash can interfere with the capacity of mink to digest fats. This can happen if limestone (calcium carbonate) is used, as it forms insoluble calcium soaps such as calcium stearate. Finally, higher ash diets may contribute to urinary calculi and/or kidney stones.

Ranchers should also be concerned about high ash mink feedstuffs such as chicken feed (19% ash) wherein the availability of the proteins to the digestive capacity of the mink is only 52% compared to 78% for chicken heads (5% ash) and 89% for chicken entrails (4% ash).

It is important to point out that in the case of both vitamin and mineral requirements, sub–optimum nutritional states must be classified as primary or secondary nutritional deficiency. In a primary deficiency, sub–optimum levels of the specific nutrient are provided by the dietary program. In a secondary deficiency, ample quantities of the mineral are present in the diet, but certain specific factors are responsible for binding the nutrient in a structure unavailable to the digestive processes of the animal. This point is well illustrated with the use of raw eggs in mink diets wherein avidin binds biotin and conalbumin binds iron.

Using chelated minerals may minimize the possibility that mink will develop a secondary deficiency of specific trace minerals such as copper, iron, or zinc. The word "chelate" originates from the Greek word for claw—*chele*; thus, chelated minerals are metallic cations combined with specific chelating agents such as amino acids to yield a ring–like structure. Chelation shields the mineral from external influences and thereby affects intestinal absorption as well as interference from other minerals on absorption. For example, calcium may bind zinc, decreasing its absorption; however, chelated zinc is protected from binding by calcium.

Mineral imbalances are another factor that mink nutritionists must consider. Calcium/phosphorus ratios as low as 0.75/1.0 and as high as 1.7/1.0 are satisfactory for the mink (Rime-slatten, 1966b). Ca/P ratios higher or lower than these specific recommendations may result in reduced absorption of these ions and bring about a secondary mineral deficiency. Some antagonism is recognized among magnesium, calcium, and phosphorus thus excesses of calcium or phosphate in the diet may decrease the absorption of magnesium, and vice versa (Glem–Hansen, 1978).

Van Campen and Scaife (1967) and Van Campen (1969) showed that copper and zinc compete for absorption in the intestine of rats and that an excess of either element reduced the absorption of the other. Studies by Aulerich et al. (1983) indicated that with high zinc levels in the mink diet, the melanin content of the hair collected from the kits in the zinc–supplemented group was significantly lower than that of the control mink. This achromatrichia was thought to result from to the failure of melanin formation, a process requiring the copper–containing enzyme tyrosinase. This same study indicated that supplementing the control diet with zinc resulted in an 18.2% reduction in liver copper concentration.

High levels of molybdenum can also induce a copper deficiency. In contrast, low levels of forage molybdenum can be associated with copper toxicity. Arrington and Davis (1953) showed that a high level of molybdenum induced copper deficiency in rabbits, with typical signs such as anemia. Studies at Michigan State University (Aulerich et al., 1983) with mink kits indicated that supplementing the control diet with molybdenum resulted in a 24.9% reduction in liver copper concentration.

224

Calcium/Phosphate

Since calcium and phosphorus are so closely interrelated, both nutritionally and physiologically, it makes sense to discuss them together. Calcium and phosphorus in the form of hydroxyapatite crystals ($3 Ca(PO_4)_2$ and $Ca(OH)_2$) are the major minerals involved in the structural rigidity of bones and teeth. Calcium is also involved in blood clotting and in impulse transmission at the neuromuscular junction. Phosphorus is involved in bone and teeth formation and, in addition, is an important structural element for the soft tissues including membrane phospholipid components lecithin and cephalin, nucleic acids, and in proteins as illustrated by the phosphorylated protein of milk, casein, and the vital role of ATP (Adenosine Tri–Phosphate) in energy transfer in animal physiology.

Physiology

A study by Newell (1999) indicated that only 10% of dietary phosphorus is excreted in the urine. Earlier studies by Leoschke (1960) on diets containing horsemeat, 50%; whole ocean whiting, 30%; liver,10%; and oatmeal, 10% supplemented with limestone or dicalcium phosphate or with the horsemeat being replaced with raw calf bone at levels of 10, 20, and 30% of the diet, the kidney excretion was calcium 0.9% and phosphorus 23.4%. It is of interest to note that phosphorus excretion decreased in sequence to the higher levels of raw calf bone replacing horsemeat, since raw bone is not a good source of phosphate compared to raw horsemeat. Very likely the studies of Newell in Nova Scotia, Canada, involved relatively high levels of bone and hence reduced loss of phosphorus via the urine. Another study by Leoschke (1962b) indicated an average phosphorus excretion of 35% of dietary intake with a range of 30–46%. In this experimental work, the diet consisted of fortified cereal, 20; beef liver, 10; and old hens and entrails at 70% of the ration. Both experimental studies indicated relatively low urinary calcium as only 1–2% of dietary intake.

Requirement

Initial experimental studies on the calcium and phosphate requirements of the mink were conducted in 1951. These studies indicated that the calcium requirement of the mink is about 0.3% of the diet when the diet contains 820 IU vitamin D/kg and the Ca/P ratio is in the range of 0.75–1.7 (Basset et al., 1951). However, all factors considered, the Basset data may be skewed because mink today are larger than the small animals employed in these studies. A review by Rimeslatten (1966) stressed that the 1951 work may have been low in the actual calcium requirements of the mink as due to the small size of the animals involved and thus too low for the mink genetics of the 20th century with faster growth rates. He recommended 0.4–1.0% calcium with the 820 IU vitamin D/kg and Ca/P ratios of 0.75 to 1.7. Studies by Argutinskaya (1954), while they predated the Rimeslatten commentary, nonetheless were in accord with his work. The 1954 study was reviewed in the Russian handbook (Penelaik, 1975) which led to a recommendation of 0.5–0.6 % calcium for growth and 0.8% during lactation.

Resources

An excellent calcium and phosphate resource is dicalcium phosphate with a Ca/P ratio of 1.1/1.0. Inferior resources include steamed bone meal with a Ca/P ratio of 2.4/1.0 and tricalcium phosphate with a Ca/P ratio of 2.0/1.0. Limestone (calcium carbonate) is a poor resource for calcium inasmuch as it contributes to a relatively alkaline urinary pH and thereby is supportive of urinary calculi (struvite) formation in the mink. Phytic acid, a hexaphosphoric acid ester of inositol, is found mainly in the bran portion of cereals. Phytic acid is not only a poor resource for phosphate in animal nutrition inasmuch as the acid is bound in an indigestible form, but also it may combine with free calcium in the intestinal lumen to yield insoluble salts which are not available for absorption. Studies on the employment of phytase enzyme to release phosphate in a digestible form from phytic acid have not proven to be of any value in decreasing the phosphate content of mink feces (Rouvinen, et al., 1996 and Bursian et al., 2003).

Nutritional Deficiency
Primary Nutritional Deficiency

A deficiency of calcium or phosphate results in the nutritional disease of rickets in young animals. Sub–optimum calcification of the skeleton leads to lameness, bent bones, fractures, and abnormal skeletal development. In adults, a decalcification of the skeleton results in osteomyelitis which causes weak and brittrle bones. An extremely low level of calcium in the mink's diet, achieved via boneless meat sources such as pork lungs and uteri, can lead to *ontogenesis imperfecta*, the effects of which were severely deformed limbs and abnormalities in the spinal cord, particularly in the breast and lumbar regions (Wenzel and Arnold, 1971).

Secondary Nutritional Deficiency

Even with ample calcium and phosphates in the ranch diet, multiple factors can lead to a secondary deficiency of these critical minerals. Absorption of calcium from the intestinal tract depends upon both vitamin D levels and Ca/P ratios.

Vitamin D is required for the absorption of calcium and phosphates from the intestinal canal and is essential for bone formation. With Ca/P ratios within the range of 0.75 to 1.7/1, the mink's requirement for vitamin D will be covered by 100 IU of vitamin D/day (Joergensen, 1985).

Ca/P ratios higher than 1.7/1 or lower than 0.75/1 can bring about sub–optimum performance of the mink. A study with Ca/P ratios of 0.9, 1.3, and 1.9 determined that the higher Ca/P ratio of 1.9 resulted in reduced weight gains and that the pelts were significantly shorter than those of the animals on the dietary programs with lower Ca/P ratios (Hansen et al., 1992).

For almost 50 years, phosphoric acid at a level of 2.0% (75% feed grade acid) dry weight basis has been used in mink ranch dietary programs throughout the world as a prophylactic medical approach to urinary calculi and as a mink feed preservative "on the wire." However, serious problems can occur if close attention is not given to optimum Ca/P ratios. Using phosphoric acid in nutrition programs where the diet contains a minimum of 35% "bone–in" fresh/ frozen feedstuffs has never resulted in the Ca/P deficiency of rickets with, to my knowledge, one exception (Leoschke, 1980). In that instance, the ranch diet with phosphoric acid provided only about 28% "bone–in" fresh/frozen feedstuffs. As a result a sub–optimum Ca/P ratio and a significant incidence of rickets occurred.

pH of the intestinal tract

The absorption of calcium is favored by a low intestinal pH which facilitates keeping the calcium salts in solution. Thus, normal gastric hydrochloric acid secretion is necessary for efficient absorption of calcium (Heinz, 1959).

Phytic Acid/Fatty Acids

A number of factors in the ranch diet may interfere with calcium absorption and thereby create a secondary deficiency of this mineral. Phytic acid, a hexaphosphoric acid ester of inositol which is found in cereal seeds, may form insoluble calcium salts with free calcium in the intestinal contents, thus making some calcium unavailable for absorption (Harrison and Mellanby, 1939). Specific fatty acids such as stearic acid yield insoluble salts (soaps) that simply pass through the intestinal tract and are excreted in the feces (Ahman, 1974; Rouvinen and Kiiskinen, 1990, 1991).

Toxicity

Within 10 days after they were placed on a rachitogenic diet high in calcium and low in vitamin D and phosphorous, mink kits experienced difficulty in walking (Smith and Barnes, 1941). They tended to crawl, and as the condition became more severe, they were unable to stand. Enlargement of the ribs at the costochondral junctions was evident. The spinal column in the thoracic region became concave (lordosis). The leg bones were bent and enlarged at the ends. The ash content of the dry fat–free femurs was 22 to 30%, compared to 60–64% for normal animals.

Chloride

Chloride anion functions in the acid–base balance of the body and is vital for proper osmotic pressure within and without cellular structures wherein chloride and bicarbonate are the major anions of the extracellular fluids. Gastric secretion includes hydrochloric acid and chloride salts. The nutritional requirements of chloride anion are closely associated with sodium cation as salt, thus chloride requirements of the mink will be discussed under the topics of sodium and salt requirements.

Chromium

Chromium has been shown to be an essential nutrient for animals as it is involved in the functioning of insulin in the regulation of carbohydrate metabolism. Chromium deficient animals have impaired glucose tolerance, that is, a reduced capacity to metabolize glucose. Chromium deficiency symptoms are similar to those of insulin insufficiency, or diabetes. A chromium deficiency in mink or fox has not been reported.

Cobalt

No scientific data indicates that cobalt as a mineral is an absolute requirement for the mink. At the same time,. it must be acknowledged that mink require vitamin B_{12}, cobalamin, which contains 4.3% cobalt. The primary role of cobalt is to serve as a nutrient resource for the bacterial synthesis of vitamin B_{12} in the rumen of cattle and in the cecum of horses and the fur–bearing herbivores, rabbits and chinchillas. Studies at the University of Wisconsin (Leoschke, 1952 and Leoschke et al., 1954) indicated that the intestinal flora of the mink are capable of synthesizing vitamin B_{12} from cobalt ions provided in purified diets. However, this amount of intestinal microbial synthesis is inadequate to meet the vitamin B_{12} requirement of the animal for growth. With mink fed purified diets, this intestinal flora synthesis of vitamin B_{12} is directly related to the carbohydrate source. The synthesis is five times as high on diets containing sucrose (table sugar) as on diets employing dextrins (smaller weight glucose polymers obtained by the hydrolysis of corn starch). This observation is exactly opposite to that noted in studies with rats (Lewis, 1951). This difference is of interest inasmuch as multiple studies indicate that intestinal flora synthesis of vitamins by rats is consistently higher when the animals are provided dextrin polymers of glucose as compared to sucrose, a dimer of glucose and fructose.

Copper

Copper is a constituent of several metalloenzymes such as cytochrome oxidase, which functions in cellular respiration, and lysyl oxidase, important in connective tissue formation. Copper is also a constituent of tyrosinase, an enzyme involved in melanin (the black pigment of skin, hair, and fur). Studies conducted by Aulerich et al. (1982) observed that darker fur occurred in pelted males fed the higher level of copper. In terms of the synthesis of blood components, copper is associated with the metalloprotein, ceruloplasmin, which functions in iron absorption. It follows that a deficiency of copper may result in an induced iron deficiency. Copper is also necessary for the synthesis of hemoglobin and red blood cell maturation. Thus a copper deficiency may result in an anemia which is really an iron deficiency, that is, a secondary iron deficiency directly related to a copper deficiency in the animal's diet. Obviously, copper is a key trace mineral in mink nutrition in terms of its role in hemoglobin formation and normal pigmentation of fur.

228

Physiology

Studies by Mejborn (1989) on copper sulfate supplementation of mink diets indicated that 75–90% of the copper intake was excreted in the feces and that urinary copper excretion was elevated with increased copper intake. However, the excretion constituted only 0.3 to 1.5% of the total copper intake.

Requirements

Kangas (1976) has reported that the mink ration in Finland was Cu:Zn:Fe at 1.5/5/25, whereas a similar analysis by the Danish Fur Breeders Association varied from 1/8/36 to 1/1/2 with minimum and maximum total contents of copper 5–34 ppm, zinc 27–60 ppm, and iron 75–203 ppm in the wet feed (Brandt, 1983) wherein 4.5–6.0 ppm is considered adequate (Glem–Hansen, 1978). In general, the copper requirements are adequately met by typical mink diets containing fish (Kiiskinen and Makela, 1977). Danish mink feed normally contains 25–35 ppm copper. Studies by Mejborn (1989) with a control diet containing 5.1 ppm copper (wet weight basis) and experimental diets containing 39 and 116 ppm copper indicated no effect of copper supplementation on growth or fur production. However, earlier studies by Aulerich and Ringer (1976) indicated superior weight gains in male kits but not in female kits fed supplemental copper at the 50 ppm level. In two other studies, Aulerich et al. (1982) and Bush et al. (1995) it was noted that darker fur color occurred in male kits but not female kits with supplemental copper at the 200 ppm level. It is of interest to note that in the 1982 research work, the control diet contained 60 ppm copper, 330 ppm zinc and 330 ppm iron.

Resources

Copper sulfate and copper chelates are considered to be excellent copper resources. On the other hand, experimental studies indicate that copper oxide has a copper availability of only about 10% at the very most.

Nutritional Deficiency
Primary Nutritional Deficiency

Copper deficiency is characterized in most species by diarrhea, anemia, achromatrichia (depigmentation of the hair), alopecia (loss of hair), dermatosis, and growth retardation. Studies at Michigan State University (Aulerich et al., 1983) employing high levels of supplemental zinc during the reproduction/lactation phase showed that mink kits at 3–4 weeks exhibited graying of the fur around the eyes, ears, jaws, and genitals with concomitant loss of hair and dermatosis in these areas. During the next few weeks the achromatrichia and alopecia spread over much of the body. The melanin content of the hair collected from the kits was significantly lower than that of the control mink. This achromatrichia was thought to be due to the failure of melanin formation, a process requiring the copper–containing enzyme tyrosinase. There was a profound immunosuppressive effect within the kits on the zinc–supplemented diet.

Secondary Nutritional Deficiency

High levels of zinc in the diet inhibit copper absorption through competition for binding sites of the protein metallothionein (Van Campen & Scaife, 1967 and Van Campen, 1969). This point is well demonstrated by the experimental studies of Aulerich et al. (1983). The research involved employing high levels of supplemental zinc during the critical reproduction/lactation period. The mink kits at 3–4 weeks of age showed graying of the fur around the eyes, ears, jaws, and genitals with concomitant loss of hair and dermatosis in those areas. During the next few weeks the achromatrichia and alopecia spread over much of the body. The melanin content of the hair collected from the kits was significantly lower than that of the control mink. This achromatrichia was thought to be due to the failure of melanin formation, a process requiring the copper containing enzyme tyrosinase. There was profound immunosuppressive effects within the kits fed the zinc supplemented diet. The Aulerich studies noted above found, more specifically, that high levels of zinc supplements reduced the copper in the liver by 18% and that supplementing with molybdenum reduced it by 25%.

Toxicity

Studies by Aulerich et al. (1982) have demonstrated that mink are among the more copper tolerant species.

Acute Toxicity

The studies by Aulerich et al. (1982) indicated that oral dosing (by gavage) of mink was not feasible because the animals consistently vomited immediately after dosing, thus requiring the alternate method, interperitoneum injection. Acute toxicity (21 days) LD–50 concentration of copper sulfate or copper acetate in adult mink were 7.5 and 5.0 mg/kg respectively.

Chronic Toxicity
Reproduction/Lactation

Studies by Powell et al. (1997) indicated that breeder mink fed conventional ranch diets with 20% copper–treated raw eggs at a level of 1,000 ppm had no adverse effects on female mink reproduction performance or kit growth and survival. Studies by Aulerich et al. (1982) wherein the control diet contained copper, 60 ppm; zinc, 330 ppm; and iron, 330 ppm (dry weight basis) with supplementation of the diet for a year with 200 ppm copper indicated that although the reproduction performance of the mink was not adversely affected, greater kit mortality and reduced "litter mass" was noted.

Kit Performance

Studies by Brandt (1983) indicate that copper levels of 320 ppm wet feed (about 1,000 ppm copper dehydrated basis) provided significant toxicity. There was 50% mortality which, on autopsy, revealed fatty liver and kidney metamorphosis, liver cirrhosis, icterus, hemolysis, and wasting. The hematological values showed a shift towards a normo/microcytic, hypochromic anemia, high total copper, low iron in blood plasma, low total protein, high ASAT and ALAT—all of which fit the pattern of copper poisoning. One surprising finding was the high plasma zinc, which might have been expected to be as low as the plasma iron, due to competition for binding sites. But, since haemolysis was prevailing and as the high ASAT and ALAT and alkaline phosphatase values indicated both muscular and liver damages, the high plasma zinc was due to cellular leakage. Brandt's detailed observations are of interest: "The present investigation shows the strong interrelationship between the absorption or utilization of copper and zinc, depending on the relative and total content of the feed. Surprisingly, 300 ppm copper/kg wet feed was highly toxic to the mink. Whether this is due to uptake facilitated (chelating properties of the diet being rich in high digestible fish proteins), or just the unphysiological ratio of Cu:Zn:Fe in the feed, future investigations will reveal."

Fluoride

Fluoride is important in the nutrition of animals inasmuch as it is normally present in bones and teeth and a proper intake is essential for achieving maximum resistance to dental caries.

Physiology

No specific biochemical role for fluoride in animal physiology has been found.

Requirements

No specific experimental data is available on the requirement of the mink for fluoride although a physiological need for fluoride has been demonstrated in other animals.

Resources

Fish are an important source of fluoride, and most meat and cereal products contain significant quantities of fluoride.

Nutritional Deficiency

Although no experimental data is available on fluoride deficiency in mink, it is known that deficiency signs in other animals include reduced rate of growth, infertility, and increased susceptibility to dental problems.

Toxicity

Fluoride sources are cumulative; so when quantities are ingested which exceed the critical level for a species, fluoride toxicosis results. In this regard, studies by Aulerich et al. (1986, 1987) are of interest.

The basal diet contained 35 ppm fluoride (104 ppm dehydrated basis) and the water supply provided 0.34 ppm fluoride. Supplemental fluoride at levels of 0, 33, 60, 108, 194 ppm, and 350 ppm provided no significant differences in growth, fur production, breeding, gestation, whelping or lactation. However, at the 350 ppm level, a high kit mortality occurred, and the skulls of males fed this level of fluoride were invariably crushed when placed in the clamping device used to hold the mink carcasses for pelting.

The only signs of toxicity noted in the mink were a general unthriftiness accompanied by hyperexciteability. A few days before death, the animals became lethargic and reluctant to move. No signs of lameness, as reported in other species exposed to toxic concentration of fluoride, were noted. No signs of dental lesions were observed in adult mink at necropsy, although several kits whelped by females fed and reared on the 350 ppm fluoride diet showed a dark mottling of the teeth, particularly the canine teeth.

Shupe et al. (1987) have recommended tolerance levels in the feed of not more than 50 ppm, wet weight basis (135 ppm dry weight basis) for breeding stock and 100 ppm, (wet weight basis) for animals being raised only for their pelts.

Iodide

Iodine is a component of the thyroid hormones thyroxine and triiodothyronine involved in the regulation of the rate of cellular metabolism.

Requirement

Wood (1962) has suggested 0.2 ppm of iodide for breeder and growth diets as an adequate level. Normal fish–containing mink ranch diets provide approximately 2.4–6.4 ppm iodide (Kiiskinen and Makela, 1977). In the published literature, recommended iodide levels for cats range from 1.4 to 4.0 ppm dry diet basis (Scott, 1964). Conflicting reports on levels of iodide supportive of the top performance of the mink can be found in reports from Michigan State University. A report by Jones et al. (1982) with a control diet containing about 6 ppm iodide (dry weight basis) indicated that supplementation of this diet with 10 ppm or 20 ppm iodide appeared to have beneficial effects on reproduction and lactation. A later study by Aulerich and Napolitano (1986) working with a control diet containing 15 ppm iodide (dry weight basis) indicated neither beneficial nor detrimental effects on reproduction or kit growth with iodide supplementation as 25 ppm of potassium iodide.

232

Resources

The best mink nutrition resources for iodide are seafood products, iodized salt, and trace mineral mixtures commonly used in commercial mink cereal fortification.

Nutritional Deficiency

Although there have been no scientific reports of a nutritional iodide deficiency in mink, clinical signs in cats include thyroid hypertrophy, alopecia, abnormal calcium metabolism, and death (Greaves et al., 1959); Scott (1960) has reported that iodine deficiency caused fetal resorption, while estrus and libido were unaffected.

Toxicity

Studies by Aulerich et al. (1978) and Jones et al. (1982) indicated that supplementation with 80–100 ppm iodide resulted in reduced reproduction and lactation performance and increased kit mortality. At a level of 1,000 ppm iodide supplementation, 100% reproduction failure occurred.

The 1982 report noted that the thyroid glands of kits whelped and nursed by dams fed more than 20 ppm supplemental iodide showed hypertrophy marked by follicular cell hyperplasia and a decreased amount of colloid. Similar histopathologic lesions were observed in the thyroids of adult mink that received 80 ppm or more supplemental iodide. Also observed were numerous lesions in the gall bladders.

It is apparent that mink are less tolerant of dietary iodine than many other species. Ranchers should take care not to use iodine solutions to clean out watering systems prior to the reproduction/lactation phase of the mink ranch year, as such a practice can severely undercut the performance of the mink. This warning is based field observations by a Pennsylvania mink rancher (Leoschke, 1985).

Iron

Iron is a key element in biological pigments including hemoglobin, an oxygen carrier in the blood which can also act as an iron storage resource, and myoglobin, an oxygen storage resource in the muscles It is also involved with the cytochromes and certain enzymes such as catalase and peroxidase. Of especial interest to mink nutrition scientists is that iron is an essential element for melanin pigments. Studies by Stout et al. (1968) with radioactive iron (Fe–59) indicated that 70% of the iron in dark mink fur is associated with melanosomes.

233

Physiology

Studies involving radioactive Fe–59 and Fe–55 hemoglobin indicated that the radioactivity of urinary output never exceeded 0.5% of the dose administered. These same studies found it likely that common steps governed iron absorption from ferrous sulfate and hemoglobin iron, since large doses of ferrous sulfate tended to induce a diminution in absorption of hemoglobin iron. Of interest is a study from Oregon State University (Adair et al., 1962) wherein physiological studies indicated that mink fed aneamogenic fish excreted about 50% more urinary iron and 20% less fecal iron, possibly indicating a factor in aneamogenic fish providing significant interference in the retention and utilization of iron.

Requirement

Purified diet studies at Oregon State University (Stout et al., 1968) indicated that at a level of 10 ppm iron, an iron deficiency achromotrichia was prevented, while at a level of 5 ppm iron, there was significant white underfur. Ahman (1966) has recommended 20–30 ppm iron for mink nutrition, and in a later report (1974), concluded that 60 ppm iron would meet the iron requirements of mink on a regular ranch diet. This level is also recommended by Jorgensen (1972). The earliest reccomendation (Wood, 1962) was an amount of iron that was equivalent to 88 ppm and 79 ppm for breeder and growth diets respectively. Most recently, Treuthardt (1992) recommended 20–90 ppm with no inhibitory factors present. In the case of mink diets containing anaemogenic fish, he recommended a level of 300–400 ppm to prevent iron deficiency anemia. Obviously, the nutritional content of the diet and/or the physiological environment of the intestinal tract will have a very significant effect on iron requirements because bioavailability may or may not be optimum. In fact, the bioavailability of iron may be the key factor in the wide variation in levels of iron recommended for the top performance of the mink. Another obvious factor is the genetics of the mink, for it has been observed that iron absorption in mink is subject to marked individual variation.

Typical Scandinavian mink ranch diets contain 156–352 ppm iron (Glem–Hansen, 1978). Thus, with most practical mink ranch diets, there should be no iron deficiencies unless dietary factors interfere with iron absorption. The hemoglobin percentage in mink blood has been used as a parameter for the iron status of the mink in multiple studies (Skrede, 1970; Rothenberg and Jorgensen, 1971; Fletch and Karstad, 1972).

Resources

Excellent iron resources include ferrous sulfate, used in human medicine for the treatment of iron deficiency; iron chelates with amino acids/proteins as illustrated by ferrous iron as iron glycinate (Stout et al., 1960); Fe–EDTA; ferric glutamate marketed as "Hemax" (Skrede, 1986a); and ferrous fumarate. An iron compound of interest is ferroceron ($C_{10}H_{13}$ FeONa$_3$. 4 H$_2$O (Rapoport and Golushkova, 1991). Another iron compound of interest is Ferroanemin (Fe–diethyl aminopentacetic acid), a complex iron compound inert to the TMAO compounds present in specific aneamogenic fish (Perel'dik and Perel'dik, 1980).

Unfortunately, some inferior iron resources are being employed in fortified cereals, including ferric sulfate, ferrous carbonate, and ferric oxide. The ferric sulfate used in water purification programs yields, on addition to water, ferric hydroxide. This insoluble product used to precipitate particles in water is obviously not an optimum iron resource for the mink. Ferrous carbonate has only 50% of the value as ferrous sulfate for cats (Rogers et al., 1986).

Ferric Oxide or iron oxide, has very little bioavailability for animals; in fact, it has been used for many years in animal nutrition research as a marker in digestibility studies, that is, as a product 100% non–biologically available. Iron oxide adds to the red color of the final mink ranch mix; perhaps it impresses the rancher but not the mink. Again, it is logical to "Listen To The Mink" in determining the biological value of any nutrient resource.

Experimental studies have noted that a number of nutrients have the capacity to enhance iron absorption from the intestinal tract, including vitamin B_{12} and the amino acid cysteine.

For vitamin B_{12}, experimental studies involving mink fed an anaemiogenic diet, that is, high levels of fish of the cod (*Gaddae*) family (Tauson and Neil, 1993a and 1993b) found that oral vitamin B_{12} supplementation—but not injections—had some effect in preventing anemia, indicating an influence on intestinal iron absorption.

For cysteine, other experimental studies on mink fed high levels of the anaemogenic fish of the cod (*Gaddae*) family, several authors (Skrede 1986a, 1987, 1988 and Skrede and Ahlstrom, 1987) showed that supplementation with L–cysteine at a level as high as 0.6% resulted in a 15–fold increase in iron absorption from ferrous fumarate. The cysteine–containing tripeptide glutathione also improved iron absorption but not at the same level as L–cysteine. These same studies indicated that L–cystine, the disulfide dimer of cysteine without free sulfhydryl functional groupings had no effect on iron absorption. Very likely the value of cysteine supplementation was directly related to the fact that the amino acid cysteine forms several complexes with ferrous cations and thus must influence the bioavailability of iron for the mink.

It is of interest to note that studies by Skrede (1970b, 1970c, 1974, and 1986b) indicated that the use of blood meal and blood increased the absorption of ferrous sulfate iron when the diet contained raw blue herring, an anaemoiogenic fish of the *Gaddae* family. Studies by Taylor et al. (1986) indicated that the value of blood or meat to non–heme iron absorption is directly related to the iron–chelating effect of cysteine and cysteine–containing peptides released during the animal's digestion of meat resources.

Nutritional Deficiency
Primary Nutritional Deficiency

The classic iron deficiency in animals is a microcytic–hypochromic anemia, which causes in mink emaciation, growth retardation, and rough pelage (Stout et al., 1960a and 1960b). If the anemia is present during the critical early phases of fur development, "cotton fur" is very likely to develop: the earlier and more severe the anemia, the more pronounced the fur defect.

Secondary Nutritional Deficiency—Non–Fish

The large number of studies on iron deficiency in the mink find that "cotton fur" is directly related to the use of high levels of anaemogenic fish of the *Gaddae* and *Merluccius* families. As this factor deserves special consideration, a unique sub–section is included in this report on specific fish resources as a significant factor in secondary iron deficiency in the mink.

Multiple factors have been shown to impede iron absorption in the intestinal tract of the mink including phosphate anions, cadmium, copper, manganese, and zinc cations, as well as conalbumin in egg whites. High levels of peroxided fats in the mink ranch diet inhibit the bioavailability of iron, and experimental studies indicate that an excessive alkaline environment in the mink intestinal tract does not favor iron absorption.

Phosphate Anion

High levels of phosphate anion in the mink's diet as provided by bone meal (Brown, 1995) or meat–and–bone meal (Anon, 1981) can yield an iron deficiency in mink as exhibited by anemia and "cotton" mink.

Copper Cation

Studies by Bush et al. (1995) involving copper sulfate supplementation of mink ranch diets noted that with increasing copper levels, the net result was higher copper levels and lower iron levels in the livers of the mink. At the same time, it is of interest to note that copper is involved in the activity of the metalloprotein ceruloplasmin which functions in iron absorption. Hence a deficiency in copper cation can result in an induced iron deficiency anemia.

Egg White

The white portion of raw eggs contains the protein conalbumin (ovo–transfurin) which inhibits iron absorption from the intestinal tract. Studies with rats by Stout and Adair (1970) with radioactive Fe–59 showed that almost 6 times as much iron was absorbed from the rats' intestinal tracts when the diet provided cooked eggs rather than raw eggs. Obviously, heating the egg resulted in denaturation (i.e., destruction of biological activity) of the conalbumin in the raw egg.

Leoschke (1973) noted that, with pregnant mink and pre–weaning kits on a pellet program with a significant level of powdered eggs, the performance of the kits was excellent at weaning (4weeks of age); however, when at 4 weeks the kits transferred from mother's milk to a pelleted mash program containing high quality powdered eggs, their growth ceased and they became, essentially, runts. The experimental work that followed with a trace mineral chelated copper, iron, and zinc, resulted in outstanding performance of the dark mink kits but minimal performance of the pastel kits. That is, the performance indicated that dark kits are more sensitive to iron deprivation via conalbumin in egg white than pastel kits.

In further studies at the National Mink Research, Leoschke's work determined that ferrous sulfate at a level of 9 grams/100 kilogram of whole raw eggs prevents the iron deficiency in kits. Mink ranch diets including raws eggs from a "spun"method that produces a high percentage of egg white would require higher levels of iron supplementation to prevent "cotton mink."

Peroxidized Fats

Numerous reports from the scientific literature indicate that feeding oxidized fatty fish or fish oils to mink can lead to an anemia characterized by a decrease in erythrocyte number and hemoglobin content and reduced iron levels in the liver (Havre et al., 1973; Tauson and Neil, 1991; Tauson, 1993; Borsting et al., 1941; Engberg and Borsting, 1994; Frindt et al., 1996). The exact chemical basis by which peroxidized fats yield an iron deficiency anemia is unknown, but a good possibility is that the peroxide structures convert ferrous iron to ferric iron, which in turn, in an alkaline environment, would yield insoluble ferric hydroxide. Years ago in field observations, Scandinavian mink farmers noted using high levels of fish liver led to the development of iron deficiency anemia and cotton pelts (Rimeslatten, 1958). Very likely the high level of fish oils with high PUFA content produced fat peroxidation and reduced bioavailability of the iron.

Secondary Nutritional Deficiency—Anaemogenic Fish

Mink are unique among domesticated animals in providing an obvious evidence of iron deficiency without specific blood changes. Examination of the fur reveals "cotton fur," a flimsy textured underfur usually drab or white in color and closely resembling the appearance of cotton. The guard hairs are unchanged.

Without question, there is a genetic sensitivity to this nutritional disease. Within a given population of mink on a specific dietary program a very small percentage of mink with show "cotton fur." Specifically, according to Michels (1980) in observations at the National Research Ranch, among 800 breeder mink on a pellet program, there would be a litter or two with "cotton fur." Yet, interestingly, kits with "cotton fur" in June provided top quality pelts at the end of the year, showing that iron requirements may be lower in the fur development phase than in the early growth phase (June). These observations on the part of Michels had also been noted at Oregon State University

(Adair et al., 1959; Oldfield et al., 1960), both of whom found an obvious genetic factor in mink with "cotton fur." Studies by Helgebostad (1966) indicated that mutation mink, specifically Aleutian gene mink, sapphires and sapphire shadows, are more liable to develop the anemia than standard dark mink. These results clearly posit that some mink have a higher requirement for iron and/or a limited capacity for iron absorption.

"Cotton mink" have been a part of the North American fur industry for most of the 20th century. Mink ranchers have observed it as early as 1939 on ranches in Oregon, Washington, and Idaho, and as early as 1924 in Iowa and Illinois (Hummon and Bushnell, 1943). "Cotton fur" has also been observed in wild mink (Seton, 1929).

For the last half–century, it has been associated with the use of high levels of aneamogenic fish in practical mink ranch dietary programs. Commonly used aneamogenic fish include:

Coal fish	*Gadus virens*
Blue Whiting	*Gadus poutassou*
Atlantic Whiting	*Gadus merlangud*
Haddock	*Gadus aeglefinus*
Atlantic Hake	*Merluccius vulgaris*
Pacific Hake	*Merluccius merluccius*
Silver Hake	*Merluccius bilinearis*
Alaskan Pollock	*Pollachius virens*

In the mid–fifties North American mink ranchers across the midwest experienced a major outbreak of "cotton mink." The event was precipitated by the discovery that di–ethyl–stilbesterol, DES, a synthetic chemical with female hormone activity responsible for reproduction failure in mink, was being fed to beef and dairy cattle. To replace this feed resource, ranchers immediately replaced the tripe portion of their ranch diets with readily available Atlantic whiting, precipitating major economic loss directly due to "cotton fur."

The earliest report in the scientific literature from North America on the development of "cotton fur" in mink was Watt (1951) who observed it in mink on a ranch diet containing two thirds silver hake (*Merluccius bilinearis*). Her report was followed in Scandinavia by work from Helgebostad (1957) and Helgebostad and Martinsons (1958) with mink fed three quarters of the diet as raw coal fish (*Gadus virens*).

In addition to genetics and nutrition, age may also be a factor in "cotton fur." Only young mink develop "cotton fur," which suggests that depigmentation occurs only when the demand for iron coincides with the stress of both fur growth and cell proliferation.

Solutions to the problem were varied. Early on, it was noted that cooking the silver hake at 93° Celsius (200° Fahrenheit) provided superior growth and no "cotton fur" (Stout et al., 1960a). Also, there was no "cotton fur" with the cooking of raw coalfish (Helgebostad and Martinsons, 1958;

Helgebostad et al., 1961; Havre et al., 1967; Helgebostad and Ender (1961). Apparently a heat labile factor was responsible for the "cotton fur" syndrome. Studies by Stout et al. (1960b) indicated the value of intramuscular injections of organic iron (ferric hydroxide as Armidexan, Armour Vet Labs). In addition, experimental work by Helgebostad et al. (1961) indicated that the administration of perenteral iron supplements during the growth phase ameliorated the anemia. However, the employment of dietary iron supplements has met with mixed success. Scandinavian researchers Ender et al. (1972) and Skrede (1974) reported that ferric glutamate and ferrous fumarate were satisfactory supplements to prevent the dietary iron deficiency related to *Gaddae* fish with high levels of Tri–Methyl–Amine Oxide (TMAO). These same iron resources were able to restore essentially normal hemoglobin and hematocrit levels in` the "cotton fur" mink. But, at Oregon State University, in studies by Wehr et al. (1976) and Wehr et al. (1977) with ferric glutamate supplementation, negative results were noted. Studies with ferrous fumarate reduced but did not eliminate the anemia and "cotton fur" caused by feeding formaldehyde or 55% Pacific hake (*Merluccius merluccius*) (Adair et al., 1974). It is obvious that both experimental groups were working with the same nutritional disease but the resources employed to create the "cotton fur" syndrome involved two distinctly different fish species, coal fish with high TMAO content in Scandinavia and Pacific Hake with high formaldehyde content in the Oregon trials.

In terms of our present understanding of the "cotton fur" syndrome in mink, it is generally acknowledged that both TMAO and formaldehyde are the factors in fish which bind iron in a structure with minimum bioavailability for the mink (Rouvinen et al., 1997). Studies by Amano and Yamada (1964) indicate that TMAO is broken down by enzymatic action in the fish digestive tract to yield several products including formaldehyde.

In Scandinavia the emphasis on the dietary factor involved in "cotton fur" in mink has centered on TMAO (Ender and Helgebostad, 1968), intraperitoneal injections of which yielded a severe anemia in the mink (Ender et al., 1972; Helgebostad, 1976). Oregon State University studies have emphasized formaldehyde as the causative factor (Costley, 1970; Wehr et al., 1976). The Scandinavian studies indicated that fish with high TMAO and low formaldehyde consistently yielded "cotton fur" while the Oregon studies found that mink fed formaldehyde on a non–fish diet containing no TMAO exhibited severe anemia and "cotton fur."

The Oregon State University studies on the "unknown" factor in Pacific Hake responsible for "cotton fur" in the mink led initially to a patent application for the discovery of a new anti–bacterial chemical. "F.M. Stout, J.E. Oldfield, and J. Adair have found that Pacific Hake contains a wide spectrum antibacterial substance. The substance shows promise in the treatment of diseases and in food preservation. Ground up Pacific Hake was extremely resistant to bacterial spoilage stored at room temperature for several weeks without decomposing. This antibacterial substance has effectively prevented the growth of every pathogenic and nonpathogenic bacteria tested" (Anon., 1968).

239

Toxicity

For the most part, animals have a well regulated physiological program for iron absorption. Thus iron toxicity in animals and humans is rare, with the exception of specific genetic disorders.

Magnesium

Although about 70% of an animal's body magnesium (Mg) occurs in the skeletal structure, magnesium is also a key component of the soft tissues. Cardiac and skeletal muscle, as well as nervous tissue depend on a proper balance between calcium and magnesium cations for normal function. Magnesium is an essential co–factor for the phosphate–transfering enzymes myokinase, diphosphopyridine nucleotide kinase, and creatine kinase. It also is a co–factor for pyruvic carboxylase, pyruvic oxidase, and the condensing enzyme of the citric acid cycle. Magnesium has a significant role in the diseases of mink inasmuch as it is a key component in mink urinary calculi, struvite (magnesium ammonium phosphate hexahydrate) (Leoschke et al., 1952).

Physiology

Studies by Leoschke (1960, 1962b) indicate that about 9 to 18% of the magnesium intake of the mink is excreted via the kidneys. The initial study indicated that magnesium present in bone is not readily available to the digestive processes of the mink.

Requirement

Mink nutrition scientists vary widely in their recommendations for optimum levels of magnesium. Wood (1962) recommended 440 ppm and 396 ppm in breeder and grower rations, respectively, while the data of Warner et al. (1964) and McCarthy et al. (1966) indicated 625 ppm as adequate for normal growth and fur development of mink on a purified diet. The National Research Council sub–committee on furbearer nutrition (Travis et al., 1982) tentatively recommended the level of 440 ppm for mink ranch diets in the absence of more definitive data. This recommendation is consistent with the 400 ppm level recommended by the National Research Council sub–committee on cat nutrition (1986).

Resources

Fish are excellent magnesium resources for the mink, and cereal grains provide good resources.

Nutritional Deficiency
Primary Nutritional Deficiency

Although there are no experimental reports on magnesiuim deficiency in the mink, Chausow et al. (1985) provides specific information on signs of a magnesium deficiency in cats, a closely related carnivore. These include poor growth, muscle weakness, hyperirritatbility, convulsions, anorexia, reduced bone and serum magnesium concentration, and calcification of the aorta.

Secondary Nutritional Deficiency

Kangas (1976) emphasized that high levels of calcium and phosphate can decrease the absorption of magnesium from the intestinal tract. In studies with rats (Forbes, 1963) indicated that a magnesium deficiency could be produced only via the addition of both calcium and phosphate to diets already low in magnesium. These additions produced interference in magnesium absorption leading to high calcium/magnesium ratios in the tissues, and led to severe kidney calcification.

Toxicity

Magnesium is antagonistic to calcium and phosphorus, thus excess levels of magnesium in the ranch diet could lead to an inhibition of calcium and phosphate from the intestinal tract (Glem–Hansen, 1978).

Manganese

Manganese functions primarily in the formation of the organic mucopolysaccharide matrix of bone. Manganse is a co–factor for a galactotransferase enzyme that participates in the formation of the mucopolysaccharide. It is also involved in critical functions of amino acid metanbolism.

Requirement

The minimum requirement for optimum performance is not known. Wood (1962) recommends levels corresponding to about 40 ppm for breeder and grower diets. This level was obtained by analysis of adequate diets commonly fed to mink on the West coast of the United States and Canada.

The only specific study on the manganese requirement of the mink was reported by Erway and Mitchel (1973), whose study involved a specific color phase of mink, pastel mink with "screw necks" or head tilting. This unique characteristic of some pastel mink was directly related to a birth defect in which the otoliths (gravity receptors in the inner ear responsible for maintenance of equilibrium) are reduced in size or are absent. Animals displaying this defect have extreme difficulty in swimming and, depending upon the extent of the defect, may be completely unable to maintain equilibrium and consequently drown. The syndrome can be prevented by 1,000 ppm manganese supplementation to the mother during embryonic development.

241

Resources

Excellent resources for manganese include manganese sulfate and manganese chelates. Manganese oxide is considered to be an inferior resource for manganese.

Molybdenum

Molybdenum is an essential nutrient, being a constituent of the enzyme xanthine oxidase, a biological catalyst involved in the catabolism of nucleic acids and the formation of uric acid. Molybdenum is a trace element of major significance in animal nutrition in terms of its relationship with copper: elevated concentrations of molybdenum can induce a secondary copper deficiency while, in contrast, low levels of molybdenum can be associated with copper toxicity. Studies by Aulerich et al. (1983) showed that the addition of 200 ppm molybdenum as MoO_3 to a control diet resulted in a 25% reduction in liver copper concentration.

Requirement

At the present time, there is no experimental data provided in the scientific literature on the molybdenum requirement of mink. Studies at Cornell University with mink on purified diets indicated that 1.5 ppm molybdenum was satisfactory for growth and fur production (McCarthy et al., 1966). All factors considered, neither a deficiency of molybdenum nor a toxicity of molybdenum is very likely.

Potassium

In contrast to sodium, potassium occurs mainly in the intracellular fluid. The cellular elements of the blood contain about twenty times more potassium, as does the plasma; skeletal muscle contains about six times more potassium as sodium. Potassium influences the contractility of smooth, skeletal, and cardiac muscle and profoundly affects the excitability of nerve tissue. Potassium is also important for the fluid and ionic balance of the body.

Physiology

The digestibility of potassium is about 90% with normal mink ranch dietary programs (Enggaard–Hansen et al., 1985).

Requirements

In the absence of precise requirement data, Wood (1962) has suggested an amount equivalent to 0.3% potassium for breeder and grower diets. Studies at Cornell University with mink on purified diets indicated that 0.58% potassium was satisfactory for growth and fur development (McCarthy et al., 1966). These levels are consistent with the recommendation of 0.4% potassium for cats (NRC cat bulletin, 1986).

242

Resources

Good sources of potassium include cereal grains, skimmilk powder, fish, and meat products.

Nutritional Deficiency
Primary Nutritional Deficiency

Although no experimental data is available on potassium deficiency in mink, signs in the cat include anorexia, retarded growth, emaciation, lethargy, locomotive problems, unkempt fur, and hypokalemia (NRC cat bulletin, 1986). Hypokalemia may be present in the case of nursing sickness in the mink, a condition often seen in lactating females with more than seven kits. Histological examination of these mink show changes in the kidneys in the formation of small blisters which could indicate a potassium deficiency which sometimes occurs in the event of a negative energy balance, as when the muscles break down to yield energy for milk production (Jorgensen, 1985).

Secondary Nutritional Deficiency

Usually about 90% of the potassium content of a normal ranch diet is absorbed. However, with the use of relatively high levels of beet pulp or potato pectin, this absorption can be significantly reduced, leading to a secondary deficiency of potassium (Enggaard–Hansen et al., 1985). In a later report, Moller (1986) it was noted that the excretion of sodium and potassium via the feces was correlated to the increasing amount of feces. The molar concentration of sodium and potassium was constant when beechwood or sugar beet pulp was used as a supplement, while wheat bran had little effect on these ions.

Toxicity

It is essentially impossible to produce hyperpotassemia in mink with normal circulation and renal function.

Selenium

Selenium is a constituent of the enzyme glutathione peroxidase which catalyzes the breakdown of hydrogen peroxide and other peroxides formed in metabolism. In animal physiology, selenium and vitamin E (alpha tocopherol) are a working pair. Vitamin E, as a natural antioxidant, acts to prevent the development of peroxide structures, while selenium, as a co–factor for glutathionine peroxidase, functions in the destruction of peroxides that may occur. In terms of fur animal nutrition, Mertin et al. (1991) provided experimental data indicating that supplementing a silver fox diet with a selenium salt had a significant effect on fur development, lengthening undercoat hairs in males and even more markedly in females.

243

Requirements

Kiiskinen and Makela (1977) have reported on the selenium content of Finish mink ranch diets with a range of 0.05 to 0.42 ppm selenium dehydrated basis. Studies by Aulerich and Napolitano (1986, 1993) indicated that 0.8 ppm dietary selenium (dry matter basis) supported satisfactory growth, fur production, and reproduction/lactation, and that supplementation of this dietary program with up to 1.6 ppm selenium as sodium selenite produced no significant increase in weight gains or fur quality. Studies at Cornell University with mink on purified diets indicated that 0.25 ppm selenium was satisfactory for growth and fur development (McCarthy et al., 1966).

Resources

Fish products are good sources of selenium; inorganic salts of selenite and selenate are sound sources often employed in commercial fortified cereal formulations.

Nutritional Deficiency

Selenium and vitamin E are interrelated in animal nutrition and physiology such that a deficiency of either of these nutrients will be intensified by a simultaneous deficiency of the other. Likewise, a deficiency of either selenium or vitamin E can be minimized by the significant presence of the other. Stowe and Whitehair (1963) determined that 0.1 ppm selenium added as sodium selenite to a vitamin E–deficient basal diet prevented all but minor vitamin E lesions in the mink.

With sub–optimum levels of selenium and/or vitamin E in an animal's diet, peroxide formation can damage cell membranes; thus, a dietary deficiency of selenium or vitamin E can cause extensive tissue damage. A selenium deficiency in calves and lambs results in skeletal and heart muscle degeneration termed "white muscle disease," also called "nutritional muscular dystrophy." Mink nutritional muscular degeneration syndrome has been reported in mink fed high levels of fatty fish (Knox, 1977; Brandt, 1983, 1984; Henriksen, 1984).

Toxicity

Studies by Aulerich and Napolitano (1986, 1993) showed that the addition of up to 3 ppm selenium to a conventional mink diet (as fed basis) was not detrimental to reproduction or early kit growth. However, supplementation with 10 ppm or higher levels of selenium caused a marked reduction in feed consumption, at which point the experimental feeding trials were ended.

Studies by Aulerich and Napolitano (1993) showed that, on a basal ranch diet with a selenium level of 0.8 ppm on a dry weight basis, supplementing with selenium levels as high as 3.0 ppm had no adverse effect on the reproduction/lactation performance.

244

Sodium Salts

Since the nutritional requirements of the chloride anion are closely associated with the sodium cation, the chloride requirements are being discussed at this point. Sodium cations and chloride anions are important in the regulation of the acid–base balance of body fluids and determine to a large extent the osmotic pressure of the extracellular fluids. Chloride anions are concentrated by gastric cells in the stomach lining and secreted as hydrochloric acid, important for protein and carbohydrate digestion. Studies by Eriksson et al. (1983) indicate that with some ranch diets, supplemental salt increases feed volume significantly.`

Physiology

In humans under normal conditions, 90% of ingested sodium is excreted in the urine, usually in the form of sodium chloride and sodium phosphates. In the mink, because of the very short passage time through the alimentary tract, the digestion of sodium is significantly lower than in other monogastric species. Digestibility of sodium is only about 65–77% (Enggard–Hansen et al., 1985; Lund, 1979; Eriksson, 1983).

An experimental study of Clausen et al. (2002) is of interest. The Control Diet provided 0.17 grams of salt/ 100 kcal and salt supplementation yielded experimental diets with 0.25 and 0.70 grams of salt/100 kcal. Judging from the concentration of sodium in the urine, it seemed that mink kits cannot concentrate urine more than to a final content of 250 nmole sodium/ liter of urine (plus other anions). This level of maximum capacity is achieved when the sodium chloride content in the feed is no more than 0.50 grams/100 kcal. Further additions of salt will increase the water needs for the kits. Aldersterone in the plasma of lactating mink at 42 days was at a normal level of <200 pg per ml when sodium chloride content of the feed was 0.50 grams/100 kcal. or above. In earlier studies, Clausen et al. (1996a) showed normal plasma aldersterone levels when the feed content was 0.42 grams of salt/100 kcal. From these investigations, it appears that 0.42 – 0.50 grams of salt/100 kcal is able to support the top performance of both lactating mink and their kits. With ranch diets providing 120 calories/100 grams in this investigation, salt levels corresponded to about 1.5% salt on a dry matter basis.

Requirements

Inasmuch as sodium and chloride ions are not deposited in body tissues in significant quantities, the mink's requirements must therefore be covered by daily intake. 0.5% salt (dehydrated basis), the equivalent of 0.2% sodium cation and 0.3% chloride anion, should be ample for mink throughout the ranch year (Travis et al., 1982). Danish mink ranch diets contain from 0.12 to 0.24% sodium, except during the lactation period when a level of 1.3 to 1.5% salt is recommended (Glem–Hansen, 1978) for the prophylactic treatment of "nursing sickness." This recommendation of Glem–Hansen is confirmed in later studies by Clausen et al. (1996) and Clausen et al. (2002).

Resources

For the most part, the natural content of sodium and chloride in mink feed ingredients should cover the mink's requirements for the ranch year with the exception of the lactation period. Commercial fortified cereals provide ample salt except for the lactation phase where extra salt supplementation is recommended at 0.25%–0.5% of the ranch diet; the rancher is advised to work with the lower level before working with the relatively high level of 0.5%.

Nutritional Deficiency
Primary Nutritional Deficiency

Although no experimental studies have been reported on salt deficiency in mink, studies with another carnivore, cats, indicate deficiency signs as weight loss, severe alopecia, and dryness of skin. A chloride deficiency may cause a severe alkalosis (NRC cat bulletin, 1986). All factors considered, in terms of the salt content of most mink feed ingredients, a serious salt deficiency is unlikely except for the special case of the lactation requirements of a nursing female with a large litter.

Secondary Nutritional Deficiency

In a normal mink ranch dietary program, about 65% of the sodium content of the diet is absorbed. However, this absorption can be cut significantly in the presence of relatively high levels of beet pulp or potato pectin, leading to a potential secondary deficiency of sodium provided that the initial dietary plan was relatively low in sodium (Enggaard–Hansen et al. (1985). In a later report, Moller (1986), it was noted that the excretion of sodium and potassium via the feces was correlated to the increasing amount of feces. The molar concentration of sodium and potassium was constant with beechwood or sugar beet pulp supplementation, while wheat bran had little effect on these cations.

Toxicity

Many factors can influence an animal's need for salt including the concentrations of salt in feed ingredients and drinking water, ambient temperatures and humidity, the concentration of other ions (potassium and sulfate) in the diet, and the genetic differences between animals. At the same time, too much salt can have adverse effects, especially if drinking water is limited.

Salt toxicosis in mink is usually directly related to reduced water intake via water deprivation. A number of experimental studies indicate that mink ranch diets providing the mink a level of salt as high as 6.0% dehydrated basis is quite satisfactory for the mink over a short time period (Ericksson et. al., 1983, 1984, 1986; Makela and Valtonen, 1982) and over a long time period (Fjeld–Pedersen, 1974; Adair et al., 1979) provided that ample water was provided the animals. Experimental studies by Aulerich et al.(2000) indicated that up to 3% salt added to a typical conventional (wet) mink diet with unlimited drinking water does not have a detrimental effect on mink breeding, gestation, or whelping; nor does that level adversely affect litter size, kit body weights through 3 weeks of age, or kit mortality through weaning.

Experimental studies have showed that mink tolerated water with 1% salt as the only source of drinking water (Fjeld–Petersen, 1974; Makela and Valtonen, 1982). Studies by Gorham and Dejong (1950) indicated that withholding water from mink for three days will kill mink under average weather conditions.

Acute Toxicity

Studies by Gorham and Farrell (1962) indicated that a single feeding of a 3% (9% dehydrated basis) ranch diet provided visible salt toxicity signs and 2/9 mink died. Single doses of 7.2, 9.0, and 18.0 grams of salt produced death within 20 to 72 hours in 7/8 mink. At the 27.9 gram level (45% salt dehydrated basis), the mink simply refused to eat their feed. In another short term study, mink were given diets containing 0, 1, 2, or 4% supplemental salt for a period of seven days with water *ad libitum* (Restum et al., 1995). Restricted water availability at 50% and 25% *ad libitum* were required over the next 14 days to produce symptoms of salt toxicity.

The clinical signs of salt toxicity/water restriction in this study were increased thirst, mild dehydration, decreased feed consumption, decreased body weight, rough coat, crusty nose and eyes, irritability in the early stage, and lethargy in the later stages. In the Gorham and Farrell study, the mink were weak and depressed. Incoordination and excitability, along with stiffness and paralysis were observed. Their eyes were sunk in their sockets and glued shut with a clear fluid. A characteristic nervous syndrome usually followed within a day and continued until death. A recumbent mink, when touched by a stick, suddenly convulsed, clamping its jaws.

In the Restum et al. (1995) study, brain sodium concentrations were not affected by salt supplementation when drinking water was provided *ad libitum*. However, restricting drinking water generally resulted in increased brain sodium concentration. Mild to moderate micro or macro vesicular changes were observed in the livers of some mink fed each higher level of salt, but these were especially prominent in mink on restricted water.

Chronic Toxicity

When excessive salt is contained in a ration, mink usually will detect it quickly and refuse to eat the feed. Unfortunately, in some cases, the mink will eat the highly salted food and die as a result. Accidental salt poisoning killed thousands of mink in Finland in spring, 1982 (Eriksson et al., 1983), and more recently in Canada (Oswald, 1995). A manufacturing error produced a fortified cereal product with 13.2% salt (5.2% sodium) and consequently a ranch diet containing about 5% salt. When mink were fed this diet during the reproduction phase, the net result was a loss of about 10% of the breeding herd within a few days.

247

Zinc

Zinc is a key component of multiple metalloenzymes including plasma alkaline phosphatase, alcohol dehydrogenase, and thymidine kinase. Thymidine kinase is involved in the ATP (Adenosine TriPhosphate) activation of thymidine, one of the initial steps in DNA polymer construction. A zinc deficiency causes limited cell division which results in dramatic effects on tissues with a rapid growth rate, such as the skin. Dermatitis is a classic sign of zinc deficiency in animals.

Physiology

In zinc balance studies, Mejborn (1986, 1988, and 1989) noted that the fecal excretion of zinc was rather constant, 60% in adult mink and 80% in growing kits receiving normal to high dietary levels of zinc (40 to 340 ppm dry matter). Urinary excretion on these diets was 2–15%, at the highest level of zinc intake. Mejborn's employment of radioactive dilution techniques to measure the endogenous excretion of zinc into the digestive tract is of interest (1986, 1988, and 1990). At a very low level of zinc intake (< 12 ppm), the endogenous excretion was low but accounted for an important part (57%) of the total amount of fecal zinc. At very high levels of dietary zinc intake (500 ppm), the absolute amount of endogenous zinc was higher but in percent of total fecal zinc excretion it was less than 7%. At higher intake, the homeostasis was regulated via absorption from the digestive tract.

Mejborn's studies deteermined that zinc concentration of the liver was not a good indication of zinc status. More reliable was the zinc concentration in the pancreas and bone (femur).

In another study by Mejborn (1987) involving relatively low levels of dietary zinc (5 to 25 ppm dry matter), fecal zinc accounted for 30–95% of the zinc intake while urinary losses of zinc were only 1–2% of intake. Also in this study was a tendency for mink on the highest protein intake to have a higher zinc balance in mg/day.

Requirement
Reproduction/Lactation

Studies by Pingel et al. (1992) indicated that a daily zinc intake of less than 6 mg by female mink before and during the breeding season, compared with an intake of 10 mg zinc/day, tended to reduce litter size. Assuming a feed intake of about 60 grams dry matter/day for a breeding female, a level of about 100 mg/kg (100 ppm dry matter zinc) would be sub–optimum for the reproduction/lactation period. Obviously, the presence of relatively high dietary calcium levels typical of Scandinavian mink ranch diets may be a factor in this relatively high zinc requirement for the reproduction/lactation phase. This very point is raised by Pingel and associates. For several animal species a high level of dietary zinc (50 to 100 ppm dry matter) is required for optimum fetal development; lacking this level, a zinc deficiency causes severe congenital abnormalities (NRC cat bulletin, 1986).

Growth

In his book *Mink Production* (1985), Jorgensen recommends 20 ppm zinc for a ready–mix feed, about 60 ppm dry matter basis. He also comments that Danish feed centers provide 25–60 ppm in their ready–mix feed, actually one to three times the recommended norm. For cat nutrition, 15 ppm dry matter is recommended for kittens fed diets containing low quantities of compounds known to decrease zinc bioavailability, phytates and fiber. With dietary programs providing significant levels of phytates and fiber, 50 ppm zinc in the dry matter is considered to be sufficient (NRC cat bulletin, 1986). It is of interest to note that Brandt (1983) found that a level of 90 ppm dry matter was satisfactory for ensuring the measured activity of plasma alkaline phosphatase.

Fur Development

A study by Lohi and Jensen (1991) indicated that dietary levels of zinc were not reflected in the hairs. On the other hand, studies by Pingel et al. (1992) indicated that a very low dietary zinc levels was reflected in the zinc content of the hair of growing mink. Tjurina and Tjutjunnik (1982) reported the zinc content of mink hair to be in the range of 238–275 ppm dry matter.

Resources

Excellent zinc compounds for mink nutrition include zinc sulfate and zinc chelates. Chelates of trace minerals are special combinations with amino acids and/or proteins which provide high bioavailability for the mink. Zinc oxide is a water insoluble product considered sub–optimum for animal nutrition.

Nutritional Deficiency
Primary Nutritional Deficiency

No experimental studies have been noted relative to a primary nutritional deficiency of zinc. On common mink ranch dietary programs, a zinc deficiency is very unlikely, since commercial fortified cereals provide ample levels of zinc.

Secondary Nutritional Deficiency

Multiple factors including the cations cadmium, copper, calcium, and ferrous iron may compete at receptor sites for zinc absorption. Organic inhibitors of zinc absorption include phytate (myoinositol hexaphosphate) and components of dietary fiber including hemicellulose and lignin. The inhibition of zinc absorption by phytate appears to be through co–precipitation with calcium and phytate to form an insoluble complex. It is likely that products of the Maillard reaction (Lykken et al., 1982) and amino acid–phytate (from soybean oil meal) compounds formed during food processing can also inhibit zinc absorption, (Erdman et al., 1980).

Toxicity

Reproduction/Lactation

A study by Bleavins et al. (1983) on the toxicity of zinc for mink is of major interest inasmuch as the experimental work involved both pregnant mink and their offspring through 22 weeks of age. The level of zinc was initiated at 1560 ppm dry matter in early January but increased to 3060 ppm dry basis on March 17. The exposure of the pregnant mink to 3060 ppm zinc on a dry matter basis did not result in grossly observable zinc toxicity or zinc induced copper deficiency. Gestational length, litter size, kit birth weights, and kit mortality to weaning was not affected by the zinc supplementation. However, the high level of zinc supplementation did produce achromatrichia, alopecia, lymphopenia, and a reduced growth rate in the offspring of the zinc–treated females. These mink kits also exhibited profound immunosuppression, but this was not a permanent defect inasmuch as a normal lymphocyte response was seen approximately 14 weeks after weaning and being placed on a zinc unsupplemented basal diet. Body weights of male kits were significantly lower at 12 weeks of age, but no significant difference in weight was noted at 22 weeks of age.

At 3–4 weeks the kits exhibited graying of the fur around the eyes, ears, jaws, and genitals with a comcomitant loss of hair and dermatosis in those areas. During the next few weeks, the achromatrichia and alopecia consistent with a copper deficiency spread over much of the body, even though the anemia and neutropenia associated with a copper deficiency were not found. A reduced hematocrit value without corresponding decrease in red blood cell count was also observed. Mean Corpuscular Volume was significantly reduced at p less than 0.10 but not at p less than 0.05.

A recent experimental study (Clausen and Sandbol, 2005b) on zinc toxicity in mink during the weaning period from May 24 to June 24 is of interest. The experimental data indicated that at a level of supplemental zinc oxide of 1,000 ppm with a conventional mink ranch diet, only one observation of negative performance was made: the normal weight increaase was reduced.

A standard ranch diet with 40% solids prior to water addition supplementating at 1,000 ppm would provide about 2,500 ppm of zinc oxide on a dry weight basis. Considering the level of zinc already provided by the ranch diet, it is logical to assume that the actual level of zinc provided in the experimental diet exceeded 3,000 ppm dry weight basis. Obviously, Blevins and associates found more severe signs of zinc toxicity, the difference being the chemical sources of zinc— ZnO in the case of the Clausen and Sandbol (2005b) and $ZnSO_4$ in the case of the Bleavins study. As previously noted zinc oxide is a water insoluble product considered sub–optimum for animal and human nutrition in terms of limited bioavailability to the digestive process. For human nutrition, zinc sulfate is the trace mineral resource of choice (Olson, 1984).

An interesting case of zinc toxicity occured when breeder mink were placed in new zinc––coated cages and the feed was placed directly on the wire. Within a few days symptoms of anorexia, chronic diarrhea, weight loss, rough hair coats, and sunken eyes were observed. Within a week to ten days the colony of six thousand breeder mink lost more than a thousand mink. (Westlake and

Poppenga, 1988). Nephrosis and hepatic lipidosis, along with elevated tissue zinc levels and relatively low copper levels, were similar to lesions found in ferrets with zinc toxicity. In normal mink liver and kidney tissue, levels of copper were in the range of 19–40 ppm; while at the ranch with the new zinc– coated cages, the copper levels were in the range of 3 to 5 ppm. On the other hand, while zinc levels of normal mink liver and kidney tissues were in the range of 21–40 ppm, the levels at this ranch were in the range of 187–205 ppm.

Late Growth/Fur Development

Although mink may be relatively sensitive to zinc toxicity during the reproduction/lactation phase, they are relatively tolerant of high levels of zinc during the period from early July to pelting in early December. A study by Aulerich et al. (1991) involving a conventional ranch diet supplemented with 4,500 ppm zinc on a dry matter basis for 144 days noted no marked adverse effects on feed consumption, body weight gains, hematologic parameters, fur quality, or survival. Zinc concentrations in the liver, kidney and pancreas increased in direct proportion to the zinc content of the diet. Histopathologic examination of the livers, kidneys, and pancreas revealed no lesions indicative of zinc toxicosis. This work parallels that of Bleavins et al. (1983): a ranch diet providing about 3,060 ppm zinc on a dry matter basis did not result in grossly observable zinc toxicity or zinc induced copper deficiency in adult mink, with one difference. Major signs of zinc toxicity were noted in the offspring of female mink provided high levels of zinc as late as 12 weeks of age.

In another early study (Brandt, 1983), a ranch diet providing about 3,000 ppm zinc dry matter basis, only one mink kit out of 8 on the experimental program exhibited fatty liver and kidney metamorphosis, liver chirrosis, icterus, haemolysis, and wasting similar to mink on toxic levels of copper sulfate. At the same time, the author felt that the level of 3,000 ppm dry matter basis was not close to the actual toxic level of zinc. It is logical to anticipate that within a worldwide mink population some mink kits would be especially sensitive to zinc toxicity.

As, Ni, Si, Sn and Va

With animals on purified diets in a sterile, dust–free environment, arsenic. nickel, silicon, tin, and vanadium must be provided to animals for top growth performance and hair or feather development (Frieden, 1972).

5

Disorders Related to Nutrition Mismanagement

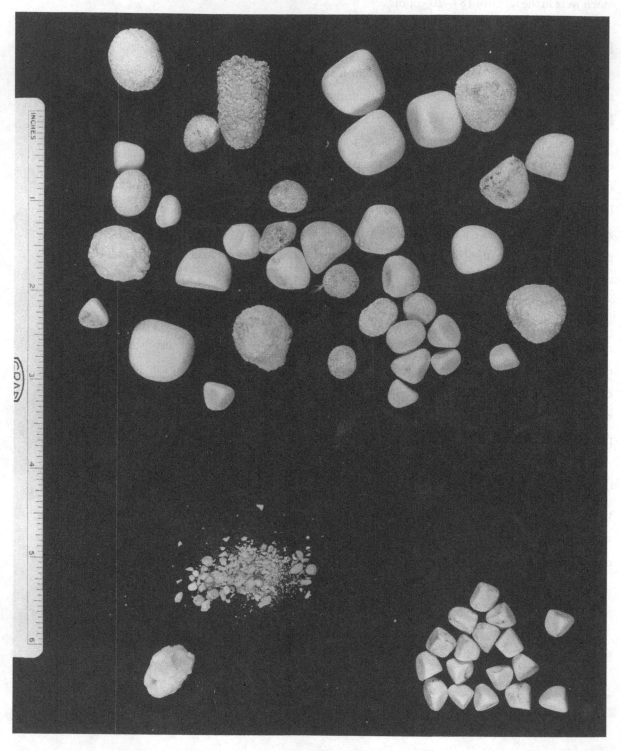

Acid/Base Imbalance

The mineral composition of the mink's diet essentially determines the acid/base balance of the diet. Feedstuffs containing an excess of inorganic cations including calcium, magnesium, potassium, and sodium relative to inorganic anions including carbonate, chloride, phosphate, and sulfate contribute to an alkaline environment in the feed and, physiologically, to a relatively high urinary pH. Obviously, vice versa, an excess of inorganic anions would contribute to an acidic feeding regimen and a relatively low urinary pH.

Large quantities of carbonic acid are produced via the catabolism of simple sugars, glycerol, fatty acids, and amino acids. This component does not cause a problem inasmuch as the lungs provide a route for loss of carbonic acid via carbon dioxide. The metabolism of simple sugars, glycerol, and fatty acids does not produce a net yield of acid for animal physiology. In the phospholipid catabolism such as that of lecithin, acidic dihydrogen phosphate anion is released. Catabolism of amino acids yields ammonia, which is converted in the liver to urea, a relatively weak base. The sulfur–containing amino acids cysteine, cystine, and methionine yield a bisulfate anion which has a pKa of 1.99, that is, an acidity value similar to that of phosphoric acid. Thus with a high animal protein intake, carnivores excrete an acidic urine with a normal pH range of 6.0 to 6.8.

Special supplementation of mink ranch diets can affect the acid/base balance. Fortified cereal containing calcium carbonate, limestone, has an alkaline effect inasmuch as calcium carbonate is a salt resulting from the combination of calcium hydroxide, a relatively strong base, and carbonic acid, a relatively weak acid. Ranchers are advised not to employ limestone in their mink diets because of its distinct potential to encourage the formation of bladder stones (urinary calculi) (Hartsought, 1951). Phosphoric acid has been used in ranch diets throughout the world since 1956 (Leoschke, 1956) as a palatable feed preservative and as a prophylactic to prevent urinary calculi in both mink and fox. Phosphoric acid is relatively easy to work with inasmuch as its acidity is only about 10% that of sulfuric acid (battery acid). North American mink ranchers have used it for many years to preserve raw eggs without refrigeration. The addition of 1–2% phosphoric acid (75% feed grade) provides a final pH of 3.5 to 2.4. With a pH of 3.5 or below, after 3 to 5 days, the bacteria count is zero (Zimbal, 1975).

Both sulfuric acid and hydrochloric acid have been effectively used in the preservation of fish as silage: after maturing, the silage has a pH in the range of 2.5–3.5 (Jorgensen et al., 1976; Poulsen and Jorgensen, 1976). Enggaard–Hansen and Glem–Hansen (1980) reported that the use of 2.5–3% sulfuric acid with fish and fish offal yielded a pH of about 2 in the initial raw material. Ranchers have also used glacial acetic acid (organic acid present in vinegar) as a feed preservative, but the value of this acid is minimal inasmuch as its acidity is only about 10% of phosphoric acid or about 1% of that of sulfuric acid.

253

Mink Feed pH Levels

Excessive acid levels in mink feed, for instance in the use of high levels of fish silage, can lead to reduced appetite, loss of weight, and a higher rate of mortality according to Enggaard–Hansen and Glem–Hansen (1980) whose studies attempted to determine optimum pH levels in practical ranch diets. In a study initiated in late July after a 20 day transition period using sulfuric acid silage at 1 to 2% per day, the control diet without fish silage had a pH of 6.6, the 20% silage diet had a pH of 5.3, and the 40% silage diet had a pH of 4.7. In the initial 4 weeks of the experiment researchers noted that in the two experimental groups the deposition of calcium, magnesium, and phosphorus was very limited; however, by mid–September they found no differences in the mineral deposition of the three groups. Mink feed pH levels as low as 4.7 and 5.3 did not lead to a reduction in growth over the nine week period of the research work. Other authors, including Jorgensen et al. (1976) and Poulsen and Jorgensen (1976), commented that mink feed pH below 5.0 (low values of base excess) yields significant acidosis resulting in loss of appetite, weight loss, and sub–optimum reproduction/lactation performance. Poulsen and Jorgensen (1977) commented that if the pH of a complete mink ration is lower than 5.5, it may cause changes in the acid–base balance of the animals. However, it appears that this particular acid–base imbalance is not of significant physiological concern inasmuch as, according to Jorgensen (1985), the blood of the mink can maintain optimum pH with a rather acidic feed (pH 5.0–5.3). If on the other hand, large quantities of fish silage preserved with sulfuric acid (giving a pH feed value of 4.5 or less) are used, the mink cannot itself maintain optimum blood pH and will die very quickly with even a slight change in the acidity of the blood. It is of interest to note that for a period of more than five years, National Fur Foods provided to 12 ranchers year round a mink redi–mix product with a pH range of 5.1–5.2. This feed resulted in top performance in all phases of the mink ranch year (Michels, 2001). The pH range of this feed was achieved with the employment of phosphoric acid (75% feed grade) at a level of 2.5% of dry matter.

Anemia

Anemia is defined as a blood disorder related to the quantity and/or size of red blood cells and/or the level of hemoglobin present in the red blood cells. A report by Kangas (1968) indicates that in male kits the hemoglobin levels rise from about 11.5 in early July to as high as 18.5 in early November. Studies by Jorgensen and Christensen (1966) indicated a close relationship between the level of hemoglobin in full grown mink kits and the pigmentation degree of the underfur, skin length, and the calculated fur value. Low hemoglobin levels in the kits during the period of fur formation (August 14 into December) resulted in pelts of poorer quality, especially pelts with more or less depigmented underfur. They recommended that, in order to obtain maximum fur development, nutrition be managed so that hemoglobin values rise gradually after mid–August and reach a maximum when the mink are full grown, about November 15.

254

Nutritional anemia in the mink can be related to primary nutritional deficiences of specific minerals and/or vitamins as noted in Table 5.1. It can also be found as a secondary nutritional deficiency of iron directly related to the employment of specific species of anaemogenic fish providing high levels of TMAO (Tri–Methyl–Amine Oxide) or formaldehyde, both of which bind iron in a structure with minimum bioavailability.

TABLE 5.1. Nutritional Deficiences and Anemia in Animals

Deficiency	Anemia Characteristics
Iron, Copper/Vitamin E	hypochromic, microcytic
FolicAcid, Ascorbic Acid, and Vitamin B_{12}	macrocytic (immature large red blood cells)
Vitamin B_{12}	pernicious anemia (macrocytic anemia with low levels of red blood cells)

It should be noted that mink can develop methaemoglobinaemia with nitrite poisoning a condition similar to carbon dioxide poisoning (Jorgensen, 1985).

Bladder Stones—Urinary Calculi

History

Bladder stones—urinary calculi or urolithiasis—is relatively common in ranched fur bearers. The condition seen in mink and foxes is similar to LUTD (Lower Urinary Tract Disorders) in domestic cats, dogs, and ferrets.

Bladder stones have been responsible for multiple deaths of mink since the very beginning of the mass production of mink as a domestic animal in North America. As early as 1936, a Virginia mink rancher (Harman, 1954) experienced significant losses of mink with urinary calculi when, unfortunately, he made a decision to move from his "homemade" mink cereal which did not contain bone meal or limestone to a commercial fortified mink cereal with 3% steamed bone meal and limestone.

A Canadian mink rancher, as reported by Harris (1939), first experienced the loss of pregnant mink with bladder stones as the direct result of changing his mink water resource from rain water to well water. A University of Wisconsin professor, Steenboch (1939), also commented on the occurrence of urinary calculi in mink and requested that mink ranchers send specimens of bladder stones to his laboratory. The mink rancher response was minimal, inasmuch as no concerted effort was made by a specific individual to provide follow up. However, in 1950, in his first year of graduate study, W. L. Leoschke did take up that challenge, obtaining the bladder stones from Wisconsin and Illinois ranchers. Through his research travels and studies, he determined the chemical composition of sixty (60) mink bladder stones.

As early as 1947, Chaddock, an American veterinarian, reported that his examination of

255

more than 4,400 mink indicated that more than 4% of the mink deaths were due to cystic calculi. Other reports indicated that the loss of pregnant mink with urinary calculi was as high as 8–10% (Harris, 1970; Hartsought, 1951). The Hartsought data shows how serious the situation was in 1951, that the loss of pregnant mink with urinary calculi represented a loss of a mininum of $1,000,000 for the 5,000 mink ranchers in the United States.

The incidence of bladder stones was seasonal, with major losses of pregnant mink in the spring just prior to whelping and later losses with male in the late growth phase of the ranch year, July–August.

A wide variety in the chemical composition of mink bladder stones has been noted including calcium oxalate, cystine, struvite, and urates.

Struvite Calculi
Composition

Struvite uroliths are the predominant form of urinary calculi reported in mink (*Mustela vison*), ferrets (*Mustela putorius furo*), and other weasels of the *Mustelidae* (Edors, Ulrey and Aulerich, 1989), as well as carnivore companion animals. In more recent years, calcium oxalate stones have become more common (Zeller, 1996).

From gross appearance, mink urinary calculi can be divided into two groups: hard, smooth, white and crystalline and friable, rough, colored calculi. The size varies from sand–like particles to stones weighing a full ounce.

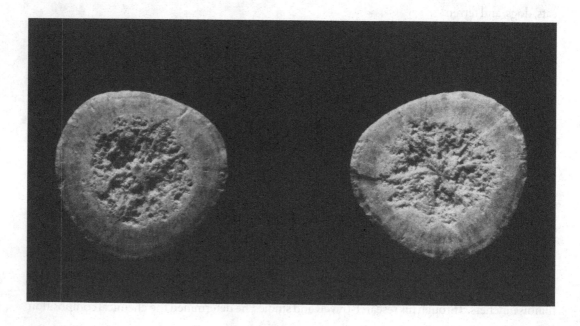

256

Chemical analyses indicate that the two types of mink urinary calculi are essentially identical in composition, the main constituent being magnesium ammonium phosphate hexahydrate, $MgNH_4PO_46H_2O$. There are also traces of calcium and magnesium phosphates (Leoschke, Zikria, and Elvehjem, 1952).

One of the earliest studies on the composition of mink bladder stones was that of Smith and Hodson (1941) which indicated that the primary salt present was magnesium ammonium phosphate with lower levels of calcium. The ash level of 82% indicated that the salt was a hexahydrate. Later studies (Bassett, Harris, Smith, and Yeomen, 1947; Sompolinsky, 1950; Konrad, Hanak, and Mouka, 1973; Nguyen, Moreland, and Shield, 1979) confirmed the earlier analytical work.

Etiology

Struvite salts crystallize from the urine above the pH range of 6.0–7.5; the lower the pH of the urine, the greater proportion of the salt present in solution. In vitro studies have shown that the compound is quite soluble in water in pHs below 6.0; however, there is a sharp drop in the solubility of the salt in a pH range of 6.0 to 7.5 (Vermeulen, Ragins, Grove, and Goetz, 1951).

Very specific conditions of pH, urine concentrations of magnesium, ammonium cations, and phosphate anions, and a nucleus such as cellular debris or bacteria are required for stone formation in the kidneys or bladder. With an optimum urinary environment, such as a urinary tract infection, calculi formation may take place in a matter of days; it may also happen over a matter of weeks as seen in the mink pregnancy period. In terms of the physiology of pregnancy, the final weeks of pregnancy sustain the concentration of ammonium cations at its highest level, providing an environment most supportive of struvite formation.

With proper mink nutrition and ranch management and with normal kidney phsyiology, the magnesium, ammonium, and phosphate ions remain in urinary solution. Without satisfactory nuclei, they may precipitate out of solution as a fine powder which is easily flushed out of the mink's urinary system.

All factors considered, the formation of struvite urinary calculi in the mink requires the concurrance of multiple predisposing factors including:

Urine Concentration

Obviously, restricted water intake in winter weather and/or limited water resource in the mink feed could lead to a relatively high concentration of the cations and anions required in the urine for calculi formation. An increased concentration of the urine reduces the frequency of urination and thereby the opportunities for flushing out small struvite crystals from the bladder and urethra. This comdition supports a urinary environment likely to ensure calculi formation (Neil, 1992).

257

Ammonium Cation Concentration

Normally a high proportion of the nitrogen in the urine is in the form of urea synthesized in the liver via the Krebs Urea Cycle involving carbamyl phosphate and ornithine (Wamberg and Tauson, 1998). However, during the pregnancy of mammals, the urinary concentration of urea decreases and the concentration of ammonia and/or ammonium cation increases due to physiological changes especially in late pregnancy. These are directly related to fetal growth, as carbamyl phosphate is being diverted to the synthesis of RNA and DNA and as the amino acid, ornithine, is being diverted to the synthesis of the polyamines, spermine and spermidine, for fetal cellular growth. These significant changes in the physiology of pregnant mink and the subsequent modification of urine composition in late pregnancy provide a logical basis for two observations of Hartsought (1951).

He found no urinary calculi in 500 non–pregnant mink, but 400 out of 5,000 pregnant mink were lost through urinary calculi prior to whelping. He concurred with rancher observations that the mink so lost were carrying six or more fetuses. The large litters involved the females in a physiological crisis which reduced urea concentration and enhanced concentration of ammonia/ammonium cation, thus supporting the formation of magnesium ammonium phosphate hexahydrate crystals or urinary calculi.

Acid–Base Balance

Inasmuch as the struvite salt, magnesium ammonium phosphate hexahydrate, is soluble in a urinary environment of pH below 6.0, it is obvious that the pH of mink urine is an important factor in the formation of struvite calculi. This specific point has been known as early as 1941 (Smith and Hodson, 1941; Kubin and Mason, 1948).

Akaline Feedstuffs

From the earliest days of mink ranching in North America, mink ranchers have added alkaline feedstuffs to their ranch diets, inadvertently contributing significantly to losses of their animals through urinary calculi.

Hard Water

Harris (1970) has reported field observations on the Marena Fur Farm, St. Mary's, Ontario, Canada, yielding specific ranch data on the correlation between the use of hard (well) water and the incidence of bladder stones in pregnant mink. In the seven years 1933–1938 and 1940, rain water was used, with no bladder stone incidents. In the two years 1939 and 1941, well water was used with the loss of 10% of pregnant mink with urinary calculi.

258

Limestone and/or Steamed Bonemeal

Rancher observations as early as 1936 and 1946 indicated a direct correlation between the incidence of urinary calculi and the employment of as much as 3–4% steamed bone meal and limestone in commercial or homemade mink cereals (Harman, 1954; Hartsought, 1951).

Studies at the University of Wiscosnin (Leoschke, 1960) provided important information on the relationship between limestone (calcium carbonate) in the mink's diet and higher urinary pHs: 0.5% limestone with Ca/P ratio of 1/1 yields pH 6.9, and1.5% limestone with Ca/P ratio of 3/1 yields pH 7.8.

Sadly, despite published commentary on the relationship between calcium carbonate in the ranch diet and the formation of urinary calculi, as late as 1969, 2.5% chalk (calcium carbonate) was a typical component of mink cereals in Great Britan (Rice, 1969). As late as 2000, a Wisconsin mink rancher (Kurhajec, 2000) observed, with the use of raw eggs with shells (almost pure calcium carbonate), a significant increase in the incidence of bladder stones in male kits.

Magnesium Oxide

Magnesium oxide, like calcium salts, produces for a rise in animal urinary pH and potential for an increased production of struvite crystals and stones, inasmuch as both magnesium and calcium are members of the Alkaline Earths family of the elements (Clausen and Wamberg, 1998; Clausen, 1999). These studies provided an equation for a linear correlation between BE (Base Excess) and the urinary pH:

$$pH = 5.04 + 0.087BE \text{ (in nml/100 g. of wet feed)}.$$

Vitamin A Deficiency

In studies with fox, it was noted that the incidence of calculi in the kidney and bladder was associated with the levels of vitamin A provided in the experimental diets. No calculi were found in the group receiving 100 IU of vitamin A/kg (Bassett,Harris, Smith, and Yeomen, 1947). This amount precludes a vitamin A deficiency. With such a deficiency, epithelial tissues (skin, urinary tract) are sloughed off, possibly leading to cells which could act as nuclei for struvite stone formation.

Bacterial Infections

Most animals affected by uroliths have a history of chronic infections of the urinary tract with Staphylococcus, Proteus, or E. coli. In ferrets, bladder stones can form in these conditions overnight (Bell, 1996).

Without question, bacterial infections of the urinary tract are a primary factor in urinary calculi, especially in male kits. The earliest detailed studies on the involvement of urinary tract bacterial populations in mink bladder stone formation were conducted by Iver Nielsen (1955, 1956).

He was able to isolate micrococci from infected mink and then induce calculus formation in experimentally infected animals. The isolated micrococci included Staphylococcus (micrococcus pyrogenes var. albus) and Proteus mirailis. These microorganisms possess a urease enzyme that hydrolyzes urea to ammonia, thereby raising both the urinary pH ammonium ion concentration and promoting the formation of struvite stones, Magnesium ammonium phosphate hexahydrate. The bacteria may act as a nucleus for calculi formation, but Nielsen also found that nuclei resources other than bacteria may be responsible for urinary calculi creation. Renal plaques may yield calculi where no bacteria culture was isolated; in fact, in 188 healthy mink, he found plaques consisting of calcium deposits in 43, nearly 40%. A plaque may increase in size and break in contact with urine in the renal pelvis, then act as a nucleus for urinary calculi formation.

Later studies by Lauerman (1962, 1965) provided new insights into those factors predisposing mink kits to urinary tract infections and subsequent bladder stone formation followed by deaths because of "bloody bladders."

These include a vitamin A deficiency, DES (DiEthylStilbesterol), estrogens, trauma, and an unsanitary pen. Improperly administrated phosphoric acid can actually lead to struvite formation inasmuch as sub–optimum levels of phosphoric acid produce phosphate anions for calculi formation but a urinary pH too high to solubilize the struvite components.

The bacteria that cause "bloody bladder" in male kits are found in the prepuce (the preputial sac) of apparently normal healthy mink. When the mink become susceptible, the bacteria in the prepuce ascend the urinary tract and cause the disease that leads to the animal's death via a toxemia resulting from obstruction of the urethera by struvite calculi and/or an inflammatory fibropurulent exudate.

DES, a synthetic estrogen, produces both a change in the kinds of cells found in the epithelial lining of the urinary passage and a sloughing away of patches of epithelium. This sloughing encourages bacterial invasion of the urinary tract, possibly by Staphylococcus aureus and Proteus mirabilis, normal components of the preputial flora in healthy mink (Lauerman and Berman, 1962, 1962c). Lauerman found that metaplasia induced by estrogens may occur without the development of infection, lacking significant bacteria carried in from the prepuce.

Prophylactic Treatment—Bladder Stones–Urinary Acidification

As early as 1940, the use of a urinary acidifier was recommended as useful in the prevention of bladder and kidney stones in the mink (Koch, 1940):

"It was suggested recently in one of the monthly fur farm publications that the addition of two drops of muriatic acid (hydrochloric acid, HCl) to each pint of drinking water continuously would be useful in eradicating bladder and kidney stones. Upon close check–up of this theory, there is no evidence in the literature pertaining to scientific research work that would indicate that muriatic acid has any beneficial effect upon bladder stones at all. Surely it could do no harm, so there would be nothing wrong in trying." Ironically, this same individual recommended, "And add a little finely ground limestone to the mix."

In my experience of more than 50 years involvement in mink nutrition research studies, I have seen repeated cases where rancher experience and wisdom (anecdotal observations) were simply ignored by the scientific community. Witness for instance, the Oregon mink rancher's comment to Dr. John Gorham that wheat germ meal (a relatively rich source of vitamin E) was useful in the prevention of "Yellow Fat" disease of the mink. This nutritional disease is brought about by oxidized PUFA (Poly–Unsaturated–Fatty–Acids) in salmon waste with their destruction of the vitamin E content of the mink ranch diet, in short, by second rate mink nutrition management. The problem was finally resolved by supplementating ranch diets with vitamin E.

Ammonium Chloride

Early in his graduate experience at the University of Wisconsin, W. L. Leoschke was given a new project in addition to his initial work on the unknown vitamin requirements of the mink. The research area of concern was that of bladder stones in pregnant mink just prior to whelping, a problem that was costing the United States mink industry a million dollars annually. This was clearly an opportunity of a lifetime, a chance to resolve a major mink industry problem of nutrition mismanagement and/or misunderstanding of mink physiology during late pregnancy which had led to very significant losses of mink. Fortunately, with the cooperation of two local mink ranchers, Tony Werth and Elmer Christiansen, the problem was resolved.

Analytical data on 60 mink urinary calculi collected throughout Wisconsin and Illinois (Leoschke, 1951) determined that the primary component of the stones was struvite (magnesium ammonium phosphate hexahydrate). A literature search resulted in an examination of a paper by Verculen et al. (1951) who reported creating urinary calculi in rats and then decreasing the size of the stones by using a urinary acidifier, ammonium chloride. This substance, when metabolized by the animal's liver, is converted from the relatively weak base, ammonia, to a weaker base, urea, with resulting acidification of the urine output.

The prophylactic dosage of one gram of ammonium chloride/mink/day was calculated by comparing the weights of the rats on the Vermeulen et al.(1951) study and the weight of mink breeder females (Leoschke, 1954; Leoschke and Elvehjem, 1954).

Data from the initial study follows.

Group	Number	Animals Lost with Cacculi
Control	400	16
Experimental	200	1*

* Death of this pregnant female occurred within one week of the start of the experiment, indicating that the urinary calculi were already present at a size large enough to bring about obstruction and death

One of the immediate responses from mink veterinarians was that Leoschke was "out of his mind" to use such a high dosage of ammonium chloride. Thus the following year, field studies with 50,000 mink on ranches throughout North America were initiated with a dosage of only 1/2 gram of ammonium chloride/mink/day. However, these studies were disappointing inasmuch as the overall loss of pregnant mink with urinary calculi on some farms ran as high as 3%.

Ammonium chloride at a level of one gram/mink/day provided an average mink urinary pH below 6.0, a pH likely to prevent any formation of struvite calculi.

Experimental studies in Denmark (Clausen, 1999) on the use of ammonium chloride as a prophylactic treatment, provided the mink with 0.35% of the ranch diet, 1% on a dehydrated basis which yielded a urinary pH below 6.6. This trial was effective even though the level of ammonium chloride is significantly lower than that used by Leoschke in 1952. The basic point is simple: "Listen To The Mink." If the lower level of ammonium chloride is effective in the prevention of mink bladder stones, use that level.

Ammonium Chloride and DL–Methionine

Recently, Westlake and Newman (1985) conducted a field study using a combination of ammonium chloride and DL–methionine for the prevention of urinary calculi in male kits, a program that had been used for many years for the prevention and/or cure of cats with struvite stones. The prophylactic medicinal level included 75 mg of ammonium chloride and 300 mg of DL–methionine/mink/day, equivalent to one pound of ammonium chloride and four pounds of DL–methionine per 6,000 mink/day. The recommendation of Westlake (1988) for the reproduction phase of the mink ranch year, 0.5 pounds of ammonium chloride and 2 pounds of DL–methionine, have been lowered in terms of the relative weights of male and female mink.

Methionine is a urinary acidifier inasmuch as the animal body converts a significant quantity of methionine to sulfate anions which are excreted by the kidneys. Note that sulfuric acid is 10 times as strong as phosphoric acid. The advantage of the ammonium chloride/DL–methionine approach is that, since the combination product is less bitter than ammonium chloride, it increases palatability and may be a better choice for feed economics.

Phosphoric Acid

Since 1956, phosphoric acid has been the urinary acidifier of choice for most North American mink ranchers because of its specific advantages. It has excellent feed preservation "on the wire," providing a final feed pH as low as 5.1–5.2 at the recommended level of 2.5% of the 75% feed grade acid on a dehydrated feed basis. Leoschke (1996) recommends this level not only for better feed preservation but also as a superior prophylactic to treat bladder stones. It has higher palatability than ammonium chloride for the mink, and it has a specific potential to enhance the availability of calcium present in bone products in the ranch diet, as the phosphoric acid is able to convert the tricalcium phosphate of bones into dicalcium phosphate, a more digestible source of calcium for the mink.

262

In human medication, phosphoric acid has historically been used in the diet to prevent urinary calculi (USA Dispensatory, 1948 and later issuues).

It has also been found to be useful for treatng struvite stones in both ferrets and foxes (Edifor et al., 1989 and White et al., 1992).

Mono–Ammomium Phosphate

This so called "dry phosphoric acid" is not as effective as phosphoric acid in preventing bladder stones and/or preserving feed "on the wire" (Wanachek, 1996).

Sodium BiSulfate

An alternative product for feed preservation and as a urinary acidifier is sodium bisulfate. This product has proven to be safe and field observations in the United States and Canada have shown that this product is safe and effective for mink (Leoschke, 2004). Sodium bisulfate is also a useful urine acidifier to prevent struvite calculi in cats (Kneuven, 2000). Experimental studies reported by Jones–Hamilton Co., Walbridge, Ohio (2002) with cats on a pellet program indicated that 0.9% sodium bisulfate yielded an average urinary pH of 6.4, considered effective in minimizing struvite stone formation in cats and mink. This urinary pH is similar to that obtained in studies with mink at the National Research Ranch with sodium bisulfate at a 1% level dry matter basis (Michels, 2002).

Calcium Oxalate

As early as 1950, researchers in Denmark reported that one out of 28 calculi contained calcium oxalate (Sompolinsky, 1950). In more recent years, the incidence of calcium oxalate stones in mink has increased dramatically, as high as 25% in Utah (Durant, 2000). Both Durant and Rouvinen–Watt (2000) have suggested that this major increase may be due to excessive levels of acid in mink ranch diets, resulting in a lower urinary pH supportive of calcium oxalate formation.

It has been suggested that changing the urinary pH merely manages the disease and does not really resolve the problem. The use of acids or acidifying agents may only change the urinary environment favoring one type of urolith problem over another. For while dietary acidifiers effectively reduce struvite crystals, they lead to enhanced formation of calcium oxalate stones.

Urate Nephrolithiasis

An inherited defect may lead to excess metabolic production of uric acid with resultant hyperuricemia and an overloading of the kidney resorption mechanism. With this defect, a net renal excretion of uric acid has been reported by Tomlinson, Perman, and Westlake (1982) and Tomlinson, Perman, and Westlake (1987). Uric acid is the end product of the metabolism of endogenous and exogenous (dietary) purine bases in human beings, some non–human primates, Dalmatian dogs and birds. Other species normally process purines to products other than uric acid, such as allanton. The hyperuricemia in mink may be due to defects of an enzyme function in purine catabolism or a deficiency of urate–binding protein. The uric acid level in the affected mink was 23 mg/dl compared to a level of only 5 mg/dl in one unaffected related mink. The renal pelves were dilated and filled by irregular pale green–yellow calculi composed of calcium and ammonium urate. Urate crystals were noted in the urine sediment. Precipitation of uric acid crystals in the skin around the muzzle and foot pads may have contributed to bleeding. Those sites would be highly irritating and could cause bleeding wounds directly or via self–mutilation and hemorrhage secondary to the irritation. That all the affected mink were of the same sex and from related sires is consistent with an inherited metabolic or physiological defect. This point was supported by the fact that the problem was eliminated by removing all affected animals and their parents from the breeding colony.

Cystine Calculi

Oldfield, Allen, and Adair (1956) have reported the presence of 140 creamy white calculi in a male mink which were 99% cystine.

Cannibalism

Cannibalism is a common problem on many mink ranches in June of each year when the kits, and even occasionally, the mothers will kill and eat the kits one by one. A half–century ago, researchers thought that the problem was easily prevented by raising the energy level of the dietary regimen in late May, via fat levels as high as 28–30%. In more recent years, West coast ranchers (Basal et al., 1994) have noted that high energy dietary programs are ineffective and that lower energy and higher protein feeding programs were required to minimize the problem in cases of specific genetics of the mink. It seems that, here again, the answer to cannibalism in the mink is simple: "Listen To The Mink".

"Cotton" Mink

Issues and solutions to this problem are discussed in the previous chapter.

Fatty Livers

All factors considered, observations of enlarged fatty livers in mink at autopsy are signs either of sub–optimum nutrition management or a disease problem wherein the disease is manifested in part by mink being "off feed" for a significant period of time prior to their death: with self imposed starvation, adipose tissue fat deposits are mobilized and result in fatty liver degeneration. The classic example of a nutritional deficiency leading to enlarged fatty livers is a vitamin B_{12} deficiency (Leoschke, 1952) in which the animals may be "off feed" for as long as ten days to two weeks prior to death. Similarly, in laboratory rats starved for as little as 24 hours, there is significant increase in the liver fat content as adipose tissue fat depots are mobilized.

"Fire Engine Red" Dark Pelts

The term "Fire Engine Red" dark pelts was coined many years ago when horsemeat was the major ingredient in commercial mink ranch diets in North America. In 1939, the Midvale Co–Op, Utah, provided a ready–mix product containing 45% horsemeat (Erikson, 1960). This diet, when it contained fresh horsemeat, produced no color problems in dark mink. However, with rancid horsemeat (stored at just below freezing temperatures from pelting in late November to early July during a period of minimum feed requirements for the mink), "Fire Engine Red" dark fur was common at pelting.

Clearly, there is a direct correlation between the level of polyunsaturated acids in a fur animal's diet and the loss of "sharp color" in dark mink in the late fall. This observation makes "common sense": when fats with significant levels of polyunsaturated fatty acids on the surface of the fur of mink and fox undergo oxidative rancidity in the final weeks prior to pelting, these are responsible for the loss of "sharp color" in dark mink and "Fire Engine Red" in darks fed rancid horsemeat. Many ranchers are familiar with the observation that too often dark mink with wonderful dark color in early November (breeder selection assessment) will yield "off color" dark pelts for marketing. It is unfortunate that some ranchers raising dark mink have felt required to pelt "early" (prior to full underfur development) to minimize the market loss associated with "off color" dark pelts. Observations of fur ranchers and fur auction people over the years have noted the correlation between "off color" dark pelts and the employment of mink feed resources with a fat content providing relatively high levels of unstable polyunsaturated fatty acids. These include: linolenic acid, 18:3 omega–3: trienoic (eighteen carbon fatty acid with three unsaturated (alkene) bonds initiated with the 3rd carbon from the tail end) found in linseed oil and rarely in animal fats except for horses (Brooker and Shorland, 1950; Gupta and Hilditch, 1951); Arachidonic acid, 20:4 omega 6: tetraenoic, found in limited but significant levels in animal and poultry fats; eicosapentaenoic acid, 20: 5 omega 3: pentaenoic; and Docosahexaenoic acid, 22: 6 omega – 3: hexaenoic, found in fish and other marine animals.

As early as 1937, Martin (1987) noted that feeding cod liver oil during the summer

265

months tended to make fox fur "brown" in color. Later in 1942–1943, Helgebostad and Ender (1955) carried out feeding experiments with herring and round coalfish for Silver fox cubs which demonstrated that pelts in these cubs lagged behind in color and gloss when compared to pelts of cubs on animal feed products. In 1960, experimental studies at Oregon State University (Stout et al., 1960) noted that mink provided rancid sardine oil, commercial herring oil, or rancid horsemeat yielded pelts with a brownish cast. A very detailed report by Stout et al. (1967) provided Table 5.2.

TABLE 5.2. Mink Fur Color as Influenced by Ration Composition

Dietary Program	Fur Color Score*
Purified Diet – 100%**	3.00
87% Purified Diet 13% Chicken Offal***	2.25
87% Purified Diet 13% Fish Offal***	2.25

 * Fur Grade Evaluation: 3–Dark, 2–Medium, 1–Brown

 ** providing vegetable fats stabilized by natural anti–oxidants.

 *** dehydrated basis.

Practices of modern mink nutrition management emphasizing the use of a high percentage of stabilized animal fats have generally resolved this problem and regained "sharp color" in dark pelts marketed. Still, unfortunately, 21st century mink ranchers are experiencing "off color" dark mink directly related to marine fats provided in a second quality salmon meal (Jones, 2001).

For mink ranchers interested in guaranteed "sharp color" in dark mink with no chance of "Fire Engine Darks" or "off–color" pelts, the decision is simple—provide the mink with a high quality meal or pellet program. As early as 1969, Professor A. Hoogerbruge, Department of Veterinary Science, University of Utrecht, The Netherlands, working with Trouw & Co., Amsterdam, exhibited pelts from mink that had been raised on Pelsifood meal to a group of North American mink ranchers (Clay, 1969). The ranchers were impressed with the sheen and sparkle in the guard hair, very obvious in contrast to pelts from mink that had been raised on conventional ranch diets. The nutritional key to the "sharp color" was a mink feeding regimen providing 100% of stabilized animal fats with a minimum quantity of marine fats.

Fur Chewing

Fur chewing or "clipping" in the mink varies from being a minor problem (chewing only the tip of the tail) to a major one to total elimination of the guard hair in all parts of the body available to a single animal. Multiple factors may be responsible for fur chewing including a genetic weakness leading to a poor temperament, nervousness, and aggressive behavior or environmental factors such as limited bedding and a confined living space. All factors considered, the major reason for fur

266

chewing is genetics, a problem which can be resolved via careful breeder selection. For instance, a Wisconsin mink rancher (Meyer, 1995) observed that one male had sired almost 75% of the mink that were fur chewing. In order to save the pelt value of such mink, feeding mild tranquilizers such as Acepromazine granules may be useful (Hunter and Lemieux, 1996).

Some experimental studies involving raw eggs have discoveered a biotin deficiency which produces fur chewing in addition to the other usual signs of a biotin deficiency (Helgebostad et al., 1952; Stout and Adair, 1969; Wehr et al., 1980; Aulerich et al., 1981; Michels, 1998). In the 1998 study, with seven week old kits on a high quality powdered egg dietary regimen, fur chewing began within ten days after starting the experimental nutrition program.

"Hippers" and "Weak Hips"

The terms "hippers" in mutation mink, especially pastels, and "weak hips" in dark mink involves male kits showing a particular anatomical phenomenon: the fur over the hips is less dense and appears to be unprime due to incomplete shedding of the summer fur (moulting) and sub–optimum growth of the winter fur. Certainly, a major factor responsible for the condition is excessive fattening of the male kits in the groin region, a condition which may result in insufficient blood circulation to the hair follicles in the moulting phase.

Experimental Design—"Hipper" Development

A classic case of nutrition mismanagement which accidentally led to greater production of "hippers" in mink was played out at the Elcho Mink Ranch in Northwest Wisconsin in the summer and fall of 1962. Even though experimental studies of 1956 at the University of Wisconsin (Leoschke, 1958) had shown the value of higher fat dietary programs for modern mink nutrition, the ranch owner was of the "old school" and had no faith in the new concepts of higher energy diets for the early and late growth phases of the mink ranch year. In the years 1961–1962–1963, I conducted a ranch survey of mink male kit weights at 10–14–18–22 and 26 weeks of age and covering five states—Michigan, Wisconsin, Illinois, Indiana, and Ohio. In early July, 1962, I arrived at the Elcho Mink Ranch with my Ford Falcon station wagon and a kilogram scale. While setting up the scale and checking the weighing cages for identical weights, I asked the owner to bring in male kits from litters of six born exactly 10 weeks before in early May. Paul and his men started to bring in the kits in the special weighing cages. I turned from the scale, looked at the kits, and said "Paul, you do not understand; we only weigh male kits." The kits were so very small that I immediately assumed that they were female kits. When Paul compared the 10 week weights of his male kits to the data I had obtained in earlier years, he admitted that maybe the ranch diet was too lean and modified the late growth diet (July–August) to provide a modern higher energy program, 26–28% fat. By early September, the Elcho

Mink Ranch had the highest incidence of "hippers" ever seen in this history of the ranch operation.

Another experimental study is also of interest. It is well acknowledged that certain mutation mink such as Blue Iris and Wilds have higher energy requirements for top performance than do pastels or darks. In an experiment on the Falcon Mink Ranch in summer/fall of 1986 using a special pellet with an extra 2% fat, raising the level of fat to 24%, the net result was a very significant increase in the percentage of "hippers," another "brilliant" idea that did not work. Later it was found that with Blue Iris mink the best approach was a leaner program (20% fat) for the period from early July through September, followed by a 24% fat pellet in early October, yielding extra size on the mink without "hipper" problems (Sandberg, 1990).

Prevention of "Hippers" and "Weak Hips"

Both nutritional physiology and exercise provide an explanation for "hippers" in mutation mink and "weak hips" in dark mink: with modern mink ranch operations, the mink have it "too good" relative to their distant cousins in the wild. Mink kits in their natural habitat are expending considerable energy to survive on limited feed resources, while the daily routine for mink kits on commercial mink ranches consists of sleeping and eating. The net result is that domesticated mink male kits end up with excessive adipose(fat) tissue development; that is, by early September they are "beer belly" mink.

The problem can be solved in two ways—by genetic management wherein careful breeder selection can gradually eliminate the problem within a few years; and/or careful nutrition management to bring about "trim" male kits coming into the fur production phase. A too common recommendation is to restrict the feed intake of the mink in August by "The Man With The Spoon." All factors considered, this is not wise nutrition management. Too often the net result is "thin" mink coming into September, who are never able to develop the optimum adipose tissue in the fall months required for maximum pelt length at marketing.

Superior ranch management for the prevention of "hippers" and "weak hips" in male kits in early September is to provide dietary programs less palatable to the mink, including higher protein commercial cereal products that can be fed at levels as high as 35–70% of the ranch diet. These programs provide a unique combination of minimum "hipper" and "weak hips" problems, enhanced fur development directly related to higher protein intake by the male kits during fur development, and for some unknown physiological reason, "lighter leather" pelts. Too many ranchers feel that these higher cereal programs will undercut the size of the mink at pelting. The experimental data provided in Table 5.3. indicates equal if not superior size of the male kits at pelting.

TABLE 5.3. Male Kit Growth—National GnF–70 Program*

Color Phase	Weight Gains — Late Growth and Fur Development		grams
	July–August	September–November	Totals
Darks			
Ranch Mix	920	310	1230
GnF–70	820	450	1270
GnF–70 gains relative to Ranch Mix with GnF–20	–100	+140	+40
Pastels			
Ranch Mix	970	450	1420
GnF–70	870	580	1450
GnF–70 gains relative to Ranch Mix with GnF–20	–100	+130	+30

* National Research Ranch (1979) Unpublished Data.

Table 5.4. indicates the significant value of dietary programs that support "trim" male kits coming into early September for enhanced fur development.

TABLE 5.4. Are Your Mink Equalling Their Genetic Potential For Fur Development?

Correlation Between Male Body Weights at Different Ages And Fur Quality of Pelts Marketed

Age	Fur Quality Correlation*
Birth	+0.16
14 Days	**
28 Days	**
42 Days	–0.16
June	–0.25
July	–0.24
August	–0.24
September	–0.40
October	–0.41
December	–0.34

* Hansen et al. (1992),

** No significant correlation.
  Note: With female pelts, a negative correlation between body weights and fur quality is found only for body weights in December (–0.18).

Nursing Sickness

For many years, North American mink ranchers have referred to the nursing disease of the mink as nursing anemia; however, this is a misnomer inasmuch as no anemia is present (Hartsought, 1955). Among all domesticated animals, the milk output of the lactating mink with 6–8 kits is unique in terms of quantity of production and quality of product, especially energy content (Brody, 1945; Wamberg and Tauson, 1998a, 1998b, 1998c). Anatomical/physiological support of this observation lies in the fact that no other domestic animal doubles its birth weight within 4 days (Hartsought, 1955). Wamberg and Tauson (1998c) estimated that a 1,100 gram lactating female mink produces more than 3,000 grams of milk during the four week lactation period. The very high energy demands of lactation may, under specific circumstances of sub–optimum ranch and/or nutrition management, lead to a severe negative energy balance of the dam which may lead to the metabolic disorder termed "nursing sickness."

Without question, nursing disease is the major cause of the loss of female mink over the whole ranch year. The economic loss is high since top mink performing in reproduction/lactation are lost from the breeding colony.

Etiology

Nursing sickness is believed to develop from a complex of multiple metabolic, nutritional, and environmental factors which simply undercut the capacity of the nursing dam to meet the extreme demands of lactation. Because lactating mink with large litters are unable to maintain a feed intake meeting their energy requirements, they experience a significant loss of body weight via the loss of adipose (fat) tissue as well as muscle structures. Korhonen et al. (1991) have shown a direct positive relationship between litter size and female weight loss over the lactation phase. The majority of these lactating mink are able to compensate for the energy depletion and maintain a normal physiological state. However, a significant number of these animals will develop nursing sickness due to a variety of factors including large litters (Clausen et al.,1992; Wamberg et al., 1992), high environmental temperatures, and sub–optimum nutrition management. Females of 2 year and 3 years are especially at risk because they raise larger litters with higher kit weight at weaning compared with younger females (Clausen, 1992).

The nursing sickness syndrome is seen in lactating mink with large litters about the time of weaning (six weeks of age) (Hunter and Schneider, 1991; Clausen et al., 1992; and Schneider et al., 1992). The disease is characterized by extreme weight loss, sudden anorexia (lack of appetite), dehydration as evidenced by an increase in the number of red and white cells per unit of volume, progressive weakness, and finally energy exhaustion and a high mortality rate. Serum biochemical abnormalities include increases in osmolality; the concentration of creatinine, phosphosphate, urea nitrogen, glucose, and potassium; and a decrease in sodium and chloride concentrations (Clausen et al., 1990, Henriksen and Elling, 1986, and Schneider and Hunter, 1993). In addition, the animals were acidotic (Brandt, 1988, Wamberg et al., 1992, and Schneider and Hunter, 1993). The presence of high serum glucose levels has been confirmed by Wamberg et al. (1992) and Schneider and Hunter (1993). A study by Wamberg et al. (1992) noted that the peripheral cells of mink suffering from nursing sickness develop lower responsiveness to insulin, resulting in a reduced cellular uptake of glucose. Studies by Borsting and Damgard (1995) and Damgard and Borsting (1995) indicated that nursing mink may have relatively high levels of blood insulin. Their work showed that the blood insulin concentration in the nursing female two hours postprandially was higher than that observed after an intravenous infusion of glucose in glucose tolerance tests.

The urine is almost devoid of sodium and chloride, urinary potassium concentration is diminished by approximately 50%, and the concentrating ability of the kidneys is reduced to less than 1/3rd the maximum value (Wamberg et al., 1992). (Clausen and Hansen (1989) emphasized the superiority of urine analyses for the assessment of nursing sickness.) Body deterioration is rapid and in the final stage the animals may exhibit a staggering gate, vomiting, and black–tarry droppings. Finally, the eyes appear sunken and the animals are nervous and ill at ease; stupor or coma usually precedes death. The ultimate cause of death can be attributed to a combination of many factors, but cardiovascular collapse, as a result of the hypovolemia, hyperkalemia (in the face of muscle potassium depletion), and uremia is likely the major process involved (Scheider and Hunter, 1993). Post mortem observations include the complete absence of any fat on the carcass, a symptom which may also be considered as a diagnostic aid. Gross necropsy findings include emaciation, dehydration, and variable degrees of hepatic lipidosis (Henriksen and Elling, 1986 and Seimiya et al., 1988).The only histologic lesions reported are hepatic lipidosis and vacuolization of renal tubular epithelium (Brandt, 1988 and Seimiya et al., 1988). Often the kits will be seen licking the saliva from the open mouth of their mother. Without treatment, death usually occurs within a few days after onset of the disease.

It has been suggested that nursing sickness is linked to a disruption in glucose homeostasis (Borsting and Gade, 2000). On the basis of their observations and the studies of Wamberg et al. (1992), Borsting and Damgard (1995) and Damgard and Borsting (1995) on insulin in the physiology of mink, Rouvinen–Watt (2002), Rouvinen–Watt (2003), and Rouvinen–Watt and Hynes (2004) have proposed a new hypothesis on the pathogenesis of nursing sickness –– that the underlying cause of nursing disease is the development of an acquired insulin resistance wherein a history of obesity (lipodystrophy), high protein oxidation rates in the lactation phase, and n–3 fatty acid deficiency are identified as key contributing factors. Genetic susceptibility, nutrition management, energy and fluid deficit as well as stress are associated factors which may further contribute to the development of nursing sickness.

The three major factors that may contribute to the development of nursing disease are as follows:

1. History of obesity: In obesity, the ability of adipose tissue to buffer the daily influx of nutrients is overwhelmed (or absent), first interfering with insulin–mediated glucose disposal, and finally leading to insulin resistance;

2. High protein oxidation rates– lactation phase: A lowering of dietary protein levels reduces oxidative stress and heat stress (due to the Specific Dynamic Action of the metabolism of amino acids) and improves water balance in nursing mink via reducing water requirement for voiding urea;

3. n–3 polyunsaturated fatty acid (PUFA) Deficiency: PUFA of the n–3 family play an important role in modulating insulin signaling and glucose uptake by peripheral tissues.

Prevention

Nursing sickness is a complex disease wherein a combination of practices is required to minimize the problem, including genetic, ranch, and nutrition management programs.

Genetic Management

The incidence of nursing disease varies markedly along genetic lines for fur color on some ranches, which implicates a genetic component in the disease (Schneider et al., 1992). For instance, at the National–Northwood Mink Ranch in the 1970s, it was observed that violet mothers committed "suicide" via their lactation efforts. The female with nursing sickness should be removed from the breeder colony unless she possesses unique genetics.

Ranch Management

At whelping, large litters should be broken down if possible to a maximum of eight kits for experienced mothers and seven kits for mothers experiencing their first pregnancy and lactation. In the case of sensitive color phases such as violets, an arbitrary weaning at five weeks may be required. In the case of darks and other color phases, any nursing female weighing less than two pounds at 5–6 weeks should be isolated with a single male kit. The fact that some cases of nursing sickness occur after weaning of the kits adds to the value of an early weaning program prior to excessive weight loss of the lactating mink. Tauson (1988) has reported a delay of at least one full week before the weight of females begins to increase after weaning.

A major effort may be required to provide a cool environment for the mink including spraying systems, shade, and sheds with reflecting roofs.

A survey of North American mink ranches by Rouvinen–Watt and Hynes (2004) found the following factors on ranches with minimal nursing sickness problems: breeder selection focusing on body length rather than body weight; initiation of breeder female conditioning beginning in the fall and keeping the potential breeders active; weaning of the kits prior to seven weeks.

272

Nutrition Management

In past years, nursing disease was considered to be related to a salt deficiency; however, in more recent years the factor of sub–optimum energy levels in the lactating dietary regimen has become more significant.

Salt Depletion

In the mid–fifties, Hartsought (1955, 1960) considered that nursing sickness was a simple salt deficiency directly related to the loss of substantial quantities of sodium cations leaving the mother's body via the milk flow, with resultant negative sodium balance and the development of typical symptoms of an extracellular dehydration. He advocated the employment of 0.3 to 0.5% salt supplementation of the ranch diet (before water addition) depending on the amount of salt in the cereal mixture. His observations noted a 5–12% loss of breeding females in the United States with nursing sickness in the period of 1948–1954 with no losses of mink with the disease in 1955. At the time of Hartsought's first publication on the relation between salt depletion and nursing sickness, vigorous opposition was expressed by Dr. W. L.Roberts, at the time the top mink nutrition scientist in the world. He felt that the levels of salt supplementation recommended by Hartsought were excessive and that levels as low as 0.5 to 0.7% dehydrated basis were quite adequate (Roberts, 1955).

Hartsought's initial observations on salt depletion as a factor in nursing sickness have been supported by subsequent experimental studies in the four decades following his initial publication (Kacmar et al., 1980; Hansen et al., 1993; Clausen et al., 1995); and Clausen et al., (1996a, 1996b). In one Clausen experiment, there was one group with 0.22 grams NaCl/ 100 Calories and a second group with a higher level of salt intake, 0.52 grams/100 Calories. In the group with the lower salt content in the feed, the average female weight loss at weaning was 100 grams more than in the group with the higher salt content. The incidence of nursing sickness (100 females/group) was 22% for the lower salt group relative to 7% for the higher salt group. Based on these studies, it was recommended that the salt content of the diet during the lactation period be 0.40 – 0.45 grams salt/100 Calories Metabolic Energy, equivalent to 1.6 to 1.8% salt on a National Research Council dietary program of 4,080 Calories/kg dry matter. This is significantly higher than the NRC (1982) recommendation of 0.5% salt for the lactation phase and certainly higher than Roberts'1955 recommendation.

In the studies of Kacmar et al. (1980), the liver salt content of deceased lactating mink was lower than the physiological standard of 2.5 to 3.6 grams/ kg.

273

Lactation Energy Exhaustion

Without question, factors other than salt depletion are responsible for nursing sickness, as well illustrated in an observation of Schneider et al. (1992) wherein nursing disease occurred on ten of eleven ranches that had supplemented their mink diets with 0.5% salt. These observations are consistent with those reported by Durant (1992).

Multiple experimental studies over the period of the last two decades support the basic thesis that progressive energy depletion is a major underlying pathologic process in nursing disease of the mink. That is, these lactating mink simply lack the capacity to ingest adequate energy to meet their needs, resulting in severe negative energy balance. This perspective is consistent with obseervations of severe dehydration and emaciation directly due to heavy losses of energy and body mass brought about by increasing milk production (Brandt, 1988; Clausen and Olesen, 1991; Hunter and Schneider, 1991; Durant, 1992; Schneider and Hunter, 1992, 1993, and Wamberg et al., 1992).

Brandt (1988) was able to induce dietary stress in lactating mink via restricted feeding or provision of less palatable feed, a clear nutritional basis for weight loss. The weight loss, in turn, induces the negative energy balance responsible for the development of nursing disease. Clausen and Olesen (1991) noted that a higher risk of nursing disease was related to ranch diets containing a high proportion of carbohydrates, dietary regimens that in themselves provide a lower level of energy concentration.

Protein Levels
Quality

Hunter and Lemieux (1996) recommend a high quality feed with increased energy content during lactation, as well as optimum salt supplementation. The "quality" diet would emphsize protein resources with high biological value for the mink including liver, cooked eggs, and muscle meat resources; it would also attend to the factor of diets with high palatability for the mink.

These recommendations are consistent with specific recommendations provided within the National Program of Mink Nutrition for the past twenty years: rendered fat supplementation at 1, 2, 3, and 4% on May 1st, 10th, 20th and 30th to achieve fat levels of 20–22, 22–24, 24–26 and 26–28% on each of those dates, the exact level to vary with the genetics of the mink and the individual farm environment. This program has also included a 0.25% salt addition about May 10th with an increase to 0.50% recommended only if required and provided that ample water resources are available.

Quantity

Rouvinen–Watt (2002) recommended lowering the level of dietary protein and elevating the level of carbohydrates. The lower protein levels would alleviate both the oxidative stress and heat stress (Specific Dynamic Action of proteins in animal nutrition physiology) and provide superior water balance via minimizing water required for voiding urea.

274

These specific recommendations are consistent with the National Program of Mink Nutrition. That program recommends of a minimum of 15% fortified cereal in the critical lactation period — higher than many ranch diets provide. In short, finally, for the lactation phase, the recommendation is a level of protein of 35% of ME as opposed to the level of 40% of ME as recommended by The Nordic Association of Agricultural Scientists.

Wet Belly Disease
Clinical Signs

"Wet belly" (WB) disease is a urine induced peri–preputial fur defect wherein the fur around the urinary orifice of males is stained and appears to be wet as a result of dribbling of urine components into the fur. See Photograph 1 (Leoschke, 1962a).

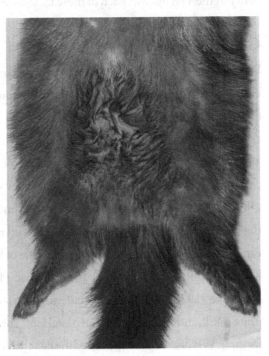

At pelting and fleshing, the leather side of a "wet belly" mink exhibits an obvious discoloration in the inguinal area; it also appears that the maturation of the skin and fur in the "wet belly" disease area is slower than normal. The dark color noted on the pelts is due to unprimeness related to the fact that the fur follicles here are still located deep within the dermis and most of the pigment is in the base and lower portions of the hair shafts which results in a dark discoloration directly related to melanin granule deposition on the underside of the skin. Normal prime mink pelts have a smooth, creamy–white appearance: during priming of the pelt, the follicles become shorter and retreat closer to the outer surface of the skin; at prime, no melanin pigment is seen on the leather side (Aulerich et al., 1962). Grossly and microscopically, the area of WB discoloration resembles the skin of an unprime pelt except that there is mild hyperkeratosis in the stratum cocneum in the skin of the pelts (Bostrom et al., 1967).

It is of interest to note that Schaible et al. (1962) classified the WB disease of the mink as two separate manifestations. He separated urinary incontinence, (UI), wherein the mink fur is soaked and matted with urine around the urinary orifice and without pelt discoloration in the inguinal area from WB disease, wherein a dark discoloration of the pelt appears in the inguinal area. This classification of the disease was based on multiple observations that the UI designation did not necessarily yield WB discoloration of the pelts about the urinary orifice.

275

Etiology

A careful examination of the urinary system of male kits affected with WB disease revealed no pathologic differences that could be responsible for yielding the disease (Borstrom et al., 1967). Multiple factors including genetics, nutrition, and ranch management are involved in the incidence and severity of the WB disease. Each of these factors deserves careful consideration.

Genetics

Without question, heredity is the major factor contributing to the incidence and severity of WB disease of mink. Thus, careful breeder selection should resolve the problem. As early as 1956, Leoschke (1959) noted that with an experimental design of 300 genetically matched males distributed to a control diet and two experimental diets, 70% of the males on the control diet with WB disease had a brother with the same signs on one of the experimental diets. This initial observation on the role of genetics in WB disease was confirmed a few years later at Oregon State University (Adair and Oldfield, 1988) where it was noted that the same sire produced 45% of the WB observed mink in the 1960 matings and 75% of the WB observed mink in the 1961 matings. Other reports throughout the world mink industry support the role of heredity in the incidence of WB disease including Kuznecov and Diveeva (1970) and Pastirnac (1977). At the National Northwood Mink Ranch Leoschke (1978) noted a high incidence of WB disease in dark mink, a moderate incidence in sapphire mink, and almost no signs in pastel kits. The genetic threshold of susceptibility is an important factor in WB disease accounting for the fact that, even with the identical dietary program and similar feed intake, significant WB disease may occur in one color phase and not in others.

Dietary Programs

The key nutrient active in the incidence and severity of WB disease of the mink is fat. See Figures 5.1 and 5.2 based on studies by Leoschke (1960).

GRAPH 5.1. Correlation Between Dietary Levels and Incidence of "Wet Belly" Disease

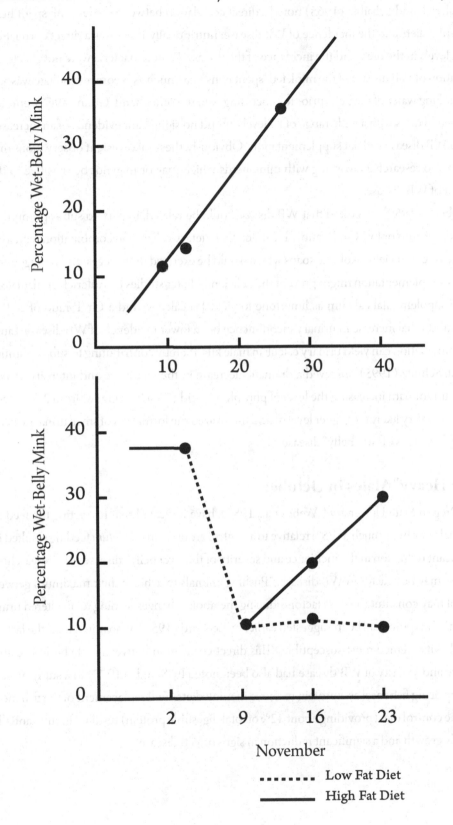

November

········· Low Fat Diet

──────── High Fat Diet

Disorders Related to Nutrition Mismanagement

These observations were confirmed by studies at Oregon State University (Wehr et al., 1983). A study by Aulerich and Schaible (1965) noted a direct correlation between the level of "spent hens" in experimental diets and the incidence of WB disease: nutritionally, there was a direct correlation between fat levels in the diets and the incidence of the disease Proteus bacteria were not a factor in this observation of WB disease of the mink fed "spent hens" inasmuch as Neomycin Sulfate was added to the drinking water of the hens prior to processing. On the other hand, Gunn (1966), employing experimental diets with a wide range of fat levels, found no significant evidence of an increased incidence of WB disease with fat supplementation. Obviously, these inconsistent observations may be due in part to researchers working with mink herds which may or may not be sensitive to the development of WB disease.

Roberts (1959) suggested that WB disease could be related, in part, to sub–optimum dietary levels of calcium or low Ca/P ratios. He noted that dietary calcium in combination with long chain fatty acids could yield insoluble soaps which would be excreted in the feces, and he suggested that calcium supplementation might prevent the deficiency. Later studies by Aulerich et al. (1963) with dietary supplemental calcium as limestone to yield 1% calcium and a Ca/P ratio of 2/1 resulted in a substantial increase in urinary incontinence but a lower incidence of WB disease. Limestone supplementation can yield urinary calculi in male kits, thereby contributing to sub–optimum urine output. Schultz (1995) observed a dramatic decrease in the incidence and intensity of WB disease concurrent with increasing the level of phosphoric acid (75% feed grade) from 2.0 to 2.5% dry weight basis. Very likely the higher level of acid minimized the formation of small urinary calculi, resulting in fewer signs of "wet belly" disease.

Stress—"Heavy" Males in October

Studies at Oregon State University (Wehr et al., 1983), have indicated that restricting the feed intake of the male kits by as much as 75% relative to a control group with *ad libitum* feeding resulted in a very significant reduction in the incidence and severity of the "wet belly" disease. The researchers felt that stress may be a factor in WB disease: "Pushing animals to achieve their maximum genetic size potential may constitute a stress factor inducing metabolic changes leading to an altered urine composition." This point had been suggested earlier by Leonard (1959), who noted that "the fastest growing male kits seemed most susceptible." This direct correlation between male body size and the incidence and severity of WB disease had also been noted by Skrede (1977) in a study on soybean meal replacing fish meal as a protein resource in mink diets. Full replacement of the fish meal content of the control diet (providing about 42% of total digestible protein) resulted in substantially reduced body growth and a significant reduction in signs of WB disease.

Role of Hormones

A study on the potential role of male hormone production in mink as a factor in the development of WB disease was initiated in the fall of 1960 by Leoschke (1962a). A number of mink ranch observations led to the experimental studies, including observations in the fall of 1957 of a group of white male kits with a high incidence of WB disease in which many of the kits had well developed testicles in early October. A few of the males in this group of WB kits were "chuckling," a phenomenon normally observed only in the March breeding season. Many reports from mink ranchers detailed a high incidence of WB disease in the male breeder herd during March, generally, among the more aggressive males.

The experimental study indicated a direct correlation between the level of methyl testosterone supplementation and the incidence and severity of WB disease. Cessation of the hormone treatment resulted in an immediate decrease in the external manifestations of the disease. It should also be noted that a synthetic male hormone derivative provided by G. D. Searle Co. was also employed in the experimental study and that this compound significantly enhanced signs of the disease in the experimental mink. These early observations on the significant role of male hormones in WB disease were confirmed by experimental studies at Oregon State University (Anon, 1964). In these studies, male kits receiving testosterone orally showed a higher incidence of WB disease than a similar group not receiving the male hormone; female kits intramuscularly injected weekly with methyl testosterone showed a significant increase in signs of the disease (Wehr et al., 1983). It is of interest to note that the studies at Oregon State University indicated that with castrated mink, significant incidence of WB disease was observed, indicating that hormones of gonadal origin are not essential for WB development.

The exact role of male hormones in enhancing the incidence and severity of WB disease is obviously unknown. It is possible that physiologically enhanced levels of male hormones may enhance blood levels of specific biochemicals that, when excreted into the urine, may change its characteristics to yield a lower surface tension. This change may increase adherence of the urine components to the fur adjacent to the urinary orifice.

Female hormones may also have a role. Intramuscular injection of estrogen (ECP, estradiolcyclopentylproprionate) in male kits, yielded extensive WB symptoms (Wehr et al., 1983); however, the signs of WB disease were secondary to severe urinary tract infections by Proteus species.

Ranch Management

As early as 1965, Leoschke (1965) had observed that the early separation of mink kits in July would minimize the problems of WB disease when it occurs in litters. This observation was confirmed by Zimmerman (1976) and Pastirnac (1977) and in a series of studies by Wehr et al.(1983), wherein WB disease in single caged sapphire males was reduced dramatically (as high as 85%) in comparison with paired (male–female) kits. The advantage of single caging of the male kits might be related to greater access to water and/or less precocious sexual development. Studies by Aulerich et al. (1963) indicated that later pelting of WB mink did not diminish the size of the affected area of the pelt.

Urinary Tract Infections and/or Bladder Stones

Studies by Gunn (1964, 1966) indicated that the WB disease may be a monovalent or polyvalent infection caused by proteus organisms and/or other potential pathogens present in chicken waste, fish, or beef tripe. Young, growing mink especially may a have a relatively low resistance to such potential pathogenic bacteria. He also noted that cooking mink feed or freezing it at -10 to -20° F. decreased the incidence and severity of WB disease by 50%. In a later study, Gunn (1964b), by using drastic cold and drafty housing conditions and low temperatures (0° to -20° F.) with "wet belly" diets was able to induce WB disease in 20 and 50% of kit females and males respectively. Changing from cold to warmer housing for a period of two weeks frequently brought about recovery from the disease. Thus, controlled lowering of body resistance by chilling stress or an environment of warm housing can respectively induce or protect mink against "wet belly" disease. To Gunn, these observations confirmed previous experimental data relating to the infectious nature of the disease, that is, chilling or stress lowers resistance to bacterial infections.

Lauerman (1965) associated Proteus mirabilis and Streptococcus fecalis with "wet belly" disease, noting that the preputial sac was an ideal area for bacteria to reside. Proteus and Streptococci can grow in a medium of high mineral concentration and cause very active protein decomposition under aerobic conditions. However, multiple studies at Oregon State University (Wehr et al., 1983) with a wide variety of anti–bacterial products proved ineffective in minimizing the type of WB disease they were encountering.

It is possible for bladder stones (urinary calculi) to yield manifestations of WB disease directly related to a sub–optimum urine excretion pattern. A Minnesota rancher, Schultz (1995) observed a significant reduction in the incidence and severity of WB disease in his herd by simply raising the level of phosphoric acid (75% feed grade) from 2.0 to 2.5% of dry matter. Enhancing the urinary acidity very likely dissolved small urinary calculi responsible for the signs of "wet belly" disease.

All factors considered, in terms of urinary tract infections and/or bladder stone formation in mink, it appears that the disease signs exhibited are a secondary factor related to the presence of urinary calculi and/or urinary bacterial infections which impede normal urine flow.

Male Urine Composition

Some researchers including Schaible et al. (1962) have stated that the WB disease is a manifestation of urinary incontinence. However, hundreds of observations over a period of many years of mink with uncomplicated WB disease (not a manifestation of a urinary infection and/or urinary calculi) indicate that the urinary flow is perfectly normal. Although urinary flow of mink with WB disease is apparently normal, it has been repeatedly observed that the last few drops of urine tend to infiltrate into the fur about the urinary orifice. A number of studies have indicated that, indeed, the biochemical composition of urine from WB mink is unique relative to that from normal healthy mink without the disease manifestations. These studies show that a common characteristic of WB urine is a significantly lower surface tension (Leoschke, 1959; Pastirnac, 1977; Sorfleet and Chavez, 1980). Studies at Oregon State University (Wehr et al., 1983) did not include experimental data on surface tension but did note that urine from WB mink had a higher viscosity, and thus more capacity to adhere. The lower the surface tension of the urine, the greater the capacity for the urine to adhere to the fur and infiltrate the area about the urinary orfice. Studies by Wehr et al. (1983) indicated that distilled water would not penetrate the fur to soak the skin. On the other hand, it was considerably easier to wet the fur by using urine with or without WB characterisrics. Moreover, the mink fur seemed to wet more easily after repeated wetting.

The studies conducted in 1966 involved wetting the fur of female kits with urine concentrates from both WB and non–WB male kits. In addition the urine concentrates were applied to two different strains of dark mink, a WB–susceptible group and a WB–resistant group. Data from these experimental trials is presented in Table 5.5.

TABLE 5.5. "Wet Belly" Spotting in Mink Treated with WB and Non–WB Urine–1966

	Non–WB Urine		WB Urine	
	Incidence–%	Severity*	Incidence–%	Severity*
Strain				
WB–Susceptible	50	1.2	80	2.0
WB–Resistant	10	1.0	60	1.3

* Severity grading on a sale of 0–3 with 3 being the most severe.

It appears that skin sensitivity as well as urine composition is involved in the development of WB disease of the mink. It is apparent from the data provided in Table 5.5 that the urine from both normal healthy mink and WB disease mink contain detergent–like components which change the surface tension of the urine and intensify the penetration of the urine into the normally water resistant mink fur. Too, these detergent–like biochemicals likely have a higher concentration in the urine of WB–diseased mink. The exact chemical nature of these biochemicals is, at the present, unknown, but they are lipid in nature. In studies by Wehr et al. (1983) involving freeze–dried urine samples from mink with and without WB disease, the "normal" mink urine residue was distinctly granular with comparatively low apparent viscosity while the "Wet Belly" urine was buttery, quite smooth and viscous, with a semisolid consistency. Additional support for the lipid nature of the unique components of WB mink urine is seen in an observation of Leoschke (1962b) who noted that the feeding of Sudan IV, a fat–soluble red dye, to mink with WB disease yielded a direct correlation between the severity of WB disease signs and the intensity of the red color in 24 hour collections of urine. In another supportive study (Leoschke, 1962b), when mink urine was extracted with petroleum ether (fat solvent) the surface tension rose dramatically, non–WB urine from 45–49 and WB urine from 35 to 55. The study showed that the unique component in the urine of WB mink responsible for lowering urine surface tension is fat–soluble in nature since a fat–soluble solvent is able to remove it from the urine. Another fascinating observation emerged: in the process of extracting mink urine with petroleum ether a very stable oil/water emulsion was produced which was difficult to break. A similar petroleum ether extraction of human urine did not yield a stable o/w emulsion, indicating that mink urine contains a biochemical component enhancing the formation of o/w emulsions, a chemical constituent with a detergent–like nature having both hydrophilic and hydrophobic characteristics.

One potential component of such WB urine (detergent–like characteristics that would lower the surface tension of the urine and also encourage o/w emulsions) would be fatty acids which at a WB urinary pH of 6.1–6.4 would yield "soaps" (sodium salts of medium and long chain fatty acids) (Leoschke, 1956). Studies by Pastirnac (1977) and Adair et al. (1988) indicated higher levels of fatty acid levels in the urine from WB mink at the level of micrograms/liter level. This observation is of interest, but, all factors considered, it is likely that additional components of mink urine unique to WB mink in presence and/or concentration will be found. One of the likely prospects would be lipoproteins. Studies by Schweigert et al. (1991) indicated that carnivores—relative to herbivores, omnivores, and rodents—are unique in terms of excretion of vitamin A and vitamin E (fat soluble vitamins). Vitamin A levels (retinol equivalents) in the urine of canines were between 423 ng/ml (dog) and 6,304 ng/ml (silver fox). Their studies after precipitation and ultracentrifugation indicated that carrier proteins in the urine exist for vitamin A and may exist for vitamin E. In the dog the total excretion of vitamin A represented 15–63% of the total uptake of vitamin A while less than 4% of vitamin E was excreted.

"Wet Belly" Disease–Treatment

Inasmuch as the "wet belly" disease of the mink is a recessive trait, the best approach is that of genetic selection against this disease in the herd. A temporary prophylactic approach is to add salt to the ranch diet at levels of 1.0 to 1.5% (Leoschke, 1962c; Stout and Adair, 1971; Adair et al., 1977). The recommendation was based on multiple reports that mink with WB disease exhibit a significant reduction in 24 hour urine volume (Leoschke, 1962; Adair et al. 1977; Sorfleet and Chavez, 1980).

282

Reduced water consumption would decrease the daily urinary output of the mink and thereby cause a concentration of minerals, metabolic by–products, and those specific biochemicals responsible for the manifestation of WB disease. Supplementing ranch diets with salt resulted in an increase in urine volume, a reduction in urine viscosity, and a significant decrease in the incidence and severity of WB disease.

Another nutrition management approach to minimize WB disease is to follow the specific recommendations of the National Program of Mink Nutrition for the fall months, that is, to raise the level of fortified cereal starting in October for five weeks and concurrently to drop that of fatty fresh/frozen feedstuff. The net result of this program is leaner dietary and slightly lower palatability ratings. These changes in the October diet can minimize excessive weight gains in the final weeks of fur development and thereby minimize the incidence and severity of WB disease. The program also contributes to sharper color in dark pelts marketed, as polyunsaturated fat resources are replaced with carbohydrates which are metabolized by the mink to yield stable saturated fatty acids.

"Yellow Fat" Disease—Steatitis

This disease has been extensively discussed in the chapter on Vitamin E, secondary deficiency, "Yellow Fat" Disease—"Steatitis".

6

Toxic Substances in the Feed Supply

Aflatoxins

Aflatoxins are potent hepatotoxic and carcinogenic compounds produced by *Aspergillus parasiticus* and *Aspergillus flavus*. These molds grow on a wide variety of feedstuffs and foods.

Mink are among the most sensitive species to aflatoxin poisoning. Koppang and Helgebostad (1972) found that daily doses as small as 5 micrograms of aflatoxin administrated for four weeks caused fatty degeneration of the liver and catarrah in the stomach and intestines. Chou et al. (1976a) noted that the single LD–50 dose of aflatoxin for mink was estimated to be 500 to 600 micrograms/ kilogram of body weight. A marked increase in plasma cholesterol and alkaline phosphatase appeared before the mink died. The liver from animals that died of aflatoxicosis showed prominent pathologic changes which included hemorrhages and the appearance of pink–yellow spots. Histopathologic examination of the liver of dead mink revealed fatty infiltration, bile duct proliferation, and bile stasis (Chou et al., 1976b).

The earliest recorded case of aflatoxin poisoning in mink occurred on a mink ranch in Great Britain in 1962 (May, 1970) and was found to be directly related to the use of peanut flour containing aflatoxins. Peanuts close to the ground are very often infected with a fungus termed *Aspergillus flavus*, a fungus in itself comparatively harmless, but during its lifetime producing aflatoxins. At this ranch, usually the animals which died had been in good condition externally and had eaten well up to a day or two prior to death. Signs of impending death were passing a black tarry product, and bleeding from the mouth or rectum. Indeed, death was often immediately preceded by a larger than usual hemorrhage of this kind.

Aneamogenic Agents

These diseases have been extensively discussed in the section on Iron, secondary deficiency, aneamogenic agents.

Arsenilic Acid

Arsenilic acid (AA) increases the blood supply to the skin by dilation of the peripheral blood vessels with the potential for a greater deposition of fat and pigment in the skin and a contribution to feather growth in chickens. AA also possesses bacteriostatic properties.

Initial studies on the nutritional value of AA were conducted at Oregon State Agricultural Experiment Station (Davis, 1954). The addition of 0.01% AA dietary dry matter did not improve color or fur quality; moreover, the animals were extremely nervous as noted at biweekly weighings.

In 1956 studies on the tolerance of mink for AA were conducted by Spruth et al. During pregnancy and lactation, levels as high as 0.02% AA dietary dry matter were employed. Normal litters were born 3–10 weeks after AA supplementation was initiated. After whelping, the level of AA was increased to 0.05% of dietary solids, and after 2.5 months, the mink were returned to the ranch and fed a regular ranch diet without AA supplementation. The adults and kits were observed 3 months later to be in excellent health with prime coats. In a separate concurrent experimental study, mink were fed 0.05 and 0.1% dietary solids without adverse effects. It is apparent that mink show a high tolerance for AA—10 times normal feeding levels, similar to chickens and rats. This degree of tolerance is higher than that of dogs, turkeys, or swine.

Botulism

Putrefactive organisms in slaughterhouse offal, poorly processed whale meat, and occasionally fish, including the anaerobic bacterium *Clostridium botulism*, can produce toxins fatal to the mink. The botulism bacteria form spores that are very heat resistant, whereas the bacteria and the toxin have a very low resistance to heat.

Botulism affects the nerve system, with subsequent paralysis. The mortality is characterized by many deaths the first day followed by a steadily diminishing number in the following days. Clinical signs of botulism poisoning may appear in mink within a few hours after ingestion of the toxin. Early symptoms include abdominal breathing (heaving of the flanks), muscular incoordination, stiffness and paralysis of the hind legs. These signs are usually followed by paralysis of the front legs and neck, excessive salivation, convulsions, coma, and death. The cause of death is usually due to respiratory paralysis and there are generally no consistent post–mortem changes other than hyperemia of the lungs (Quortrup and Gorham, 1949; Gorham, 1950).

The obvious answer to botulism is kit vaccination against the toxin. However, this prophylactic program does not cover the critical June period between maternal immunity and complete effectiveness of the botulism vaccine. There is a simple answer to this problem. Ranches with a history of botulism should feed a quality mink pellet product as a wet mash for the month of June.

Pesticides and Chlorinate Hydrocarbons

An excellent review of experimental studies on the toxicity of various pesticides and chlorinated hydrocarbons is provided by an article by R. J. Aulerich and S. J. Bursian in the book *Mink: Biology, Health and Disease*, Hunter and Lemieux (1996), Graphic and Print Services, University of Guelph, Guelph, Ontario, Canada N1G 2W1.

Di–Ethyl Stilbesterol

The decision of the poultry industry in North America to implant di–ethyl–stilbesterol (DES), a synthetic estrogen, in the heads of cockerels to enhance the growth rate created a major problem for the mink industry. When poultry with implanted DES were used in ranch diets, there was an absolute failure of the reproduction/lactation performance of the mink. Yet, interestingly, there were no reproduction/lactation problems with the mink when unwashed tripe (beef rumen) from DES fed cattle was used in experimental diets at the University of Wisconsin (Shackelford, 1959).

Reproduction/Lactation Studies

Experimental studies with specific levels of DES in mink rations indicated that relatively low levels of DES, 2 to 1.5 micrograms/mink/day, actually resulted in an enhanced reproduction/lactation performance (Travis et al., 1955, 1956). But another study by Bassett et al. (1957) indicated reduced reproduction/lactation performance at 2 micrograms/mink/day. At a level of 5–10 micrograms/ mink/day a complete reproduction failure occurred (Travis et al., 1957 and Shackelford and Cochran, 1962).

Of major interest is the fact that even a late introduction of DES in the mink ranch diet—as late as April 23, after ova have been implanted but before the dam had whelped—can result in almost complete failure of the normal reproduction process (Travis and Schaible, 1962). Reproduction anomalies included resorption of litters, lowered kit production, reduced kit weights, and increased kit mortality. Significant loss of kits between birth and 14 days indicated sub–optimum lactation performance and/or inadequate strength of the kits to nurse.

Finally, it is important to point out that the studies of Shackelford and Cochran (1962) indicated complete recovery from DES treatment one year after cessation of the study.

Growing and Furring

The use of 10 micrograms of DES/day to mink kits reduced body length and weight gains during a 23 week period (Travis et al., 1956).

Urinary Calculi Formation

DES is known to predispose male mink kits to a urinary tract infection, subsequent bladder stone formation, and death of the animals from "bloody bladders." DES produces a change in the epithelial lining of the urinary passage of the mink as well as a sloughing away of patches of epithelium, which processes encourage bacterial invasion of the urinary tract by *Staphylococcus auereus* and *Proteus Mirabilis*, normal components of the preputial flora in healthy mink (Lauerman and Berman, 1962a, 1962c).

287

DiMethyl-Nitrosoamine (DMNA)

Much more than other species, mink seem to be sensitive to the acute toxic effects of DMNA. Young rats, for example, may survive for 26–40 weeks on diets containing 50 ppm DMNA (Barnes and Magee, 1954); mink kits receiving diets containing 2.5 or 5.0 ppm die within 23–34 days (Carter et al., 1969). In mink the LD–50 of DMNA given by subcutaneous injection is 7 mg/kg/bw, and in those animals which have a short life span the cumulative hepatotoxic and carcinogenic doses of DMNA are the same. A daily exposure of less than 0.2 mg/kw/bw will usually cause toxic hepatotosis, and a lower dose may be carcinogenic. The carcinogenic action of DMNA on mink appears to be nearly 100% if the exposure time is long enough and if the dose is near to a hepatotoxic level. A threshold level below which no toxicity occurs has not been established in mink (Koppang and Rimeslatten, 1976).

Mink exposed to DMNA yield a hepatotoxicosis which leads to death via massive abdominal hemorrhage stemming from one or more ruptured blood–filled sacs or hematomas on the liver as well as hemorrhages from gastric and duodenal erosions (Stout et al., 1968). For more information on fish yielding DMNA toxicity, see the section on Iron—secondary nutritional deficiency—anaemogenic fish.

Ergot

Ergot is the common term given to a sclerotic produced by fungal species of the claviceps that infect grasses (rye grass, wheat grass, blue grass, and fescue) and cereal grains (wheat, rice, corn, sorghum, rye, barley, oats, and millet). The sclerotic contains numerous biological active alkaloids that are responsible for the toxicity of the mycotoxin. Ergot is one of the earliest recognized mycotoxins. It has been responsible for the deaths of thousands of humans and animals including mink (Cameron et al.,1988).

A specific study on the effect of ergot alkaloids on the reproduction/lactation performance of mink is of interest (Sharma et al., 2000). The experimental trial indicated that ergot alkaloids as low as 3 ppm wet weight of mink feed resulted in sub–optimum reproduction/lactation performance wherein hydrocephaly was the predominant deformity. With current improved grain handling procedures and sanitation employed by millers and feed manufacturers, ergotism is seldom a problem. At the same time, mink ranchers who use cereal grains raised on their own property should be aware that the consumption of very low concentrations of ergot alkaloids by mink can result in serious negative reproduction results.

288

Ethylene Glycol

A report of a Louisiana mink rancher (McWilliams, 1990) is of interest. The rancher was using an old water pail with ethylene glycol residue from his repair garage. The mink would get thinner and thinner and just gradually die. Typical crystalline crystals (oxalate) were found in the kidneys, substances which indicated antifreeze poisoning.

Gossypol

Gossypol is a phenolic substance in cottonseeds. It can cause various adverse effects on animals including depressed appetite, weight loss, lung and heart lesions, and anemia.

Vilas Klevesahl (1983), a Georgia mink rancher, experienced a loss of almost 20% of his breeder females and significant lactation failure directly related to his use of cottonseed hulls as bedding for his breeder mink. He had restricted the feed intake of his mink to obtain proper "trim" condition for breeding and whelping, but the mink consumed a significant amount of cottonseed hulls containing 0.53% gossypol, with subsequent loss of breeder mink and kits.

Histamine

The increasing use of acid–preserved raw materials in mink feed increases the potential risk of the formation of histamines in mink feedstuffs inasmuch as the optimum acidity for histidine decarboxylation by certain bacteria (*Clostridium, Proteus, Salmonella* and *Escherichia coli*) is within the pH range of 5.0 – 5.5.

Fish silage has very significant levels of free lysine, histidine, and arginine while poultry by–products and most mink rations have insignificant levels. However, it is important to note that the storage of mink feed for even as little as 24 hours can increase histamine levels significantly. In fact ordinary levels in mink feed from 0–30 ppm histamine can increase to 100–120 ppm.

Woller (1977) has reported on the effects of high levels of histamine in the mink ranch diet where the control diet contained a normal level of histamine (15 ppm) and the experimental diets contained 118, 203, 677, and 847 ppm histamine. Increasing histamine content gave increasing diarrhea and all experimental animals showed very dilated stomachs. With the highest level of histamine, vomiting was noted at the beginning of the experimental feeding tests but only for the first three days. Daily weight gains of the kits fell with increasing histamine concentration.

Zhong (1986) has reported major losses of mink due to histamine poisoning on rations containing as much as 20–30% fish meal imported from Peru with a histamine content in the range of 3.5 to 5.0 mg/gm.

Lead

Lead poisoning in mink is usually associated with the use of red lead paint or paint containing lead on cages, feed, water containers, or feed equipment (Gorham et al., 1972). Another potential source of lead in the mink's diet is acid water which has been carried in lead pipes or lead treated valves or even, as one Canadian rancher experienced, lead painted kitchen walls wherein the scraped paint was used as mink bedding (Pridham, 1967).

The clinical signs of acute plumbism in mink are anorexia, musclar incoordination, stiffness, trembling, dehyration, convulsions, and microprint discharge around the eyes. Levels of lead in an acute outbreak of plumbism were about 5 times higher than those in the chronic outbreak, according to Pridham (1967). Treatment recommended by Gorham (1949) includes adequate amounts of calcium as dicalcium phosphate or calcium gluconate, with ample levels of vitamin D. The purpose of the vitamin D and calcium supplementation is to tie up the lead circulating in the system and store it in the bones where it will do little harm. After it is stored in the skeletal structure, it may be slowly eliminated from the mink's body.

Mercury

Mercury contamination resulting from industrial pollution of lakes and rivers is widespread; its relevance to practical mink nutrition has been reported by Wobeser et al. (1975a). Wobeser and Swift (1976) have reported an incidence of mercury poisoning in wild mink, and it would appear that with diets high in fish, mercury poisoning could be a potentially serious problem in ranch mink.

Studies by Aulerich et al. (1974) have shown that mink are quite sensitive to methyl mercury but comparatively tolerant of mercury in an inorganic form. Mink fed diets containing 5 ppm methyl mercury showed clinical signs of mercury poisoning within 25 days, with death occurring as early as the 30th day. The degradation of methyl mercury by mink has been investigated by Jernelov et al. (1976).

Swedish investigators Ahman and Kull (1962) reported 0.24 and 2.4 mg mercury as magnesium–brom–alkyl–mercuric chloride, per kilogram of feed to be highly toxic to mink and noted a high correlation between the mercury content of the feed and the mercury concentration in the organs. Borst and van Lieshout (1977) investigated phenyl mercuric acetate intoxication in mink related to a ranch diet contaminated by PMA with a mean mercury content of 27 ppm. This diet was fed for more than a year, with a high mortality in the animals older than one year; yet reproductivty and fur quality of the surviving mink were not affected by PMA intoxication. Their studies indicated that clinical symptoms began with a progressive loss of appetite and a decrease in activity. The most important pathologic changes were observed in the kidneys. Tubulonephrosis, primarily localized in the proximal convoluted tubules, was most severe in mink dying 2–3 weeks after feeding of the contaminated diet. Lesions in the central nervous system were not observed. Thus, the effects of aryl mercurial compounds such as PMA differed strikingly in the mink, as they do in other animals and man, from those of the alklylmercurial compounds.

Blood urea nitrogen levels increased up to six times and proteinuria were observed up to 5 weeks after exposure to PMA. Tubulonephrosis, less severe than in the mink that died spontaneously, was found up to 5 weeks after exposure to PMA. Mercury residues in brain, liver, kidney, and skeletal muscles were analyzed. The half–time of the disappearence of mercury was 10 days in skeletal muscle, 22 days in the kidneys, 27 days in the brain, and 36 days in the liver.

Other clinical signs of mercury poisoning observed by Aulerich et al. (1974) and Wobeser et al. (1975b) include incoordination, anorexia, weight loss, tremors, ataxia, paralysis, paroxysmal convulsions, and high–pitched vocalizations. When suspended by the tail, mercury–treated mink show the typical limb–crossing phenomena indicative of mercury poisoning in other species.

Nitrates

Nitrate and nitrite contamination of water is sometimes of concern. These contaminants enter water supplies via sewage, animal waste, and fertilizer use. Ground water in the vicinity of feed lots or confinement livestock operations including mink ranches often contain excessive levels of nitrates and nitrites. Levels of 200 ppm nitrates in water are potentially hazardous and levels as high as 1500 ppm cause acute toxicity.

The only report within the worldwide fur industry on nitrate toxicity in mink was from Adair et al. (1981). The experimental studies initially involved nitrate levels of 1.06 to 3.20% as fed basis (providing 0.15, 0.30, and 0.45% of nitrate nitrogen (N). Then final levels were reduced to 0.05, 0.10, and 0.15% nitrate N. There was some early growth depression in male kit growth on each of the higher nitrate levels; however, these differences tended to disappear as the kits gained extra weight during fur development. A reproduction study with 8 females involving 0.15% nitrate N indicated that at 3 and 8 weeks of age, nitrate–fed kits were smaller than non–nitrate kits, a phenomenon directly related to the relatively low feed volume of the mink kits from 4 to 6 weeks of age.

Nitrites

Since the 1950s, it has been a common practice in Norway to preserve herring with an aqueous solution of sodium nitrite. Sodium nitrite, $NaNO_2$, retards the conversion of trimethyl oxide present in fresh fish tissues to dimethylamine and trimethylamine, a process which is catalyzed by bacteria and autolytic enzymes. But, if sodium nitrite is added to stale fish in which dimethylamine and trimethylamine are already present, dimethylnitrosoamine (DMNA) may be formed on subsequent heating when the herring yield a fish meal. To a lesser extent, monomethylamine and trimethylamine oxide, which are present in a variety of marine teleosts, give rise to the formation of DMNA when brought into reaction with sodium nitrite and heat processing (Ender, 1966). Studies by Koppang (1974) indicate that DMNA can also form in production of fish meal from fresh and frozen formaldehyde–and benzoate–preserved catches.

Propylene Glycol

Initial studies on the toxicity of propylene glycol (PG) were conducted in the USSR by Maksimov and Nikolaevskii (1986). The experimental study lasted only 45 days and involved the use of a product containing a mixture of 50% PG, 35% glycerol and 15% sorbitol at 2.5, 5.0, 7.5 and 10.0% of feed solids. The additive caused dystrophy of the liver and kidney stones, and it induced aggression. There was no effect on the growth of the mink but the additives caused histological alterations in liver, spleen, kidneys, and intestines.

A later study by Weiss et al. (1994) involved a higher level of PG, 12% of feed solids for a period of only one week. The net result was a 17% decrease in hematocrit, a 21% decrease in RBC count, and a 4.8 fold increase in reticulocyte count. A marked increase in Heinz body and eccentrocyte numbers was consistent to oxidative injury to RBC. Because of the high feed intake, the mink ingested approximately twice the quantity of PG/kg/bw as domestic cats fed diets containing 12% PG. Therefore, the severity of the hematolytic dyscrasia in mink may be the result of a greater intake of PG rather than a unique sensitivity of mink to RBC oxidative injury. However, the high food intake and the mink's position at the top of the food chain may increase its exposure to environmental contaminants.

Sulfathiazole

In 1990, a mink rancher in British Columbia, Canada, experienced complete reproduction/lactation failure in the process of providing his mink sulfathiazole during March and April (Leoschke, 1990).

Thiaminase

The toxicity of this enzyme is discussed in the section on Thiamiine, vitamins.

Thyroid Glands

After a New York state mink rancher experienced major reproduction/lactation failure with the use of beef gullet trimmings in his ranch diet, Travis et al. (1964, 1966) initiated experimental studies on this effect. The gullet trimmings consisted of the trachea, larynx, esophagus and adhering muscle, and glandular (thyroid–parathyroid) tissue of calves about 10 days old. Each pound of gullet trimmings contained 5.5 grams of dried glandular tissue wherein each gram of dried tissue contained thyroid activity equivalent to 0.16 mg of tri–iodo–thyroxin and 1.0 mg of sodium L–thyroxin pentahydrate. The control diet contained 15% of beef lungs, replaced with 15% gullet trimmings in the experimental diet. The following chart indicates mortality rate.

Group	#Females	Total Kits		
		Birth	14 Days	Weight–gms
Control	14	64	56	57
Gullets	16	22	5	38

Later experimental studies of Travis and Duby (1969) indicated reproduction/lactation failure in mink fed calf heads with the thyroid glands attached.

The initial work of Travis and co–workers was confirmed by Kangas and Makela (1972). Their experimental study indicated no adverse effects from the gland material feeding in the period prior to breeding in March, but they observed 80% loss of kits within two weeks of whelping when the gland material was fed during the period from March through April. They also noted retarded growth of male kits. It is very important to point out that Kangas and Makela felt that the parathyroid effect was minimal, inasmuch as no clinical signs of parathyroidism appeared in the experimental group.

293

7

Modern Mink Nutrition Management

Fortified Cereals

At the beginning of commercial mink ranching in North America, practical mink diets contained very little or no cereal grains, as the mink farmers tried to mimic the natural diet of the mink in the wild. However, as the years passed, higher and higher levels of fortified cereals were employed as the direct result of pertinent experimentation and practical experience at ranch sites. These higher levels of fortified cereals were used for one primary reason—feedstuff economics. As resources of metabolic energy per the ranchers' feed dollars, only rendered animal fats, such as choice white grease (pork fat) provides a more economical caloric product. A secondary nutrition factor is that practical mink ranch diets without fortified cereals or with relatively low levels of cereal provide excessive protein levels, with resultant physiological stress on the liver and kidneys of the mink. Over the period of the last half century, ranchers on the National Program of Mink Nutrition have noted top performance of their mink via the employment of fortified cereals at the 15% level (National XX–15) for the reproduction/lactation phase and at the 20% level (GnF–20) for the late growth and fur production periods. Higher levels of simple fortified cereals, formulations without dehydrated animal or fish protein products, can be employed at even higher levels during periods of the ranch year when the protein requirements of the mink are minimal—November through February.

A high quality fortified cereal formulation should provide the mink with a highly palatable product which provides, in addition, highly digestible carbohydrate resources, especially cooked grains; ample vitamin fortification; and trace mineral resources readily available to the digestive/absorption processes of the mink.

Simple Cereal Formulations For Mink

Table 7.1 provides an insight into the design of fortified cereal formulations that have been employed at Oregon State University, the University of Wisconsin, and the Central Fur Foods Co–Op of Wadsworth, Illinois.

295

TABLE 7.1. Simple Mink Cereal Formulations—Late Growth and Fur Production

Ingredients	Oregon(1)	Wisconsin (2)	Illinois(3)
Ground Oat Groats	65	—	—
Steam Rolled Oats	—	55	43
Wheat Germ Meal*	10	20	—
Soybean Oil Meal	10	8	15
Cooked Wheat	—	—	17
Cooked Corn	—	—	12
Alfalfa Leaf Meal*	10	2	2
Brewers Yeast*	4.3	10	—
Wheat Bran	—	5	10
Salt	—	—	1
Fortification	0.7	—	—

(1) Adair, Stout and Oldfield (1963), (2) Leoschke and Rimeslatten (1954), and Krieger (1962).
* Low taste appeal ingredients for the mink.

The relatively low palatability of the simple cereal formulations which employ alfalfa leaf meal, wheat germ meal, and brewers yeast is seen in experimental studies conducted at Oregon State University (Stout and Oldfield, 1963). Researchers noted that replacing the OSU 52 simple cereal formulation with ground oat groats resulted in enhanced growth and longer pelt length in male kits but not in female kits.

Currently, low palatability ingredients are not used in quality commercial fortified cereals. These low taste appeal ingredients such as alfalfa leaf meal, brewers yeast (B–vitamin resource) and wheat germ meal (vitamin E resource) have been replaced with stabilized vitamin A, B–vitamins and stabilized vitamin E products..

Cereal Ingredients

Alfalfa Leaf Meal

Svenden (1995) showed that adult mink demonstrated a significant preference for mink feed without alfalfa. This important observation had been noted earlier. W. Roberts, Federal Fur Foods, was the first mink nutrition scientist to exhibit mink fed pellets at the International Mink Show in Milwaukee in 1952. The mink were small and their fur had a greenish tinge, possibly related to the alfalfa leaf meal content of the Federal Fur Foods new mink pellet formulation.

Barley

This excellent carbohydrate resource is similar to other cereal grains such as oats and wheat. Cooking increases both palatability and carbohydrate digestibility for the mink. Carbohydrate digestibility is somewhat lower for both raw and cooked barley, inasmuch as barley has a higher level of fiber and also a relatively high level of beta–glucans which are poorly digested by animal alpha amylases.

Brewers Yeast

Without question, brewers yeast has a low palatability rating for mink. Studies at the University of Wisconsin in the 1950s on the value of brewers yeast as a B–vitamin resource had to resort to "debittered" brewers yeast before the study could be completed since the mink would simply not accept a regular brewers yeast composite sample provided by the Brewers Yeast Council (Leoschke, 1953). This observation was confirmed by later studies in Alaska by Leekely et al. (1962) wherein the deployment of 2% brewers yeast in a mink ration resulted in kit growth depression.

Corn Germ Meal

Field observations by Leoschke (1985) indicated that within certain dietary programs, corn germ meal in mink and fox rations can bring about significant diarrhea.

Corn (Maize) Gluten Meal

Corn (maize) gluten meal (60% protein) has a relatively high content of sulfur amino acids, cysteine, and methionine, that is, sulfur amino acids at a level of 5.0 g./16 g. N. Serious attention must be given to the basic fact that the product has relatively low levels of lysine and arginine. Experimental studies in Denmark by Lund (1975) and Jorgensen and Hansen (1978) indicate that the product can be fed up to 20% of the total protein in the mink feed with good results. It is very important to note that these specific experimental diets did not include another high protein plant resource, soybean meal. The value of combining two high protein plant resources in modern mink nutrition must be carefully studied prior to marketing.

Peas–Extruded

Experimental studies by Ahlstrom et al. (2003) indicate that extruded peas as the only carbohydrate resource at a level of 15% of the ranch diet provided satisfactory performance of the mink. The product contained protein, 23%; ash, 3%; fat, 1%; and carbohydrates, 65%. In another study, Clausen et al. (2001), it was noted that levels of 9–10% heat treated peas and 3–4% heat treated soybeans was satisfactory alone or in combination. It is important to note that peas must be treated to inactivate (denature) trypsin and chymotrypsin inhibitors which undercut protein digestion. Heating is also important to provide optimum carbohydrate digestibility.

Rapeseed

Rapeseed meal is the residue remaining after oil is extracted from seeds of rape (Brassica and B. campestris). The meal contains glucosinolates, compounds that inhibit the activity of the thyroid gland. Canadian plant breeders have developed cultivars low in glucosinolate content, producing non–toxic rapeseed meal, and have given it a special name, canola meal, to distinguish it from the high glucosinolate rapeseed meal.

Belzile et al. (1974) conducted studies on the value of rapeseed flour (RF) with 67% protein and one–tenth the quantity of glucosinolates as found in commercial meal. Levels employed were 6.5, 13.4 and 20% dry matter representing 4.3, 9.0 and 12.4% of total protein in the diets. At the higher levels of 13.4 and 20% of the protein as RF there were reduced weight gains and hence shorter pelts. At the higher levels there was also noted a decrease in digestibility of dry matter, gross energy, and N and N–retention. At 20% of protein as RF there was significant thyroid hypertrophy, but at slaughter, serum–bound iodine level was the same for all groups.

The relative sub–optimum nutritional value of unhusked rape seeds for the mink was reported by Lorek et al. (1996) in experimental studies involving 10% of the total weight of fresh feed as flaked "00" rape seeds with kits 6–14 weeks of age and at a 5% level from 14 to 24 weeks of age. Protein levels in the control and experimental diets were similar but the kits on the rapeseed regimen exhibited poor weight gains and relatively short pelts with inferior fur quality.

In a more recent study by Bjergegaard et al. (2002), water–extracted rapeseed at a 3% level in the mink feed yielded unchanged or slightly better fur quality. However, at the 7.5% level, significantly reduced weight gains were noted.

Soybean Hulls

Studies at the National Research Ranch (Leoschke, 1980) indicated that when ground soybean hulls replaced beet pulp as a fiber resource in practical mink nutrition, second–rate droppings were noted in all of the dark kits and a high proportion of the pastel kits. These same mink kits had excellent good form droppings within a few days after being placed on National Early Growth pellet formulations with beet pulp as a fiber resource.

Soybean Products
Introduction

Properly processed soybean products have the potential to make a significant contribution to the modern nutrition of the mink. The clear advantage of soybean products for the rancher businessman is the relatively low cost per protein unit compared to quality dehydrated animal and fish products. The disadvantages of using soybean products for modern mink nutrition are multiple.

(1). Protein Digestibility: Ratings for the mink in some cases are relatively low, as noted in the scientific literature: raw soybeans, 62% (Ahman, 1959); cooked soybeans, 67% (Ahman, 1958 and Glem–Hansen, 1977); soybean oil meal, solvent extracted and heat treated, 80% (Glem–Hansen and Eggum, 1977; Kiiskinen et al., 1974; Jorgensen and Glem–Hansen, 1975a, 1975b; Skrede, 1977;Glem–Hansen, 1978; Skrede and Herstad, 1978; Glem–Hansen and Mejborn, 1985; Kiiskinen et al., 1985);

(2). Carbohydrate Digestibility: Ratings for the mink are significantly lower than most cereal grains, as follows: commercial 44% protein soybean oil meal with hulls (solvent extracted), 49% (Smith, 1927); commercial 44% protein soybean oil meal with hulls (solvent extracted)—cooked, 57% (Smith, 1927 and Ahman, 1958, 1959); commercial soybean oil meal, 50% protein without hulls (solvent extracted and heat treated), 58% (Leoschke, 1965).

The relatively low digestibility of the carbohydrates found in soybean oil meal is related in part to the oligosaccharides raffinose and stachyose which are not hydrolyzed by monogastric animals due to the absence of indigenous beta–glycosidase enzymes (Skrede and Ahlstrom, 2002).

It is very important to note that multiple studies over a period of years have indicated that the digestibility of the carbohydrates in soybean products is significantly lower than that reported in the paragraph above. Specifically, the digestibility of commercial soybean meal (44% protein with hulls, solvent extracted and heat treated) is in the range of 12–35% (Seier et al., 1970; Kiiskinen et al., 1974; Rimeslatten, 1974; Jorgensen and Glem–Hansen, 1975; Glem–Hansen, 1978; Skrede and Herstad, 1978). Obviously, this wide variation in experimental data indicates that multiple factors are involved, including variation in soybean products, wide variation in heat treatment, and significant differences in analytical procedures employed.

In terms of chemical analytical procedures, it is well known that the most accurate experimental studies on carbohydrate digestibility involve examining the content of the glucose in the mink feed and in the feces. In other cases the reports do not involve any intensive studies on the glucose content of the mink feed and feces, but simply take the less laborious and less expensive approach of calculating the glucose content of the mink feed and feces through Nitrogen Free Extract (NFE) data. In this calculation, analytical data on protein, fat and ash is simply subtracted from 100: NFE = 100% – (% protein + % fat + % ash). With this simple program all errors that have occurred in the analysis of protein, fat, and ash are combined, leading to potentially inaccurate data on the actual carbohydrate content of the mink feed and feces;

(3). Amino Acid Pattern: For the most part, soybean products are relatively unique as a plant protein for animal nutrition with a good content of lysine and limited only in methionine content.

Anti–Trypsin Factor

Raw soybean products contain constituents which inhibit the action of trypsin and chymotrypsin enzymes in animal digestive physiology. These anti–trypsin anti–chymotrypsin factors in raw soybean products can be inactivated (denatured) by optimum heat treatment.

An excellent experimental study on the value of optimum heat treatment of soybean oil meal to be employed in mink nutrition was reported by Hillemann (1977). See Table 7.2.

TABLE 7.2. Mink Kit Growth and Heat Treatment of Soybean Oil Meal

Group	Darks	Pastels	Feed Efficiency– Kcal/g. Growth
Control	100%	100%	100%
Experiments with 20% of protein as			
SOM (solvent extracted)			
Moderate Heat Treatment	77	76	121
Normal Heat	93	89	108
Extra Heat	89	92	103

Mink Performance

All factors considered, experimental studies with mink over a period of a half–century indicate that "Not all soybean products are created equal."

Growth Depression

As early as 1955, Rimeslatten (1974) noted that the replacement of 32% of animal protein resources with soybean cake resulted in sub–optimum kit growth and lower pelt values. Later experimental studies confirm this initial observation of second rate growth with any employment of soybean products in mink diets (Seier et al., 1970; Hillemann, 1972; Belzile and Poliquin, 1974; Alden and Johansson, 1975; Belzile, 1976; Johansson et al., 1977; Skrede, 1977; Hillemann, 1978; Narasimhalu et al., 1978; Skrede and Herstad, 1978; Hejlesen and Clausen, 2001, and Skrede and Ahlstrom, 2002).

Relative to mink kit growth, comment is warranted on experimental studies with full fat soybean products wherein there is a consistent pattern of sub–optimum kit growth performance, as noted by Hillemann (1976, 1977, 1978) and experimental studies at the National Fur Foods Research Ranch (unpublished data).

Fur Quality—Sub–Optimal

A number of experimental studies have indicated that the employment of significant levels of soybean oil meal in mink diets may have an adverse effect on the fur quality of the mink in terms of longer guard hair (Alden and Johansson, 1975; Johansson et al., 1977; Skrede, 1977; Skrede and Herstad, 1978; and Skrede and Ahlstrom, 2002); and less silky fur (Hillemann, 1977 and Jorgensen, 1978).

Fur Quality—Superior

Finnish experimental studies wherein 25% of the animal protein was replaced with soybean meal proteins indicated similar pelt quality relative to the control group without soybean meal (Kiiskinen et al., 1974). However, with lysine and methionine supplementation of the soybean meal regimen, significantly better fur quality was noted (Kiiskinen and Makela, 1975).

Studies by Hillemann (1976) and Johansson (1977) noted that, although with the use of soybean oil meal there was kit growth depression, the final pelt lengths were similar and the pelts marketed were better in both color and quality.

Experimental studies by Skrede (1977) and Skrede and Herstad (1978) indicated that with higher levels of soybean meal, there was a significant reduction in the incidence and severity of "wet belly" disease resulting in higher pelt prices. Very likely the reduction in "wet belly" disease signs with soybean oil meal experimental diets was directly related to reduced kit weights as of September 1, followed by enhanced feed volume and weight gains during the fur production phase.

Late Growth—Fur Production – Summary

All factors considered, it is apparent that high quality soybean products with optimum heat treatment have a potential to make significant contributions to the 21st century nutrition of the mink. In a recent study, Skrede and Ahlstrom (2000) employed modern enhanced quality soybean products at levels of 15% and 30% of total protein, replacing Norwegian fish meal for the fall months. The overall finding from the study indicated only minor effects on fall weight gains, pelt size, and fur characteristics. It would be of interest to see whether future experimental studies using the same products and adding lysine and/or methionine supplementation would fulfill the potential for enhanced fur quality as noted in the Kiiskinen and Makela (1975) studies.

Reproduction/Lactation

A number of experimental studies (Kiiskinen et al., 1974; Kiiskinen and Makela, 1975; Belzile, 1976; Hillemann, 1977) have indicated normal reproduction/lactation performance of the mink with the employment of soybean meal. However, there was reduced kit growth and increased weight loss in lactating females. A study by Hillemann (1977) wherein 20% of the animal protein in the control diet was replaced with soya protein from a full fat product noted a reduced number of kits per female in both standard and pastel mink, kit growth being very depressed, and increased weight loss by the lactating mothers of both color phases.

Wheat Germ Meal

For many years, mink ranchers in the USA have used wheat germ meal in the basic cereal composition for the reproduction/lactation phase. Wheat germ fat contain components with estrogenic properties supportive of the breeding activity of the mink. All factors considered, there is no scientific basis for use of wheat germ meal or wheat germ oil for enhanced performance of the mink for reproduction/lactation. Without question, wheat germ meal has a low palatability rating for the mink as noted in experimental studies at the National Research Ranch (see Feedstuff Palatability) wherein the taste appeal of wheat germ meal rated only 55%. Confirmation of this low rating is also seen in studies conducted in Alaska (Leekly et al., 1962) providing experimental data indicating kit growth depression with the use of 2% wheat germ meal. At Oregon State University, the use of 10% wheat germ meal yielded depressed mink kit growth (Watt, 1952).

Wheat Gluten

Wheat gluten is a good source of plant protein for modern mink nutrition, valuable for increased water binding capacity and improved feed consistency. Wheat gluten protein, like soybean oil meal protein, is relatively low in the critical amino acid for fur production, methionine. Special supplementation with DL–methionine may be required to yield top performance. The first experimental study published on the value of wheat gluten for the mink was that of Kiiskinen (1984). He found that true protein digestibility was 95.8% and the true amino acid digestibility was 97.3%. Additional research by Laue et al. (1997) provided data on the digestibity of protein, 95%; fat, 94%; and carbohydrates 38%; as well as the amino acids cystine, 91%; methionine, 96%; and threonine, 86%.

Fresh/Frozen Feedstuffs

Liver

Animal, fish, and poultry livers are wonderful nutrient resources for mink. They are rich in vitamins, trace minerals, essential fatty acids; as a protein source of high biological value for the mink they are exceeded only by whole egg and whole milk proteins. It must be acknowledged, however, that fish livers are especially potent in vitamin A and polyunsaturated fatty acids to the point that high levels of fish liver may undercut the performance of the mink.

Rancher experience over many years supports the view that liver is nearly an essential for top performance diets during the critical reproduction/lactation phase of the mink ranch year. The major pellet programs in North America—National Fur Foods of New Holstein, Wisconsin, and XK Mink Foods of Plymouth, Wisconsin—provide specific liver products in their pellet formulations for this period.

High quality liver meals, that is, those marketed at $1 to $2/pound for cat nutrition, may replace fresh/frozen liver in modern mink diets; however, the average liver meal on the world market is not recommended since "economical" processing programs provide inferior mink nutrition. Among other failures there is a significant loss of the biological value of the protein content and a potential significant loss of other nutrients, including the essential fatty acids critical for mink nutrition. Further, these "economical" liver meals on the market are unpalatable for the mink (Howell, 1951).

To date, experimental studies with the goal of replacing liver with potent trace mineral and vitamin products have yielded only sub–optimum performance of the mink (Belzile, 1982).

Quality Protein Feedstuffs

Egg Products

Without question, whole egg proteins are the global standard for quality of protein resources. These provide the highest biological value for all animals including humans and mink in terms not only of amino acid pattern but also of availability to the digestive processes of animals. This statement has been confirmed in multiple studies at the National Research Ranch facilities including a detailed study on an experimental design to evaluate the biological value of dehydrated protein resources for modern mink nutrition (Leoschke, 1976).

The combination of optimum mink nutrition management and sound mink ranch business management suggest that expensive egg products should not be employed during the late growth period (July–August) or the maintenance period (December–February), as these periods have minimum protein requirements in both quality and quantity of protein. Reserve these products for more critical phases of the mink ranch year, unless whole egg protein is available at a cost per unit of protein lower than that of other quality muscle protein resources.

Cooked Eggs

These are an excellent protein resource for the mink. Egg white components binding biotin and iron have been denatured by the cooking process and salmonella bacteria have been killed.

Raw Eggs

These are also an excellent protein resource for mink provided that serious consideration has been given to the potential problems that exist relative to salmonella populations, the avidin protein component binding biotin, and conalbumin component binding iron.

303

Salmonella Component

Without question, the most serious problem with raw eggs is that of salmonella bacteria which can undercut the reproduction/lactation. These bacteria in raw eggs can be destroyed by using 65 pounds of phosphoric acid (75% feed grade) per 4,000 pounds of raw eggs to achieve a final pH of 3.5 or below and holding the mixture for 3–5 days to achieve a zero bacteria population (Zimbal Jr., 1996). Another method of preserving raw eggs at room temperature is to use sulfuric acid or phosphoric acid at levels as high as 4% (Powell et al., 1997). This program suppresses bacterial growth but leads to the hydrolysis of the proteins to yield polypeptide structures and a semi–solid gel state. This final product is difficult to handle at the ranch site.

A new approach to preserving raw eggs is to use a copper sulfate solution yielding a copper concentration of 1,000 ppm which minimizes both bacterial and fungal growth for up to one week at 50–79 degrees F. This product has been used at a level of 20% of the mink ration throughout the breeding, gestation, lactation, and early growth periods with no adverse effects on reproduction/lactation performance or kit growth (Powell et al., 1997).

Avidin Protein Component

Egg whites contain avidin, a protein which binds biotin, a B–vitamin, in a structure which is unavailable to the digestive processes of animals. It is recommended that each 100 pounds of raw eggs be supplemented with 3 to 5 pounds of a biotin concentrate containing 100 milligrams of biotin per pound. The higher level of supplementation is especially recommended for raw egg products with a high proportion of egg whites, such as "spun eggs."

Conalbumin Protein Component

Egg whites contain a specific protein, conalbumin, which binds ferrous cations in a structure unavailable to the digestive processes of mink. A few years ago, a Wisconsin mink rancher lost more than 400 kits in June with a classic iron deficiency anemia (hypochromic, microcytic) through using raw eggs as the "water" component of the mink feed mixture (Trimberger, 1993). To prevent iron deficiency anemia in mink fed raw eggs, the rancher is advised to supplement each 100 pounds of raw eggs with 20 grams of ferrous sulfate (Leoschke, 1973). If the raw eggs contain a high proportion of egg whites, the level of iron supplementation should be doubled.

Powdered Eggs

Powdered eggs are an excellent protein resource for modern mink nutrition provided proper supplementation with biotin and ferrous sulfate are provided. Inasmuch as raw eggs contain about 25% solids, the biotin and iron supplementation for powdered eggs should be four times that recommended for raw eggs, that is, 12 pounds of a biotin concentrate (100 mg biotin/pound) and 80 grams of ferrous sulfate per 100 pounds.

Milk Products
Whole Milk

Ranch observations have indicated that milk products are excellent feedstuffs for the mink, especially during the reproduction and early growth phases (Makela et al., 1967,1968).

All factors considered, cow's milk is an excellent feed resource in terms of protein quality (second only to whole eggs in terms of biological value for the mink as determined by essential amino acid pattern), calcium, phosphate, trace minerals, and vitamins. However, mink are intolerant to milk sugar, lactose, a fact which simply limits the level of milk and milk products that can be provided in a ranch diet without serious diarrhea problems. At the same time, mink can survive on a sole whole milk diet (supplemented with iron, copper, and manganese salts) for as long as two years. A study at the University of Wisconsin (Leoschke and Elvehjem, 1956) was designed to ascertain any potential differences between raw and pasteurized milk. In this experimental work, six male kit brothers were divided into two groups and three kits were placed on a raw milk program while the other three kits were placed on a pasteurized milk regimen. In terms of the known composition of cow's milk, a special trace mineral supplement was required. The milk was supplemented with a solution which provided 3.0 mg of iron, 0.15 mg of copper, and 0.15 mg of manganese per 100 mls milk. The special trace mineral supplement used salts known to be readily absorbed by animals including iron pyrophosphate, copper sulfate, and manganese sulfate.

The mink kits moving into adulthood over a two year period were able to survive without any unique requirement of unidentified factors present in liver, hog mucosa, or lard.

Cheeses—Whole Milk

Whole milk cheeses including cheddar and Swiss are wonderful mink feed products in terms of nutrient content and, in most cases, mink feed economics with feed solids as high as 60%.

In terms of biological value for the mink, whole milk cheeses are second only to whole eggs as protein resources, a fact related to high digestibility rating and amino acid pattern. They provide stable fats with a relatively low level of polyunsaturated fatty acids and thus strongly support "sharp color" in dark mink pelts. Processed cheeses are an excellent resource for digestible calcium and phosphate; however, too often they are relatively high in salt as a result of manufacture. Ranchers (Zimbal, 2000) have experienced the loss of hundreds of mink kits in June when they have used cheese with excessive levels of salt. Salt levels above 3% of mink feed can be toxic even when ample water is available.

High levels of lactose (milk sugar) can bring about significant diarrhea in mink kits in June; adult mink are especially susceptable inasmuch as their physiology provides sub–optimum levels of the lactase enzyme required for the conversion of lactose into galactose and glucose simple sugars.

Cottage Cheese

Cottage cheese provides levels of feed solids as low as 25% with a protein level of 20%. The protein provided in cottage cheese is casein, which has a sub–optimum amino acid pattern relative to the amino acid requirements of the mink for fur development and reproduction/lactation, especially with a relatively low level of arginine.

Muscle Meats
Introduction

For the most part, the protein quality of the muscle portion of cattle, fish, horses, pigs, poultry, and sheep is essentially similar. However, brief commentary is warranted on individual muscle meat resources.

Fish–Whole

In terms of amino acid pattern, all whole fish are, for the most part, essentially equal; however, special care must be given to fish species that are aneamogenic (see section on iron nutrition) or contain the thiaminase enzyme (see section on vitamin B_1). In addition it should be noted that certain fish species including salmon and tuna contain high levels of polyunsaturated fatty acids which can lead to "yellow fat" disease of the mink on dietary programs providing sub–optimum levels of vitamin E.

Certain species of fish—cod, haddock, silver hake, and whiting—store fat in their livers and not within their muscle structures. On the other hand, other fish, such as herring and mackerel, store fat in their muscles rather than their livers. In practical terms, the fat content of fish viscera may vary considerably from species to species.

Horsemeat

Horses are unique among all domesticated herbivores in terms of adipose tissue (fat) with a relatively high content of linolenic acid, a polyunsaturated fatty acid very sensitive to oxidative rancidity. The combination of rancid horsemeat and sub–optimum dietary levels of vitamin E can lead to "yellow fat" disease of the mink.

Spent Hens

"Spent Hens" are 18–20 months old hens no longer profitable for egg production or as food for human consumption. Unbled, un–eviscerated, defeathered hens at this age have a proximate analysis of % basis of moisture, 58; protein, 16; fat, 20; fiber, 0.5; NFE (carbohydrates), 1.8; ash, 3.3; calcium, 1.0; phosphorus, 0.6. (Aulerich and Schaible, 1965a).

Multiple studies at Michigan State University indicate that "spent hens" should be properly processed, that is, they should be treated with Neomycin Sulfate (325 mg/pound) at level of 2 mg

Neo Mix/ml in the drinking water for 20 hours prior to processing to "sterilize" the digestive tract (Aulerich and Schaible, 1965b). The "spent hens" were employed in mink diets at levels as high as 75% for the late growth and fur production phases of the mink ranch year (Schaible et al., 1966) and as high as 50% for the more critical reproduction/lactation phase of the mink ranch year (Aulerich and Schaible, 1965c). Performance of the mink in all of these studies was satisfactory.

Additional studies at Michigan State University indicated that "spent hens" with feathers were of practical value in mink nutrition (Aulerich & Schaible, 1966). This new method of processing "spent hens" for mink feed involved birds that were not defeathered. The presence of feathers in the new poultry product did not adversely affect the consumption of the mink during the late growth and fur production phases of the mink ranch year: the mink simply ate around the coarser feather particles. Growth rate of the mink fed the non–defeathered "spent hens" was superior to that of the control group and the fur quality was about equal.

Animal By–Products
Beef By–Products
Beef Ears

Used by Russian mink nutrition scientists in experimental diets designed to yield a poor amino acid pattern for studies on the amino acid requirements of the mink.

Beef Gullet Trimmings

After a New York state mink rancher had major reproduction/lactation losses with the employment of beef gullet trimmings in his mink ranch diet, Travis et al. (1964) initiated experimental studies on the effect of gullet trimmings on the reproduction/lactation performance of mink.

The gullet trimmings consisted of trachea, larynx, pharynx, esophagus and adhering muscle and glandular (thyroid–parathyroid) tissues of calves about 10 days old. Each pound of gullet trimmings contained 5.5 grams of dried glanular tissue, each gram of which contained thyroid activity equivalent to 0.16 mg of tri–iodo–thyroine and 1.0 mg of sodium L–thyroxine pentahydrate.

The control diet contained 15% of beef lungs which were replaced with 15% gullet trimmings in the experimental group. The third group was given tri–iodothyronine and sodium thyroxin pentahydrate equivalent to that found in the 15% gullet trimmings.

Group	# Females	Kit Count	Birth	14 days Kit Weight–gms
Control	14	64	56	57
Gullets	16	22	5	38
T&T	16	51	11	39

High kit mortality in both experimental groups was very likely due to sub–optimum milk production.

307

Beef Lips

Russian mink nutrition scientists used this protein source in experimental diets designed to yield a poor amino acid pattern for studies on the amino acid requirements of the mink. Note: Studies by the Oscar Mayer Co., Madison, Wisconsin (1950) with rats indicated that the protein quality of beef lips was inferior.

Calf Heads

Calf heads contain about 35% ash–dehydrated basis. Used without brains and tongue by Russian mink nutrition scientists as a very poor protein resource for the study of the amino acid requirements of the mink, calf heads are low in tryptophan, methionine, isoleucine, and histidine as ascertained via amino acid supplementation required to achieve good mink performance.

Calf heads may, however, have significant levels of thyroid and parathyroid tissue, depending on how they are cut.

Poultry By—Products
Baby Chicks

Day old chicks may be employed in mink ranch diets at a level as high as 25% in all phases of the year provided that full attention is given to the proportion of quality protein feedstuffs to meet the essential amino acid needs of the animals. High levels of baby chicks may yield loose droppings.

Entrails

This feedstuff has high protein digestibility, 87%, (Leoschke, 1959a) and excellent palatability for the mink. When using poultry entrails from old hens, ranchers must be concerned about the content of egg residues which contain both avidin protein and conalbumin. The first of these substances can bind biotin and the second, iron. The avidin content of turkey eggs is especially potent, leading to a biotin deficiency in the mink and yielding an abnormal gray underfur coloration (Adair et al., 1961).

Feet

This feedstuff has a relatively low protein digestibility of only 52% (Leoschke, 1959a); it also has a high collagen content and hence a second rate amino acid pattern with low levels of cystine, tryptophan, and tyrosine as well as a low palatability for the mink. The experimental study on the digestibility of the proteins present in chicken feet conducted at the University of Wisconsin involved 20 adult male mink. With a research diet consisting of chicken feet, fat, and sugar, and fortified with vitamins and trace minerals, only half of the animals consumed enough of the diet over the period of a week to allow valid data on protein digestibility ratings.

Heads

The protein digestibility of chicken heads for the mink is 77% (Leoschke, 1959a).

308

Fish By–Products
Fish Bones

Fish bones provide a second rate protein resource for the mink: it has a relatively low digestibility rating and a high content of collagen protein which has low levels of cystine, trytophan, and tyrosine. Helgebostad (1973) first noted that fish bones may impose a severe strain on the mink digestive system. Studies by Skrede (1978c) noted a linear negative relationship between ash content in fish dry matter and resulting nitrogen digestibility. A 1% unit increase in ash content on a feed solids basis depressed protein digestibility by a factor of 0.6%.

Other experimental studies by Skrede (1978a) involving high levels of filleting scrap—frames with no heads or viscera—resulted in poor performance of the mink in terms of growth and fur development. The study by Skrede (1978c) indicated that fish backbone (vertebral column not including the ribs) had a protein digestibility of only 84% compared to 96% protein digestibility for fish fillets. It is of interest to note that cooking of the backbones altered the structure in such a way as to improve the protein digestibility.

Fish Fillet Cuttings

Fillet cuttings represent discarded fractions from fish fillets marketed for human consumption. Diets using these fillet cuttings, according to studies by Skrede (1978a), tended to yield a heavier pelt than experimental diets containing filleted scrap. As expected, fillet cuttings, with a superior protein digestibility and amino acid pattern, provided a higher underfur density of male skins..

Fish Heads

Experimental studies by Skrede (1978a) involving high levels of fish heads indicated that this protein resource is second rate in terms of essential amino acids as evidenced by a relatively low nitrogen balance. The partial replacement of fish scrap with fish heads negatively influenced general fur quality. Cod fish heads could exert a negative effect on fur quality indices because of the increased fatness of the animals coming into the fur development period. Although 29% of ME as protein was satisfactory for growth of the male kits on experimental diets, it was not so for the mink fed the high level of fish heads. In fact, replacimg filleting scrap with fish heads caused a significant reduction in protein digestibility, as would be expected with the relatively higher ash content of the fish heads.

Fish Scrap (Filleted Fish)

In the 21st century, for the most part, the employment of fish scrap as a significant proportion of mink ranch diets is an example of second rate nutrition management and unwise business management. These high ash products with low protein digestibility ratings and low protein biological value due to minimum muscle and maximum collagen content do not support top fur development of the mink.

309

Special attention must be given to salmon by–product, a fish protein resource with the unique fat content of a relatively high level of polyunsaturated fatty acids (PUFA). This feedstuff has a high potential for causing "yellow fat" disease unless supplemented with high levels of vitamin E (Leekley, Cabell, and Damen,1962) and undercutting the quality of fur development through yielding "off color" dark pelts. Even a commercial salmon meal containing powerful anti–oxidants can still yield "off color" dark pelts (Taylor, 2000).

It should be noted that, more than twenty years ago, National Fur Foods stopped using high ash whitefish meal in their mink feed products after experimental studies on the biological value for mink kits indicated second rate performance of animals. In terms of sound business management, high quality whole fish meals on the market provide better mink nutrition economics, offering more pounds of digestible protein for the rancher's feed dollar.

Fish Skin

Fish skin has a large component of collagen protein and thus a relatively low biological value for the mink related to low levels of cystine, tryptophan, and tyrosine. Although relatively low ash skin has high digestibility ratings for the mink, its poor amino acid pattern undercuts its value for mink nutrition. Studies by Skrede (1978a) noted that replacing half the filleting scrap protein with fish skin resulted in a striking depression on the growth of the kits. The poorest fur elasticity was noted in both male and female pelts in experimental diets containing high levels of fish skin.

Fish Viscera

Fish viscera accounts for 6–12% of the total weight of the fish if the liver is excluded and about 50% of the weight of the fish if the liver is included. The variable fat content of fish viscera is directly related to its liver content. True digestibility of nitrogen in fish viscera products averages about 92%. Studies by Skrede (1981a) indicated that fish viscera at levels of 15–30% of apparent digestible protein in the ranch diets had no adverse effects on reproduction, kit viability or kit growth, or fur quality.

Miscellaneous
Crustacean Waste
Crab

Watkins et al. (1982) have provided experimental data on the use of a crab protein concentrate from king crab (*paralithodes camschatica*) waste with a protein level of 67%. The experimental diets were planned to replace 10–20% of the protein provided by a standard mink ranch diet. Weight gains of both males and females were lower with greater feed consumption compared to the control group. The major effect was on the male kits where it appeared that the level of dietary fat was a critical factor. Pelt characteristics were not appreciably affected. It was concluded that crustacean waste products could be satisfactory protein resources for mink provided that the protein and energy concentration of the diets are maintained at optimum levels and that dietary calcium does not become excessive.

Shrimp

Watkins et al. (1982) have provided experimental data on the use of untreated shrimp waste, shrimp meal, and sieved shrimp meal. The experimental diets were designed to replace 10–20% of the protein provided in a standard mink ranch diet. Weight gains of both male and females were lower with greater feed volume, compared to the control group. The major effect was on the male kits where it appeared that the level of dietary fat was a critical factor.

No attempt was made in this experiment to balance fat levels between treatments. Mink given untreated shrimp waste sorted out larger fragments of shrimp with resultant higher fat intake.

Shrimp meal at 20% of total protein resulted in final weights that did not conform with the dietary fat level, suggesting an additional size limiting factor which could be related to a high calcium/phosphorus ratio of 3.5/1.0, one which would yield a reduced fat absorption.

Squid

Experimental studies on the use of squid in practical mink ranch diets were reported by Skrede and Koppang (1982). Raw squid (*Todarodes sagittatus Lamarck*) at 34% of the mink's diet caused reduced litter size, increased kit mortality, and reduced body weights of the kits and their mothers.

Mink kits fed raw squid in the summer and fall at 18% of the diet showed impaired growth and excessive mortality within one month. Necropsy revealed that poor performance and death was caused by severe gastric ulcers, in some cases leading to perforation.

Dehydrated Protein Resources

Bacterial Proteins

Bioprotein, a bacterial protein meal produced from natural gas (Skrede et al., 1998) contains 70% protein, with an amino acid composition similar to that of fish meal except for less lysine and more tryptophan. Protein digestibility is 79%, with lysine at 89% the highest and cystine at 47% the lowest digestibility. The low digestibility rating for cystine may be due to several factors including a low cystine level in BPM, a high cystine–level content in endogenous secretions of the mink (Skrede, 1979), and a secondary disulfide cross–linking during production of fish meal as shown by Opstvedt et al. (1984). N–balance studies demonstrated that the amino acid quality of bioprotein was similar to fish meal (Ahlstrom and Skrede, 2002). Their experimental studies also indicated that

311

the bioprotein product may account for as much as 40% of total digestible protein provided the mink (when provided at 8% of the ranch diet). At this level of protein there was reduced feed intake and lower weight gains for the male kits but not for the female kits. With 40% of digestible protein from bioprotein, there was significantly reduced hair length for both male and female kits. Increasing the level of bioprotein to 60% of total digestible protein resulted in slightly lower ME intake relative to metabolic weight, with subsequent slightly lower weight gains of the kits (Hellewing and Tauson, 2002).

Bloodmeals

One of the earliest studies on the role of whole beef blood and beef blood meal in practical mink nutrition was conducted by Adair et al. (1961). These studies reported on the replacement of 20% sole in the mink's diet with 20% whole beef blood, observing that the fur quality of the female pelts marketed was relatively low. At the same time, in a similar study with beef blood meal, the fur color of the male pelts marketed was superior. Obviously, factors of antioxidant stabilization of the fats present in the two products cannot be ignored.

A later study on whole blood conducted in Denmark (1981) indicated that levels above 6% can have a negative dietetic effect and can undercut the color quality of dark mink. At the same time, it should be noted that the employment of 30% blood in the diet did not show any negative effects on the growth of the mink.

Another study conducted on the role of flash dried blood meal and cooker dried blood meal conducted at Michigan State University (Aulerich, 1984) is, all factors considered, too supportive of the employment of blood meals in modern mink nutrition. The summary statements claim: "both flash dried and conventional cooker dried blood meals can be used to supply a moderate portion of the protein in mink ranch diets and that blood meals are high in lysine and have a good a good amino acid profile." These statements can mislead mink ranchers on the value of flash dried and conventional cooker dried blood meals in the modern ranch diet because of the following points. Bloodmeals do have a relatively high level of the essential amino acid lysine, but no laboratory data was provided on the specific biological availability of the lysine present in the blood meals employed and no commentary was given on the basic fact that the availability of lysine in any given dehydrated protein feedstuff will vary from one commercial product to another on the market. The key modern animal nutrition fact is that access to laboratory procedures for lysine content and available lysine content have been available for many years;

The protein level of the experimental diets was 36% of ME, a level of protein 20% in excess of the protein requirement of the mink for top growth and fur development. Thus within the experimental design, second rate blood products could provide top growth and fur development of the mink.

Feather Meals

Feather meals have a high content of sulfur amino acids, especially cystine. However, Eggum (1970) has reported that the protein digestibility in feather meal is often very low because of incomplete hydrolysis in preparation. Furthermore, he showed that the various treatments which increase the digestibility coefficients to 79–84% for total protein content did not give rise to the same high digestibility of the sulfur amino acids present, as their digestibility ranged from 44–78%.

Chavez (1981) noted a negative effect of feather meal on palatability of mink rations. When feather meal was fed at 5, 10 and 20% in place of fresh animal by–products, at the 20% level, although pelt quality was improved, the size of the pelts was greatly reduced, giving a pelt return 17% below that of the control group.

In an earlier study, Rietveld (1976) also observed the low palatability of feather meal. He employed feather meal on top of a conventional ranch control diet at levels of 1.5 and 3.5%. There was essentially no significant increase in female pelt quality at 1.5% and decreased pelt size at the 3.5% level of feather meal.

Fish Meals

Quality fish meals with protein resources of high biological value for the mink are a major asset in modern mink nutrition management. Worldwide standards relative to the quality of fish meals for modern mink nutrition are provided in Table 6.1.

TABLE 6.1. Fishmeal Standards—Mink and Fox Production

Chemical Analysis	Norway (1)	Denmark (2)	Soviet Union (3)
Crude Protein—min. %	70	73	60
Crude Fat—max. %	10	10	10
Water—%	6–10	6–8	6–10
Ash—max. %	16	13	22
Ammonia Nitrogen—max. %	0.2	—	0.2
Salt, NaCl—max. %	3	—	3
DiMethyl Nitrosamine	NOT DETECTABLE		
Total Volatile Nitrogen—max. mg/100 g.	—	120	—
Free Fatty Acids—max. % Crude Fat	—	10	—

(1) Ugletveit (1975), (2) Konsulenternes (1976), and (3) Kuznecow (1976).

Observations at the National Mink Research Ranch over a period of 50 years indicate that "All Fish Meals Are Not Created Equal." More than 80% of the fish meals on the North American market are inferior for mink nutrition due to multiple factors including (1) sub–optimum handling prior to processing, (2) overheating during production to the point of destruction of amino acids and/

or bonding of other amino acids in structures unavailable to the digestive processes of the mink, (3) low palatability for the mink, and (4) high levels of polyunsaturated fatty acids which undercut "sharp color" in dark pelts.

Ranchers selecting a fish meal for their mink ranch diets should request specific experimental data on the biological value of the protein content. Relative to fish meal purchases, ranchers should consider experimental studies—"Listen To The Mink In His/Her Cage" and not simply accept proximate analyses from "The Chemist In His/Her Laboratory."

A study on the nutritional value of herring meal as a mineral resource is of interest (Hansen, N.E., 1974). The data indicate that herring meal makes substantial contributions of calcium, phosphorus, sodium, iron, and zinc. Levels of magnesium were satisfactory, but levels of manganese and copper were sub–optimum.

Meat and Bone Meals

Experimental studies on the biological value of dehydrated protein resources for modern mink nutrition were conducted by W.L. Leoschke over the period 1970–2000 at the National Fur Foods Co. Research Ranch. Of the dozen or more meat and bone meals assayed in this research, not one single product was satisfactory.

Experimental studies in Europe reported by Ahlstrom et al. (2000) confirmed the observations of Leoschke in terms of research work involving meat and bone meals from condemned cattle and pig carcasses, fox and mink carcasses, and inedible slaughterhouse by–products. None of the products exhibited protein biological value satisfactory for mink nutrition. The products had a proximate analysis of protein, 50%; fat, 15%; and ash, 24%. These protein resources failed to provide levels of digestible sulfur amino acids required for the top performance of the mink for growth and fur development.

A study by Skrede and Ahlstrom (2001) involved the use of meat–and–bone meal at 22.5% and 45% of digestible protein, replacing fish meal during the critical reproduction/lactation period. Reproduction/lactation was satisfactory, but kit weaning weights were significantly lower when 45% of digestible protein was provided by meat–and–bone–meal.

A more recent article, on the use of meat and bone meal in mink diets during the summer and fall is of interest (Clausen et al., 2003). The experimental diets involved meat and bone meals at levels of 3–6–9–12%. At levels of 6% and higher, the kit weight gains for July and August were sub–optimum resulting in "trim" mink kits coming into September and "extra" gains in the fur production period. However, the net result was pelt length reduction in all groups receiving more than 3% meat and bone meal. The experimental group with 12% meat and bone meal yielded pelts of dark genetic origin with reddish fur. The authors suggested that the reddish color was directly related to sub–optimum dietary levels of phenylalanine/tyrosine leading to inadequate synthesis of melanin pigment.

In cat studies, black kittens develop red fur on low phenylalanine/tyrosine experimental diets. The normal black fur color can be restored in new fur development provided that the animals are provided with extra tyrosine supplementation (Morris et al.,2002).

In meat and bone–meal products, it is of interest that there is a good correlation between the level of glycine in the products and the ash/protein ratio. This observation makes common sense in terms of the fact that glycine is a key amino acid in collagen connective tissue which is a major protein in skeletal structures. It should also be noted that hydroxy–proline level of a dehydrated protein product is a good measure of the proportion of skeletal tissue present inasmuch as hydroxy–proline is unique to collagen protein.

Poultry Meals

In the United States poultry meals are considered to be a dehydrated product wherein the poultry resource represents whole bird and/or poultry products employed for human consumption with no significant content of poultry feathers, feet, heads, or entrails.

Initial experimental studies at the National Research Ranch on the biological value of poultry meals relative to poultry by–product meals (those containing poultry entrails, feet, and heads) indicate that the biological value of poultry meals may not be superior for the mink. These studies may suggest that the conditions of processing poultry products to yield poultry meal or poultry by–product meal may be more critical than the basic ingredients employed.

Poultry By–Product Meals

Poultry by–product meals may contain, in addition to whole birds, discarded bird parts intended for human consumption and poultry by–products including poultry feet, heads, and entrails but not feathers. It is of interest to note that in Europe, the poultry–by–product meals do contain poultry feathers, thus undercutting their nutritional value for mink significantly. Evidence for this viewpoint is seen in experimental studies reported by Kiiskinen (1984): Poultry By–Product Meal – true protein digestibility, 56% and true amino acid digestibility, 72%.

All factors considered, it would be logical to assume that the biological value of poultry by–product meals provided in North America would be similar to fish meals of comparable ash content. However, multiple studies at the National Research Ranch over a period of thirty years (Leoschke, 1970–2000) indicate that a very high majority of poultry by–product meals marketed in North America are inferior to fish meals with comparable ash content due to multiple factors. They may have low moisture levels (below 6.0%), the Maillard Reaction can bring about the destruction of lysine and other amino acids and the bonding of amino acids including arginine in structures unavailable to the digestive processes of the mink. Because they may also have an amino acid pattern which is significantly sub–optimum.

315

Potato Protein Concentrate

A number of experimental studies have been conducted on the nutritional value of potato protein concentrates for the growth and fur development of the mink. One specific product, Protamyl PF, Avebe, Holland, contained 86% protein, 4% fat, 0.3% ash, and 9% moisture. A consistent pattern was found of decreased kit growth at the higher levels of the potato protein products (Glem–Hansen, 1979; Tausen and Alden, 1980; Lund, 1981; Aulerich, 1984; Hillemann, 1984). In a study at the National Research Ranch, wherein half of the total protein provided to the mink was by potato protein concentrate, Leoschke (1980) noted that the reduced weight gains of the kits was not related to palatability of the product for the kits. It has been reported by Simova et al. (1982) that the sulfur amino acids cystine and methionine may be the limiting amino acids in potato protein concentrates. Studies by Lund (1981) indicated that the depressed growth of kits was due to solanin, a poisonous alkaloid present in potatoes.

It is of value to note that replacement of fish meal protein with potato protein product up to 20% of total protein resulted in more efficient feed conversion and superior fur development.

Single Cell Proteins

For thousands of years, humans have enjoyed the fruits of fermentation—witness bread, beer, and wines! In more recent years, the fermentation has employed yeasts and bacteria to yield protein feedstuffs from petroleum products and methanol. These processes have a major potential for protein production, as illustrated by the fact that the rate of protein production by a yeast may be as much as 100,000 times as great as that by a yearling steer per unit weight of the yeast and steer respectively, (Shacklady, 1972).

Studies on Prina yeast from British Petroleum were conducted at the National Research Farm facilities at the Northwood Fur Farm in 1974. This n–paraffin petroleum product yeast had a proximate analysis % of water, 4; protein, 65; fat, 8; and ash, 6. Relative to fish meal, this Prina yeast had an equal or higher content of the essential amino acids except for histidine (91%) and methionine (70%). Low methionine levels is characteristic of all yeast proteins and the level of methionine in the Prina yeast product was higher than that present in soybean oil meal. In addition, the arginine level is relatively low, with an excessive lysine/arginine ratio where a 1/1 ratio would be preferable. There were also questions about the capsule coat, which is a polyuronide quite different from the polysaccharides normally found in the diet of the mink. Another interesting point was the high proportion of the fatty acids present, with an odd number of carbon atoms compared to the even number commonly found in animal, fish, and plant fats.

The experimental plan for the June early growth period was a 20, 40, 60, and 80% replacement of fish meal protein in the control diet mash which provided 31% protein and 30% fat, a dietary regimen designed to assess the biological value of proteins for the mink. The clear result was an inverse relationship between kit growth and the level of yeast provided in the experimental diets. Even

316

at the 20% level of total protein with good feed volume the kit growth was sub–optimum, indicating that factors other than palatability were responsible for the poor performance. In a longer range summer/fall experiment with 38% protein and 24% fat the results were more favorable, with good weight gains in July–August and superior weight gains in the fall months directly related to the "trim" mink coming into September fur production period.

Danish studies conducted a few years later on single cell protein products were not especially encouraging. (Jorgensen et al., 1975) noted a protein digestibility of only 79% and relatively low palatability of the product. Another study with a penicillin secondary product yielded an even lower protein digestibility rating (Jorgensen et al., 1976). A later study by Kiiskinen (1984) involved a single cell protein product termed Pekilo with a more favorable true protein digestibility of 83%, and a more recent experimental study by Bjergegaard et al. (2002) noted that with an experimental design of 2.5–10% of single cell protein product in a ranch mix, only the 2.5% level provided satisfactory performance.

Skim Milk Powder

Without question, whole milk proteins provide a quality of protein for mink nutrition unequalled by any other protein resource except whole egg protein in terms of biological value as ascertained by digestibility ratings and amino acid pattern. However, all factors considered, the milk sugar (lactose) content of skim milk powder limits the level that may be employed in a mink ration without significant diarrhea problems. That content is usually 2–4% on a ranch mix basis equivalent to 5–10% dehydrated basis, depending upon the genetics of the mink.

Experimental studies by Howell (1951) indicated 100% mortality within 30 days after mink kits on an experimental dry diets were involved in a transition from 15% skim milk powder to a 26% level.

Experimental studies by Leoschke (1970) indicated that dark kits were much more sensitive to skim milk powder diarrhea than pastel kits. The pastel kits would accept a 10% skim milk powder level (dehydrated basis) without significant diarrhea problems.

317

Supplementation

Fat Resources

Raw Animal Fats

Fresh/frozen animal feedstuffs can provide palatable and economical fat resources with high digestible energy for the mink. However, a primary concern for the rancher must be the quality of the initial fresh product and especially the quality of the frozen product in terms of freezer temperatures and length of storage time. With second rate storage conditions and long term storage these products can undercut the vitamin E nutrition of the mink as they undergo a process termed "oxidative rancidity" and too often yield dark pelts with sub–optimum "sharp color." The basic nutritional problem with raw animal fats, fresh or frozen, is that the level of natural antioxidants including the tocopherols as well as vitamin E (alpha–D Tocopherol) present in these products is simply sub–optimum. This basic point is well illustrated in Table 7.2.

TABLE 7.2. Mink Fur Color as Influenced by Ration Composition*

Program	Fur Color Score**
100% Purified Diet***	3.00
87% Purified Diet, 13% Chicken Offal (dehydrated basis)	2.25
87% Purified Diet, 13% Rockfish (dehydrated basis)	2.25

* Stout et al. (1967),**Fur Grade Evaluation: 3–Dark, 2–Medium, and l–Brown, ***containing lard (stabilized with anti–oxidants) and vegetable oils containing natural vitamin E.

Rendered Animal Fats

These are excellent energy resources for modern mink nutrition since the commercial products are stabilized by antioxidants.

"Spent" Oils

Rendered animal or vegetable fats that have been subjected to many hours of high temperatures in the preparation of "fast food" products for human consumption undergo both oxidative rancidity and hydrolytic rancidity. These yield products that provide sub–optimum kit growth as noted in experimental studies at the National Research Ranch (Rietveld, 1973).

Fish Oils

Fresh Fish Oils

Fish oils contain a high proportion of polyunsaturated fatty acids (PUFA) and hence are very sensitive to oxidative rancidity with the development of hydro–peroxide, peroxide, and free radical structures which can undercut both vitamin E and iron nutrition. Even with fresh fish oils, nutrition for the mink may be sub–optimum. In a study by Damgaard and Borsting (2002), fresh fish fats were employed in experimental diets to levels as high as 70% of total fat during the lactation phase. There were no adverse effects on nursing mink body weights, but the experimental diets with the highest levels of fish oil yielded sub–optimum kit weights at weaning. These observations were confirmed by Bjergegaard et al. (2003) and Damgaard et al. (2003). In the latter study, it was noted that a high intake of PUFA resulted in decreasing numbers of blood platelets for both the lactating mink and their kits.

Fresh Fish Oils and Vitamin E Nutrition Status

The National Research Council, USA, Bulletin Number 7, Nutrient Requirements of Mink and Fox, 2nd Revised Edition (Travis et. al., 1982) recommends 27 milligrams of vitamin E/kilogram of feed solids. Experimental studies involving 95% fatty fish products and fresh fish oil (Peroxide Number < 10 ME/kg) (Tauson, 1993) indicated that this recommended level was inadequate, as noted in the loss of pregnant mink and inferior kit growth due in part to reduced feed intake. It was apparent that the fish oil provoked an oxidative stress that was not fully counteracted by the level of vitamin E supplementation.

Fish Oils and Iron Nutrition Status

A number of studies (Tauson, 1993 and Engberg and Borsting, 1994) have noted that the inclusion of fresh fish oils in mink diets, even with high levels of vitamin E, can result in lower hemoglobin and hematocrit levels as well as low liver iron content in kits. It is very likely that the development of an iron deficiency was due to impaired iron absorption, as ferrous cations are oxidized to ferric cations which in turn yield an insoluble ferric hydroxide. All factors considered, significant levels of fresh fish oils in modern mink ranch diets should be used only in combination with special vitamin E and selenium supplementation to minimize adverse effects. (Brandt et al., 1990).

Rancid Fish Oils

When rancid fish oils are included in mink rations, the nutrition of the animal is undercut, resulting in reduced feed intake and sub–optimum kit growth. Studies by Engborg and Borsting (1994) found elevated plasma activities of creatine kinase, aspartic acid amino transferase, and alanine amino transferase, indicating degerative processes in the liver and skeletal muscles. With increasing oxidative rancidity, the mink's capacity for fat digestion is reduced significantly (Borsting et al., 1994). A study by Damgaard et al. (2003) concluded that moderately oxidized fish oil could be employed in mink diets through the winter maintenance and reproduction periods without any negative effects on the health or performance of the mink. However, high levels of the rancid fish oil resulted in lower kit weights at weaning. The high intake of PUFA from rancid fish oil also resulted in a decrease in the number of blood platlets in both the lactating mink and their kits.

Vegetable Oils

Vegetable oils including corn oil, peanut oil, and soybean oil are excellent energy resources for the mink as well as key resources for the essential fatty acid, linoleic acid, which has relatively low concentrations in animal, poultry, and fish fats.

Rapeseed Oil

Rapeseed oil is marketed in North America as "Canola Oil," a name related to its primary source, Canada. Tauson and Neil (1991) reported on experimental studies with rapeseed oil as the main source of fat in mink diets in the growing–furring period. Diets with 3 or 6% OO–variety rapeseed oil provided mink with significantly higher T–4 values and elevated ME intake compared with the control group fed slaughterhouse offal and poultry waste. Fur characteristics were superior in the rapeseed group compared to the control group, a fact which could be related in part to the relatively stable fat resources in a vegetable oil compared to those in fresh/frozen animal fats. A more recent study (Bjergegaard et al., 2003) employing rapeseed oil in the gestation and lactating periods noted superior performance of the mink, compared to performance on a conventional ranch diet without rapeseed oil (Bjergegaard et al., 2003).

Antioxidants

Antioxidants are specific chemicals which work individually or as a group to prevent oxidative deterioration of foods and feedstuffs. The net result of their employment is higher nutritive value, increased palatability, and more attractive appearance of foods and feedstuffs. There are multiple types of antioxidants including (a) naturally occurring compounds such as vitamin E (alpha–D–tocopherol) and other tocopherols, as well as other compounds including flavonols and tannins; (b) synthetic compounds including BHA (butylated–hydroxy–anisole), BHT (butyrate–hydroxy–toluene), and ethoxyquin with the commercial name of Santoquin; and (c) organic and inorganic acids such as citric, tartaric, ascorbic, phosphoric, and phytic.

The natural and synthetic antioxidants such as vitamin E and BHT contain phenolic groups which inhibit the process of oxidative rancidity by reacting with the free radical structures which arise at an early stage of oxidative rancidity.

The antioxidant activity of organic acids such as citric or the inorganic acid phosphoric is termed synergistic, since these do not have antioxidant activity per se, but are able to enhance the antioxidant activity of the natural or synthetic antioxidants. These acids can form stable complexes with such trace mineral cations such as copper, iron, or nickel which may actually act as pro–oxidant catalysts to enhance oxidative rancidity.

Oxidative rancidity of fats is accelerated by moisture, storage at elevated temperatures (human milk is stored at –70° Fahrenheit to minimize oxidative rancidity), and exposure to light. Inasmuch as oxidative rancidity involves the initial formation of free radical structures, the fatty acids most sensitive to oxidative rancidity are the polyunsaturated fatty acids (PUFA) with multiple alkene (double bonded). Antioxidants can work in both in vitro (outside of the body) in the case of animal feedstuffs in storage and in vivo (inside the body) where high levels of vitamin E in the blood circulation support higher levels of vitamin A, for example, as vitamin E anti–oxidative activity protects the vitamin A from oxidation by fish oils. Experimental studies of Cabell and Leekley (1960a) on the value of BHT for the prevention of yellow fat disease (steatitis) in mink, indicated that BHT did not work in vitro; for example, the addition of BHT to salmon waste prior to freezing and storage did not prevent the development of steatitis. Apparently the BHT had been destroyed during the storage period by the PUFA of the salmon fat. To prevent this disease,the BHT had to be added to the mink feed via the cereal component on the day the mink feed was mixed, thus acting in vivo.

BHT

BHT and BHA have been used to preserve rendered fats such as lard and tallow, as well as in multiple commercial food products for many years. The initial experimental work on BHT in mink rations as a prophylactic to minimize the loss of mink kits with yellow fat disease was conducted by Cabell and Leekley (1960b). BHT supplementation of the ranch diet at 112 grams/ton of feed was very effective in preventing steatitis when used from breeding through the lactation and early growth periods. Importantly, the antioxidant had no adverse effects on the reproduction/lactation performance of the mink.

Studies by Travis and Schaible (1961a) indicated that 0.2% of BHT on a dry matter basis had no adverse effects on the late growth, fur development, or reproduction/lactation performance of the mink. This level of BHT is 10X the levels of the antioxidant which is allowed in feeds by the FDA (Federal Drug Administration).

DPPD

DPPD (diphenyl–p–phenylene diamine) is a very effective antioxidant for the prevention of yellow fat disease in mink (Leekley et al., 1959). However, DPPD has a major adverse effect on the reproduction/lactation performance of the mink.

Ethoxyquin

Ethoxyquin (dehydroethoxy trimethylquinoline) with the commercial name of Santoquin is a very useful antioxidant employed in many fish processing plants in the preparation of quality fish meals. Leekley et al. (1962) have shown that a level of 123 mg/kg of ranch diet was very effective in preventing steatitis in mink fed diets with high levels of salmon waste. Travis and Pilbean (1978) showed that Santoquin was effective in minimizing vitamin E losses and oxidative rancidity in fish in frozen storage for five months.

Travis and Schaible (1961a) noted no adverse effects on the reproduction/lactation, growth, or fur quality of mink with the feeding of 0.125% (1,250 ppm) ethoxyquin on a feed solids basis, a level 10X that allowed in feeds by the FDA. However, in a more recent study, Clausen and Dietz (1999) noted that the use of 167 ppm (about 500 ppm solids basis) ethoxyquin in a ranch diet significantly reduced kit weights as of September and produced shorter pelt lengths. Their experimental work also indicated significant changes in the liver and kidneys of the male kits. Another recent experimental study (Bjorgegaard et al., 2002) noted that the use of 300 ppm (about 1,000 ppm solids basis) of ethoxyquin on a control diet already containing some Santoquin had no adverse effects on the lactation performance. Here, the ethoxyquin was transferred from the feed of the mother to her milk. Appreciable differences were seen also in the types of protein and carbohydrate (17 different oligosaccharides) components present in the milk.

Miscellaneous
Acids
Mink Feed pH

The pH of mink feed "on the wire" is of major importance for optimum nutrition and physiology of the mink, as well as for their capacity for top performance. Both high and low pH feed mixtures can lead to significant physiological stress and sub–optimum performance of the animals.

Relatively alkaline (higher pH) feeds (related to an alkaline water supply and/or the employment of limestone, calcium carbonate, in the ranch diet) can lead to the formation of struvite bladder stones. On the other hand, relatively acidic (low pH) feeds can lead to the acid–base imbalance termed metabolic acidosis, a low value of base excess in terms of optimum blood physiology of an animal.

There are inconsistent reports about metabolic acidosis related to a relatively low pH of the feed. At the 6th International Congress In Fur Animal Production, Warsaw, Poland, (1996), Christian Borsting, one of the leading Scandinavian mink nutrition scientists, stated that the pH of mink feed could be as low as 4.7 and still provide top performance of the animals. On the other hand, earlier reports by Jorgensen et al. (1976) indicated that mink feed at a pH below 5.0 yielded significant metabolic acidosis; Paulsen and Jorgensen (1977) concurred that a mink feed pH less than 5.5 could lead to metabolic acidosis.

It is obvious that multiple factors determine that pH of mink feed which brings about sub–optimum performance, including the nature of the dietary program and the genetics of the animals involved.

Dietary Composition

Without question the nutrient composition of Scandinavian mink ranch diets is distinctly different from that of commercial mink ranches in North America. Thus, one can anticipate significantly different recommendations from the respective mink nutrition scientists.

At the International Mink Show in Milwaukee, Wisconsin, January 1998, Rouvinen (1998)

322

provided a Finnish perspective, recommending relatively high pH levels for mink feed "on the wire" for each specific period of the ranch year: lactation, 6.0; early growth, 5.8; and furring, 5.5. At the same time, Weis (1998) reported on Danish mink diets providing pH levels of 5.7 (January–May), 5.3 (June–August), and 5.0–5.1 (September–pelting). All factors considered, it does appear that the mink are less sensitive to a lower feed pH in the period from July to December.

Weis (1998) noted that in mid–August, Danish ranches switched from phosphoric acid to sulfuric acid (10 times more acidic), a decision directly related to better mink feed economics. The recommendation for higher pH levels in mink ranch diets in Europe may be related to the experience of Scandinavian ranchers who used sulfuric acid silage in ranch diets. The use of silage containing feed with a pH lower than 5.5 during the breeding season and the early growth period provided fatal results (Jensen and Jorgensen, 1975; Lund, 1975).

The lower pH of North American mink ranch feeding programs has the specific advantage of lower bacteria populations as the feed remains on the wire for a number of hours.

National Fur Foods redi–mix programs in the years 1996–1998 provided a feed pH in the range of 5.1–5.3 throughout the reproduction/lactation, early growth, late growth, and fur production periods (Michels, 1998) with top performance of the mink. The lower pH levels were achieved through using phosphoric acid (75% feed grade) at a level of 2.5% feed solids basis. At the same time, the company was marketing mink and fox pellets with a similar phosphoric acid concentration.

The sharp differences between European and North American recommendations for optimum pH in mink feed "on the wire" may suggest that other factors besides mink feed pH are responsible for the relatively poor performance of European mink with lower pH feeding programs. One factor, for example, might be the final levels of sulfate anions in the ranch diets.

Mink Genetics

While a majority of mink in North America with mink feed programs providing feed "on the wire" at a pH range of 5.1–5.3 exhibit top performance, it is quite possible that specific mink genetics such as "wilds" and "mahogany" are more sensitive to lower pH levels in the feed (Brown, 2000). When this South Dakota rancher used mink feed below a pH of 5.3, a significant problem occurred: mink kits with protruded rectums.

Acetic Acid

Acetic acid is a relatively weak acid with a net yield of hydrogen cations only about 1% of that of sulfuric acid in a one normal solution. Acetic acid, an organic acid, is thus metabolized by the mink to yield water and carbon dioxide. Essentially, it has no effect on urinary pH and is thus not effective in preventing struvite bladder stones.

Yet, apple cider vinegar, a weak solution of 5% acetic acid, may contain significant levels of potassium salts and may thus actually raise urinary pH via the kidney excretion of potassium cations.

323

When Loftsgard et al. (1972) used 0.5% acetic acid in the process of cooking cereal grains, the net result was to convert the normal starch polymeric structure of the grains to smaller polymers, dextrins, which had greater water adsorption qualities. Improving the water–binding properties of the cereal grains decreased losses of feed "on the wire."

A number of studies have been reported on the use of acetic acid in fish silage preparation, including those of Jensen and Jorgensen (1975) and Sando (1975). The fish silage was prepared with a mixture of sulfuric acid (96%) at 3% together with 0.5 – 1.0% acetic acid or formic acid, lowering the pH of the silage to between 2.5 and 3.2.

Formic Acid

In addition to the work reported above, recent studies (Polonen et al., 1997; Polonen et al., 1999) established that when formic acid is used in mink rations, extra folic acid is required for its optimum metabolism.

Phosphoric Acid

For more than 50 years, phosphoric acid has been a key component of mink ranch dietary programs around the world. Not only does it preserve feed "on the wire," it acts as a dietary urinary acidifier for the prophylactic treatment of struvite bladder stones in mink and fox.

Phosphoric acid was first employed in mink ranch diets in North America in 1954 with the marketing of a special commercial by–product by Wilson & Co., Chicago, termed "Drum Pak." This was a mink feed product provided in 55 gallon drums filled with beef by–products and 2.6% phosphoric acid (75% feed grade) as a feed preservative which did not require refrigeration (Hallinan, 1958). About the same time, Vitamins, Inc. (1960), Chicago, was marketing a mink feed product also using phosphoric as "Fish Pak Condensed whole fish – protein > 22. Ingredients – whole fish, 2% phosphoric acid, sodium proprionate and BHT. 100# plastic bag = 150# of fresh fish. Store without refrigeration. Store refrigerated after opening."

Within a few years, Northwood Fur Farms, Cary, Illinois, was employing phosphoric acid as a feed preservative in its mink feed "on the wire," (Gross, 1955).

Earlier studies at the University of Wisconsin had resulted in the use of ammonium chloride as a urinary acidifier for the prevention of struvite bladder stones in pregnant mink (Leoschke (1954; Leoschke and Elvehem, 1954). With the observation that phosphoric acid was being safely employed on a mink ranch, Leoschke (1956) initiated studies to ascertain that level of phosphoric acid (75% feed grade) that would provide the same degree of urinary acidity (pH 5.7–5.8) within the ammonium chloride program for the prophylactic treatment of mink struvite bladder stone formation. It was ascertained that a level of 0.8% phosphoric acid (75% feed grade), that is, 16#/ton of mink feed prior to water addition (equivalent to 2.0% of phosphoric acid (75% feed grade) on a dehydrated basis) would provide optimum mink urinary acidity for the prevention of struvite blad-

324

der stones.

Studies in the spring of 1956 (March 1st to June 15th) at the Christiansen Mink Ranch, Cambridge, Wisconsin, and summer/fall (July 1st to pelting) at Associated Fur Farms, New Holstein, Wisconsin, indicated that the employment of phosphoric acid (75% feed grade) at the 2.0% level—dehydrated basis—had no adverse effects on the mink in terms of reproduction/lactation, growth, or fur development. Later observations at the National Research Ranch (Klevesahl, 1968) indicated that phosphoric acid could be fed the full 12 months of the mink ranch year without adverse effects on the mink. Other field observations over a period of years, including the performance of the mink on the National Fur Foods redi–mix program, indicated that phosphoric acid could be fed at the recommended level of 2.0% (solids basis) with optimum performance of the mink in all phases of the mink ranch year.

Studies reported by Witz (1984) indicated that the use of phosphoric acid in mink feed programs enhanced fur quality and reduced feed waste at the ranch.

Two specific advantages of using phosphoric acid rather than ammonium chloride to prevent struvite bladder stones in mink are definite feed preservation value in the mink feed "on the wire" and higher palatability of the mink feed "on the wire." Fuhrman (1990), a mink rancher who had used phosphoric acid in his mink feed for many years, reported that one morning a pail of water was accidentally used in the ranch mix instead of the usual pail of phosphoric acid, causing the whole ranch to be "off feed." In a field study at a Wisconsin mink ranch Dietrich (1995) provided specific data that phosphoric acid caused a significant reduction in bacteria populations of the mink feed "on the wire".

Ranch Feed Mix	Bacterial Count/gram
Ranch Diet	2,800,000
Ranch Diet w/phosphoric acid	400,000

Recommended Levels

Field observations from ranchers on the National Fur Foods redi–mix program, the National Gro–Fur mink pellets, and National fox pellets with 2.0% phosphoric acid (75% feed grade) feed solids basis indicated too many cases of bladder stone formation. Apparently changes in fur animal genetics and/or mink and fox dietary programs had resulted in a fur animal urinary pH environment favorable to struvite stone formation even in the presence of 2.0% feed solids of phosphoric acid (75% feed grade). Hence the company made a decision to raise the level of phosphoric acid in its redi–mix program and mink and fox pellets to 2.5% (75% feed grade) solids basis (Leoschke, 1996). This significant increase in the phosphoric acid level without experimental studies was made in part on the basis of experimental reports at two other research sites. Relatively higher levels of phosphoric acid were being used in the diets of ferrets at Michigan State University, as high as 2.9 to 4.7% dehydrated basis (Edors et al., 1989) and at Nova Scotia Agricultural College, fox diets at 2.5% dehydrated basis (White et al., 1992). This increase has resulted in (a) superior feed preservation in mink feed "on the wire," (b) a significant decrease in the incidence of bladder stones in mink and fox on the National Fur Feeds pellet programs, and (c) a significant reduction in the incidence of "wet–belly" disease on a Minnesota ranch (Schultz, 1995).

Sub–Optimum Levels of Phosphoric Acid

Two examples of the effect of sub–optimum levels of phosphoric acid are pertinent here. The accumulated evidence of field observations over many years and on multiple ranches and on one specific example of a commercial fox pellet at half the recommended level (Mosshammer, 1985) lead to the conclusion that sub–optimum levels can actually contribute to the formation of struvite bladder stones in mink and fox. The scientific basis for this apparent inconsistency in the value of phosphoric acid for the prophylactic treatment of struvite bladder stones in mink is obvious. The composition of struvite stones is magnesium ammonium phosphate hexahydrate (Leoschke, 1952), a salt that is relatively soluble at a urinary pH below 6.0 but increasingly insoluble at a urinary pH above 6.0 and that carries a significant potential to form struvite crystals (bladder stones). The use of phosphoric acid at levels less than recommended results in a urinary environment which (1) has a relatively higher pH with an environment favorable to struvite formation and (2) enhanced levels of phosphate anion, encouraging struvite formation.

Wanachek (1996), a West coast mink rancher, noted a high incidence of mink bladder stones and lower mink feed preservation "on the wire" with the employment of ammonium phosphate. Obviously he was using a mink supplementation program with a major potential for struvite bladder stone formation through enhancing urinary ammonium and phosphate ions and increasing urinary pH, thereby contributing to the formation of struvite crystals.

Excessive Levels of Phosphoric Acid

Without question, excessive use of any mink feed ingredient creates the potential for sub–optimum performance, including drowning (excessive water intake). Field observations over a period of years indicate that when excessive levels of phosphoric acid have been employed in mink ranch diets, significant cases of mink malnutrition have occurred.

Rickets/Osteoporosis

For many years, National Fur Foods programs have specifically recommended that for mink diets employing phosphoric acid, a minimum of 40% of the diet should contain "bone–in" feedstuffs to prevent rickets. Specifically, that dicalcium phosphate can be used to attain that goal, as a level of

1/4% of the salt has a calcium content equivalent to 10% whole fish. With ranch diets providing relatively low Ca/P ratios, it has been shown that moderate levels of phosphoric acid can lead to osteoporosis (Durant, 2000).

Protruded Rectums

As early as 1968, Pridham (1968) reported a number of cases of prolapsed rectums in mink kits which he attributed to excessive levels of phosphoric acid in the June diet. Later observations at the Walter Brown Mink Ranch (2000) indicated a relative sensitivity of wilds and mahogany kits to phosphoric acid supplementation leading to a feed pH below 3.5 with resultant protruded rectums. Another mink rancher in Minnesota (Sonnenberg, 2000) reported protruded rectums in mink kits with the employment of 3.5% phosphoric acid (75% feed grade) solids basis, obviously an excessive level.

Urinary Tract Infections

Lauerman (1965) reported that one of the factors promoting urinary problems in mink was improperly administered phosphoric acid.

Calcium Oxalate Urinary Calculi

While dietary acidifiers in mink ranch diets may reduce the incidence of struvite bladder stones, they can also lead to enhanced formation of calcium oxalate calculi (Durant, 2000).

Propionic Acid

Propionic acid has been employed as a preservative in fish silages, as reported by Johnsen and Skrede (1981). Silages were produced at industrial scale processing plants with 0.75% formic acid and 0.75% propionic acid followed by autolysis at 30–33° Celsius for 3 days and then heating to 96°Celsius and removal of fat by centrifugation.

Sulfuric Acid

A number of studies have been reported on the use of sulfuric acid in a fish silage. These include the experimental work of Jensen and Jorgensen (1975) and that of Sando (1975). The fish silage was prepared with a mixture of sulfuric acid (96%) at 3% together with 0.5 – 1.0% acetic acid or formic acid. The pH range of the silage was 2.5–3.2.

N. E. Hansen (1979) has expressed concern about the effect of acid silage on calcium and phosphomeostasis. In response to acid loading, bone salts are dissolved and blood ionized calcium increases. Along with the increased excretion of acids, the urinary calcium and, especially, phosphate excretion increases. The mineralization of bone in growing animals is disturbed with a potential for ossification in the mink.

327

Bone meal

In North America, bone meal has been employed for many years in mink ranch diets and even in commercial fortified cereals. In theory, from the perspective of a chemist, steamed bone meal with 30% calcium and 12% phosphorus should be an excellent resource for these two vital nutrients. However, from the viewpoint of a biochemist involved in the science of the nutrition of fur bearing animals, a basic question must first be answered. What is the biological value of this mineral resource for the mink. That is, what is the actual availability of the calcium phosphate ions present in this mineral resource to the blood circulation of the mink via digestive action?

One approach to examining the biological value of a given feedstuff such as a mineral resource or a vitamin supplement is to feed the product to the mink and assay the kidney excretion of the ions or vitamins provided by the product. In the case of calcium, this is not practical inasmuch as only about 1% of dietary calcium is found in the urine. However, in the case of the phosphate anion, the average urinary excretion is about 24% of dietary intake (Leoschke, 1960b). These studies at the University of Wisconsin indicated that even the soft bones of calves were a relatively poor resource for phosphate compared to the phosphate anions present in horsemeat. Even though raw calf bones contain 15 times the level of phosphate of boneless horsemeat, the 24 hour urinary excretion of phosphate actually decreased as 10, 20 and 30 % of boneless horsemeat in the experimental ration was replaced with raw calf bones. Apparently, the mink have a low capacity for the digestion of bone products.

Not only are bone products a poor calcium and phosphate resource for mink nutrition, studies indicate that high levels of steamed bone meal in a mink's ration can lead to iron deficiency anemia—that is, hypochromic, microcytic anemia—and signs of "cotton" (white underfur) in mink kits (Anon, 1981 and Brown, 1995). The bone meal promoted a more alkaline pH in the intestinal tract and thereby created an environment less conductive to iron absorption.

DiCalcium Phosphate

Dicalcium phosphate has been employed in North American mink ranch diets for many years. It has been the primary calcium resource in multiple commercial fortified cereals for modern mink nutrition, in commercial dog food formulations, and even in commercial breakfast cereals such as "Total." In the early 1950s, Dr. Willard Roberts, the top nutrition scientist within the world fur animal industry, recommended the employment of dicalcium phosphate in mink ranch diets with sub–optimum calcium content (Roberts, 1955). Of all the common calcium supplements used in mink ranch diets, only dicalcium phosphate provides the Ca/P ratio of 1/1 recommended as optimum for animal nutrition.

Calcium/Phosphorus Resource	Ca/P Ratio
Steamed Bone meal	2.4
Tricalcium Phosphate	2.0
Dicalcium Phosphate	1.1
Monocalcium Phosphate	0.6

Studies at the University of Wisconsin (Leoschke, 1960b) indicated that dicalcium phosphate is a superior calcium and phosphate resource for modern mink nutrition; that is, it has higher biological value than the bone products available for practical mink nutrition.

Larvadex

When larvacides, or IGR (Insect Growth Regulator), are provided in mink feed, they travel through the digestive tract and are excreted in the feces. There they are toxic to fly larvae as these hatch from eggs and enter into the larval stage of development.

Aulerich et al. (1981), Aulerich et al.(1984), and Adair et al. (1986) conducted experimental studies with Rabon (Diamond Shamrock Corp. Rabon Oral Larvacide ROL) at levels ranging from 2.5 ppm (recommended) to 50 ppm fed throughout the ranch year from breeding through fur development. These studies indicated no adverse effects on the performance of the mink. Decreased feed consumption was observed at the 100 ppm level of the larvicide. Studies by Aulerich et al. (1981) in an acute short term high level exposure program employing levels of Rabon at 200, 1,000, and 2,000 ppm for three days indicated no adverse effects and plasma acetyl cholinesterase was not affected adversely.

Limestone

Even though limestone (calcium carbonate) contains 38% calcium, it is a very poor calcium resource for modern mink nutrition inasmuch as it can enhance urinary calculi formation and can decrease the digestive capacity of the mink to process fats and proteins present in feedstuffs.

Urinary Calculi Formation

Studies at the University of Wisconsin (Leoschke, 1960b) indicated that the addition of calcium carbonate to a mink ranch diet resulted in a rise in the urinary pH and led to the formation of struvite urinary calculi. Earlier field observations by Hartsought (1951) indicated that the use of limestone in mink diets contributed to losses of animals with bladder stones.

Reduced Fat Digestion

As early as 1974, Ahman (1974) reported that even small amounts of limestone (as little as 1% of the diet) depressed fat digestion by as much as 10% whereas the addition of dicalcium phosphate, tri–calcium phosphate, or bone meal did not have an adverse effect on fat digestion. This initial study was confirmed by Hansen and Glem–Hansen (1980) and later by Rouvinen and Kiiskinen (1991). The alkaline nature of limestone raises the pH of the gastro–intestinal contents and thereby

329

enhances saponification (hydrolysis of fats to yield glycerol and free fatty acids). The free fatty acids released, especially saturated fatty acids such as stearic acid, react with the calcium ions present to form essentially insoluble soaps which simply pass through the mink's body as part of the feces.

Reduced Protein Digestion

Makela and Polonen (1978) reported that limestone grist added for neutralizing acid–preserved slaughterhouse offal has been shown to lower protein digestion in the mink.

Potassium Sorbate—Mold Inhibitor

In a study by Rouvinen et al. (1997), potassium sorbate at 200 ppm was found to prevent mold growth in silage involving silver hake and Atlantic herring.

Propylene Glycol—Mink Feed Anti–Freeze

Propylene glycol (PG) is a clear, colorless liquid with a sweetish taste. It is relatively non–toxic with an oral LD–50 of 32.5 gm/kg for rats. Because of its low chronic oral toxicity, PG is considered to be safe for use in human foods. Glycogenic, that is, converted to glucose in body metabolism, it is an emulsifier, preservative, and freezing point depressant.

The first reported use of PG in practical mink nutrition was that of Gillies (1966) where the product was employed on a number of mink ranches in Alberta, Canada, during the spring and fall. In the spring the product was fed at a level of 1% of the dietary dry matter for the prevention of pregnancy disease of the mink. Treatment for 3–4 weeks prior to whelping continued for 1–2 weeks after whelping. In the fall, PG was fed at a level of 10% of the dietary dry matter to prevent freezing of the feed "on the wire" to maintain body weights, and thereby, maintain pelt quality.

The experimental observations of Gillies included using propylene glycol in both the feed "on the wire" and in the drinking water. The mink feed tests involved PG at 5% of dietary solids. At 10 degrees F., the PG feed took up to two hours longer to freeze than the control diet. After overnight freezing in a deep freeze, the control feed was as hard as a rock, so hard that a nail could hardly be driven into it, while the PG feed could easily be penetrated with the point of a pencil. The control feed could not be easily removed from the wire, whereas the PG feed was readily removable.

In water tests, PG was used at a 5% level at a temperature of 10 degrees F.. The water with PG took over three hours to become thick while the control water had a hard layer of ice in about 1.5 hours; Moreover, the control water froze hard and fast while the treated water did come to a final solid state, but it remained flaky rather than hard and similar to frozen orange juice.

In 1966 the Kellogg Co. started marketing "Kel–Prime," a product containing animal fat, 57% and 20–30% PG in an emulsion with water, lecithin, and polysorbate (preserved with BHA, citric acid, and sodium propionate). The product provided about 3,000 Calories/pound; and at the 1–3% level of employment in the mink ranch diet, reduced bacterial growth in the feed in direct

330

proportion to the level added. At the 3% level, the product provided a maximum of 2% PG in final dietary solids (Howell, 1966).

Studies reported by Witz (1984) indicated that the use of PG at a level of 2% of dietary dry matter involving 100 mink kits during the fall period provided equal weight gains of the kits with superior fur quality and no significant effect on the color of the pelts marketed.

One adverse effect of PG in practical mink nutrition was reported by DeHart (1981) where the use of PG on a number of mink ranches in Wisconsin in 1979 and 1980 resulted in an increased incidence of wet–belly disease. This increase may have been directly related to a higher feed intake (and weight gains) of the male kits on the PG supplemented diets. It coould also be attributed changes in the composition of the mink urine, which would increase the incidence and severity of the disease.

Sodium Bisulfate

The pKa (the chemist's mathematical basis for designating the relative acidity of a specific acid) for phosphoric acid, 2.16, and for bisulfate anion, 1.99, are very similar; thus, their use as a mink feed preservative at equal levels of acid concentration should yield about the same final pH of the feed as noted in experimental data on involving a fish slurry, is provided by Jones–Hamilton, Co., Walbridge, Ohio, its product which contains 75% bisulfate anion, similar to the level of phosphoric acid present in commercial phosphoric acid (75% feed grade).

However, all factors considered, it appears, in terms of field observations by United States mink ranchers employing sodium bisulfate, that this acid resource is not quite as effective as phosphoric acid (Leoschke, 2004). Ranchers Zimbal (1998) and Brown (2000) have indicated that sodium bisulfate was not quite as effective as phosphoric acid (75% feed grade) for the preservation of raw egg products at room temperature. In the case of the Walter Brown Mink Ranch, the use of 3% phosphoric acid (75% feed grade) achieved a raw egg mixture with a pH of 3.5 and zero bacteria population after 3–5 days storage time. The use of sodium bisulfate at a level of 2.0% yielded a raw egg mixture of 3.6 but a sub–optimum lowering of bacteria populations.

Observations of Durant (1999) on the urinary pH of mink fed 0.8% phosphoric acid (75% feed grade) in the ready–mix program prior to water addition or 0.9% sodium bisulfate indicated a higher mink urinary pH with sodium bisulfate supplementation. Durant employed sodium bisulfate levels as high as 2.5% dehydrated basis in the ready mix program without observing any negative effects on feed consumption.

Sodium BiSulfite

Sodium bisulfite, $NaHSO_3$, is an efficient agent for preservation of fish and has a stabilizing effect on the fat fraction. Multiple studies have been conducted on the use of sodium bisulfite as a preservative for mink feed silage (Ahman, 1966; Jorgensen, 1965; Jorgensen et al. 1971; Quist and Akela, 1961, 1963, 1964). However, there is one very serious drawback to the use of sodium bisulfite as a

331

mink feed preservative: it destroys thiamine, vitamin B$_1$ (Moller, Jensen and Jorgensen, 1975; Wehr et al. 1977). These studies have demonstrated that almost all the thiamine in a mink diet is destroyed within 24 hours when bisulfite is present. To prevent this vitamin deficiency, fish silage preserved with sodium bisulfite has to be fed immediately after preparation of the diet or it must be used alternately with non–sodium bisulfite silage to prevent a thiamine deficiency in the mink.

8

Modern Mink Nutrition—Pellets

Pellet Development—Mashes

Mink Nutrition Concepts

I first became involved in experimental studies on mink nutrition at the beginning of graduate school studies at the University of Wisconsin in June, 1950. To my surprise, I learned that, as late as 1900, chickens were being fed fresh meat as a means of nutritional survival. I believed that as poultry research had enabled farmers to progress beyond the requirement for fresh meat in poultry rations, mink nutrition research would do the same. At that time I felt that it was only logical to believe that major progress in understanding mink nutrition would result in the development of mash and pellet formulations to provide superior performance of the mink. These would certainly surpass current ranch diets containing 15% fortified cereal and 85% of fresh/frozen horsemeat, fish, poultry, and beef by–products.

It is of interest to note that as early as 1953, Harris et al. (1953) in the National Research Council's first bulletin on the nutrient requirements for foxes and minks recommended, "Research to develop a dry diet for minks should have a high priority."

Today in the 21st century that vision of a young graduate student has become reality. For the reproduction phase, pellet programs are supporting performance of the mink which equals that of common mink ranch diets containing primarily fresh/frozen animal, bird, and fish products. In all other phases of the ranch year, they support superior performance. In lactation, "nursing anemia" rarely exists, and in the early growth phase there are minimal problems of "June Blues" (excessive losses of kits and nursing mothers).In late growth, pellet programs provide "trim" mink coming into September, resulting in enhanced feed intake during fur development. At pelting, pellet programs provide unique pelt quality with lighter leather, silkier fur, and the sharpest color ever achieved in the mink ranching industry.

Initial studies at university experimental stations and commercial research ranch facilities leading to the development of pellet programs for the mink involved mash formulations. A dry powder was mixed with an equal weight of water to yield a mixture capable of supporting the top performance: "Open The Bag, Add Water, and Raise Mink". The early studies with "dry" diets led to insights into specific factors required to achieve top growth of the mink, including:

1. dry mink feed ingredients with high palatability for the animals;

2. optimum energy levels, that is, higher fat levels; and

3. dehydrated animal protein resources with high biological value for the mink, protein resources with (a) an optimum amino acid pattern and (b) high digestibility ratings. Major progress in re

335

search studies on "dry" diets was achieved only after a fourth critical factor was acknowledged:

4. the age of the kits at the initiation of experimental feeding trials.

Most of the earlier studies with "dry" diets were initiated in August, by which time specific taste preferences have already developed. In late June or early July, the kits are eager for any mink feed made available to them; however, by August these same kits are highly prejudiced against any significant dietary changes. With young, eager kits in early July, a given "dry" diet formulation provided as a wet mash may be readily accepted, but the identical wet mash product presented in August may receive a very negative reaction. This observation comes from my personal experience at the National Research Ranch, Eagle River, Wisconsin, in the summer of 1959. I assure you, I heard the mink respond, "Do you expect us to eat this crap?"

Palatability Of Dry Mink Feed Ingredients
Initial Research

Initial experimental studies on the taste appeal of dehydrated feedstuffs for the mink were conducted at the National Research Ranch, Menasha, Wisconsin, in the summer of 1962 (Leoschke, 1976). The research planning involved "dry" formulations fully fortified with vitamins and trace minerals and in which the major nutrient divisions were provided as follows:

Nutrient Class	Mink Feedstuff
Carbohydrates	bakery steamed rolled oats
Fats	lard (rendered pork fat)
Proteins	high quality fish meals

For the initial experiments, the plan was to provide each mink of the 20 adult male palatability panel with two bowls of the wet mash. However, this plan proved fruitless, for with identical mash in each butter bowl, the mink in one cage would prefer the bowl on the left while another mink would prefer the bowl on the right. Such unequal feed intake from two bowls with identical mash proved nothing. It was decided to begin an alternative day feeding program wherein the ten mink with odd numbers received Control Diet #1 on Monday, Wednesday, and Friday and the Experimental Diet A on Tuesday, Thursday, and Saturday while the even numbered mink received an opposite feeding pattern. As a net result, each of the 20 mink on the palatability panel over the six-day feeding period, had the choice of Diet #1 for 3 days and Diet A for 3 days. Careful measurement of feed intake allowed the calculation of a "Palatability Rating" for each specific mink feed product tested compared with the standard carbohydrate, fat, or protein resource.

This alternate day feeding program received approval from a Danish mink researcher,

Vindys(1968), who referred to the program as one of "surprise and indirect self selection." He commented also that, as wasted feed was readily identifiable, the experiment is more exact. The absence of the opportunity for direct self selection does not seem to me a disadvantage. Experimental data from these initial studies are as follows:

Palatability Ratings—Dehydrated Mink Feedstuffs

Carbohydrates		Fats		Proteins	
Unique Carbohydrate	118	Turkey Fat	106	Quality Fishmeal	100
Toasted Wheat Flakes	103	Lard	100	Poultrymeal	90
Rolled Oats	100	Corn Oil	86	Livermeal	77
Soybean Oil Meal	85	Vegetable Oil	81	Torula Yeast	54
Wheat Middlings	74			Low Quality Fishmeal	47
Corn Hominy	60			Meatmeal	40
Wheat Germ Meal	55			Bloodmeal	35
				Whalemeal	23
				Fish Solubles	17
				Brewers Yeast	7

These observations on the wide variation in the palatability of dehydrated mink feedstuffs are confirmed by multiple research reports including studies indicating the low taste appeal of wheat germ meal (Watt, 1952); the relatively low palatability of soybean oil meal (Watt, 1952 and Belzile, 1976); and the low palatability of commercial liver meals and meat meals (Howell, 1951, Rouvinen et al., 2002). In addition Kifer and Schaible (1956) and Leekley et al. (1962) confirmed these observations on the relatively low taste appeal of torula yeast and the especially low taste appeal of brewers yeast for the mink. My own experience at the University of Wisconsin (1956) in an experimental study designed to compare the urinary excretion of B–vitamins (niacin, pyridoxine, and thiamine) and/or their metabolites with mink placed on ranch diets with 2% brewers yeast or 2% live yeast product indicated that the mink simply do not like the bitter taste of brewers yeast even when it is a relatively small proportion of the animal's diet (2% of the dietary regimen "on the wire" equivalent to about 5% of the total dietary program dehydrated basis). Initial studies with a composite sample of brewers yeast from the Brewers Yeast Council had to be terminated inasmuch as the adult mink would simply not accept the experimental feeding program. Finally, the experimental work was completed with a special "debittered" brewers yeast product.

Although these were not included in the palatability testing, cooked potatoes have high taste appeal for the mink (Ericksson, 1954).

Mink do not enjoy the presence of alfalfa leaf meal in a regular ranch diet or within a "dry" experimental product fed as a wet mash. This was noted in studies at the Christiansen Mink Ranch in Wis-

337

consin sponsored by the VioBin Corporation (Leoschke, 1958) and later confirmed by Svenden (1995).

As to the palatability of fish oils versus plant fats, studies reported by Polonen et al. (2000) demonstrated that the linseed oil diet provided smaller weight gains than did fish oil supplementation.

A classic example of the naivete and inexperience of a young "fresh" Ph.D. mink researcher shows is the simple Wisconsin Cereal formulation by Leoschke and Rimeslatten (1954). which contained rolled oats, 55; wheat germ meal, 20; brewers yeast, 10; soybean oil meal, 8; wheat bran, 5; and alfalfa leaf meal, 2. A mink cereal formulation provided by two university staff members with 32% of the mink feedstuffs having low palatability ratings –– wheat germ meal, alfalfa leaf meal, and brewers yeast. Obviously, research experience gained over a period of many years is a major asset.

Palatability Enhancement–Flavoring Products

Experimental studies at Cornell University (Warner, 1966) have yielded little positive data on the value of flavoring products such as MSG to enhance feed consumption. More recent studies by Rouvinen–Watt et al. (2002) confirm these earlier findings. Flavorings including aspartame, dextrose (glucose), fructose, kelp, liver powder, liver–whey powder, molasses, and salt or xylose did nothing to enhance palatability; in fact, the liver–whey powder product actually reduced feed intake of the male mink. Still, high quality liver powders and liver digests do enhance taste appeal of pellet formulations as noted in studies at the National Research Ranch facilities.

Some mink ranchers have observed the positive value of molasses (Langenfeld, 1955) and liquid vanilla extract (Lang, 1996). One gallon/ton of mink feed proved of definite value in keeping mink "on feed" at whelping.

However, while mink may enjoy raspberry and strawberry gelatin mixtures, they do not particularly appreciate chocolate puddings (Zimbal, 2000).

Optimum Energy Levels–Higher Fat Levels

Without question, a key factor in the successful development of top performance mash and pellet formulations is the level of fat provided. As carnivores, mink have much higher energy requirements than omnivores or herbivores. The diet of the mink in the wild involves animals and fish with body composition low in carbohydrates and high in protein and fat. In addition, fat contributes to the palatability of any animal's diet, humans included.

My first direct insight into the unique value of fat as a palatability factor in mink mash formulations was in 1956 at the University of Wisconsin with a research project on the digestibility of rendered fats (pork lard, beef tallow, and poultry fat) for the mink. To enhance the value of the experimental data, I decided to add an additional facet to the program by ascertaining the metabolic fecal fat excretion of the mink, that is, fat excretion limited to the basic physiology of the animal and not to the fat resources in the mink's diet. With this experimental data, I could convert data on "Apparent Fat Digestibility" to "True Fat Digestibility" simply by subtracting the daily fecal metabolic fat excretion data from the data on daily fecal fat excretion on the experimental diets containing a specific rendered fat.

To achieve an experimental diet designed to yield data on the metabolic fat (fat excretion via the feces directly related to the metabolism of the animal and not to dietary fat intake) excretion of the mink, I required a dietary program essentially free of fat and which, at the same time, met the daily nutritional requirements of the mink for protein, carbohydrates, vitamins and trace minerals. This (I thought) would be easy to do: simply delete any significant fat resource from the original formulation for the study.

Experimental Diets–Fat Digestibility Studies–University of Wisconsin

Ingredients	Orignal Diet	"Fat Free" Diet
Fortified Cereal	46%	57%
VioBin Fishflour*	35%	43%
Rendered Fat	19%	

* A fat solvent extracted fish meal providing a minimum level of fish oils.

Studies on the digestibility of lard, tallow, and rendered poultry fat with the original diet, found the acceptance of the adult male mink quite satisfactory. However, the mink's lack of enthusiasm for the "fat free" diet completely surprised me. In nutrition theory, the mink should have consumed more of the "fat free" diet to achieve daily energy requirements; however, I had simply forgotten about the key role of fats in the palatability of any diet for animals, including human animals.

My initial observations on the key role of fat levels in the formulation of meal diets for the mink have been confirmed by mink nutrition scientists throughout the world. Michigan State University researchers Kifer (1956), Kifer and Schaible (1956), Travis and Schaible (1961), and Aulerich and Schaible (1965) have concurred. The last authors stated succinctly, "The addition of fat to dry diets increases palatability." Oregon State University's Stout et al. (1962) remarked "A significant finding in this brief study was that the higher energy ration, that is those containing higher levels of fat, consistently resulted in improved growth." A 1978 Danish review said "Success with dry diets are for instance to a high degree depending on the energy density of the feed." Kumeno et al. (1970) of Japan specified "24% fat level provided superior growth to 20% on a meal formulation study."

Biological Value of Protein Resources

The biological value of a given protein resource for an animal is related to the amino acid pattern and availability of those amino acids to the digestive capacity of the animal. The latter point is of unique importance for the mink with its relatively short gastro–intestinal tract and limited time of feed passage from mouth to anus.

339

In developing "dry" diets using dehydrated protein feedstuffs, the challenge was to obtain protein resouces with high biological value for the mink. The processes required to convert fish, fish scrap, poultry by–products, blood, and other fresh/frozen mink feedstuffs into fish meals, poultry meals, and blood meals often result in a significant Maillard Reaction wherein lysine, cystine, and tryptophan are destroyed and arginine is bound in a structure unavailable to the digestive processes of the mink (Allison, 1949). In the production of fish meals, secondary disulfide (cystine units) cross–linking may occur, leading to a reduced digestibility of cystine in these proteins (Opstvedt et al., 1984). Experimental studies at the National Research Ranch facilities over a period of three decades (1970–2000) indicate that a maximum of only 10–20% of the fish meals, poultry meals, blood meals, and meat and bone meals available in North America are of high biological value for the mink (Leoschke 1976, 1995). A consistent pattern has been noted in this failure: a majority of dehydrated protein feedstuffs with a moisture content less than 6.0% produce sub–optimum performance of the mink.

It is important to appreciate the basic fact that the protein destruction processes of the Maillard Reaction are observed during both the dehydration processes yielding high protein meals and extrusion of mink feed ingredients to yield pellets with less than 6.0% moisture. In the later process, the temperatures employed in the drying segment may affect the biological value of the total protein content of the pellet. Experimental studies by Ljokjel et al. (2004) involving extruder temperatures of 100, 124, and 150° C. indicated reduced true digestibility of crude protein and total amino acids present. With pellets processed at 125 and 150° C. there was reduced availability of the essential amino acids – arginine, histidine, isoleucine, leucine, lysine, and valine–and lowered digestibility of the non–essential amino acids – alanine, aspartic acid, cystine, and glutamic acid. The greatest loss of amino acid availability was that of cystine, a critical amino acid for fur development.

Mashes—Academic Studies

Lyophilization–Freeze Drying

In an early study, Gunn (1960) carried out the dehydration of a practical mink ranch diet via a steam jacked dryer at 250° F. with one atmosphere vacuum. However, with this experimental diet, feed intake was limited, with reduced weight gains. Gunn suggested that a vacuum freeze program might be more effective.

Another method of converting a standard mink ranch diet to a powder is through the very gentle but expensive process of lyophilization, or freeze drying under a vacuum. A study at Cornell University (Travis et al., 1967) indicated that the freeze dried powder mixed with water to yield a mash provided excellent growth of mink kits with reduced feed efficiency (grams of feed consumed/grams weight gain). The digestibility of the protein and fat in the freeze dried powder was identical to that of the ranch diet containing 20% cereal and 80% fresh/frozen mink feedstuffs. Reproduction/lactation performance on the experimental mash was somewhat sub–optimumm, indicating a possible loss of specific nutrients via the lyophilization process.

340

A similar program at Oregon State University (Adair et al., 1962) involved an azeotropic process to convert a typical ranch diet to a powder which was then rehydrated just prior to feeding. Unfortunately, significant fat rancidity developed during the processing, to the point of significantly lower growth performance and fur development.

Michigan State University

As early as 1949 experimental studies on the development of "dry" diets for the mink were initiated in North America (Travis et al., 1949). This experimental diet provided the following mink nutrition management program:

Experimental Dry Diet–1949

Ingredients	%
Cooked Yellow Corn	30
Rolled Oats	20
Corn Flakes	15
Soybean Oil Meal	10
Fish Meal	5
Meat Scraps	5
Wheat Germ Meal	5
Homogenized Fish	2
Blood Meal	2
Brewers Yeast	2
Alfalfa Leaf Meal	2
Ground Limestone	1
Fish Oil	0.5
MnSO4	0.5
Proximate Analyses: Protein, 23; Fat,3; and Ash; 7	

Obviously, the diet was not patterned after the mink ranch diets current in North America at that time, which contained 40% protein and 24% fat. However, for the very first time in dry diet experimental work, these studies used amino acid supplementation in the fur production period to enhance kit growth and obtain superior fur development. The addition of DL–methionine, DL–lysine, and DL–tryptophan to this formulation provided extra weight gains during October and November. Nevertheless, since this initial formulation was low in fat and placed too great an emphasis on plant protein resources, it could support only sub–optimum growth and fur development. Additional research studies on "dry" diets at Michigan State University are detailed in the following reports from Kifer and Schaible (1955a, 1955b, and 1956).

One of the better "dry" diet formulations from these studies is noted here:

Experimental "Dry" Diet—1955-1956

Ingredients	%
Cooked Grains	18
Torula Yeast	7.5
Brewers Yeast	7.5
Soybean Oil Meal–50 Protein	20
Fish Meal–55 Protein	20
Meat Scraps–60 Protein	10
Liver Meal	3
Cottonseed Oil	6
Vitamins/Trace Mineral Mix	3
Proximate Analyses: Protein, 45 and Fat, 13	

During the fur production period this experimental diet provided growth equal to that of male kits on a ranch diet.

In retrospect, it must be noted that the basic guidelines for the development of higher quality "dry" diet formulations were unknown at that time. Low palatability ingredients included meat meal, liver meal, brewers yeast, and torula yeast. The research work was sponsored in part by the Brewers Yeast Council and the Sulfite Pulp Manufacturers Research League, torula yeast producers. Obviously the fat level was sub–optimum and the only higher quality protein resources were the fish resources employed, creating an excess emphasis on plant protein resources. Moreover, the experimental studies were not initiated until late August when the kits were already 16 weeks of age.

In later studies the best kit growth performance came from formulations providing 25% crude casein and a higher protein fish meal. That is, experimental diets, in particular emphasized protein resources with higher biological value, crude casein in particular, in which production is less likely to suffer protein damage during production than are fish meals, liver meals, and meat meals.

The first studies on "dry" diets for the mink during the critical reproduction/lactation phase of the mink ranch year were conducted at Michigan State University (Travis et al., 1949 and Travis and Schaible, 1962). The mink on the "dry" diets responded with sub–optimum reproduction/lactation performance due to a higher mortality as well as lower kit weights at 21 days. Partly accounting for this failure was the fact that the formulations were identical to those employed from late August to early June, with no unique formulations for the reproduction/lactation period, the most critical facet of mink nutrition management. As a result, the scientists recognized that "...some factor or factors necessary for adequate milk production was missing from the dry rations."

342

Norges Institute for Fjorfe or Pelsdyr

An early European "dry" diet formulation is provided in the following table (Rimeslatten, 1954)

Norwegian Dry Diet Formulation

Ingredients	%
Fishmeal	45
Soybean Oil Meal	5
Wheat Meal	20
Oatmeal	10
Wheat Germ Meal	5
Grass Meal	2.5
Brewers Yeast	2.5
Cod Liver Oil	1.25
Lard or Tallow	5
Dry Skim Milk	3.75

Reproduction/lactation and early growth (June) was sub–optimum, very likely due, in part, to the low fat content, 10–14%. The palatability of the mash was undercut by significant levels of wheat germ meal, grass meal, brewers yeast, and the low level of fat.

University of British Columbia

The following "dry" diet formulation, supplemented with 5% liver was supportive of good growth and re-production/lactation (Wood, 1962). The liver enhanced the palatability of the ration as well as increasing its total nutritional value in terms of essential amino acids, trace minerals, essential fatty acids, and vitamins.

UBC Mink Ration No. 144–625

Ingredients	%
Herring Meal	25
No. 1 Feed Screening	40
Stabilized Fat	27.5
Brewers Yeast	2
Iodized Salt	0.5
Vitamin Pre–Mix	5

Oregon State University–1956

The following "dry" diet formulation was supportive of excellent growth of male kits, with slightly lesser weight gains of the female kits. Fur quality was equal to that of the control group fed modern anti–oxidants or natural anti–oxidants such as vitamin E (Adair et al., 1966)

Ingredients	%
Oat Groats	35
Blood Meal	14
Herring Meal	20
Lard	14
Soybean Oil	2
Brewers Yeast	2
Molasses	2
Beet Pulp	2
Dicalcium Phosphate	3
Wheat Bran	6
Proximate Analyses	**%**
Protein	34.8
Fat	21.4
Fiber	2.4
Ash	6.6
NFE	34.8

Pennsylvania State University

The following "dry" diet formulation produced size and fur quality nearly equal to a conventional mink ranch diet (Cowan et al., 1972).

Pennsylvania State "Dry" Diet

Ingredients	%
Poultry By–Product Meal	57.5
Cereal	25
Fish Meal	15
Vitamin/Mineral Mix	2.5

The formulation is very simple and very economical, as on the world market today, poultry by–product meal is much lower in cost than fish meals. However, I totally disagree with the authors' viewpoint which emphasized feed economics over top performance of the mink "Dry feed @ $200/ton can provide a mink with size and fur quality nearly equal to mink receiving diets costing $400/ton." It seems, however, that "nearly equal" has a major pelt economic cost. Pelt prices for mink on the Penn State "dry" diets were 20% below those of pelts achieved through a conventional mink ranch diet.

The goal of any serious research program on "dry" diets should be formulations that produce superior performance especially in fur quality and that allow North American mink ranchers to compete within the world fur market.

Mashes—Commercial

Trouw Pelsifood

Without question, the first successful "dry" dietary program in the worldwide fur industry was marketed in 1963 by Trouw & Co., Amsterdam, Holland. By 1969, more than 200 ranchers were using Trouw meal with 100–150% water addition to yield a mash (Broyles, 1969). Chief components of the product marketed in the United States included fish meal, animal fat, spray-dried blood meal, gelatinized corn, dried meat product, skimmed milk, oat groats, wheat middlings, inactivated yeast, soybean meal, dehydrated alfalfa meal, soybean oil, lecithin, and dicalcium phosphate, with vitamin and trace mineral supplementation. Guaranteed analyses data included protein, 34; fat, 22; crude fiber, 3; and ash, 10 (Anon 1969).

The success of Pelsifood in Europe was related, in part, to the basic fact that the research studies preceeding marketing provided insights into low ash dehydrated protein feedstuffs with high biological value for the mink(Van Limburgh et. al., 1969). American mink ranchers visiting the Netherlands and observing pelts from Pelsifood mink were impressed with the silkiness of the pelts (Anon, 1969) and the sharp color (sheen and sparkle) compared to pelts of mink raised on conventional ranch diets (Clay, 1969).

After a very successful decade in Europe, Trouw Pelsifood was produced and marketed in North America, but it proved a complete failure in terms of the performance of the mink. All factors considered, it seems that locally available dehydrated protein feedstuffs proved to be inferior to those available in the Netherlands. Apparently, no effort was made to assess their biological value for the mink. The naive perspective that "All Fish Meals Are Created Equal," obviously equally applied to blood meals and poultry meals available in North America, proved to be a downfall for Trouw Pelsefood. In fact, studies at the National Research Ranch over the period of 1970 – 2000 indicated that a high majority of fish meals and other dehydrated protein feedstuffs from both South and North America provide proteins with low biological value for the mink. These same studies showed that fish meals being produced in Scandinavia had unusually high biological ratings for the mink, indicating that European fish meal producers had higher standards, very likely for the high protein quality required for aquaculture. This basic point is witnessed by a minimum moisture standard of 6.0% fishmeal specification for Special A and Lt–999 (Esbjerg Fiskeindustri a.m.b.a., Denmark,1998).

345

National GnF–100 Meal–1955-1963

National Fur Foods has been a somewhat reluctant bride to fur animal nutrition research studies. However, over the period of the last half–century, National Fur Foods has provided more labor and financial support to mink, fox, and ferret nutrition research studies than any other fur animal feed company within the worldwide fur industry. I am proud to have been a part of that story.

In the period from June 1954 to June 1959, I was employed as a post–doctoral research associate at the University of Wisconsin at Madison. As of February 1, 1955, I was hired as a mink nutrition consultant for National Fur Foods by Paul Langenfeld, Chairman of the Board, National Fur Foods Co., Fond du Lac, Wisconsin. Naively, I thought that I was being employed as a mink nutrition scientist with the research goal of enhancing the mink nutrition management program of the company (which at the time marketed only a simple fortified cereal, XX–15) and ultimately with the futuristic goal of the developing a high performance mink pellet which would produce superior performance of the mink in all phases of the mink ranch year relative to the conventional ranch feeding programs containing 15% cereal and 85% fresh/frozen feedstuffs. As I found out later, Paul Langenfeld had no interest whatsoever in research on the development of mink pellets (he said "they would never work"), but wanted Dr. W. L. Leoschke on the payroll simply for public relations, since "the mink ranchers trust Leoschke." Over the period of the next eight years, he was able to undercut progress on research studies at the National Research Ranch in many ways. He formulated the initial "dry" diet formulations inasmuch, as far as he was concerned, it was "common sense" to assume that an uneducated millionaire knew more about mink nutrition in 1955 than a Ph.D. University of Wisconsin fur animal scientist. He limited access to mink kits for experimental studies until late August (to minimize the loss of pelt size) when the mink already had a mind–set established about the palatability of a given mink nutrition regimen. In one case he stopped the experimental feeding trial in early September after he saw high levels of feed residues "on the wire". In fact, the research manager had "overfed" the mink on Saturday to eliminate the Sunday feeding.

VioBin, Inc. Research Studies –1956-1957

Making no progress with research studies under the leadership of Paul Langenfeld, I welcomed an initiative by the VioBin Foods, Inc., to conduct studies on "dry" diets employing VioBin fishflour, a solvent (ethylene dichloride) extracted fish meal with minimum content of fish oils. The initial two "dry" diet formulations were as follows:

346

VioBin "Dry" Diet Formulations–1956

Ingredients	#1 – Unpalatable	#2 – Palatable
Proteins		
VioBin Fishflour	45	35.5
VioBin Liver Meal	1	1
Carbohydrates		
Cereal Mix	25	35
Glucose Hydrate	13	—
Molasses	—	5
Fats		
Lard	13	18
Lecithin	2.5	5
VioBin Wheat Germ Oil	0.5	0.5
Proximate Analysis		
Protein	40	35
Fat	18	26
Carbohydrates	25	20
Growth Performance		
Grams/Day	2.5	10

Formulation #1 was designed by Ezra Levin, CEO, VioBin Foods, Inc., which provided financial support for the experimental studies at the Elmer Christiansen Mink Ranch, Cambridge, Wisconsin. "Dr." Levin seemed to believe that a CEO of a major corporation with essentially no knowledge of mink nutrition was more qualified to formulate a "dry" diet than a Ph.D. mink nutrition scientist. Yet, I felt obligated to follow "Dr." Levin's formulation since I was enthusiastic about this new opportunity to conduct research studies.

The mink kits placed on experimental mash #1 in early July did not readily accept the dietary regimen. Thus with limited feed intake, at the end of three weeks they had gained only about 1/6th of the growth achieved by their matched brothers on the conventional ranch diet. I well remember that hot day in late July, 1956, when I observed the weight data. I was very discouraged at that point but I was also determined to continue my personal fight for the development of a 100% "dry" diet for the mink. The thought that I would have to wait another year before testing new "dry" dietary formulations could not be taken seriously. I discussed the matter with Elmer Christiansen and requested an opportunity for an immediate "on the spot" modification of the research formulation to yield Diet #2, with the hope that the new formulation would achieve greater acceptance. Fortunately, Elmer gave me a second chance at a a critical point in the history of mink nutrition research on "dry" diets and pellet formulations.

The important changes in formulation resulted in a much greater acceptance of the mash and unusually fine growth in the weeks to follow. There were two basic changes in the formulation enhanced palatability.

1. The use of 5% molasses, a tasty feedstuff for the mink. I had gleaned this information from conversations with Ed Langenfeld, Associated Fur Farms, New Holstein, Wisconsin, who had been working on the National Fur Foods "dry" diet formulations prior to my association with the company;

2. An increase in the level of fat, that is, an increase in caloric density and energy concentration. This decision was based on earlier observations I had made at the University of Wisconsin in studies on the digestion of fats by mink (Leoschke, 1959). I had noted that while the adult mink had accepted an experimental diet with 22% fat content (cereal mix, 46; VioBin fish flour, 35; and a rendered fat resource, 19), the mink had not readily accepted a special research diet with only 2% fat but providing the same proportion of cereal and fish flour (cereal mix, 55 and fish flour, 45). This dietary regimen was specifically designed to provide data on the metabolic fat excretion of the animals. It was obvious from this experimental observation that fats play a major role in mink nutrition, not only as a rich energy resource but also as an ingredient enhancing the taste appeal of the dietary mixture.

The VioBin experiments of 1956 were an important turning point in my basic understanding of the modern nutrition of the mink and led to more palatable "dry" diet formulations. In the weeks to follow, the growth of the mink kits on the new diet actually exceeded that of their brothers on the ranch mix. However, at pelting, the final size and fur quality was sub–optimum. Additional studies on the VioBin formulations were continued in the summer and fall of 1957 and thus, in spring 1958, I was able to show "Dr." Paul Langenfeld that the mink kit growth on the VioBin formulations was 5 times that of the animals on the Langenfeld formulations. The net result was that after three years of fruitless experimental studies at the National Research Ranch, Paul Langenfeld reluctantly gave me the opportunity to play a major role in the formulation of "dry" diets for mink in the summer of 1958 and the years ahead.

National Research Ranch Eagle River–1959

After completing five years of post–doctoral studies at the University of Wisconsin in June, 1959, I was given the opportunity to work full time for two months at National Research Ranch facilities. The location of the ranch was a critical factor in the successful studies that followed: I was able to conduct serious studies on "dry" diet formulations without the inhibiting oversight of Paul Langenfeld. The two major achievements of that summer were:

1. The discovery of a new carbohydrate resource with high water absorption and high taste appeal for the mink;

348

2. The observation that while mink kits in early July readily accepted an experimental "dry" diet formulation as a mash, these same animals in late August were reluctant to accept the identical mash. By late August the mink kits were less eager to eat and had developed a set pattern of taste.

Completely disregarding the summer's major accomplishments, National Fur Foods Co. management decided to emphasize "bottom line" economics, cutting my consultant fee in half, and discontinuing my role as "biochemist in residence" at the National Research Ranch facilities. Without question, this essentially foolish decision set the research program back a full 4 years. I was able to resume in–depth work at the National Research Ranch again only in the summer of 1963, since there was new management at New Holstein.

National Research Ranch, Neenah-Menasha–1963-1969

With the appointment of Grant Barrett as manager of National Fur Foods Co., a whole new world of research opportunities arrived. Grant Barrett had an enlightened viewpoint on mink feed marketing. He spearheaded the development of special antibiotic products for the mink, Power–Pak and Dyna–Pak, and showed an enthusiasm for research studies that would lead to the marketing of new mink nutrition products. The net result was the purchase by National Fur Foods Co. of a mink ranch at Neenah–Menasha, Wisconsin, which was placed under the management of Vilas Klevesahl, a wonderful gentleman who was very helpful and encouraging.

The years as "biochemist in residence" in the summer months at the mink ranch at Neenah–Menasha were very fruitful in terms of two major advances in my basic understanding of modern mink nutrition "dry" diet formulations. These were:

1. The initiation of the basic mink nutrition management concept "At National, We Listen To The Mink" as illustrated by a study on the palatability of a wide variety of dehydrated mink feedstuffs wherein a taste appeal panel of 20 mink made the important decisions. The net result was new "dry" diet formulations with higher taste appeal.
2. Major progress in the discovery of "dry" formulations supportive of a sound reproduction/lactation performance of the mink. In the spring of 1967, the ranch achieved an average of 4.0 with dark mink, a respectable average at the time.

However, in the spring of 1968, problems in ingredient quality control arose with a second–rate fish meal from a Canadian plant that had provided first-rate fish meals in past years. This lapse led to sub–optimum reproduction/lactation performance of the mink. Apparently there were changes in the processing of fish to yield fish meals at the plant, engineering changes that resulted in a fish meal with a protein content of significantly lower biological value for the mink. It would take four years of intensive studies before high performance reproduction/lactation pellets could be marketed, the new formulations involving new higher quality fish meals chosen by the mink in studies on the biological value of dehydrated protein protein resources, the introduction of a high quality but very expensive liver meal (two dollars per pound) and the use of a new dehydrated protein feedstuff of high biological value for the animals.

National–Northwood Research Facilities–1970-1979

Major progress was made in this decade in our basic understanding of the wide variation in the quality of fish meals and other dehydrated protein feedstuffs on the market. Intensive experimental feeding trials on the biological value of dehydrated protein feedstuffs involved as many as 300 mink kits in early June (Leoschke, 1976). Tony Rietveld, the new manager of the National–Northwood Ranch with 20,000 breeders at Cary, Illinois, encouraged not only these, but studies that extended over the next three decades.

With enhanced research activity at National–Northwood, by 1978 National Fur Foods was marketing a high quality GnF–100 mash mink nutrition program with good performance of the mink in all phases of the mink ranch year, especially in terms of unique quality pelts with sharp color and lighter leather. Finally, mink ranchers could discard their conventional ranch dietary programs which had involved the intensive labor of preparing fresh/frozen feedstuffs. They could discard their expensive freezers and grinding equipment and simply "Open The Bag, Add Water, and Raise Mink."

The mink pelt crop in 1978 at National–Northwood consisted of more than 100,000 pelts from mink raised from early July to pelting on National GnF–100 "dry" feed fed as a mash. Unfortunately, the concept of a mink mash program did not receive support within the North American fur industry since the rancher still had the labor-intensive work of feeding the mink twice a day, seven days a week, and scraping the feed boards the following morning. Thus the need for a pellet program, with it's specific advantage of reduced labor.

National Research Ranch Oshkosh–1979-1991

During these years of research studies on an enhanced pellet program Mark Michels, Paul Mauer, and Bruce Voss provided major assistance. Finally, the research team had developed a pellet supporting top performance of the mink in all phases of the mink ranch year. With a combination of superior (a) genetic management, (b) nutrition management, and (c) ranch management, over a period of ten years the reproduction/lactation performance of dark mink was raised from about 4.0 to a solid 5.0 ranch average (50–50 old and virgin breeders). In addition, progress was made in replacing bitter brewers yeast with a unique vitamin and mineral formulation with higher palatability for the mink. With the nutrition insights of Dr. Dan Chausow, major progress was made on the special amino acid supplementation required for the growth and fur development of the mink. Adding to that, a unique trace mineral supplementation program resulted in the sharpest color ever achieved in dark mink within the world fur industry.

National Research Ranch Facilities–1992-2006

Important research on the modern nutrition of the mink through a pellet program was continued with the cooperation of Denney Patton, of Friendship, Wisconsin, and Tom Geiger, of Mosinee, Wisconsin. These experimental studies led to new fish meal resources; that discovery led, in turn, to the marketing of a special pellet — Gro–Fur Gold — supportive of even sharper color in dark pelts. With new insights into the critical value of higher carbohydrate levels in the fur development period provided by Dre Sanders from the Netherlands, the degree of sharp color in darks was enhanced further.

Pellet Development–Academic

Michigan State University

The first studies on pellets for the mink conducted at an academic institution within the worldwide fur industry took place at Michigan State University (Travis and Schaible, 1961). The mink kits on the pellet program grew as well as the kits on the "dry" diet formulations.

France

As early as 1971 pellet formulations supportive of good growth and fur development similar to that of conventional ranch feeding programs had been developed (Rougeot, 1971). Detailed "dry" diet formulations of later years are shown. See table below:

French Pellet Formulations

Ingredients–%	1979	1980
Proteins		
Fish Meal	25	20.6
Meat Meal	15	5
Soybean Meal–50	5	5
Blood Meal	5	5
Skimmilk Powder	—	4
Brewers Yeast	5	5
Carbohydrates		
Cooked Cereals	20.6	30
Toasted Wheat	2	2
Lucerne Meal	2	2
Beet Pulp	2	3
Fats		
Tallow	12	6
Lard	4	6
Fortification – Vit/TM	2.4	5.4

Proximate Analysis	1979	1980
Protein	44	35.5
Gross Energy, Cal/kg	5,460	5,470
Metabolic Energy, Cal/kg	—	3,860

Charlet–Lery et al., 1979 and 1980.

Nova Scotia Agricultural College

Ingredients	NSAC Dry
	%
Proteins	
Corn Gluten Meal	21
Fish Meal	14
Meat Meal	12
Poultry Meal	6
Soybean Oil Meal	6
Whey Powder	6
Carbohydrates	
Extruded Wheat	21
Fats	
Fish Oil	8
Lard	6
Fortification	1

Rouvinen et al., 1996.

This formulation is unique in terms of providing a wide variety of protein resources.

Pellet Development–Commercial

Kellogg Co.

As early as 1950, Kellogg Co. had offered and sold to a selected group of mink ranchers a complete dry diet in the form of a pellet which was given to the mink kits from July 1 to pelting. This product was the result of considerable research and feeding trials with mink under the direction of Roland Howell. The initial results were good: the mink were well furred with clear and sharp color, but they were about one inch shorter than mink on conventional ranch diets. Thus the product was withdrawn from the market (Borsum, 1969).

By 1953, Kellogg Co. was marketing "Mink–etts Kellogg," a new pellet form of meat extending as a free–choice supplement. "Especially practical for Sunday feeding. Water them and take the day off when you have Mink–etts in every pen. Mink–etts and water provide adequate holiday fare for your herd" (Anon, 1953). Howell (1953) noted that the lactating mink were less restless, an observation made many times since with mink raised on quality pellet programs.

Federal Foods–1948-1970

Without question, the "Man of the Century in Mink Nutrition" and the "father" of pellet development in the worldwide mink industry is Dr. Willard L. Roberts, Federal Foods Co., Ross–Wells Co., and XK Co. After a prolific career, he finally ceased his efforts on mink nutrition research at the age of 96.

He began his research studies on the formulation of pellets for modern mink nutrition in 1948 with the cooperation of the Fromm Brothers Nieman Mink Ranch in Wisconsin. The first public exhibit of mink kits raised on pellets was in January 1953, at the International Mink Show, Milwaukee, Wisconsin. The mink were relatively small with a greenish cast over their fur, a feature directly related, no doubt, to Dr. Roberts' enthusiasm for alfalfa leaf meal in Federal mink cereals and pellet formulations. Dr. Roberts' 1953 commentary is of interest: "Ideal pellet–maximum benefits of minimum costs and feeding effort–adequate taste appeal." He felt that his current pellet did a good job from August to pelting with promise for whelping but was inadequate for early growth (Roberts,1953). After many years of intensive research effort, Dr. Roberts felt that his Complete Mink Food Pellets had been improved to a point of warranting large scale production.

Nippon Formula Feed Mfg. Co.–Japan

An interesting study in mink pellet formulations was published by Kumeno, (1970) with one of the formulations as follows:

High Protein/High Fat Diet

Ingredients	%
Proteins	
Whitefish Meal	44
Torula Yeast	2
Carbohydrates	
Cooked Wheat	9.4
Cooked Potato Starch	18.6
Alfalfa Meal	4
Fats	
Stabilized Extra Fancy Tallow	21
Fortification	**1**
Proximate Analysis	
Protein	34.3
Fat	24.0
Ash	9.1
Crude Fiber	1.1
Moisture	4.2*

* Without question, the low moisture content undercut the protein quality of the formulation and may have been responsible for a relatively high protein level required for top fur development.

353

The experimental studies involved protein levels of 25–35–45 and tallow levels of 11–16–21–24. The 35% protein and 24% fat diet provided the best weight gains while the 25% protein diet produced very slow growth regardless of fat level. The best fur was produced with the diet containing 45% protein and 20% fat, that is, a diet with a relatively low caloric density, providing the highest level of protein intake for the critical fur development phase of the mink ranch year.

The pellet engineering was unique. Pellets were prepared by pelleting a whole formulation without tallow into a 0.48 cm pellet, with some meal set aside. Pellets were then placed into a mixer with a steam jacket and heated to 60° C., at which point tallow heated to 80° C. was added and blended. The product was removed and cooled. The original meal previously saved was added to the surface to prevent the pellets from sticking to each other.

Ross-Wells (Beatrice Foods)–1970-1990

The most popular mink pellet ever marketed in the 20th century throughout the worldwide fur industry in terms of distribution in both North America and Europe and the total number of mink ranchers employing the product was sold under the name of two salesmen, Bill Ross and Loyal Wells. These individuals had not only no basic knowledge of mink nutrition themselves, they failed to acknowledge the people responsible for the concept of a mink pellet—the Kelly brothers, Paul Iams, and one single individual, Dr. Willard L. Roberts, as noted above, who provided the mink nutrition expertise and engineering required for the marketing of a top performance pellet.

As early as 1949, the Kelly brothers had patented the concept of a simple pellet for mink nutrition involving freezing an extruded pellet consisting of a cereal mix and fresh animal feedstuffs (Kelly and Kelly, 1949).

> "A preferred feed in accordance with the present invention comprises individually compact, frangible moist pellets of which the main ingredients are essentially dried cereal particles and fresh meat particles in about equal proportions. Passed through a grinder and extruded to yield pellets ¾ inch in diameter and 1/2 to 1.25 inches in length. Pellets normally frangible. Pellets immediately frozen –10 to –20 degrees F. and placed on a feeding tray. Moisture content limited to natural fluids of the meat. Present feed much preferred by dogs and foxes."

In later years, Paul Iams and Bill Kelley were about to give up on the development of dry feed formulations for the mink as pellets, due to limited availability of high quality dehydrated protein resources. Fortunately, they were able to team up with Ross–Wells (Beatrice Food Co.) who installed their own dehydrating equipment to produce low temperature, high quality protein meals. Ross–Wells were able to control the quality of ingredients, which consisted largely of reject whole broilers, some

354

poultry offal, and also some fish meals (which they actually produced in their own dryers). The result was a really good product. It had good meat and fish meal proteins, adequate fat levels and carbohydrates as grain resources that were fully cooked and dried (Hartsought, 1974). However, all factors considered, the reproduction/lactation performance of the mink on the Ross–Wells pellet was definitely sub–optimum. The product had no future until Dr. Willard L. Roberts (1979) became involved with his unique nutritional and engineering expertise Within a few years, the Ross–Wells pellet had earned a reputation for good reproduction, good size, and top fur development. The secret to the success of this mink pellet was summarized by Dr. G. Hartsought:

> "I think the secret of success, if you can call it success at this point, is the manner in which the protein products are produced in the dry feed and the means of control before they go into the drier, the control of temperatures in the drier, so that you are not decharacterizing those proteins in the production." (Hartsought, 1974).

The word "decharacterizing" is actually a layman's expression for a processing program that involves modifying the protein structure of the ingredients via the Maillard Reaction and other processes which undercut the biological value of the product for the mink. The Maillard Reaction which can occur during dehyration procedures; it results in the destruction of lysine, methionine, cystine, and tryptophan and bonding of arginine in a structure unavailable to the digestive processes of animals. The procedures employed in the production of the Ross–Wells pellet after Roberts' association were so gentle that a significant number of ranchers reported "hairless" kits in the early weeks after whelping, direct evidence of a biotin deficiency brought about by an un–denatured avidin protein present in the egg whites provided in the formulation. The Ross–Wells pellet program engineering was so gentle that the egg white proteins were not denatured and the avidin protein unmodified to the point of inactivity relative to the bonding of biotin.

Since the mink is a carnivore, the biological value of the proteins provided in terms of (1) amino acid pattern and (2) availability of those amino acids to the digestive processes is the key to the success of any pellet program.

Until 1990, the Ross–Wells pellets had the highest market of any pellet in the history of the worldwide mink industry. However, quality control problems arose, and a decision was made to discontinue marketing the product. The company profit margin was significantly higher with dog food processing, a rationale similar to that of Beatrice Foods when it discontinued the marketing of an excellent fox pellet.

355

Extruded Pellets vs. Steam Process Pellet Mill Production

Over the last 25 years, a minimum of 12 extruded pellet products have been marketed in North America for mink. Only three of these formulations, those of National Fur Foods, Ross–Wells, and XK Mink Foods extruded pellet, the latter two formulations indebted to the nutritional and engineering expertise of Dr. Willard L. Roberts, proved to be successful in supporting of the top performance in all phases of the mink ranch year.

Sub–optimum extruder pellet engineering can cause a significant loss of nutrients due to excessive dehydration and/or excessive temperatures. These losses include:

1. The loss of fat–soluble vitamins A and D and the loss of specific B–vitamins including thiamine, pantothenic acid, pyridoxine, vitamin B_{12}, and folic acid. This problem can be readily resolved by pellet formulations with "overage," or extra vitamin fortification, as much as 50% for the most sensitive vitamins and by a factor as high as 1,100 for the most sensitive vitamin, folic acid;

2. The loss of essential amino acids critical for fur production and reproduction/lactation performance. Extruded pellet products with less than 6.0% moisture have very likely undergone significant Maillard Reaction (Allison, 1949), with destruction and/or bonding of amino acids in indigestible structures. Evidence to support this commentary is seen in the following experimental data from Norway (Skrede, 1980).

Amino Acid Composition and Amino Acid Digestibility–Pellets

	Pellet Resource			
	#1*	#2**	#1	#2
Amino Acid	% of Protein		% Digestibility	
Arginine	5.9	5.6	92	85
Cystine	1.7	1.2	74	52
Lysine	6.9	5.8	88	71
Tryptophan	1.0	0.9	85	67

* Ross–Wells
**An extruded pellet with sub–optimum formulation and/or processing procedures.

Pellet #2 shows a significant loss of critical amino acids as well as a loss in digestibility. The loss of arginine is especially important: in mink fur, arginine is the essential amino acid in highest concentration, and the arginine content of mink milk is 1.6 times the level found in cow's milk. Unfortunately, there is no simple formulation answer to the loss of amino acids with the extruded pellet process. Although DL–methionine may be available at a reasonable price, DL–arginine is very expensive. The answer to this problem, then, lies not in amino acid supplementation but the application of superior gentle engineering processes.

356

In all the known cases of nutritionally sub–optimum extruded pellets marketed in North America in past years, the companies had maximum standards for water content (not over 11% moisture to minimize mold development) but no written standards for minimum moisture content. In the extruded pellet for mink processed in Wisconsin, the actual moisture level was 3.5%; and the pellet for mink processed in Indiana contained a moisture range from 1.2–4.7%.

Problems of the loss of methionine and arginine with excessive heat treatment of dehydrated protein feedstuffs were first noted at the University of Wisconsin with mink on purified diets containing vitamin–test casein (Leoschke and Elvehjem, 1959). Vitamin–test casein is dried cottage cheese (casein) which has been extracted with alcohol to remove all traces of vitamins and then heated to remove the alcohol residues. High quality vitamin–test casein has a light yellow color. In the summer of 1955, a Wisconsin laboratory received a shipment that was tan in color, a clear marker for the Maillard Reaction. Further studies with this product led to the discovery of the key roles of methionine and arginine in fur production.

Vassa Mills Ltd., Finland–1981

A modern European mink pellet formulation has the following composition.

Vassa Mills Ltd. Taysrehu 3 (Complete Dry Feed)

Ingredients	%
Proteins	
Fish Meal	40
Soybean Oil Meal	8
Meat and Bone Meal	5
Brewers Yeast	3
Feather Meal	3
Blood Meal	1
Carbohydrates:	
Processed Cereals	19
Fats	19
Fortification	2
Proximate Analysis	
Protein	46.1
Fat	22.1
Fiber	1.4
Ash	8.7
N.F.E.	22.1

Digestibility data on this pellet for the mink is of interest: Protein, 80; Fat, 86; Carbohydrates, 65; and Ash 38 (Kiiskinen and Makela, 1981).

357

XK Pellets–1984–Present

In 1984 Dr. W. L. Roberts retired from Beatrice Foods and became a mink nutrition consultant to XK Mink Feeds. The initial mink pellet marketed did not support top reproduction/lactation of the mink and was followed with a new XK–100 reproduction/lactation–liver product wherein fresh liver was employed as a basic ingredient for the extruder process with subsequent fine reproduction/lactation performance.

National Fur Foods Pellets–1972-Present

The history of experimental studies from 1950 to 1972 on the development of "dry" diets fed as a mash has already been discussed. However, by 1972, National Fur Foods was able to market a pellet with good reproduction/lactation performance and superior late growth and fur development (Leoschke, 1981). In 1976, considering the fact that research studies had shown uniquely different nutritional requirements for dark mink and pastels, National Fur Foods marketed Pel–Dark and Pel–Light pellets for the growing and furring phases of the mink ranch year.

In 1982 the company took a major step to enhance the reproduction/lactation performance of the mink on National Fur Foods pellets. With the specific encouragement of Jack Brennan, significant changes in pellet formulation were made relative to vitamin and trace mineral fortification, higher quality dehydrated protein feedstuffs, and changes in the amino acid supplementation program. The net result was enhanced reproduction/lactation performance. In the years to follow, new advances were made, especially in new formulations enhancing the quality of pelts produced on National Gro–Fur pellets.

By the year 2000, the dream and goal of that young graduate student–1950 research assistant, mink project, University of Wisconsin–was fully realized. Throughout the world, a pellet program was being marketed wherein the performance, without question, was superior to that achievable on conventional ranch diets containing fresh/frozen feedstuffs. All factors considered, it now appears that mink on a quality pellet program have a greater potential for reaching their genetic potential for top performance than those on conventional feeding programs. This statement is documented with the following field observations.

358

Reproduction/Lactation

National Pellets– Reproduction/Lactation Performance

Year	Location	Diet	Ranch Averages
1986	Scotland–Mundell Ranch	Ranch Mix	4.1
		Pellets	4.5
1986	Denmark – Kemovit	Ranch Mix	4.8
		Pellets	5.7
2001	US–Minnesota–Whites	Pellets	5.0*

*Denotes the highest production ever achieved with his white mink. A net loss in reproduction/lactatation occurred upon return to ranch mixes.

Early Growth–June

National Early Growth Pellets as a mash provides excellent growth of the kits with minimal problems of "June Blues," losses of kits and nursing females.

Late Growth –July–August

Unique mink kit growth pattern resulting in "trim" male kits coming into September with resultant enhanced feed volume and superior fur development.

Fur Development

Finest fur quality achieved within the history of the world mink industry in terms of (1) sharp color (2) lighter leather and (3) silkier fur.

Specific commentary of ranchers and fur graders support this statement as follows.

Unique Pelt Characteristics
Sharp Color

1981, Scotland, Mundell Mink Ranch. Jim Baxter, former mink pelt grader with Hudson's Bay – Annings, London, for twenty years prior to employment by Blake Mundell. With the conversion of the Mundell ranch from conventional mink ranch diets to National Gro–Fur pellets, Mr. Baxter commented, "It was the very first time in my life that I had seen black mink. About 10% of the dark pelts from the Mundell Ranch were black."

2001, Wisconsin, Geiger Mink Ranch. An auction house field grader remarked on observing mink fed Gro–Fur pellets, "My God, these mink are black." The sharp black color of mink raised on National Gro–Fur pellets should not be surprising since the pellets contain 100% stabilized fat resources, which best support sharp black color.

In the early years of the North American mink industry, ranch diets using rancid horsemeat yield-

359

ed "fire engine red" dark pelts, an observation verified in more recent years in studies at Oregon State University (Anon, 1956). Horse fat is unique in terms of animal fats for its high content of polyunsaturated fatty acids including linolenic acid. As early as 1950, Oregon State University reported on the "brown" fur of dark mink raised on a diet containing an oily fish, turbot (Watt, 1951). In more recent years, "fire engine red" dark pelts have been found in mink on the West coast being fed high levels of brown rock cod, an unusually high fat fish with excessive levels of polyunsaturated fatty acids (Petersen, 1982).

An objective evaluation would indicate that essentially all fresh/frozen products undercut sharp black color in dark mink because of their content of unstabilized fats. This statement is verified by experimental data from Oregon State University which notes that the replacement of 13% of a purified diet (100% stabilized fat content) dehydrated basis with poultry by–products or fish resulted in a color score loss of 25% (Stout et al., 1967). A number of mink nutrition scientists have reported on the loss of sharp color in mink and foxes when fish oils, a high polyunsaturated fatty acid fat resource, are used in experimental diets (Anon, 1956; Adair, et al., 1958; Rouvinen et al., 1991).

This summary of experimental data from the National Northwood Mink Ranch provides support for a direct relationship between sharp color gradings in mink and the percent of stabilized fat in the final ration. The higher the level of fresh/frozen products in the mink ration, the lower the fur color rating (Leoschke, 1971).

National Research Ranch–Experimental Data–1971

Dietary Program	Fur Evaluation–Color	
	Pastels	**Darks**
Ranch Mix	1.8	1.5
GnF–70	1.9	1.9
GnF–100	2.1	2.1
GnF–Pellets	2.5	—

A final note relative to sharp color with mink on Gro–Fur pellets: One should not be surprised at the fact that mink on Gro–Fur pellets have sharp color. In 1932, Purina Company marketed a pellet named Fox Chow Checkers with the following observations from the auction house (Smith, 1932).

"Outstanding feature was the color of the pelts, which was far superior to that of the other pelts on the same sale. The lustrous black color of the pelts and the absence of brown, rusty color were the interesting features of the pelts."

The biochemical basis for the loss of sharp color in mink pelts is directly related to the presence of high levels of unstabilized polyunsaturated fatty acids in the fur of the mink. These levels lead to "off color" in dark pelts since the acids are sensitive to oxidative rancidity in the final weeks of fur development.

360

Lighter Leather Pelts

Prior to the marketing of GnF–100 meal and Gro–Fur pellets, ranchers on the National "Hi–Dri" feeding programs,–the GnF–35/37 and GnF–50 products –had noted lighter leather in their mink pelts. One British Columbia, Canada, rancher working with the GnF–35 program for the first time had an immediate negative reaction to the program when the mink were pelted and fleshed: the leather of the female mink was so light that her workers were damaging too many pelts in the process of fleshing. She had to make a decision between modifying the pelt processing or going back to the heavier leather GnF–20 program of previous years. It was an easy choice to stay with the GnF–35 "Hi–Dri" program (Engebretson, 1975).

Mink ranchers throughout North America, from the Mullen Mink Ranch in Nova Scotia to the Watt Mink Ranch in Oregon, have commented on the "boardy" pelts produced on low cereal, high fish content mink dietary regimens. Buyers, too were aware of this shortcoming. For many years, Charlie Reich, a major fur merchant in New York City, had purchased female pastel pelts but ignored the male pastel pelts since they were "boardy" The first major fur merchant in New York City to note the lighter leather male pastel pelts from National–Northwood Mink Ranch – raised on National GnF–100 mash was Charlie Reich in 1975. For many years, Charlie had purchased a large volume of Northwood female pastels but simply ignored the male pastels inasmuch as they were what he considered "boardy," a heavy leather by the finer quality furriers in New York. With the decision of Tony Rietveld, the new manager at the Northwood Mink Ranch, to employ the National GnF–100 mash program from early July to pelting, there was a dramatic change in the quality of leather of the male pastel kits. Modern nutrition management had triumphed over the genetic history of the mink. Charlie Reich was one of the first to note the change and was very excited about the new lighter leather male pastel pelts from Northwood (Reich, 1975).

The pattern of lighter leather on 100% "dry" diets continued with the National Gro–Fur pellet program. Mr. Schumacher, an Oregon furrier employed by Ed Sandburg, Blue Iris mink rancher, to provide mink coats for his wife and sister commented, "I was really excited about the light leather and thought they were beautiful mink when I got them. However, after working with them, I was even more impressed." He had never worked with pellet fed mink before and was surprised at the light leather.

From the point of view of the physiology involved in mink fur and leather development, there is no specific logical biochemical basis for the obvious relationship between light leather pelts and the "Hi–Dri" mink feeding regimens GnF–35/37, Gnf–100 meal, and modern mink pellet programs. This biochemical/physiological enigma is something for a new generation of mink nutrition research scientists to discover.

"Silkier" Pelts

Ranchers have observed that mink kits raised on pellet programs have pelts with a "silkier" nature since the very beginning of the marketing of mink pellets in the worldwide fur industry. American mink ranchers visiting the Netherlands in 1969 observed that live animals and dressed pelts from mink raised on the Trouw pellet program seemed to have an unusual silkiness of fur texture (Anon, 1969). In later years, mink ranchers in Holland feeding National Gro–Fur pellets commented: "The pellet fed mink looked beautiful; they had sharp color and were extremely silky."

Here again, the biochemical relationship between mink kits raised on pellet formulations and greater "silkiness " of pelts is another enigma for mink nutrition scientists to resolve in the years ahead.

362

9

Nutrition Management–21st Century

Reproduction

Without question, the reproduction/lactation phase of the mink ranch year is the most critical, requiring a combination of sound nutrition management, in depth genetic management, and careful ranch management to achieve top performance of the mink.

The most critical nutrition factor is the quality of protein ingredients. Mink nutrition scientists recommend a minimum of 40% liver and high quality protein feedstuffs (cooked eggs, whole milk cheese products, and muscle meats) with 35–40% of ME as protein (Leoschke, 2001). The balance of energy should be provided in a ratio of two thirds from fat and one third from carbohydrate resources (Jorgensen, 1985).

The most critical ranch management factor is "The Man With The Spoon" who assures that the animals are "trim" for breeding and in optimum condition for whelping. Graph 9.1.(Moustgaard and Riess, 1957a) provides a vital insight into the nutrient requirements of the mink during gestation. Like other mammals, mink need minimal extra protein and energy requirements in the early gestation period and no significant weight gains until the last trimester of the pregnancy. It is well known that excessively heavy females at whelping may have difficulty in giving birth; they also have a reputation for poor lactation performance. It is obvious that lean reproduction diets with fat levels in the range of 14–18% support of "The Man With The Spoon" in proper conditioning of mink for breeding and whelping.

365

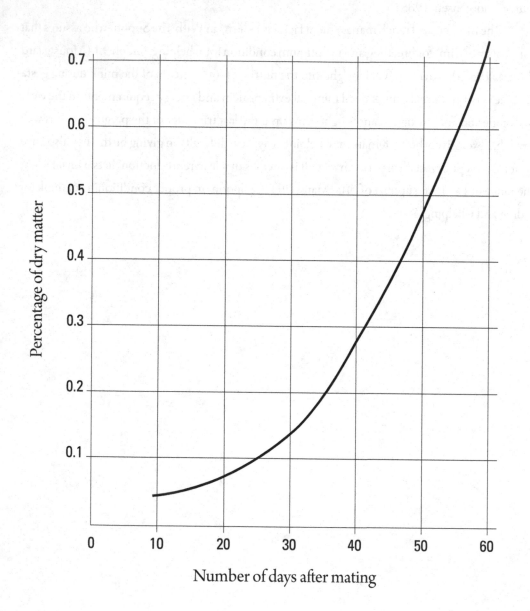

GRAPH 9.1 Protein Deposition in Pregnant Uterus of Mink

Lactation

Graph 9.2 (Glem–Hansen and Jorgensen, 1975) provides a good insight into the changing nutrient requirements of the lactating mink. A progressive increase in energy requirements must be supported by enhanced fat levels in the diet as the lactation period moves forward. Fat levels recommended include 20–22% fat at whelping and as high as 26–28% by late May.

GRAPH 9.2 Mink Milk Composition—Danish Data

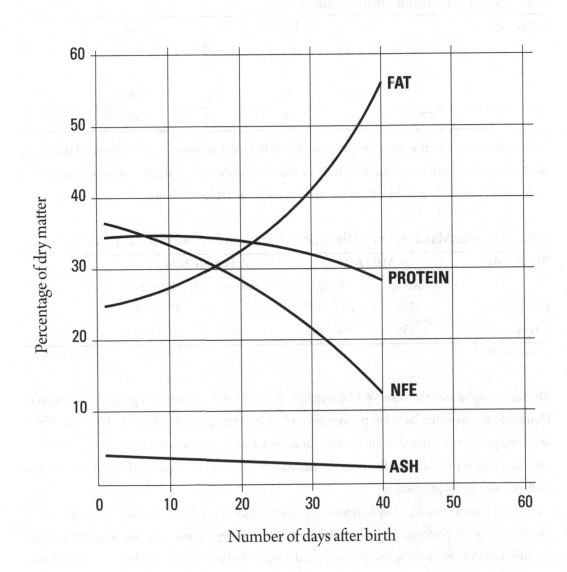

Without question, protein quality is as important, if not more important, for the lactation period than for the reproduction phase, as noted in multiple studies at the National Research Ranch,. This important point is well illustrated by the fact that a "dry" dietary program with good quality protein resources yielded a 6.0 ranch average at whelping but only a 2.0 ranch average at weaning. A superior quality protein regimen was an absolute requirement to yield a 4+ ranch average at weaning in the 1970s experimental trials.

The dam must provide the suckling kits with two grams of milk for every gram of kit growth within the litter. Table 9.1 (Jorgensen, 1985) illustrates the remarkable milk output of the lactating mink.

TABLE 9.1 Mink Milk Production Grams/Day

Litter Size			Days		Total
	3–10	11–17	18–24	25–31	
4	36	43	48	56	1275
6	52	64	72	84	1910
8	72	85	96	112	2550

The major factor affecting lactation performance of the mink is genetics. Experimental studies at the National Research Ranch facilities indicate that the lactation potential of a mink with six kits (3 males and 3 females) varies by a factor as much as 100%, as Table 9.2 shows.

TABLE 9.2 Genetic Management of the Mink–Lactation Performance at Three Weeks*

Mother Age	Male Kits		Female Kits	
	Average	Range	Average	Range
Experienced	146	103 – 172	124	98 – 138
Virgins	139	93 – 187	124	93 – 157

*Chausow (1992).

The table emphasizes the value of a kit-weighing program at three weeks, a program common in Denmark, to assess the lactation performance of each lactating mink relative to litter size. If this weighing program is limited to litters of five or more for dark kits and for six or more for mutation kits, the labor is minimal and the rewards maximal in terms of enhanced performance as seen in both pelt size and reproduction.

Within a decade, a Washington state rancher, Hackel (1980) was able to raise the reproduction/lactation performance of his mink by a full kit, using a mink-weighing program wherein the new breeder herd each spring consisted of the top half of the breeder herd in terms of lactation ratings as ascertained by litter weights at four weeks. A similar situation occurred at the National Research Ranch in the period from 1980 to 1990 wherein the pellet nutrition program was able to move from a 4.0 ranch average at weaning to a 5.0 ranch average with a combination of lactation rating selection of breeders and new understandings of the basic nutritional concepts of using modern mink nutrition pellets. However, this excellent genetic management program which had provided top reproduction/lactation performance was undercut in one single year: The breeder selection for the new year was based solely on fur quality, with complete disregard for lactation ratings. In the following year, the reproduction/lactation performance of the mink dropped from a ranch average of 5.0 to 4.5. Mink genetic management wisdom?

Early Growth

"June Blues"

All factors considered, it is very likely that the 2–4% of kit losses experienced by too many ranches in June is directly related to high bacterial populations in the mink feed "on the wire." This statement is supported by the clear observation that mink ranchers employing the National Program of Mink Nutrition pellets experience only 1% or less kit losses in June.

Bacteria populations in mink feed "on the wire" can be reduced significantly with the use of phosphoric acid and/or National Early Growth crumlets (crumbled pellets) "on top" of ranch or redi–mix programs. Table 9.3 and Table 9.4 give these results.

TABLE 9.3 Mink Feed Bacteria Populations With and Without Phosphoric Acid*

Feeding Program	Bacterial Count/Gram
Ranch Mix	2,800,000
Ranch Mix w/Phosphoric Acid	
@ 2% (75% Feed Grade)	400,000
Dehydrated Basis	

*Dietrich (1955)

TABLE 9.4 Mink Feed Bacteria Populations With and Without Crumlets "On Top"

Feeding Program	Bacterial Count/Gram
Ranch Mix*	2,800,000
Ranch Mix*	
+ 20% Crumlets	350,000
+ 25% Crumlets	290,000
+ 30% Crumlets	275,000
Crumlets as a Wet Mash**	7,800

*Wisconsin Analytical (1987), **Bacterial populations after the feed mixtures have been held at room temperature in June for two hours.

The importance of top mink nutrition management in the early growth period is seen in Table 9.5.

TABLE 9.5 June Male Kit Weights and Pelt Length*

Kit Weights – 8 Weeks – Grams	Pelt Length – Cm
450	69
500	71
550	72
600	73
670	75

*National Research Ranch (2002)

Late Growth

The late growth period, July–August, represents a phase of the mink ranch year with minimum protein requirements—as low as 25% of ME as protein with 20% of quality protein feedstuffs in the ranch diet. Thus it is recommended that the highest quality protein feedstuffs including liver, cooked eggs, and whole milk cheese products be reserved for the more critical fur development period of the ranch year if freezer inventory of these protein feedstuffs is limited. For top fur development yielding the highest pelt prices, it is better to feed 0% liver in July–August and 15% liver in the fall months than to feed 10% liver in the period from July to pelting. Likewise for cooked eggs and whole milk cheese products: their use at minimal levels in the late growth period and at maximum levels in the fur development period will yield a fur quality in pelts marketed and a greater dollar return at market. It is important to note that any contribution of liver in the late growth period in terms of enhanced taste appeal would actually undercut fur development by encouraging excessive adipose (fat) tissue development in the male kits, leading to sub–optimum feed intake in the critical fur development period.

Mid–August is a key period for modern mink nutrition management design, with a potential for enhanced fur development. Studies by Lohi and Hansen (1990) indicate that by mid–August mink have reached 94% of body length at pelting.

Studies by Jorgensen and Christensen (1966) indicated that levels of hemoglobin in mink kits as of August 14 do not show any great influence on mink fur qualities at pelting. However, during the period after August 14, the period of fur development, that level is of great importance for fur qualities achieved. Top mink nutrition management should support hemoglobin values gradually rising to maximum by November 15.

September Male Weights

Few mink ranchers fully appreciate the mink nutrition management fact that decisions made about dietary planning in July–August have a major effect on the quality of fur development in the months to follow. This point is well documented in Table 9.6.

TABLE 9.6 Correlation Between Male Body Weights at Different Ages and Fur Quality of Pelts* Marketed

Kit Age	Fur Quality Correlation
Birth	+0.16
14 days	**
28 days	**
42 days	−0.16
June	−0.25
July	−0.24
August	−0.24
September	−0.40
October	−0.41
December	−0.34

*Hansen, et al. (1992), ** no significant correlation

For female kits, a negative correlation between body weights is found only in December (−0.18). The previous table shows that a late growth nutrition management program which leads to "heavy" male kits coming into September essentially undercuts fur development because of their reduced protein intake during the critical fall fur development period. It is important for mink ranchers to appreciate the fact that male kits on commercial ranches essentially have it "too good" with a daily routine of "sleep and eat," while their distant cousins in the wild must spend the greater part of each day hunting for food. Unfortunately, the viewpoint of North American mink ranchers relative to kit growth during the summer months has not changed much in sixty years. In the 1940's their perspective on practical mink nutrition for the kits was simple, "a growing kit cannot eat too much"(McDermid and Ott, 1947). This naïve perspective led to major losses of male kits with "yellow fat" disease when the ranch diets contained high levels of rancid horse meat. In the 21st century, this same naïve viewpoint is undermining the fur development of the mink, resulting in economic loss to ranchers.

Recently, when the experimental data of Table 9.6 was presented to a group of mink ranchers, the immediate reaction from one rancher was, "We cannot afford to employ late growth diets for the kits that result in "small" mink in early September and sub–optimum pelt size at marketing." Fortunately, experimental data from the National Research Ranch indicate that a lower palatability program for July and August leading to "trimmer" male kits coming into September can cause higher protein intake in the fall months, a situation that leads to equal, if not superior, size of pelts at marketing. Table 9.7 establishes this point.

TABLE 9.7 Modern Mink Nutrition Management for "Trimmer" Male Kits in Early September

Color Phase	Diet	Male Kit Weight Gains – Grams		
		July–August	Sept–Nov	Totals
Darks				
	GnF–20	920	310	1,230
	GnF–70	820	450	1,270
Weight Gains				
GnF–70 vs. GnF–20		−100	+140	+ 40
Pastels				
	GnF–20	970	450	1,420
	GnF–70	870	580	1,450
Weight Gains				
GnF–70 vs. GnF–20		−100	+130	+30

"Common sense" dictates that high quality commercial fortified cereals with enhanced protein content designed to be fed at 30–35–50 and 70% level in combination with 70–65–50 and 30% fresh/frozen feedstuffs would have lower palatability for the mink kits than a program providing GnF–20 fortified cereal at 20% in combination with 80% fresh/frozen feedstuffs. However, all factors considered, it is seen that these summer lower palatability programs provide trim kits for enhanced fall feed intake and subsequent extra weight gains.

Mink ranchers of the 21st century are well advised to give more serious consideration to these experimental studies and overcome the naïve viewpoint that "a growing kit cannot eat too much."

November Kit Weights

Studies reported by Moller (1997) indicate that the regression of skin length on body weight varies between farms and color type. Somewhat surprising was that the weight of the kits about one month prior to pelting was of the greatest importance relative to pelt length, while weight changes during the last month before pelting did not have the same effects on a given final weight. A positive effect on skin length was found if the mink had been even bigger and thus had lost weight before pelting. Since skin does not have a muscular structure, a reduction in weight from early November to early December does not affect pelt length.

Melatonin Implantation

The pineal gland secretes the hormone melatonin in different patterns related to the photoperiods. Melatonin determines the onset of both fur development and reproduction by regulating the levels of another hormone, prolactin. Shorter photoperiods increase the secretion of melatonin, initiating summer fur development and the regrowth of winter fur in late August. Longer photoperiods reduce melatonin secretion and lead to greater activity of prolactin within the reproduction phase where prolactin enhances the implantation of embryos.

For more than a decade, melatonin implants have been used to bring about precocious prime. Silastic implants injected under the skin of the nape of the neck in early July provide a gradual release of melatonin into the blood circulation, thereby shortening the fur production phase of the mink ranch year. Pelting can thus occur a few weeks earlier; the rancher can save on labor and feed costs and take advantage of less severe weather during pelting.

As early as 1984, Rose et al. (1984) noted that melatonin implants in late June were able to induce earlier fur development of the mink. Additional studies by Rose et al. (1987) and Gregorio and Murphy (1987) provided in-depth insights into the value of melatonin implants for shortening the fur development period. Their studies and those of others indicated that the net result of melatonin implantation was enhanced size of pelts marketed but also a potential for significant "off color" in violet mink. Observations of North American mink ranchers in the years to follow agreed. The size of pelts was enhanced, but implantation created a significant problem of "off color" at pelting which also affected dark and mahogany mink (Zimbal and Zimbal, 1995). These same ranchers observed another benefit: minimal losses of mink with Aleutian disease, an observation made earlier by a Canadian mink rancher, Hugh Freeman (1994).

It is important to remember that melatonin will continue to be released from the silastic implants for many months beyond the pelting phase and into the breeding and reproduction periods. Basic animal physiology indicates that melatonin will decrease the activity of prolactin and thereby prevent embryo implantation. Thus, mink with melatonin implants can not be used in the reproduction phase of the mink ranch year.

Melatonin Implants–Mink Nutrition Management

About three weeks after melatonin implantation, mink kits show enhanced appetite as their body physiology begins to feel the simultaneous stresses of late growth weight gains and fur development. Special dietary planning is required to produce top performance from melatonin-implanted kits, including higher energy levels, enhanced protein nutrition in both quantity and quality, and special attention to the unique nutrition management required to provide "sharp color" in dark pelts. The unique energy and protein recommendations for nutrition management is well illustrated in Table 9.8.

TABLE 9.8 National Program of Mink Nutrition–Melatonin Implanted Kits

Nutrients	Regular Mink	Melatonin Mink
	September–Pelting	August – October
Protein Nutrition		
% Quality Protein Feedstuffs	30	35–40
% Quantity of Protein	35–37	40
% Metabolic Energy		
As Protein	min. 30	min. 30
Eneregy Nutrition		
% Fat	20–22	24–26

The results of experimental studies designed to produce superior color in dark mink implanted with melatonin are provided in Table 9.9

TABLE 9.9 Nutrition Management for Superior Color in Melatonin-Implanted Mink National Research–2001*

Program	Color at Pelt Grading**	
	Males	**Females**
Gro–Fur Pellets	4.9	5.5
Gro–Fur Gold Pellets	5.8	5.7
Program	**Auction Data – Black–XXD–%**	
Gro–Fur Pellets	18	26
Gro–Fur Gold Pellets	34	31

*National Research Ranch (2001), **Grading Scale of 0–7.

Fur Development

For mink fur economics, pelt size is as important as fur quality. Moller (1999) reported that skin length is closely related to body size, which can be expressed either as body weight or as body length and condition.

Top fur development requires a minimum of 30% ME as protein and 30% of quality protein feed-stuffs within a 20% simple fortified cereal program as of early September. Since fur development is maximum in September with a slower pace for the remaining weeks of the fall, a program using increasing levels of a quality fortified cereal in combination with a reduction in poultry by–products at 1% per week as of October 1, 8, 15, 22 and 29 provides a useful nutrition management program.

It yields reduced levels of fats with relatively high levels of polyunsaturated fatty acids (PUFA), in combination with higher levels of carbohydrates where the "extra" simple sugar intake converts to stable fatty acid structures, inasmuch as mink cannot convert glucose to the PUFA responsible for "off color" fur in the final weeks of fur development. The net result of this program developed by National Fur Foods combines superior mink nutrition management with enhanced ranch business management. Enhanced mink nutrition management is achieved through a lower level of PUFA which results in "sharp color" in dark pelts, and better ranch business management is attained because a quality fortified cereal provides the rancher businessman with the very "best buy" in feed solids for the rancher's feed dollar.

Brief commentary is warranted on that dietary planning required for the achievement of those pelt qualities of major concern for buyers at the world's fur auctions.

Sharp Color

Nutrition management decisions that minimize the degree of "off color" pelts include (a) an emphasis on fresh/frozen feedstuffs with a low percentage of polyunsaturated fatty acids (PUFA), including beef products and whole milk cheese products; (b) the use of high quality fish meals stabilized by potent anti–oxidants; (c) the use of higher levels of quality fortified cereals in the final months of fur development, resulting in higher carbohydrate levels. This specific recommendation is supported by the data provided in Table 9.10. Finally, the best nutrition management solution is to move to a quality pellet program providing 100% stabilized fat resources.

TABLE 9.10: National Research Ranch–Experimental Pellet Data 2006–Sharp Color in Dark Pelts–Males

Diets	Control	Experimental
% Metabolic Energy	30–50–20*	30–40–30* High Carbohydrate
Breeders	0	6
Black	10	7
Black–XXD	34	44
XXD	2	6
XXD–XD	—	1
XD	5	2
XD–DK	11	—
DK	—	—
???	4	—
$$$	$53.33	$57.76

*% ME as Protein, Fat, and Carbohydrates

Lighter Leather

Lighter leather pelts can be achieved by genetic selection or by nutrition management. For many years it has been noted that mink nutrition programs providing relatively high levels of fortified cereals containing enhanced levels of high quality dehydrated protein resources such as National GnF–35 and GnF–50 yield lighter leather in pelts marketed. However, it must be noted that at the present time, basic knowledge of the nutrition and nutritional physiology of the mink does not provide a biochemical reason for the lighter leather pelts.

Silky Fur

In almost 60 years of experimental studies on the nutrition of the mink and observations from North American mink ranchers on fur development on a great variety of ranch dietary programs, the only observation of mink fur quality expressed as "Silky Fur" was made by ranchers using pellet programs. Obviously, something is unique about the fur development of mink on these programs. It is hoped that further research studies will allow mink nutrition scientists to gain a full biochemical understanding of the physiological basis for the "silky" pelts achieved on pellet formulations.

References

References

AALAS (1972). Syllabus For The Laboratory Animal Technologist, Amer. Assn. Lab. Anim. Sc. Publ. 72-2, AALAS, Joliet, Illinois.

Aarstrand, K. and Skrede. A. (1993). Effects of different storage methods on quantities and chemical composition of manure from fur-bearing animals. Norsk Landbruksforskning 6: 339-358. ISSN 0801-5333. Scientifur 17(1): 34.

Abernathy, B. P. (1960). Studies on the nutrient requirements of mink. Ph.D. Dissertation, Cornell University, Ithaca, N.Y.

Adair, J. and Oldfield, J.E. (1988). Progress in understanding wet belly. Fur Rancher 68(9): 8-11.

Adair, J., Oldfield, J.E., Scott, R. L. and Thomson, C. (1986). Use of orally administered insect growth regulator (Larvadex R) with mink. Progress Reports: Mink Farmers Research Foundation.

Adair, J., Stout, F.M. and Oldfield, J.E. (1958) Improving Mink Nutrition - 1957. Progress Report. Misc. Paper 58, Ag. Exp. Station, Oregon State College.

Adair, J., Stout, F. M. and Oldfield, J.E. (1959). New developments in mink nutrition - 1958. Progress Report - Misc. Paper 80 (Sept. 1959).

Adair, J., Stout, F.M. and Oldfield, J.E. (1960). 1959 research in mink nutrition, Oregon State Ag. Station (1960). Misc. Paper #89, March 1960.

Adair, J., Stout, F.M. and Oldfield, J.E. (1961). 1960 progress report - improving mink nutrition (1961). Oregon State Ag. Station, Misc. Paper #103, Jan. 1961.

Adair, J., Stout, F.M. and Oldfield, J.E. (1962). 1961 progress report - experiments in mink nutrition. Oregon State University.

Adair, J., Stout, F.M. and Oldfield, J.E. (1963). The value of cereal grains in mink rations. National Fur News 35(6): 10,11, 26-30.

Adair, J., Stout, F.M. and Oldfield, J.E. (1966). A 100% dry diet for growth and furring. Amer. Fur Breeder 39(9): 13.

Adair, J., Stout, F.M. and Oldfield, J.E. (1966). Mink nutrition research, Special Report No. 207, Oregon State University.

Adair, J. Thomson, C., Scott, R. L., Hu, C.Y. and Wehr, N.B. (1988). Wet belly disease. Progress Reports. Mink Farmers Research Foundation.

Adair, J.,Wehr, N. B., Oldfield, J. E. and Stout, F.M. (1974). Measurement of avidin in fresh eggs. Progress Reports. Oregon State University.

Adair, J., Wehr, N. B., and Oldfield, J.E. (1977). Investigation of possible toxic effects of high levels of vitamin A in contributing to early kit losses. Progress Reports. Mink Farmers Research Foundation.

Adair, J., Wehr, N.B., Stout, F.M., and Oldfield, J.E.(1974). Effects of oral iron supplementation on prevention of formaldehyde induced and hake induced anemia in mink. Progress Reports. Mink Farmers Research Foundation.

Adair, J., Wehr, N.B., Thomson, C., Scott, R. L. and Oldfield, J.E. (1979). Effects of gradual feed restriction and/or salt, NaCl, supplementation during growth and furring on wet belly in standard dark mink. Progress Reports. Mink Farmers Research Foundation.

Adair, J., Wehr, N.B., Thomson, C., Scott, R. L. and Oldfield, J.E. (1981). Nitrate toxicity in mink. Progress Reports. Mink Farmers Research Foundation.

Ahlstrom, O. (1994). Fat as an energy source in fur animal diets. Effects of dietary fat:carbohydrate ratios and fat sources on physiological parameters and production performance, Ph. D. Dissertation, Agricultural University of Norway, Doctor Scientarum Theses, 1994:25.

Ahlstrom, O. (1995). Fordoyelighet av for med ulike fettniva hus blarer og mink. Norsk Pelsdrblad 3:12.

Ahlstrom, O. and Skrede, A. (1995). Feed with divergent fat:carbohydrate ratios for blue foxes (*Alopex lag opus*) and mink (*Mustela vison*) in the growing-furring period. Norwegian J. Agr. Sc. 9: 115-126.

Ahlstrom, O. and Skrede, A. (1995). Fish oil as an energy source for blue foxes (*Alopex lag opus*) and mink (*Mustela vison*) in the growing-furring period. J. Anim. Physiol. A. Anim. Nutr. 74: 146-156.

Ahlstrom, O. and Skrede, A. (1995). Comparative nutrient digestibility in blue foxes (*Alopex lag opus*) and mink (*Mustela vison*) fed diets with diverging fat:carbohydrate ratios. Acta Agr. Scan. 45(1): 74-80.

Ahlstrom, O. and Skrede, A. (1997). Energy supply for blue fox and mink in the reproduction period. Scientifur 21(4):300.

Ahlstrom, O., Skrede, A., Heggset, O.S., Mikkelsen, O. and Tangen, S.F. (2000). Meat-and-bone meals from different animal by-products as protein sources for fur animals. Scientifur 24(4) -IV-A- Nutrition, Behavior and Welfare. Pp. 63-66. Proceedings of the VIIth Inter. Sci. Congr. In Fur Animal Production, Kastoria, Macedonia, Greece, Sept. 13-15. Ed. B. N. Murphy and G. Lohi.

Ahlstrom, O. and Skrede, A. (2002). N-balance and production experiments with bioprotein for growing mink. NJF Seminar No. 347:97. Vuokatti, Finland. Oct.2-4, 2002.

Ahlstrom, O. and Skrede, A. (2003). Extruded peas as feed ingredient for mink in the growth period. NJF Seminar No. 354.

Ahman, G. (1958). Minkens van-ligaste fodermeal. Vara Palsdjur 29(11): 225-228.

Ahman. G. (1959). Smaitbarhets forsok me mink (Digestibility experiments with mink). Vara Palsdjur 30(1): 4.

Ahman, G. (1961). Minkuping Nordisk Handbok For Mink Uppfodare. Edited by A. Lund. Lts Forlag, Stockholm, 303 pp.

Ahman, G. (1962a). Whalemeat meal. Vara Palsdjur 33(9): 98-99.

Ahman, G. (1962b). Metodstudier I. Samband med smalt. Barhetsforsok med lic. Thesis, Uppsula.

Ahman, G. (1965). Eksperimetelle metoder til bestemmelse af fordojelig N aering og omsaettelig energi i. Forsog Med Mink, Nordisk Jordbruks Forskning. Suppl. 2, Del II: 72-73.

Ahman, G. (1966). Forsog me varierende maegder foder, konserveret ved tilsaetning af natriumbisulfit. Dansk Pelsdyravl 29: 469-470.

Ahman, G. (1966). Minkopdraet, 2nd ed. Dansk Pelsdyravler Forening, Glostrop, Denmark.

Ahman, G. (1967). Forsok med olika protein-fat-och kolhydratshalt I. Foderblandninger til mink av olika typ under reproductkionsperiod. Report No. 671002 from Lantbrukshogskloon, Uppsala, Sweden.

Ahman, G. (1974). Fett som naring-och energikali for mink. Bilag til mode I. NJF's Subsektion For Pelsdyr, Grena, Denmark, 1974.

Ahman, G. (1974). Fett som energioeh naringskalla for mink. Bilag til mode I NJF's Subsektion For Pelsdyr (Report to the Annual Meeting in NJF), Grena, Denmark, 1974, p. 22.

Ahman, G. and Kull, K.E. (1962). Provning av magnesium bromalkylvicksilverkloride i foder til mink (Studies on magnesium bromalkylmercurychloride in feed for mink). Int. Rep. S. P. R. Inst. Of Husdj Utf. O. Vard Palsjuravdelningen. 5 pp.

Aitken, F. C. (1963). Feeding of fur bearing animals. No. 23. Commonwealth Bureau of Animal Nutrition. Rovett Research Institute, Bucksburn, Aberdeen, Scotland. Commonwealth Agricultural Bureau, Farnham, Royal Bucks, England.

Akande, M. (1972). The food of feral mink (*Mustela vison*) in Scotland. J. Zool (London) 167: 475-479.

Akimova, T. I. (1969). Nauchnye trudy Nii pushnogo zverodstva Krolikovodstva (Scientific reports on Nii on the breeding of fur animals and rabbits).

Alden, E. (1987a). Digestibility trial with poultry meal. Vara Palsjur 58(5-6): 183-184.

Alden, E. (1987b). Digestibility trials on mink given fish meal and meat meal. Vara Palsjur 58(8-9): 276-278.

Alden, E. (1988). Digestibility trials with commercial cereal mixtures. Vara Palsjur 59(7): 256-258.

Alden E. and Johansson, A-H. (1975). Sojamjol som foder till mink. Vara Palsjur 46: 75-77.

Alden, E. and Tauson, A-H. (1981). Potentially thiaminase containing fish species in mink feeding. Vara Palsjur 52: 19-21.

Allen, M. R. P., Evans, E. V. and Sibbald, I. R. (1964). Energy-protein relationships in the diet of growing mink. Can. J. Physiol. and Pharm. 42: 733-744.

Allison, J. B. (1949). Biological evaluation of proteins. Adv. Prot. Chem. 5: 194-196.

Amano, K. and Yamada, K. (1964). A biological formation of formaldehyde in the muscle tissue of gadoid fish. Bull. Japan Soc. Sci. Fish. 30:430.

Andersen, K.E., Bjergegaard, C., Buskov, S., Clausen, T. N., Mortensen. K., Sorensen, H., Sorensen, J. C. and Sorensen, S. (1999). Glycosides in mink milk. Annual Report, Danish Fur Breeders Research Center, Holstebro, Denmark. Pp. 20-26.

Anderson, P. A., Baker, D. H., Sherry, P.A. and Corbin, J.E. (1980). Histidine, phenylalanine-tyrosine and tryptophan requirements for growth of the young kitten. J. Anim. Sc. 50: 479-483.

Anderson, P.J. B., Rogers, Q. R., and Morris, J. G. (2002). Cats require more dietary phenylalanine and tyrosine for melanin deposition in hair than for maximal growth. J. Nutr.132(7): 2037-2042.

Anon (1936). Food habits of the mink in New York. J. Mammalogy 17: 169.

Anon (1953). Advertisement, Black Fox Magazine and Modern Mink Breeder 37(5): 5.

Anon (1957). 1956 Mink Nutrition Research, Oregon State Agricultural Station, Misc. Paper 41.

Anon (1964). Investigation of factors causing "wet belly" disease in mink. Oregon State University. Progress Reports. Mink Farmers Research Foundation.

Anon (1968). A new weapon in the struggle against bacteria. Oregon's Ag. Prog. Summer. 15(2).

385

Anon (1969). New European 100% dry diet unveiled to Americans. American Fur Breeder 42(1): 10-12.

Anon (1977). Fiskeensilagens indflydelse pa passagegetiden hos mink. Stenc. Bilag Til Statens Husdybrugs-forsogs - forsogs Arsmode.

Anon (1979). Satisfactory growth and reproduction in trout fed alpha-linolenic acid as the only unsaturated fat. Nutr. Rev. 37: 298-299.

Anon (1981a). Effect of whole blood in the diet on mink color? Quantities, nutritive value and applicability to offal products from slaughter houses for fur animal feed. 518 Beretning fra Statens Husdyrbrugs Forsog Kobenhavn: p.28.

Anon (1981b). Quantities, nutritive value and applicability of offal production from slaughter houses for fur animal feed. 518 Beretning fra Statens Husdyrbrugs Forsog, Kobenhaven. p. 29.

Anon (1981). Dansk Pelsdyravl No. 5.

Anthonisen, A. C. and Loehr, B. C. (1974).The handling and treatment of mink wastes by liquid aeration processing and management of agricultural waste. Proc. 74th Cornell Ag. Waste Mgt. Conf. March 25-27, 1974, Rochester, N.Y. pp. 295-307.

Arnold, A., Kline, O. L., Elvehjem, C. A. and Hart, E. B. (1936). Further studies on the growth factor required for chicks. The essential nature of arginine. J. Biol. Chem. 16:699.

Arrington, L.R. and Davis, G. K. (1953). Molybdenum toxicity in the rabbit. J. Nutr. 51: 295-304.

Atkinson, J. (1996). Mink Nutrition and Feeding, in Mink - Biology, Health and Disease 3:12. Edited by D.B. Hunter and N. Lemieux. Graphic and Print Services, University of Guelph, Guelph, Ontario, Canada,.

Atwater, W. O. and Bryant, A. P. (1899). The availability and fuel value of food materials. Storrs Agr. Expt. Stat. Ann. Report. pp. 73-110.

Aulerich, R. J. (1984). Bloodmeals as mink feed ingredients. Fur Rancher 64(4): 16-17.

Aulerich, R. J. (1984). Potato concentrate value of its protein in mink feed. Fur Rancher 64(9): 4-8.

Aulerich, R. J. (1985). Single Cell Protein - B. P. Personal Communication.

Aulerich, R. J., Bleavins, M.R., Hochstein, J.R., Hornshaw, T. C., Napolitano, A.C. and Ringer, R.K. (1983). Effects of copper deficiency on mink production. Progress Reports. Mink Farmers Research Foundation.

Aulerich, R. J., Bleavins, M. R., Napolitano, A.C. and Ringer, R.K. (1981). Feeding spray-dried eggs to mink and its effects on reproduction and fur quality. Feedstuffs 53 (July 6th): 24-28.

Aulerich, R. J. and Bursian, S. J. (1997). Response of female mink and their litters to supplemental folic acid during the reproduction period - verification of preliminary results. Michigan State University, Fur Animal Research (1997): 68-73.

Aulerich, R. J., Bursian, S. J,. Bush, C. R., Napolitano, A. C. and Summer. P. (1995). Response of female mink and their litters to supplemental dietary folic acid during the reproduction period. Blue Book of Fur Farming - 1996:4-12.

Aulerich, R. J., Bursian, S. J., Bush, C. R., Napolitano, A. C. and Summer, P. (1997). Reponse of female mink to folic acid supplementation during the reproduction period. Michigan State University, Fur Animal Research (1997):59-67.

Aulerich, R. J., Bursian, S. J., Napolitano, A. C. and Bleavins, M.R. (1981). Evaluation of oral larvicide's with mink. Fur Rancher 62(5): 4-8.

Aulerich, R.J., Bursian, S. J., Napolitano, A.C. and Oleas, T.B. (2000). Effect of supplemental dietary salt (NaCl) on the reproductive performance of female mink and the growth and survival of their kits. Michigan State Animal Research Report 2000. Animal Science Dept. Mich. State University. East Lansing, MI 48824.

Aulerich, R. J., Bursian, S. J., Poppenga, R. H., Braselton. W. E. and Mullaney, T. P. (1991). Toleration of high concentration of dietary zinc by mink. J. Vet. Diagn. Invest. 3:232-237.

Aulerich, R. J. and Napolitano, A.C. (1986). Effects of supplemental iodine on the reproductive performance of mink. Progress Reports. Mink Farmers Research Foundation.

Aulerich, R.J. and Napolitano, A. C. (1986). The role of selenium in mink nutrition. Progress Reports. Mink Farmers Research Foundation.

Aulerich, R. J. and Napolitano, A.C. (1993). The role of selenium in mink nutrition. Fur Bearing Animal Research, Animal Science Dept. Michigan State University: 21-24.

Aulerich, R. J., Napolitano, A. C. and Bursian, S. J. (1986). Dietary fluoride - its effects on mink production. Fur Rancher 66(12):4,6, 16.

Aulerich, R. J., Napolitano, A.C., Bursian, S. J., Olson, B. A. and Hochstein, J. R. (1987). Chronic toxicity of dietary fluorine in mink. J. Anim. Sc. 65:1759-1767.

Aulerich, R.J., Napolitano, A. C. and Hornshaw, T.C. (1985).Effect of supplemental copper on mink kit hemoglobin conc. Progress Reports. Mink Farmers Research Foundation.

386

Aulerich, R. J. Napolitano, A. C., Lehning, F. J. and Ringer, R. K. (1984). Fatty liver in mink. Progress Reports. Mink Farmers Research Foundation.

Aulerich, R. J., Napolitano, A.C., Lehning, F. J. and Ringer, R.K. (1991). Fatty liver in mink. Research Report 508. Mink and Poultry Research 1990.

Aulerich, R. J., Napolitano, A. C. and Ross, R. H. (1984). Fur ranch fly control. Fur Rancher Blue Book of Fur Farming - 1985:48-51.

Aulerich, R. J., Nelson, L. J. and Braselton, W. E. (1989). Element conc. In milk of ranch mink. Scientifur 13(4): 273-276.

Aulerich, R. J., Powell, D. C. and Bursian, S. J. (1999). Handbook of Biological Data For Mink. Department of Animal Science, Michigan State University, East Lansing, Michigan.

Aulerich, R. J. and Ringer, R. K. (1976). Feeding copper sulfate. Could it have benefits in nutrition of mink. U.S. Fur Rancher 56(12): 4-9.

Aulerich, R. J., Ringer, R.K., Bleavins, M.R. and Napolitano, A.C. (1982). Effects of supplemental dietary copper on growth, reproductive performance and kit survival of standard dark mink and acute toxicity of copper in minks. J. Anim. Sc. 55(2): 337-343.

Aulerich, R. J., Ringer, R. K., and Hartsought, G. R. (1978). Effect of iodine in reproduction performance of minks. Theriogenology 9(3): 295-303.

Aulerich, R. J., Ringer, R. K. and Iwamoto, S. (1974). Effects of dietary mercury on mink. Arch. Environ.Contam. Toxicol. 2:43.

Aulerich, R. J., Ringer, R. K. and Schaible, P.J. (1962). Mink pelts affected with "wet belly". Mich. Agr. Exp. Station Quarterly Bull. 45(3): 441-443.

Aulerich, R. J., Ringer, R.K. and Schaible, P. J. (1962).Some observations on wet belly pelts. American Fur Breeder 35(5): 16-17, 36.

Aulerich, R.J. and Schaible, P.J. (1965). A semi-dry diet for mink. Amer. Fur Breeder 38(1): 20-21/

Aulerich, R..J. and Schaible, P.J. (1965a). The use of "spent" chickens for mink feeding. Q. Bn. Mich. Agr. Expl. St. 47(3): 451-458.

Aulerich, R. J. and Schaible, P. J. (1965b). Use of spent hens for feeding mink. American Fur Breeder 38(6): 14-15, 48-49.

Aulerich, R.. J. and Schaible, (1965c). A preliminary report on "spent" chickens for mink feeding during reproduction and early growth. Q. Bn. Mich. Agr. Expl. St. 48(1): 13-16.

Aulerich, R.J. and Schaible, P.J. (1966). Simple method of processing "spent" chickens for mink feeding and their use during late growth and furring. Q. Bn. Mich. Agr. Expl. St. 49(3): 24-27.

Aulerich, R. J., Schaible, P.J. and Ringer, R. K. (1963). Effect of time of pelting on the incidence of "wet belly". Mich. Agr. Exp. Sta. Quarterly Bull. 45(3): 441-443.

Aulerich, R. J., Shelts, G. and Schaible, P.J. (1963). Influence of dietary calcium level on the incidence of urinary incontinence and "wet belly" in mink. Mich. Agr. Exp. Station Quarterly Bull. 45(3):444-449.

Aulerich, R. J. and Swindler, D. R. (1968). The dentition of the mink (*Mustela vison*). J. Mammalogy 49: 488-494.

Austreng, E., Skede, A. and Eldegard, A. (1979). Effect of dietary fat source on the digestibility of fat and fatty acids in rainbow trout and mink. Acta Agr. Scand. 29: 119-126.

Baalsrud, N.I. and Kveseth, C. (1965). Forsok med okande mangder B-6 pyridoxinekloride till mink. Vara Palsdjur 36: 455-459.

Bachrach, M. (1946). Fur: A Practical Treatise, Rev. Ed. Prentice Hall, New York: 610.

Basal, T. (1990). Personal communication. Oregon mink rancher.

Basal, T., Delaney, D. and Kellow, S. (1994). Personal communication, Oregon Mink Ranchers.

Basset, C.F. (1948). The protein requirement of growing minks. American National Fur and Market J. 27(3): 5-6.

Basset, C.F. (1961). Vitamin A requirement of the ranch mink. Amer. Fur Breeder 34(l2): 18-20.

Basset, C. F., Harris, L. E., Llwellyn, L.M. and Loosli, J. K. (1951). The protein requirements of growing minks. Unpublished data. U.S. Bur. Anim. Ind. And Cornell University, Ithaca, N.Y.

Basset, C.F., Harris, L. E., Smith, S.E. and Yeomen, E.D. (1946). Urinary calculi associated with vitamin A deficiency in the fox. Cornell Vet. 36(1): 5-16.

Basset, C.F., Harris, L.E., Smith, S.E. and Yeomen, E.D. (1947). Urinary calculi may result in animal losses. Black Fox Magazine 30(11):13.

Basset, C. F., Harris, L.E. and Wilke, C. F.(1951). Effects of various levels of calcium, phosphorus and vitamin D intake on bone growth. II. Minks. J. Nutr. 44:433-442.

Basset, C.F., Loosli, J. K. and Wilke, F. (1948). The vitamin A requirement for growth of foxes and minks as influenced by ascorbic acid and potatoes. J. Nutr. 35:629-638.

Basset, C. F., Travis, H. E., Warner, R. G. and Loosli, J. K. (1957). Stilbesterol and reproduction. Amer. Fur Breeder 30(1): 10.

Barnes, A.G. and Magee, M.H. (1954)Br. J. Ind. Med. 11:167.

Barkvull, A. E. (1987). Personal communication.

Barrett, W. (1973). Personal communication, Michigan mink rancher.

Bauer, J.E. (1997). Fatty acid metabolism in domestic cats (*Felis catus*) and Cheetahs (*Acinonyx jubatas*)/ Proc. of the Nutr. Soc. 56:1013-1024.

Baxter, J. (1990). Personal communication, Hudson's Bay Co. Fur Grader.

Belcher, J., Evans, E.V. and Budd, J.E. (1959). Feeding fresh water fish to mink. Black Fox Magazine and Modern Mink Breeder 42(1): 14-15.

Belcher, J., Evans, E. V. and Budd, J. E. (1959). The feeding of freshwater Fish to mink. Fur Trade J. of Canada 36(2): 12-13, 24.

Belcic, I. (1981). A survey of the vitamin status in mink of various age, sex and pelt qualities in ranches, Yugoslavia. Scientifur 5: 33-35.

Bell, J.M. and Thompson, C. (1951). Freshwater fish as an ingredient of mink rations. Bull. No. 92, Fisheries Research Board of Canada.

Bell, J. A. (1966). Ensuring proper nutrition in ferrets. Vet. Med.:1092-1103.

Belzile, R. J. (1976a). Replacement value of soybean meal for raw meat in mink diets. 1st Inter. Sc. Congr. In Fur Anim. Prod., April 27-28, 1976. Helsinki, Finland.

Belzile, R. J. (1976b). Utilization of plant products by growing, furring and breeding mink. Proc. 1st Inter. Symp. Feed Composition, Anim. Nutr. Requirements and Computerization of Diets. July 11-16, 1976, Utah State University, Logan, Utah: 327-331.

Belzile, R. J. (1982). Evaluation of a vitamin and mineral replacement for liver in diets for mink (*Mustela vison*). Can. J. Anim. Sc. 62:1245-1247.

Belzile, R. J. and Dauphin, F. (1984). Nutritive value of enzymatically pre-hydrolyzed soybean meal for mink. 3rd Inter. Sc. Congr. In Fur Anim. Prod. Versailles, France, April 25-27, 1984.

Belzile, R. J. and Poliquin, L.S. (1974). Effects of feeding soya flour on the performance of growing furring mink. Can. J. Anim. Sc. 54(3): 385-388.

Belzile, R. J., Poliquin, L. S. and Jone, J.D. (1974). Nutritive value of rapeseed flour for mink. Effects on live performance, nutrient utilization, thyroid function and pelt quality. Can. J. Anim. Sc. 54(4):639-644.

Berg, H. (1986). Rehutietoutta Turkiselainkasvattajille (Feed knowledge for fur farmers). Finnish Fur Breeders Assoc., Fur Animal Laboratory, Painopinta Ky, Vassa, Finland. Berg, H, (1987). Use of alternative protein feedstuffs in mink feeding. Scientifur 11(2): 138-139.

Berg, H. (1989). Digestibility, chemical composition of dried protein meals for fur bearers. Fur Rancher 69(4): 4,6,8.

Berg, H., Juokslahti, J., Makela, J. and Tang, L. (1983). Forsok med lag Protein\halt i mink fodreti bilag til seminar I. NJF's subsektion for pelsdyr Malmo.

Berg, H., Tyopponen, J. and Valtonen,.M. (1985). Inverkan av fodrets proteinniva pa blodvarden jamford med skinnresultatet hos mink bilag til seminar I. NJF's subsektion for pelsdyr i, Alborg.

Berg, H., Valtonen, M., Tang, L. and Ericksson, L. (1984). Protein digestibility and water and nitrogen balance studies in mink using different protein levels. 3rd Inter. Congr. In Fur Anim. Prod., Versailles, France. April 25-27,1984: 9-1 - 9-8.

Bernard, R. and Smith, S.E.(1941).Digestion and absorption by foxes and mink. American Fur Breeder XIV (2): 22.

Bernard, R., Smith, S. E. and Maynard, L. A. (1942). The digestion of cereals by mink and foxes with special reference to starch and crude fiber. Cornell Vet. 32(1): 29-36.

Birks, J. D.S. and Dunstone, N. (1989). Sex-related differences in the diet of mink (*Mustela vison*). Holarg. Ecol. 8: 245-252.

Bjergegaard, C., Clausen, T.N., Dietz, H,H,, Mortensen, K., Sorensen, H. and Sorensen, J. C. (2002). High content of ethoxyquin in feed for mink in the reproduction, lactation and early growth period. Annual Report 2001: 133-139. Danish Fur Breeders Research Center, Holstebro, Denmark.

Bjergegaard, C., Clausen, T.N., Hejlesen, C., Mortensen, K., Ochodzki, P. and Sorensen, H. (2002). Single cell protein for mink in the growing season. Annual Report 2001: 109-117. Danish Fur Breeders Research Center, Holstebro, Denmark.

Bjergegaard, C., Clausen, T.N., Hejlesen, C., Mortensen, K. and Sorensen, H. (2002). Lupin and rapeseed for mink feed in the growing period. Annual Report 2001: 97-104. Danish Fur Breeders Research Center, Holstebro, Denmark.

Bjergegaard, C., Clausen, T.N., Mortensen, K., Sorensen, H., Sorensen, J.C. and Sorensen, S. (1999). Investigation of proteins and peptides in mink milk. Annual Report (1999). Danish Fur Breeders Research Center, Holstebro, Denmark: 145-153.

Bjergegaard, C., Clausen, T. N., Lassen, T. M., Mortensen, K., Sorensen, J.C and Sorensen, S. (2003). Growth and development of mink as a function of different plant or fish oils included in the feed used during the gestation and nursing period. Annual Report 2003: 53-66. Danish Fur Breeders Research Center, Holstebro, Denmark.

Blaxter, K. L. (1982). Fasting metabolism and the energy required by animals for maintenance. Festskrift til Knut Breirem, Mariendals Boktrykkeri A/S, Gjovik, Norway: 19-36.

Bleavins, M. R. and Aulerich, R. J. (1981). Feed consumption and food passage time in mink (*Mustela vison*) and European ferrets (*Mustela Putorius furo*). Laboratory Animal Science 31(3): 268-269.

Bleavins, M. R., Aulerich, R. J., Hochstein, J. R., Hornshaw, T. C. and Napolitano, A. C. (1983). Effects of excessive dietary zinc on the intrauterine and postnatal development of mink. J. Nutr. 113(11): 2360-2367.

Block, R. J. and Bolling, D. (1945). The amino acid composition of proteins and foods. 1st ed. Charles C. Thomas, Springfield, Ill. Bloomstedt, L. (1987). Hair structure and development of fur in young mink and silver fox, blue-silver fox and blue fox. Scientifur 11(3): 264-265.

Boisen, S. (1996). Ideal protein - and its suitability to characterize protein quality in pig feeds. Acta Agric. Scand. 46: 31-38.

Borgstrom, G. (1961). Fish as food. Academic Press. N.Y.

Borst, G. H. A. and van Lieshout, C. G. (1977). Phenylmercuric acetate intoxication in mink. Tijdschrift voor Diergeneeskunde 102: 495-503

Borsting, C. F., Damgarrd, B. (1995). The intermediate glucose metabolism in the nursing period of the mink. NJF Seminar No. 253, October 4-6, 1995. Gothenburg, Sweden.

Borsting, C. F. (1996). Personal communication.

Borsting, C. F. (1997). Amino acid digestibility of feed mixtures and digestibility of crystalline amino acids fed to mink.Scientifur 21(3): 199-200.

Borsting, C. F. (1998). Ernaeringens indflydelse pa skindkvaliteten. Intern Rapport No. lll. Danmark, Jorbrugs Forskining: 55-62.

Borsting, C. F., Bach-Knudsen, K. E., Steenfeldt, S., Mejborn, H. and Eggum, B. O. (1995). The nutritional value of decorticated mill fractions of wheat. 3. Digestibility experiments with boiled and enzyme treated fractions fed to mink. Animal Feed Science and Technology 53: 317-336.

Borsting, C. F. and Clausen, T.N. (1995a). Requirements of essential amino acids for mink. VII. Symposium on Protein Metabolism and Nutrition, 24-27 May, 1995.. Portugal, 7 pp.

Borsting, C. F. and Clausen, T.N. (1996). Requirements of essential amino acids for mink in the growing-furring period. Animal Production Review. Polish Society of Animal Production. Applied Science Reports 28. Progress in Fur Animal Science, Nutrition, Patology and Disease. Proceedings from the VIth Inter. Sc. Congr. In Fur Anim. Prod. Warsaw, Poland, 1996.

Borsting, C. F. and Damgaard, B.M. (1991). The effect of treating wheat fractions with amyloytic and cell wall degrading enzymes on carbohydrate digestibility in mink. Scientifur 15(4): 302.

Borsting, C. F. and Damgaard, B.M. (1995). The intermediate glucose metabolism in the nursing period of the mink. NJF-seminar no. 253. Goteborg, Sweden, 5 pp.

Borsting, C. F. and Engberg, R. M. (1995). Influence of oxidized fish oil in mink diets on nutrient digestibility, fatty acid accumulation, performance and health. Scientifur 19(10): 69-70.

Borsting, C. F., Engberg, R.M. Jakobsen, K., Jensen, S.K., Andersen, J.D. and Anderson, J. O. (1994). Inclusion of oxidized fish oil in mink diets. I. The influence of nutrient digestibility and fatty acid accumulation in tissues. J. Anim. Physiol. A. Anim. Nutr. 72:132-145.

Borsting, C. F., Engberg, R.M., Jensen, S. K. and Damgaard, B.M. (1998). Effects of high amounts of dietary fish oil of different oxidative quality on performance and health of growing-furring male mink (*Mustela vison*) and of female mink during rearing, reproduction and nursing. J.Anim. Physiol. A. Anim. Nutr. 79: 210-223.

Borsting, C. F. and Gade, A. (2000). Glucose homeostasis in mink (*Mustela vison*). A review based on interspecies comparison. Original Review. Scientifur 24(1): 9-18.

Borsting, C. F. and Olesen, C. R. (1993). The influence of amino acid supply on the production results of mink. Nord. Assoc. Agr. Scient. NJR Report No. 92: 88-90.

Borsting, C. F. and Riis , B. (2002).Methionine metabolism in the mink. Effect of season and the supply of methionine and betaine. Annual Report 2001: 125-131. Danish Fur Breeders Research Center., Holstebro, Denmark.

Borstrum, R.E., Aulerich, R. K. and Schaible, P.J. (1967). Histologic observations on the urinary system of male mink effected with "wet belly". Mich. Agr. Sta. Quart. Bull. 50(1): 100-105.

Bostrum, R. E., Aulerich, R.K. and Schaible, P.J. (l967). Histologic features of inguinal skin of "wet belly" and normal mink. Amer. J. Vet. Res. 28(126): 1549-1554.

Borsum, L. S. (1969). U.S. ranchers not ready for complete ration. National Fur News 40(3): 22-23.

Bowman, A.M., Travis, H. F., Warner, R. G. and Hogue, D.E. (1968). Vitamin B-6 requirement of the mink. J. Nutr. 95: 554-562.

Brandt, A. (1980). Vitamin og selen probemer hos Danske mink. Danish Vet. Idske 65: 149-152.

Brandt, A. (1983). Effect of dietary copper and zinc on the hematology of male pastel mink kits. A pilot investigation. Scientifur 7(2)51-53.

Brandt, A. (1983). Vitamin E and selenium in Danish mink. Scientifur 7(2): 52-57.

Brandt, A. (1984). Nutritional muscular degeneration syndrome in mink. Clinical-chemical studies. A preliminary report. 3rd Inter. Congr. in Fur Animal Production. Versailles. 25-27, April 1984.

Brandt, A. (1988). The effect of dietary factors on the development of nursing disease symptoms and performance in lactating mink. Proceedings of the 4th Inter. Sci. Congr. In Fur Animal Prod. Ed. Murphy, D. and Hunter, D. B. pp. 112-120. University of Saskatchewan Printing Svc., Saskatchewan, SK, Canada.

Brandt, A., Wolstrup, C. and Krogh-Nielsen, T. (1990). The effect of dietary alpha-tocopherol acetate, sodium selenite and polyunsaturated fatty acids in mink (Mustela vison). J. Anim. Physiol. A. Anim. Nutr. 64: 280-288.

Brody, S. (1945). Bioenergetics and growth, Reinhold Publ. Co., N.Y.

Brooker, E.G. and Shorland, F. B. (1950). Studies on the composition of horse oil. Biochem. J. 46: 80-85.

Brown, W. (1995 and 2000). Personal comm, South Dakota mink rancher.

Brouwer, W. (1965). Report of sub-committee on constants and factors. Proceedings of the 3rd Symposium in Energy Metabolism, EAAP. Publ. No. 11: 441-443.

Broyles, G. E. (1969). Ranchers, scientists and editors tour Europe. National Fur News 40(12):8-11.

Bruun, de Neergaard, J. K. (1972). Symposium om Forkvalitet, Roros, Norway, NJF's Subsektion Fur Pelsdyr. 62-68.

Buddington, R. K., Chen, J.W. and Diamond, J. M. (1991). Dietary regulation of intestinal brush border sugar and amino acid transport in carnivores. J. Physiology 261(4): Part 2: R793-R801.

Buddington, R. K., Malo, C., Sangild, P.T., and Elnif, J. (2000). Intestinal transport of monosaccharides and amino acids during postnatal development of mink. Am. J. Physiol. Regulatory Integrative Comp. Physiol. 279: R2287- R2296.

Buraczewski, S. (1973). Amino acid composition of the body of pigs and its implication in the amino acid requirements. Proceedings From The Symposium - On Amino Acids, Brno, Czechoslovakia - 1973.

Bursian, S. J., Hill, R.R., Mitchell, R. R. and Napolitano, A.C. (2002). The use of phytase as a feed supplement to enhance utilization and reduce excretion of phosphorous in mink. 2003 Blue Book of Fur Farming. 8-12.

Bush, C, R., Restum, J. C., Bursian, S.J. and Aulerich, R. J. (1995). Responses of growing mink to supplemental dietary copper and biotin. Scientifur 19(2): 141-147.

Cabell, C.A. and Leekley, J. (1960a). Antioxidants and mink feeding. National Fur News 32 (4): 17, 29, 54.

Cabell, C. A. and Leekley, J.(1960b). Effect of butylated-hydroxy-toluene, biphenyl-para-phenylenediamine and other additives on reproduction and steatitis in mink fed fish diets. J. Amer. Sc. 18: 1534-1535.

Cameron, J. K., Bursian, S.J. and Aulerich, R. J. (1989). Effects of zearalenone and other mycotoxins on mink. Blue Book of Fur Farming 1990: 76-78.

Carter, R. L., Percival, W. H. and Roe, F. J. C. (1969). Exceptional sensitivity of mink to the hepatotoxic effects of dimethylnitrosamine. J. Path. 97: 79-88.

Carvalho da Silva, A.R., Fried, A. R. and de Angelis, R. C. (1952). The domestic cat as a laboratory animal for experimental nutrition studies. III. Niacin requirements and tryptophan metabolism. J. Nutr. 46: 399-409.

Carver, D.S. and Waterhouse, H.N. (1962). The variation in the water consumption of cats. Proc. Animal Care Panel 12:267-270.

Ceh, L., Helgebostad, A. and Ender, F. (1964). Thiaminase in Cepelin (Mallotus villosus), an Artic fish of the sal-onidae family. Interzeitschrift Fur Vitaminforshung 32(2):189-196.

Chaddock, T. T. (1947). Veterinary problems of the fur ranch. Vet. Med. 42: 409.

Chanin, P.R.F. and Linn, I. (1980). The diet of feral mink (Mustela vison) in Southeran Britian. J. Zool. London. 192: 205-223.

Charlet-Lery, G., Fiszlewicz, M., and Morel, M-T. (1980). Energy and nitrogen balances in male mink during the adult life. 2nd Inter. Sc. Congr. In Fur Anim. Prod. Vedbock, Denmark, April 8-10, 1980.

390

Charlet-Lery, G., Fiszlewicz, M. and Morel, M-T. (1981). Variation of energy balance in male adult mink fed freely around the year. Proc. 32nd Ann. Meeting of the European Assoc. for Anim. Prod. 8/31-9/3, 1981. V-36.

Charlet-Lery, G., Fiszlewicz, M., Morel, M-T and Allan, D. (1979). Evolution de la composition corporelle du vision male in croissance a parter du sevrage (Evolution of body composition of growing male mink from weaning). Ann. Zootech. 28(4): 423-430.

Charlet-Lery, G., Fiszlewicz, M., Morel, M-T and Allan, D. (1980). Variation in body composition of male mink during growth. 2nd Inter. Sc. Congr. In Fur Anim. Prod. Vedbock, Denmark, April 8-10, 1980.

Charlet-Lery, G., Fiszlewicz, M., Morel, M.T. and Richard, J. R. (1981). Rate of passage of feed through the digestive tract of mink according to the type of diet (wet form or pellets). Ann. Zootech 30(3): 347-360.

Chausow, D. G., Forbes, R..M., Czarnecki, G. I. and Corbin, J.E. (1985). Experimental induced magnesium deficiency in growing kittens. Proc. Am. Soc. Anim. Sc. (Abstr.): 285.

Chausow, D. G. (1992). National Research Ranch, Unpublished data.

Chavez, E.R. (1980). Amino acid profile of the plasma, pelt and hair of the adult mink. 2nd Inter. Sci. Congr. In Fur Anim. Prod. Copenhagen, Denmark.

Chavez, E. R. (1981). Effect of organic and inorganic sulfate supplementation in mink diets. Can. J. Anim. Sci. 61(4): 1094.

Chou, C. C., Marth, E. H. and Shackelford, R. M. (1976a). Experimental acute alfatoxicosis in mink (*Mustela vison*). Am. J. Vet. Res. 37(10): 1227-1231.

Chou, C.C., Marth, E. H. and Shackelford, R.M.(1976b). Mortality and some biochemical changes in mink (*Mustela vison*) given sublethal doses of aflaxtonins each day. Am. J. Vet. Res. 37(10): 1233-1236.

Christensen, K., Fischer, P., Knudsen, K. E. B., Larsen, S., Sorrensen, H. and Venge, O.(1979). A syndrome of hereditary tyrosemia in mink (*Mustela vison Schreb*). Can. J. Comp. Med. 43(3): 333-340.

Christensen, K., Henriksen, P., Mortensen, K. and Sorensen, H. (1989). Metabolic tyrosine disorder in mink and PLP therapy of hereditary tyrosemia. Amino Acids: Chemistry, Biology and Medicine. 762-772.

Christensen, K., Henriksen, P. and Sorensen, H. (1986). New forms of hereditary tyrosemia type II in minks. Hepatic tyrosine amino transferase Deficiency. Hereditas 104(2): 215-222.

Christensen, K., Venge, O. and Sorensen, H. (1980). Syndrome of hereditary tyrosemia in mink. 2nd Inter. Congr. In Fur Anim. Prod.Vedbock, Denmark, April 8-10, 1980.

Chwalibog, A., Glem-Hansen, N.,Heneckel, S. and Thorbek, G. (1980). Energy metabolism in adult mink in relation to protein-energy levels and environmental temperatures. Proc. 8th Sym. on Energy Metabolism, Cambridge, England. (ed. L. E. Mount). European Assoc. for Anim. Prod. Publ. No. 26: 283-286. Butterworth, London.

Chwalibog, A., Glem-Hansen,. N., and Thorbek, G. (1982). Protein and energy metabolism in growing mink (*Mustela vison*). Arch. Tierernahrung, Bd. 32: 551-562.

Chwalibog, A., Lind, J. and Thorbek, G. (1979). Description of a respiration unit for quantitative measurements of gas exchange in small animals applied to indirect calorimetry. Z. Tierphysiol. Tierernahg u. Futtermittelkde. 41: 154-162.

Chwalibog, A. and Thorbek, G. (1980). Nitrogen and energy metabolism in growing mink fed two levels of protein. Zeitschrift Fur Tierphysiologie Tiererahrung Und Fullermittelkunde 44(1): 30-31.

Clausen, T. N. (1992). Effects of dry matter content of mink feed in the nursing period. Annual Report of the Danish Fur Breeders Organ. 118-123.

Clausen, T.N. (1999a). The effect of feed composition on the acidity of urine in mink kits. NJF Seminar No. 308. Reykjavik, Iceland. 21-24 October 1999. Cand. Med. Vet. T. N. Clausen, Pelsdyrervets Forsogs-og Radgivningsvirksomhed A/S Herningvej 112 C. DK-7500 Holstebro, Denmark.

Clauson, T.N. (1999b). Ammonium chloride fed to mink kits. Ann. Report. Danish Fur Breeders Research Center, Holstebro, Denmark. Pp. 89-92.

Clauson, T.N. (2001). Addition of phosphoric acid or ammonium chloride to mink feed in June. Influences on urine pH and body growth. Annual Report 2000: 55-57. Danish Fur Breeders Research Center, Holstebro, Denmark.

Clausen, T.N. and Damgaard, B. M. (2002). Sodium chloride in the feed on the nursing period. Annual Report 2001: 60-65. Danish Fur Breeders Research Center, Holstebro, Denmark.

Clausen, T. N. and Diet, H. H. (1999). Ethoxyquin in the feed for mink kits in the growing-furring period. Annual Report 1999 Danish Fur Breeders Res. Center. ed. P. Sandbol,.

Clausen, T.N. and Hansen, O. (1989). Electrolytes in mink with nursing sickness. Acta Physiol. Scand. 136: 24AP9.

Clausen, T.N., Hansen, M.V., Lassen, M., Mortensen, K., Sorensen, H., Sorensen, J.C. and Tauson, A-H. (2004). Fatty acid profiles of mink milk and tissues of kits as a function of different n-6:n-3 fatty acid ratios in the diet. Annual Report 2003: 67-74. Danish Fur Breeders Research Center, Holstebro, Denmark.

Clausen, T. N., Hansen, O. and Wamberg, S. (1990). Electrolyte and acid/base changes in lactating female mink subject to nursing sickness. Scientifur 14:117.

Clausen, T. N. and Hejlesen, C. (2000). Increasing amounts of toasted soybean for mink kits in the growing-furring period. 1998. Scientifur 24(2): l48.

Clausen, T.N. and Hejlesen, C. (2001). Lecithin for mink in the growing and furring period. Annual Report 2000: 61-71. Danish Fur Breeders Research Center, Holstebro, Denmark.

Clausen, T.N. and Hejlesen, C. (2002a). Ad libitum or restricted feeding of mink females in May. Annual Report 2001: 77-80. Danish Fur Breeders Research Center, Holstebro, Denmark.

Clausen, T.N. and Hejlesen, C. (2002b). Peas and soybeans for mink in the growing-furring period. Annual Report 2001: 105-l08. Danish Fur Breeders Research Center, Holstebro, Denmark.

Clausen, T.N. and Hejlesen, C. (2003). Reduced protein for scan black females during winter and lactation periods. Annual Report 2002: 23-25. Danish Fur Breeders Center, Holstebro, Denmark.

Clausen, T.N., Hejlesen, C. (2004). Cracklings to mink in the growth period. Influence on health, growth and fur quality. Ann. Rept. Danish Fur Breeders Research Center, (2003) Holstebro, Denmark. Pp. 121-124.

Clausen, T. N., Hejlesen, C. and Sandbol (2003). Protein content in feed for mink females and kits. Annual Report 2003: 75-79. Danish Fur Breeders Research Center, Holstebro, Denmark.

Clausen, T. N., Hejlesen, C. and Sandbol, P. (2004). Protein to mink in the nursing period and in the period of early growth. Annual Report 2004: 77- 81. Danish Fur Breeders Research Center, Holstebro, Denmark.

Clausen, T. N., Hejlesen, C., Sorensen, H., Bjergegaard, C., Mortensen, K. and and Christiansen, C. (2003). Carbohydrate sources for mink in the nursing period. Annual Report 2002: 27-31. Danish Fur Breeders Research Center, Holstebro, Denmark.

Clausen, T. N., Kjelgaard-Hansen, M., Mortensen, K., Sorensen, A. D., Sorensen, J. C., Sorensen, H. and Sorensen, S. (2004b). Mink proteinases; Trypsin and chymotrypsin and the effect of proteinase inhibitors. Annual Report 2004: 173-184. Danish Fur Breeders Research Center, Holstebro, Denmark.

Clausen, T. N. and Olesen, C. R. (1991). Relationship between feeding and nursing sickness in mink. In: Glem-Hansen, N. (ed.) Technical Year Report 1990, Danish Fur Breeders Assoc. 204-213.

Clausen, T. N., Olesen, C. R., Hansen, O. and Wamberg, S. (1992). Nursing sickness in lactating mink (*Mustela vison*). 1. Epideiological and pathological observations. Can. J. Vet. Res. 56: 89-94.

Clausen, T.N. and Sandbol, P. (2003). The influence of the essential to non-essential amino acid ratio to mink during the growing period - fur production results. Annual Report 2002: 43-49. Danish Fur Breeders Research Center, Holstebro, Denmark.

Clausen, T. N. and Sandbol, P. (2005). Fasting of mink kits fed different feed rations and its effect on liver fat content, plasma metabolites and enzymes. Proc. From NJF-seminar No. 377. 5 pp.

Clausen, T.N., Sandbol, P. and Damgagaard, B.M. (2002). Sodium chloride in the lactation period and early growth fase. Annual Report 2001: 91-96. Danish Fur Breeders Research Center, Holstebro, Denmark.

Clausen, T. N., Sandbol, P. and Hejlesen, C. (2003). Meat and bone meal in mink diets during the growing period. Annual Report, 2002: 33-37.Danish Fur Breeders Research Center, Holstebro, Denmark.

Clausen, T.N., Sandbol, P. and Hejlesen, C. (2004c). Methionine and other methyl donors to mink in the furring period. Annual Report 2003: 97-100. Danish Fur Breeders Research Center. Holstebro, Denmark.

Clausen, T.N., Sandbol, P. and Hejlesen, C. (2005a). Protein to mink in the nursing period and in the period of early kit growth. Continued investigations. Annual Report 2005, Danish Fur Breeders Research Center, Holstebro, Denmark. 65-70.

Clausen, T. N., Sandbol, P. and Hejlesen, C. (2005b). ZnO to mink kits in the weaning period. Annual Report 2005, Danish Fur Breeders Research Center, Holstebro, Denmark. 71-74.

Clausen, T.N., Sandbol, P. and Hejlesen, C. (2005c). Sulfur containing amino acids and methyl donors to mink in the furring period. Annual Report 2005. Danish Fur Breeders Research Center. Holstebro, Denmark. 81-88.

Clausen, T. N., Sandbol, P. and Hejlesen, C. (2005d). Protein to mink in the furring period. Importance of fat and carbohydrate. Annual Report 2005. Danish Fur Breeders Research Center. Holstebro, Denmark. 89-98.

Clausen, T. N., Therkildsen, N. and Borsting, C. F. (1998). Determination of the requirement for methionine and cystine in the growing period. Scientifur 22(1): 49.

Clausen, T.N. and Wamberg, S. (1998). Changes in the urine pH of mink kits (7-9 days old) when given different feeds. Scientifur 22(2): 43-44.

Clausen, T.N., Wamberg, S. and Hansen, O. (1995). The importance of feed salt content for the incidence of nursing sickness. Scientifur 19(3): 221.

Clausen, T.N., Wamberg, S. and Hansen, O. (1996a). Effects of dietary salt supplementation on clinical and subclinical nursing sickness in lactating mink (*Mustela vison*). Anim. Prod. Rev., Polish Soc. Of Anim. Prod., Applied Sc., Reports 28. Progress In Fur Anim. Sc. Nutr., Patology and Disease. Proceedings of the 6th Inter. Sc. Congr. In Fur Anim. Prod. Aug. 21-23. 1996, Warsaw, Poland. Ed. A. Frindt and M. Brzozowski: 87-91.

Clausen, T. N., Wamberg, S. and Hansen, O. (1996b). Incidence of nursing sickness and biochemical observations in lactating mink with and without salt supplementation. Can. J. Vet. Res. 60: 271-276.

Clay, C. (1969). Dutch complete dry mink diet achieves impressive results. Fur Trade J. of Canada. 47(5): 10-18.

Close, W. H. and Mount, L. E. (1978). The effect of plane of nutrition and environmental temperature on the energy metabolism of the growing pig. Br. J. Nutr.40: 413-421.

Cobbledeck, M. (1997). Personal communication. English mink rancher.

Coffint, D. L. and Holzworth, J. (1954). Yellow fat in two laboratory cats. Acid-fast pigmentation associated with fish-base ration. Cornell Vet. 44:63-71.

Colin, G. (1854). Traite de Physiologic Comparee des Animaux Domestique, Tome Premier: 408-410. Paris, J-B. Bailliere.

Conant, R. A. (1962). A milking technique and the composition of mink milk. Amer. J. Vet. Res. 23(96): 1104-1106.

Conwey, C. B., Cooke, D.J., Matty, A. J. and Adron, J. W. (1981). Effect of quantity and quality of dietary protein on certain enzyme activities of rainbow trout. J. Nutr. 111: 336-345.

Coombes, A. I. (1941). Nutrition and the proper feeding of foxes and minks. The American National Fur and Market Journal 20(2): 13-14, 26-27.

Costley, G. E. (1970). Involvement of formaldehyde in depressed iron absorption in mink and rats fed Pacific hake (*Merluccius productus*). Thesis, Oregon State University.

Cowan, R. L., Keck, E. and McCabe, F. L. (1972). Penn State works with dry diets. U.S. Fur Rancher 51(10): 22-24.

Crampton, E. W. (1964). Nutrient-to-calorie ratios in applied nutrition. J. Nutr. 82: 353-365.

Dahlman, T., Niemela, P., Kisskinen, T., Makela, J. and Korhonen, H. (1996). Influence of protein quantity and quality on mink. Anim. Prod. Rev. Polish Soc. Of Anim. Prod. Applied Sc. Reports 28. Progress in Fur Anim. Sc. Nutr. Patology and Disease. Proc. From the 6th Inter. Sc. Congr. In Fur Anim. Prod., Warsaw, Poland, August 21-23, 1996. Ed. A. Frindt and M. Bezozowski: 9-14.

Dahlman, T., Valaja, J., Niemela, P. and Jalava, T. (2002).Influence of protein level and supplemental L-methionine and lysine on growth performance and fur quality of blue fox (*Alopex lagopus*). Acta Agr. Scand. Sec. A. Anim. Sc. 52(4): 174-182.

Dalgaard-Mikkelsen, S., Kvorning, S. A., Momberg-Jorgensen, H. C., Haagen-Petersen, F. and Schambye, F. (1953). Studies over "gult fedt" hos mink. Nord. Vet. Med. 5:79-97.

Dalgaard-Mikkelsen, S., Momberg-Jorgensen. H. C. and Haagen-Petersen, F. (1958). Control of yellow fat disease in mink. Beretn. Forsogslab,. Kobenhan. No. 308. 31 pp.

Dam, H.(1962). Interrelation between vitamin E and PUFA in animals. Vitamins and Hormones 20: 527-540.

Damgaard, B.M.(1997). Dietary energy supply to mink (Mustela vison). Effects of physiological parameters, growth performance and health. Ph.D. Thesis, Dept. of Anim. Sc.and Anim. Health. The Royal Vet. And Agr. University. Copenhagen, Denmark.

Damgaard.B.M. (1997) Dietary protein supply to mink (*Mustela vison*). Effects of physiological parameters, growth performance and health. Ph.D. Thesis. Scientifur 21(4): 281-284.

Damgaard, B. M. (1998). The effects of dietary supply of arginine on urinary orotic acid excretion, growth performance and blood parameters in growing Mink (*Mustela vison*). kits fed low protein diets. Acta Agric. Scand. A. Anim. Sc. 48:113-121.

Damgaard, B.M. and Borsting, C. F. (1995). Glucose tolerance tests in mink. NJF-Seminar No. 253, October 4-6, 1995. Gothenburg, Sweden.

Damgaard, B.M. and Borsting, C. F. (2002). Effects of high dietary levels of fresh and oxidized fish oils on mink female performance and kit growth during the nursing period. Annual Report 2001: 87-90. Danish Fur Breeders Research Center, Holstebro, Denmark.

Damgaard, B.M., Borsting, C. F., Engberg, R. M. and Jensen, S. K. (2003). Effects of high dietary levels of fresh and oxidized fish oil on performance and blood parameters in female mink (*Mustela vison*) during the winter, reproduction, lactation and early growth periods. Acta Agric. Scand. Sect. A. Animal Sc. 53: 136-146.

393

Damgaard, B.M., Borsting, C. F. and Fink, R. (2000). Effects of dietary protein and carbohydrate supply on feed consumption, growth performance and blood parameters in mink dams during the nursing period. Scientifur 24(4): IV-A Nutrition. Proceedings of the VIIth Inter. Congr. In Fur Anim. Prod., Sept. 2000. Ed. B. D. Murphy and J. O. Lohi: 17-21.

Damgaard, B.M., Borsting, C.F., Ingvartsen, K. and Fink, R. (2003). Effects of carbohydrate-free diets on the performance of lactating mink (*Mustela vison*) and the growth performance of suckling kits. Acta Agr. Scand. Sec. A. Anim. Sc. 53(1): 127-135.

Damgaard, B.M., Clausen, T.N. and Henricksen, D. (1994). Effect of protein and fat content in feed on plasma alanine-aminotransferase and hepatic fatty infiltration in mink. J. Vet. Med. A. 41: 620-629.

Damgaard, B.M. and Clausen, T. N. (1999). Dietary protein content and health of mink. Scientifur 23(4): 285.

Damgaard, B.M., Clausen. T. N. and Borsting, C. F. (1998). Effects of dietary supplement of essential amino acids on mortality rate, liver traits and blood parameters in mink (*Mustela vison*). fed low protein diets. Acta Agr. Scand. 48(3): 175-183.

Damgaard, B.M., Clausen, T. N. and Dietz, H. H. (1998). Effect of dietary protein levels on growth performance, mortality rate and clinical blood parameters in mink (*Mustela vison*). Acta Agr. Scand. 48(1): 38-48

Damgaard, B.M., Engberg, R.M., Borsting, C. F. and Jensen, S. K. (2003). effects of high dietary levels of fresh or oxidized fish oil on performance and blood parameters in female mink (*Mustela vison*) during the winter reproduction, lactation and early growth. Acta Agr. Scand. Sec. A Animal Science 53(1): 136-146.

Danish Fur Breeders Association Voluntary Feed Control Program. Mark Munkoe B. (2003). Arsberetning Analysel Aboratoriet Dansk Pelsdyr Foder Als 7500, Holstebro, Denmark.

Danse, L. H. J. C. (1976). Early changes of yellow fat disease in mink, the involvement of the reticuloendothelial system. 1st Inter. Congr. In Fur Anim. Prod. Helsinki, Finland, April 27-28.

Danse, L. H. J. C. (1978). Pathogenic study of yellow fat disease. Thesis: University of Utrecht.

Danse, L. H. J. C. (1976). Early changes of yellow fat disease in mink fed a vitamin E deficient diet supplemented with fresh or oxidized fish oil. Zbl. Vet. Med. A. 23:645-660.

Davis, K. G. (1954). Mink Nutrition Research, Progress Report No. 4 (1954). Oregon State Agr. Exper. Station.

Davis, T. A., Nguyen, H. V., Garcia-Bravo, R., Fiorotto, M.I., Jackson, E.M., Lew, D. S., Lee, D.R. and Reeds, P. J. (1994). Amino acid composition of human milk is not unique. J. Nutr. 124(7): 1126-1132.

Day, M. G. and Linn, I. (1972). Notes on the food of feral mink (*Mustela vison*) in England and Wales. J. Zool. London 167: 463-473.

Deady, J. E., Anderson, B., O'Donnell, J. A. III, Morris, J. G. and Rogers, Q.R. (1981). Effect of level of dietary glutamic acid and thiamine on food intake, weight gains, plasma amino acids, and thiamine status of growing kittens. J. Nutr. 111: 1568-1579.

Dearfor, N. (1932). Foods of some predatory fur bearing animals in Michigan. Bull I. School of Forestry and Conservation, Univ. Michigan: 32.

De Hart, H. (1981). Personal communication, Wisconsin mink rancher.

De Ritter, E. (1976). Stability characteristics of vitamins in processed foods. Food Technology 30: 48-52.

Deshmuk, D. R. and Rusk, C. D. (1989). Effects of arginine-free diet on urea cycle enzymes in young and adult ferret. Enzyme 41(3): 168-174.

Deshmuk, D.R., Sarnaik, A. P., Mukhopadhyay, A. and Portoles, M. (1991). Effect of arginine-free diet on plasma and tissue amino acids in young and adult ferrets. J. Nutr. Biochem. 2(1): 72-78.

De Silva, A.C., Fried, R. and De Angelis, D.E. (1952). The domestic cat as a laboratory animal. Niacin requirement and tryptophan metabolism. J. Nutr. 46: 399-409.

Deutsch, H.F. and Hasler, A. D. (1943). Distribution of vitamin B-1 destructive enzyme in fish. Soc. Exp. Bio. Med. 53:63-65.

Deutsch, H.F. and Ott,. G. L. (1942). Mechanism of vitamin B-1 destruction by a factor in raw smelt. Soc. Exp. Biol. Med. 51: 119-122.

Disse, P,. H. (1957). Amino acid composition of mink fur. Wisconsin Research Foundation.

Disse, G. and Disse, K. (1995). Personal communication. Minnesota mink ranchers.

Dittrich, E. (1995). Personal communication. Wisconsin mink rancher.

Dolnick, E.H. (1959a). Histogenesis of hair in the mink and its relationship to dermal fetal fat cells. J. Morph. 105(1): 1-31.

Dolnick, E.H. (1959b). Priming-up of hair in the mink. National Fur News 31(10): 20, 66.

Dolnick, E. H. (1959c). Priming-up of hair in the mink. Amer. Fur Breeder 32(12): 22,42, 44.

Dolnick, E.H. (1961). Hair growth in the mink. National Fur News 33(10): 12-13,30,35.

Dolnick, E.H., Warner, R. G. and Bassett, C. F. (1960). Influence of diet on fur growth. National Fur News 32(4): 12-13, 56.

Dunstone, N. and Birks, J.D.S. (1987). The feeding ecology of mink (*Mustela vison*) in coastal habitat. J. Zool. (London) 212: 69-83.

Durant, G. (1992). Vet says nursing sickness related to energy exhaustion. Fur Rancher 73: 2.

Durrant, G. (1999). Personal communication. Utah veterinarian.

Edfors, C. H., Villey, D.E. and Aulerich, R. J. (1989). Prevention of urolithiasis in the ferret (*Mustela putorius furo*) with phosphoric acid. J. of Zoo and Wildlife Med. 20(1): 12-19.

Eggum. B.O. (1970). Evaluation of protein quality of feather meal under different treatments. Acta Agric. Scand. 20: 230-234.

Eggum. B.O. (1973). A study of certain factors influencing protein utilizatioin in rats and pigs. 406. Beretning fra Forsogslaboratoriet, Statens Husdyrbrugsforsog, Copenhagen, Denmark.

Eichel, H. (1965). Der vitamin B-stoffwechsel der farmgehaltenen edelpelztiere, insbesondere der thiamin, riboflavin, pyridoxine und biotin - stoffwechsel des nerzes unter besonderer Berucksightigung der darmflora. Report from Veterinar Physiologischen Institute der Karl-Marx Universitat, Leipzig. DDR: 118.

Einarsson, E. and Enggaard-Hansen, N. (2000). Different protein conc. in feed for mink kits. Nitrogen-energy, water and mineral balance during growth period. Scientifur 24(1). 70.

Elnif, J. (1987). Growth of the digestive tract in the neonatal mink. Technical Yearbook, l987, Danish Fur Breeders Assoc. Copenhagen, Denmark.

Elnif, J. and Enggaard-Hansen, N. (1987). Sammenligning afnaeringsstoffernes. Fordojelighed hos minkvalpe og udvoksedehanner (pastel type). NJE Seminar Nr. 128. Tromso, Norway.

Elnif, J. and Enggaard-Hansen, N. (1988). Production of digestive enzymes in mink kits. Biology, Pathology and Genetics of Fur Bearing Animals. 4th Inter. Congr. In Fur Anim. Prod., August 21-24, 1988, Rexdale, Ontario, Canada: 320-326.

Elnif, J. and Sangild, P.T. (1996). The role of glucocorticoids in the growth of the digestive tract in mink (*Mustela vison*). Comp. Biochem. Physiol. 115A: 37-42.

Ender, F. (1966). N-Nitrosomethylamine, the active principle in cases of herring meal poisoning. IV. Inter. Tagung der Weltgesellschaft fur Buiatrik. 4 bis 9, Aug. 1966 in Zurich: l-6.

Ender, F., Dishington, I. W., Madsen, R. and Helgebostad, A. (1972). Iron-deficiency anemia in mink fed raw marine fish - a five year study. Advances in Anim. Physiol. And Anim. Nutr. Suppl. To Zeitschrift Fur Tierphysiologie. Tiernahrung and Futtermittelkunde.

Ender, F. and Helgebostad, A. (1944). Fiskeforingens innflytelse pa forskjellige pelsegenskaper hos solvrev. Norsk Pelsdyrblad 18:73-96.

Ender, F. and Helgebostad, A. (1945). Om pavisning av en vitamin B-1 destrvernde factor; prover av sild og brisling. Antivitaminer som arsak till beri-beri hos reve. Nordisk Vet. Tidskr 2: 32-76.

Ender, F. and Helgebostad, A. (1953). Yellow fat disease in fur bearing animals. Proc. XVth Inter. Vet. Congr. Stockholm, Sweden, Aug. 9-15. IV: 141.

Ender, F. and Helgebostad, A. (1958). Further studies on the influence of various food constituents on the quality of fur in foxes and mink. Saertrykk av Norsk Pelsdryblad nr. 1.

Ender, F. and Helgebostad, A. (1959). Experimental biotin deficiency in mink and foxes. Nord. Vet. Med. 11: 141-161.

Ender, F. and Helgebostad, A. (1968). Studies on the anemiogenic properties of trimethylamine oxide, an etiological factor in fish-induced anemia in mink. Acta Vet. Scand. 9: 174-176.

Ender, F. and Helgebostad, A. (1975). Unsaturated dietary fat and lipoperoxides as etiological factors in vitamin E deficiency in mink. Acta Vet. Scand. 16. Suppl. 55, 1975: 1-25.

Engberg, R.M. and Borsting, C. F. (1994). Inclusion of oxidized fish oil in mink diets. 2. The influence on performance and health considering histopathological, clinical-chemical and hematological indices. J. Anim. Physiol. And Anim. Nutr. 72: 146-157.

Engberg, R. M., Jakobsen, K., Borsting, C. F. and Gjern, H. (1993). On the utilization, retention and status of vitamin E in mink (*Mustela vison*) under dietary oxidative stress. J. Anim. Physiol. And Anim. Nutr. 69(1): 66-78.

Engebretson, G. (1975). Personal communication. British Columbia mink rancher.

Enggaard-Hansen, N., Finne, L., Skrede, A. and Tauson, A-H. (1991). Energiforsyningen hos mink og raev. (Review of energy requirements of mink and fox. In Danish). Scand. Assoc. of Agr. Scientists. NJF Report No. 63. DSR Forlag, Landbohjskolen, Copenhagen. 57 pp.

Enggaard-Hansen, N. and Glem-Hansen, N. (1980). Digestibility of crude protein, crude fat and carbohydrates in growing mink related to feeding with sulfuric acid preserved fish. Dansk Pelsdyravl 43: 59-61.

395

Enggaard-Hansen, N. and Glem-Hansen, N. (1980). Deposition of nutrients in growing mink related to feeding with sulphuric acid preserved fish. 2nd Inter. Sc. Congr. In Fur Anim. Prod., Vedback, Denmark, April 8-10, 1980.

Enggaard-Hansen, N., Glem-Hansen, N. and Joergensen, G. (1984). Energy metabolism in mink during the period of growth. 3rd Inter. Sc. Congr. In Fur Amim. Prod., Versailles, France, April 25-27,1984: 11-1 - 11-7.

Enggaard-Hansen, N., Glem-Hansen, N. and Joergensen, G. (1992). Effects of dietary calcium/phosphorus ratio on growth, skin length and quality in mink (*Mustela vison*). Scientifur 16(4): 293-297.

Enggaard-Hansen, N., Mollerm. S. and Mejborn, H. (1985). Fiberholdige stoffer i pelsdyrfoder. NJR Seminar #85, Aalburg, Denmark, 13 pp.

Erdman. J. W., Jr., Weingartner, K. E., Mustakas, G. C., Schmulz, R.D., Parker, H.M. and Forbes, R. M. (1980).Zinc and magnesium bioavailability from acid precipitated and neutralized soybean protein products. J. Food Sci. 45: 1193-1199.

Erickson,. A. (1960). Personal communication. Utah mink rancher.

Ericksson, S. (1954). Potatoes for mink - a ranch survey. Vara Pelsdjar 25(6): 113-116.

Ericksson, L., Valtonen, M. and Makela, J. (1983). Effects of dietary NaCl on the water and electrolyte balance of mink. Acta Physiol. Scand. 118(2): 35A.

Ericksson, L., Valtonen, M. and Makela, J. (1984). Water and electrolyte balance in male mink (*Mustela vison*) on varying dietary NaCl intake. Acta Physiol. Scand. Suppl. 537(2): 59-64.

Ericksson, L., Valtonen, M. and Makela, J. (1986). Effects of salt load in mink. Scientifur 10:294.

Erway, L. C. and Mitchell, S. E. (1973). Prevention of otolith defect in pastel mink by manganese supplementation. J. Hered. 64(3): 111-119.

Esbjerg Fiskeindustri - (1998a). A. m .b. a. Fiskenhavnsgade 35, P.O. Box 1049 DK-6701 Esbjerg, Denmark.

Esbjerg Fiskeindustri - (1998b). Fishmeal Specification LT - 999. A. m. b. a.

Eskeland, B. and Rimeslatten, H. (1979). Studies on the absorption of labeled dietary alpha tocopherol in minks as influenced by some dietary factors. Acta Agr. Scand. 29: 75-80.

Evans, E. V. (1963). Progress in the evaluation of energy and protein requirements of growing mink. Fur Trade Journal of Canada 40(12):8-9,16.

Evans, E. V. (1967, 1976 and 1977). Determination of metabolizable energy in mink diets and feed ingredients. Unpublished data.

Falcon, H.(1986 and 1988). Personal communication. Wisconsin mink rancher.

Farrell, D. G. and Wood, A. J. (1968a). The nutrition of the female mink (*Mustela vison*).I. The metabolic rate of the mink. Can. J. Zool. 46(1):41-45.

Farrell, D. G. and Wood, A.J.(1968b). The nutrition of the female mink (**Mustela vison**). II. The energy requirement for maintenance. Can. J. Zool. 46(1): 47-52.

Farrell, D. G. and Wood,. A.J. (1968c). The nutrition of the female mink (*Mustela vison*). III. The water requirement for maintenance. Can. J. Zool. 46(1): 53-56.

Fink, R. (2001). Nutrient and energy metabolism of the lactating mink (*Mustela vison*). Ph. D. Thesis. Dept. of Animal Science and Animal Health. The Royal Veterinary and Agricultural University, Bulorbsvej. 13, DK-1870 Fredriksberg, C. Denmark.

Fink, R. and Borsting, C.F. (2002). Quantitative glucose metabolism in lactating mink (*Mustela vison*) - effects of dietary levels of protein, fat and carbohydrates. Acta Agr. Scand. Sect. A. Anim. Sc. 52(1): 34-42.

Fink, R., Borsting, C.F. and Damgaard, B.M. (2002). Glucose metabolism and glucose homeostasis of the lactating mink. Annual Report 2001:81-86. Danish Fur Breeders Research Center, Holstebro, Denmark.

Fink, R., Borsting, C.F. and Damgaard, B.M. (2002). Glucose homoeostasis and regulation in lactating mink (Mustela vison): Effects of dietary protein, fat and carbohydrate supply. Acta Agr. Scand. Sect. A. Anim. Sc. 52(2): 102-111.

Fink, R., Borsting, C. F., Damgaard, B.M. and Rosted, A.K. I.(2002). Glucose metabolism and regulation in lactating mink (*Mustela vison*) - effects of low dietary protein supply. Arch. Anim. Nutr. 2002. 56: 155-166.

Fink, R. and Tauson, A-H. (2000). Maikproduktion - effeck af energifordeling mellem protein, fedt or kulhydrat. Danish Institute of Ag. Sc., Foulum, Denmark. Intern Report 135: 39-44.

Fink, R. and Tauson, A-H. (2004). Growth rate in mink kits - effect of protein, fat and carbohydrate supply. Meeting of DIAS Research Centre

Foulum, 30 September, 2004 on the subject "Research in relation to practical mink production. Internal Report 2004, No. 208.
Fink, R., Tauson, A-H and Borsting, C. F. (2001). Dietary protein, fat and carbohydrate supply to lactating mink (*Mustela vison*). Effect on glucose homeostasis, energy metabolism and milk production. Proceedings from NJF-seminar No. 331. 90 pp.

Fink, R., Tauson, A-H and Hansen. N.E. (2004). Effect of protein, fat and carbohydrate supply on growth and energy metabolism in mink kits. Annual Report 2004. Danish Fur Breeders Research Center, Holstebro, Denmark: 83-87.

Fishler, J. (1955). Urolithiasis or urethral impaction in the cat. J. Am.Vet. Med. Assoc. 127: 121-123. Fiskenhavnsgade 35. P.O. Box 1049 DK-6701 Esbjerg Denmark.

Fjeld-Petersen, J. J. (1974). Eksperimentel saltforgiftning hos mink. Danks Pelsdyraul 37:153-154.

Fleming, P. (1999). Personal communication. Nova Scotia mink rancher.

Fletch, S..M. and Karsted, L. H. (1972). Blood parameters of healthy mink. Can. J. Comp. Med. 36:275-281.

Fog, A. (1974). A vitamin. Bilag til kursus om vitamin-og mineralstoffer iminkens foder, deres indflydelse pa reproduktion, vaekst og pelsudvikling. Tune Landboskloe, Denmark.

Forbes, R.M. (1963). Mineral utilization in the rat. Effects of varying dietary ratios of calcium, magnesium and phosphorus. J. Nutr. 80: 321-326.

Fraps, R. M., Hertz, R. and Sebrell, W. H. (1943). Relationship between ovarian function and avidin content in the oviduct of the hen. Proc. Soc. Exp. Biol. Med. 52: 140- 145.

Freeman, H. (1994). Personal communication, Ontario mink rancher.

Frieden, E. (1972).The Chemical elements of life. Sci. Amer.227: 52-60.

Friend, D. W. and Crampton, E. W. (1961a). The stability of vitamin E in some mink feeds stored frozen. Fur Trade J. of Can. 38(11): 8-9.

Friend, D. W. and Crampton, E. W. (1961b). The adverse effects of raw whale liver on the breeding performance of female mink. J. Nutr. 73: 317-326.

Friend, D. W. and Crampton, E. W. (1961c). The need for supplemental fat by breeding mink fed rations containing codfish products. J. Nutr. 74:397-400.

Frindt, A., Brzozowski, M., Engberg, R. M. and Borsting, C. F. (1996). Inclusion of fish fat in mink diets. Anim. Prod. Rev., Polish Soc. Anim. Prod., Applied Sc. Reports, 28. Progr. In Fur Anim. Sc., Nutr. Patology and Disease. Proc. From the 6th Inter. Sc. Congr. In Fur Anim. Prod., August 21-23, 1996. Warsaw: 51-59.

Fuhrman, K. (2000). Personal Communication. Wisconsin mink rancher.

Gallo-Torres, H.E. (1972). Vitamin E in Animal Nutrition. Inter. J. Vit. Nutr. Res. 42: 312-323.

Gallo-Torres, H.E., Weber, F. and Wiss, O. (1971).The effect of different dietary lipids on the lymphatic appearance of vitamin E. Inter. J.For Vit. Nutr. Res. 41(4): 504-515.

Garcia-Mata, R. (1980). Towards a more efficient mink production. 2nd Inter. Congr. In Fur Anim. Prod. Vedback, Denmark, Apr. 8-10.

Gershoff, S.N., Andrus, S. B., Hegsted, D. M. and Lentini, E. A. (1957). Vitamin A deficiency in cats. Lab. Invest. 6:227-240.

Gershoff, S.N. and Norkin, S. A. (1962). Vitamin E deficiency in cats. J. Nutr. 77: 303-308.

Gilbert, F. F. and Nancekivell, E. G. (1982). Food habits of mink (*Mustela vison*) and otter (*Lutra Canadensis*) in Northeastern Alberta. Canadian J. Zool. 60: 1282-1288.

Gillies, R. W. (1966). Propylene glycol in mink nutrition. Information bulletin from the supervisor, R. W. Gillies, Fur Farms Branch, Dept. of Agr., Edmonton, Alberta, Canada.

Glem-Hansen, N. (1970). Foirdojelighed af oksetlag og skagteaffald. Dansk Pelsdyravl 33: 222.

Glem-Hansen, N. (1974a). Minkens protein forsyning, normal, aminosyre - analyser og andre kvalitetskriterier. Bilag tl NJF's Mode Om Optimal Ernaering Af Mink Og Blaraeve. Grena, Denmark, 19-21, August 1974.

Glem-Hansen, N. (1974b). Minkens protein forsyning, normal, aminosyre-analyser og andre kvalitetskriterier. Lando/Konomisk Forsogelaboratorium. Roskildevej, 48 H. 3400 Hilleroed, Denmark. p. 85.

Glem-Hansen, N. (1976). The requirement for sulfur containing amino acids for mink in the growth period. 1st Inter. Sci. Congr. In Fur Anim. Prod. Finn.

Glem-Hansen, N. (1977a. 1977b and 1978). National Institute of Animal Sc. Hilleroed Denmark. Personal Communication.

Glem-Hansen, N. (1977c). The requirement of protein, sulphuric amino acids and the energy value of nutrients for mink. Ph.D. Thesis. AFD for Forsog med pelsdyr, Statens Husdyrbrugsforsog.

Glem-Hansen, N. (1978). Erfarungen mit dem soyaprotein-produkt "Nurupan". Die Muhle + Misch Futterechnik. 15: 572.

Glem-Hansen, N. (1979a). Protein requirement for mink in the lactation period. Methods for evaluation of protein requirement during lactation. Acta Agr. Scand. 27: 129-138.

Glem-Hansen, N. (1979b). Digestibility of feedstuffs determined on mink. Dansk Pelsdyravl 42(4): 164-165.

397

Glem-Hansen, N. (1979c). Digestibility of feedstuffs determined in mink. Scientifur 3(2): 23-25.

Glem-Hansen, N. (1980a). The protein requirements of mink during the growth period. I. Effect of protein intake on N-balance. Acta Agr. Scand. 30(3): 336-344.

Glem-Hansen, N. (1980b). The protein requirements of mink during the growth period. II. Effect of protein intake on growth and pelt characteristics. Acta Agr. Scand. 30(3): 345-348.

Glem-Hansen. N. (1980c). The requirements for sulphur-containing amino acids of mink during the growth period. Acta Agr. Scand. 30(3): 349-356.

Glem-Hansen, N. (1982b). Utilization of L-cystine and L and D-methionine by mink during the period of intensive hair growth. Acta Agr. Scand. 32(2): 167-170.

Glem-Hansen, N. (1992). Review of protein and amino acid requirements for mink. Scientifur 16(2): 122-142.

Glem-Hansen, N., Christensen, K. D. and Jorgensen, G. (1973). Undersoggelse af den indiviadelle varation i minkmaelkens indhold af torstof, raprotein og fedt samt indholdret af aminosyrer og fedsyrer ved brug aft forsellige foderblandinger. Lando/Konomisk Forsogslaboratoriums Arbog, Denmark. 1973: 260-267.

Glem-Hansen, N., Christensen, K. D. and Jorgensen, G. (1977). Content of different carbohydrate fractions related to digestibility of carbohydrates in diets for mink. Scientifur 1(4): 28-33.

Glem-Hansen, N. and Chwalibog, A. (1980). Influence of dietary protein, energy levels and environmental temperature on energy metabolism in adult mink. Proc. 2nd Int. Sc. Congr. In Fur Anim. Prod. Copenhagen. 10 pp.

Glem-Hansen, N.and Eggum, B. O. (1973). The biological value of proteins estimated from amino acid analyses. Acta Agr. Scand. 23: 247-251.

Glem-Hansen, N. and Eggum. B. O. (1974). A comparison of protein utilization in rats and mink based on nitrogen balance experiments. Z. Tierphysiol. Tierenahrg, und Futtermittelkde 33: 29-34.

Glem-Hansen, N. and Enggaard- Hansen. N. (1981). Amino acid deposition in mink during the growth period. Acta Agr. Scand. 31(4): 410-414.

Glem-Hansen, N. and Jorgensen, G. (1973). Proteinmae gdens og-kvalitetens indflydelse pa hvalpense tilvae kst i dieperioden hos mink. Forsogslaboratoriets Arbog Statens Husdyrbrugsforsog, Copenhagen, 1973: 267-272.

Glem-Hansen, N. and Jorgensen. G. (1973). Proteinmae gdens og-valitetens indflydelse pa hvalpense tilvae kst i dieperioden hos mink. Forsogslaboratoriets Arbog Statens Husdyrbrugsforsog,Copenhagen, 1973:267-272.

Glem-Hansen, N. and Jorgensen, G. (1973a). Description of technique used in N-balance experiments with mink. Proceedings From Symposium on Amino Acids, Brno, Czechoslovakia, 1973.

Glem-Hansen, N. and Jorgensen, G. (1973b). Determination of metabolic fecal nitrogen and the endogenous urinary nitrogen on mink. Acta Agr. Scand. 23: 34-38.

Glem-Hansen, N. and Jorgensen. G. (1974). Fat levels during the lactation period. En serie Rapporer om Forsog Med. Mink l973/1974. 422 Beretning Fra Forsogslaboratoriet 5-11. Fur Anim. Exp. Stat., Trollesminde.

Glem-Hansen. N. and Jorgensen, G. (1975a). Undersogelse af maelkens kemiske sammensaetning gennem lactations - perioden hos mink. 422 Beretning Fra Forsogslaboratoriet, Statens Husdyrbrugsforsog, Copenhagen, 1975: 12-18.

Glem-Hansen. N. and Jorgensen, G. (1975b). Forsog med forskellig energikoncentration til diegivende minkhunner. 422. Beretning Fra Forsogslaboratoriet, Statens Husdyrbrugforsog, Copenhagen 1975: 5-11.

Glem-Hansen, N. and Jorgensen, G. (1975c). Undersogelse af maelkens kemiske sammensaetning gennem laktationsperiod hos mink.422. Beretning Fra Statens Husdyrbrugsudvalg. : 12-18.

Glem-Hansen, N. and Jorgensen, G. (1976). Danes study nitrogen in feces, urinary system of the mink. U.S. Fur Rancher 56(9): 6-12.

Glem-Hansen, N. and Jorgensen, G. (1977). Fedstoffer fordojelighed og kemisk sammensaetning. Dansk Pelsdyravl 40: 253-255.

Glem-Hansen,N. and Mejborn, H. (1985). Personal communication.

Glem-Hansen, N. and Sorensen, P. B. (1981). Should grain (for mink) be ground finely? Dansk Pelsyravl 44(7): 295-297.

Gnaedinger, R. H. (1963). Problem of thiaminase in mink feed. National Fur News 35(7): 8.

Gnaedinger, R, H, (1965). Thiaminase activity in fish. An improved assay method. U.S. F. W. S., B. C. F., Fish. Ind. Res. 2 (4): 55.

Gorham.J.R.(1949). Red lead poisoning in mink. Amer. Fur Breeder 21(12):56

Gorham. J. R. (1950). Botulism in mink. Amer. Fur Breeder (June):9-34.

Gorham, J. R. (1951). Yellow fat disease in mink. American Fur Breeder 24(5): 12-27.

Gorham, J. R. (1962). Hemoglobinuria in mink. National Fur News 1962: 34(6):18-20.

Gorham, J.R. (2001). A student lecture on mistakes that I have made. Fur Rancher 82(1): 8-9.

Gorham. J. R., Baker, G. A. and Boe,. N. (1951). Observations on the etiology of yellow fat disease in mink. Preliminary Report. Vet. Med. 46(3): 100-102.

Gorham. J.R. and Dejong, D. H. (1950). Dehydration and salt poisoning in mink. Amer. Fur Breeder 23(6): 12.

Gorham, J.R. and Farrell. R. K. (1962). Salt poisoning in mink. Nat. Fur News 34(9): 17-33.

Gorham. J. R. , Hagen, K. W. and Farrell, R. K. (1972). Minks - diseases and parasites. Agr. Handbook No. 175,. U.S.D.A.

Gorham. J. R. and Nielsen, I.M. (1953). Some nutritional diseases of mink. Mich. State Coll. Vet. 13(2): 94-97.

Greaves, J. P., Scott, M. G. and Scott, P.P. (1959). Thyroid changes in cats on a high protein diet, raw hearts. J. Physiol. 148: 73pp.

Green, R.G. (1938). Chastek paralysis. American Fur Breeder XI (1): 4-8.

Green, R. G., Carlson, W. E. and Evans, C. A. (1942). The inactivation of vitamin B in diets containing whole fish. J. Nutr. 23(2): 165-174.

Gregorio, G. B. D. and Murphy, B.D. (1987). Furring in mink - the role of melatonin implants. Fur Rancher 67(1): 4,8,11.

Gregory, J. F. and Kirk, J.R. (1981). The bioavailability of vitamin B-6 in foods. Nutr. Rev. 39: 1-8.

Greig, R. A. and Gnaedinger, R. H, (1971). Occurrence of thiaminase in some common aquatic species of the United States and Canada. U.S. D. C., NOAA, Natl. Mar. Fish Serv. Spec. Sci. Rep. No. 631. Seattle, Wash.

Gross, O. (1955). Personal communication.

Gupta, S.S. and Hilditch, T. P. (1951). The component acids and glycerides of a horse mesenteric fat. Biochem. J. 48: 137-146.

Gunn, C. K. (1960). Dehydrated mink ration tests during growth and furring season. National Fur News 32(8): 10.

Gunn, C. K. (1964a). Aetiolgy (cause) of wet belly in ranch bred mink. Amer. Fur Breeder 37(1): 16-17.

Gunn, C.K. (1964b). Experimental control of wet belly disease. Amer. Fur Breeder 37(7): 14, 34.

Gunn. C. K. (1966). Wet belly disease in mink. Fur Trade Journal of Canada 38(12): 10-11.

Haarrison, D. C. and Mellanby, E. (1939). Phytic acid and the rickets producing action of cereals. Biochem. J. 33: 1660.

Hallinan, F. J. (1958). Phosphoric acid in the mink ration - Why, How, When. Nat. Fur News 30(4): 33.

Hamilton, W. J. Jr. (1936). The food of wild mink. Amer. Fur Breeder 8(12):14.

Hansen, B.K. (1997). The lactating mink (Mustela vison) - Genetic and metabolic aspects. Doctoral Thesis, Dept. of Anim. Sc. And Anim. Health, the Royal Veterinary and Agricultural University, Copenhagen, Denmark.

Hansen, B.K. and Berg, P. (1998). Mink dam weight changes during the lactation period. I. Genetic and environmental effects. Acta Agr. Scand. Sec. A. Anim. Sc. 48: 49-57.

Hansen, B.K., Lohi, O. and Berg, D. (1992). Correlation between the development of mink kits in the lactation and growth periods, correlations to fur properties and heritability estimations. Norwegian J. of Agr. Sc. Suppl. No. 9, 1992. Prog. In Fur Animal Sc. Proc. From the Vth Inter. Sc. Congr. In Fur Anim. Prod., Oslo, Norway: 87-92.

Hansen, N.E. (1974). The content of minerals in herring meal. Z. Tierphysiol Tierernahrg u. Futtermittelkde 32: 233-239.

Hansen, N.E. (1978). The influence of sulfuric acid preserved herring on the passage of time through the gastro-intestinal tract of mink. Zeitschrift Fur Tierphysiologie Tierernarung and Futtermittelkunde 40: 285-291.

Hansen, N.E. (1982). Methods and analyses in determination of mink carcass composition by slaughter techniques. Acta Agr. Scand 32(3): 305-307.

Hansen, M.V., Lassen, M., Tauson, A-H, Sorensen, H. and Clausen, T.N. (2004). Different ratio between n-6 and n-3 fatty acids for lactating mink (Mustela vison) dams - effect on milk and kit tissue fatty acid composition. Scientifur 28(3): 103-109.

Hansen, N.E. (1979). Fiskeen silagens indflydelse pa knogleuduiklingen hos mink (The effect of fish silage on the ossification in mink). NJF Meeting, Kungalv, Sveden (1979).

Hansen, N.E. and Glem-Hansen, N. (1980). Digestibility of crude protein, crude fat and carbohydrates in growing mink related to feeding sulfuric acid preserved fish. Dansk Pelsdyravl 43: 59-61.

Hansen, N.E., Finne, L.,Skrede, A. and Tauson, A-H. (1991). Energy Supply For mink and fox. Nordic Assoc. of Agricultural Scientists, NJF-Utredning rapport no. 63, DSR forlog, Den. Kgl. Veterinaer. og Landbohotskole, Copenhagen, Denmark. 59 p.

Hansen, O., Clausen, T. N. and Wamberg, S. (1993). The etiology of nursing sickness. Electrolytes in normal lactating mink towards the end of the nursing period. Dansk Veteraetid Skrift 76(20): 877-880.

399

Hargrove, D. M., Rogers, Q. R. and Morris, J. G. (1984). Leucine and isoleucine requirements of the kitten. Br. J. Nutr. 52:595-605.

Harman, R. D. (1954). Some observations on urinary calculi in foxes and mink. Personal communication, Virginia mink rancher.

Harper, R. B., Travis. H. E. and Glinsky, M.S. (1978). Metabolizable energy requirements for maintenance and body composition of growing farm-raised male pastel mink (*Mustela vison*). J. Nutr. 108: 1937-1943.

Harris, E. P. L. and Embree, N.D. (1963). Quantitative consideration of the effect of polyunsaturated fatty acid content of the diet upon the requirements of vitamin E. Am. J. of Clin. Nutr. 13: 385-398.

Harris, L. E., Bassett, C.F. and Wilke, C .F.(1951). Effect of various levels of calcium, phosphorus and vitamin D intake on bone growth. I. Foxes. J. Nutr. 43: 153-165.

Harris, L. E., Cabell, C. A., Elvehjem. C. A., Loosli, J. K. and Schaefer, H.C. (1953). Nutrient Requirements For Domestic Animals. Number VII. Nutrient Requirements For Foxes and Minks. Publication 296. Division of Biology and Agriculture, National Research Council, 2101, Washington 25, D.C.

Harris. R.S. (1970). Bladder stones - mink water quality? Am. Fur Breeder 43(4):4.

Hartsought, G. R. (1951). Personal communication. Veterinarian to the North American fur industry.

Hartsought, G. R. (1955). Nursing sickness - is it a lack of salt? Amer. Fur Breeder 28(5): 10, 42-44.

Hartsought, G. R. (1960). Nursing sickness basically a salt depletion. Fur Trade J. of Canada, Sept. 1960. 38(1): 14-12.

Hartsought. G. R. (1974). Personal communication. N. Am. Vet.

Hartsought,. G.R. and Gorham, J.R. (1949). Steatitis (yellow fat) in mink. Vet. Med. 44(8): 345-346.

Hartsought, G. R. and Mason, K.E. (1951). Steatitis or "yellow fat" in mink and its relation to dietary fats and inadequacy of vitamin E. J. Amer. Vet. Med. Assoc. 119:72-75.

Hassam, A. G., Rivers, J. P. W. and Crawford, M.A. (1977). The failure of the cat to desaturate linoleic acid: Its nutritional implications. Nutr. Metab. 21: 321-328.

Havre, G. N., Helgebostad, A. and Ender, F. (1967). Iron resorption in fish-induced anemia in mink. Nature 215: 187-188.

Havre, G.N., Helgebostad, A. and Ender, F. (1973). The influence of peroxidized unsaturated fat and vitamin E on hemoglobin formation and iron absorption. Nord. Vet. Med. 25: 79-82.

Hayes, K.C., Carey, R.E. and Schmidt, S. J. (1975). Retinal degeneration associated with taurine deficiency in the cat. Science 188: 949-951.

Heger Co. (1998). Ingman Lab. Inc., Minneapolis, Mn.

Hegreberg, G.A., Hamilton, M.J., Camacho, Z. and Gorham, J.R. (1974). Biochemical changes of muscular dystrophy of mink. Clin. Biochem. 7:313-319.

Hegreberg, G. A., Hamilton, M.J. and Padgett, G. A.(1976). Muscular dystrophy of mink. A new animal model. Fed. Proc. 35(5): 1218-1224.

Heinz Handbook of Nutrition (1959). McGraw-Hill Book Co., Inc. p. 31.

Hejlesen, C. (2002). Digestibility of nutrients by young mink (*Mustela vison*) kits and adults. NJF-Seminar No. 347: 53-61. Vuokatti, Finland.

Hejlesen, C. (2003). Digestibility of feed ingredients of mink kits and adults - A comparison. Annual Report 2002: 51-57. Danish Fur Breeders Research Center, Holstebro, Denmark.

Hejlesen, C. and Clausen, T. N. (2000). Different energy distribution in the feed for mink females in the winter and reproduction period. Scientifur: 24(4). No. 4-IV-A Nutrition. Proceedings of the VIIth Inter. Sc. Congr. In Fur Anim. Prod. Ed. B. D. Murphy and O. Lohi: 74-77.

Hejlesen, C. and Clausen, T. N. (2001a) Phase feeding mink in the growth-furring period. Annual Report. 2000: 77-80. Danish Fur Breeders Research Center, Holstebro, Denmark.

Hejlesen, C. and Clausen, T.N. (2001b). Energy distribution in mink feed in the winter and reproduction periods. Annual Report 2000: 43-50. Danish Fur Breeders Research Center, Holstebro, Denmark.

Hejlesen, C. and Clausen, T.N. (2001c). Toasted soya beans for mink in the growing-furring period. Annual Report 2000:73-76. Danish Fur Breeder Research Center, Holstebro, Denmark.

Hejlesen, C. and Clausen, T. N. (2002). Energy distribution in mink feed in the winter and reproduction periods. Annual Report 2001: 69-75. Danish Fur Breeders Research Center. Holstebro, Denmark.

Hejlesen, C., Clausen, T.N. and Therkildsen, N. (1998). Phase feeding of mink kits in the growth period (protein reduction). Scientifur 22(2): 144.

Helgebostad, A. (1955). Experimental excess of vitamin A in fur animals. Nord. Vet. Med. 7: 297-308.

Helgebostad, A. (1957). Fiskeforingens innflytelse pa pelsutvikling og sunnhetstilsand hos mink. Norsk Pelsdyrblad 31: 114-121.

Helgebostad, A. (1961). Fortsatte studier over forinsbetinget anemi hos mink. Norsk Pelsdyrblad 35: 396-409, 420-429.

Helgebostad, A. (l963). Tomme tisper hos mink ved esperimentall vitamin B-6 mangel. Norsk Pelsdyrblad 37: 211-213.

Helgebostad, A. (l966). Anemia in standard mink and mutation mink. Nord Vet. Med. 18: 347-356.

Helgebostad, A. (1967). Vitamin E problems in mink. Dans Pelsdyraul 30(12):8-10.

Helgebostad, A. (1968). Vitamin B-1 mangel hos mink og rev I. Forbindelse` (Gadiculus thori). Norsk Pelsdyrblad 42: 619-622.

Helgebostad, A. (1971). Vitamin E mangel syndrome hos pelsdyr. Norsk Vet. Tidske. 83: 11-17.

Helgebostad, A. (1973). Torskehoder og seivskjaer som fur til mink og blarev. Norsk Pelsdyrblad 47: 321-325.

Helgebostad, A. (1974). Vitamin E mangel hos mink valper for avvenning. XII. Nord Vet. Congr. 1974. Proceedings: 193-196.

Helgebostad, A. (1976). Recent aspects of vitamin E deficiency in fur-bearing animals. The lst Inter. Sc. Congr. In Fur Anim. Prod. Helsinki, Finland. April 25-27, 1976.

Helgebostad, A. (1976). The formaldehyde content in fish in relationship to anemia in mink. Nord. Vet. Med. (2) 28: 108-114.

Helgebostad, A. (1980). Embryonic death in mink due to riboflavin deficiency. Nord. Vet. Med. 32: 313-317.

Helgebostad, A. and Dishington, J. W. (1977). Thiamin-mangel hos pelsdyr som folge as antithiaminfakturer i foret. Norsk Pelsdyrblad 51: 155-157.

Helgebostad, A. and Ender, F. (1955). Effect of feeding on pelt development in fox and mink. V. Marine fat - the cause of discoloring in the pelt. Norsk Pelsdyrblad No. 8, 20 Apr. 1955, Arg. 29: 139-147.

Helgebostad, A. and Ender, F. (1961). Nursing anemia in mink. Acta Vet Scand. 2:236-245.

Helgebostad, A. and Ender, F. (1973). Vitamin E and its function in the health and disease of fur-bearing animals. Acta Agr. Scand. Suppl. 19: 79-82.

Helgebostad, A., Gjonnes, B. and Svenkerud, R. A. (1961). Nursing anemia in mink, symptoms, pathology and therapy. Nord Vet. Med. 13: 593-603.

Helgebostad, A. and Martinsons, E. (1958). Nutritional anemia in mink. Nature: 181: 1660-1661.

Helgebostad, A. and Nordstogen, K. (1978). Hypervitaminosis D in fur-bearing animals. Nord. Vet. Med. 30: 451-455.

Helgebostad, A., Svenkerud, R. R. and Ender, F. (1952). Experimental biotin deficiency in mink and foxes. Nord. Vet. Med. 4:141-161.

Helgebostad, A., Svenkerud,. R. R. and Ender. F. (l963). Sterility in mink induced experimentally by deficiency of vitamin B-6. Acta Vet. Scand. 4: 228-237.

Hellewing, A.I.F. and Tauson, A-H. (2002). Bioprotein meal, a new protein source for mink. NJF Seminar No,. 347: 103-107. Vuokath, Finland, Oct. 2-4,

Henriksen, P. (1980). The tyrosine-emia syndrome in mink. 2nd Inter. Sc. Congr. In Fur Anim. Prod., Denmark, April 8-10, 1980.

Henriksen, P. (1984). Nutritional muscular degeneration syndrome in mink (clinical and pathological observations). 3rd Inter. Sc. Congr. In Fur Animal Production. Versailles, France. 25-27 April 1984.

Henriksen, P. and Elling, F. (1986). Nursing sickness in mink. Scientifur 10:79.

Hillard, M. (1977). Personal communication. Ontario mink rancher.

Hillemann, G. (1972). Beretning over forsogens pa Nordjysk Pelsdyrforsogsfarm sommeren 1971. Dansk Pelsyravl 35: 166-168.

Hillemann, G. (1976). Forsog me sojamel. Danish Pelsyravl 39: 124-126.

Hillemann, G. (1977). Forsog med forskellige sojaprodukter. Stenciled report given at a meeting of NJF's Fur Animal Division, Hurdalsjoen, Norway. 28-29 Sept. 1977. 4 pp.

Hillemann, G. (1978a). Forsog med sojamel. Dansk Pelsdyravl 41: 173-174.

Hillemann, G. (1978b). Forsog med fiskeolie og vitamin D. Dansk Pelsdyavl 41: 245-246.

Hillemann, G. (1981a). The effect of soya protein conc. - soycomil K - on mink in the growth phase. Dansk Pelsdyrav144(7): 304-305.

Hillemann. G. (1981b). Denmark research "soycomil K" - af konsulent G. Hillemann. Nordjysk Pelsdyrforsogfarm. A.M.B.A. 1980 Studies.

Hillemann, G. (1982). NJF-Seminar No. 42, Alesund, Norge.

Hillemann, G. (1984). Trials with Danish potato protein for mink. Dansk Pelsdyravl 47(12): 695-697.

Hillemann, G. and Lyngs, B. (1989). Energifordelinger til mink i vaekstperioden. Faglig Arsberetning 1988: 70-97.

Hillemann, G. and Mejborn, H. (1983). Different types of fat for mink. Dansk Pelsdyravl 46(7): 383-385.

Hjertnes, T., Gulbrandsen, H., Mundheim, H. and Opstvedt, J. (1988). Norsemink and Norse-LT - two special qualities of fish meal for fur animals. Biology, Pathology and Genetics of Fur Bearing Animals. Ed. B.D. Murphy and D. B. Hunter. Proceedings of the 4th Inter. Sc. Congr. In Fur Animal Production. Rexdale, Ontario, August 21-24.

Hodson, A. Z. and Maynard, L. A. (1938). Digestion and metabolism studies with mink. American Fur Breeder (11) Jan: 20-24.

Hodson, A.Z. and Smith, S.E. (1942). Estimated maintenance energy requirements of foxes and mink. Fur Trade J. of Canada 19(6): 12-15.

Hodson, A. Z. and Smith, S. E. (1945). Estimated maintenance energy requirements of foxes and mink. Amer. Fur Breeder XVIII (4): 44-52.

Hoie, J. (1953a). On soyacake and soyameal as food for fur animals. Norsk Pelsdyrbland, No. 11 and No. 16.

Hoie, J. (1953b). Om soyakaker og soyamjol som for till fjorfe og pelsdjur. Norsk Lantbruk 19: 384-387.

Hoie, J. (1954). Experiments with different amounts of fat and carbohydrates in the food of mink kits. Norsk Pelsyrblad 28(10/11): 175-183.

Hoogerbrugge, A. (1968a). University of Utrech. Netherlands. Unpublished data.

Hoogerbrugge, A. (1968b). 1928-1968 Jaar NFE Symposium - Forty Years of the Dutch Fur Breeders Assoc., Rotterdam, Netherlands. Sept. 24, 1968.

Hornshaw, T. C, Aulerich, R. K., Ringer, R.K. and Martin, M.B. (1991) Mineral concentration in the hair of natural dark and pastel mink. Research Report 508. Mink and Poultry Research 1990: 46-53.

Howell, R.F. (1951). A study in dry rations for mink. Nat. Fur News 23(11): 26-37.

Howell, R. F. (1955). Kellogg Mink Research Ranch. Personal communication.

Howell, R.F. (1957). Inquisitive guests attend nutrition conference. Amer. Fur Breeder 30(7): 16.

Howell, R. F. (1966). Product Information Bulletin. Kellogg Co., Battlecreek, Michigan.

Howell, R. F. (1975). Milk of the mink - what it tells us. Blue Book of Fur Farming 1975: 26-30.

Howell, R. F. (1976). How mink kits use food intake in growing, furring. Blue Book of Fur Farming 1976: 36-41.

Hummon, O. J. and Bushnell, F. R. (1943). Studies of cotton mink with reference to heredity and blood elements. Amer. Fur Breeder XV(5):30-34.

Hunter, D. B. and Lemieux, N. (1996). Mink ... Biology, Health and Disease. Graphic and Print Services, University of Guelph, Guelph. Ontario, Canada N1G 2W1.

Hunter, D. B. and Schneider, D. (1991). A new look at nursing disease and what can be done to curtail it. Fur Rancher 71(7&8): 3-4.

Ikeda, M., Tsuji, H., Nakamura, S., Ichiyama, A., Nishizuke, Y. and Hayaish, O. (1965). Studies on the biosynthesis of nicotinamide adenine dinucleotide. II. A role of picolinic carboxylase in the biosynthesis of NAD from tryptophan in mammals. J. Biol. Chem. 240: 1395-1401.

Iverson, J. A. (1972). Basal energy metabolism of mustelid. J. Comp. Physiol. 81: 341-344.

Jarl, F. (1952). Feeding experiments with toasted (heat-treated) cereal for fur animals. Kungl Lantbrukshogsk, Statens Husdjursforsok Sartryck Forhandsmedd No. 86. Vara Palsdjur 1952. No. 4.

Jensen, L. V. and Lohi, O. (1988). Correlation between mineral content of feed and mink hair and the relation of mineral content to fur characteristics. Biology, Pathology and Genetics of Fur Bearing Animals, ed. Murphy, B.D. And Hunter, D. B. Proceedings of the 4th International Congress In Fur Animal Production, Rexdale, Ontario, Canada August 21-24, 1988.

Jensen, P. and Jorgensen, G. (1975). Fremstilling or anvendelse af fiskeensilage specielt til mink. 427 Beretning Fra Statens Husdjurbrugsforsog. 79 pp. Dept. DF Fur Anim. 48H Roskilddevej, 3400 Hilleroed, Denmark.

Jensen, S.K. (2004). Transfer of vitamin E from mink mothers to kits via the milk. Meeting at DIAS Research Centre Foulum, 30 September, 2004 on the subject of research in relationship to practical mink production. Internal Report 2004, No. 208.

Jernelov, A., Johansson, A.H., Sorensen, L. and Svenson, A. (1976). Methyl mercury degradation in mink. Toxicology 6:315.

Johansson, A-H., Alden, E. and Soderdahl, K. G. (1977). Cooked fish silage in combination with fish meal or micronized soybean meal as mink food. Vara Palsdjur 48(4): 123-125.

Johnsen, F. and Skrede, A. (1981). Evaluation of fish viscera silage as a feed resource. Chemical characteristics. Acta Agr. Scand. 31: 21-28.

Johnson, V. (1988). Personal communication. Minnesota mink rancher.

Jones, A. (2001). Personal communication. Ontario mink rancher.

Jones, R. E., Ringer, R. K. and Aulerich, R. J. (1980). A simple method for milking mink. Fur Rancher 60(9): 4, 6.

Jones, R.E., Aulerich, R.J. and Ringer, R.K. (1982). Feeding supplemental iodine to mink. Reproduction and histopathologic effects. J. Toxicology and Environmental Health 10: 459-471.

Jones, W. G. (1960). Fishery Leaflet 501,. Bureau of Comm. Fish. Fish and Wildlife Svc Dept. Inter. p. 228.

Jorgensen, G. (1960). Composition and nutritive value of mink milk. Dansk Pelsdyravl 23(2): 37-38.

Jorgensen, G. (1965a). Forsok med tilskud af vitamin B-6 (pyridoxin) til et alsidigt sammensat avlsdyrfoder. Dansk Pelsdyravl 28:132-134.

Jorgensen, G. (1965b). Forsog me ensileret sild og hvilling i minkvalpenes foder. Dansk Pelsdyravl 28: 184-190.

Jorgensen, G. (1967): Minkens energi-och protein behov. Vara Pelsdjur 38(1):20-24.

Jorgensen, G. (1972). Jernbehov og jerntilskud til mink. Dansk Pelsdyravl 35: 160-162.

Jorgensen, G.(1977a).Vitamin D content in various species of fish and its influence on vitamin D content of mink feed. Dansk Pelsdryravl 41:138-139.

Jorgensen, G. (1977b). Thiaminase activity in fish silage. Dansk Pelsdryravl 40: 166-167,

Jorgensen, G. (1978a). The Use of Soybean Products in Feeding For Fur Bearing Animals. A Review of the Recent Work. Report given to the Amer. Soybean Assoc. Nov. 1978., Vest Farm 1974. Unpublished data.

Jorgensen, G. (1978b). Personal communication. Danish researcher.

Jorgensen, G. (1979). Future prospects for soy feeds in the diets of mink and foxes. Fur Rancher 59(3): 8.

Jorgensen, G. (1985). Mink Production, Scientifur, Hilleroed, Denmark.

Jorgensen, G., Baek-Olsen, K. and Glem-Hansen, N. (1971). Undersogelse af forskellige konserveringsmetoders egnethed til konservering af sild, der skal anvendes i minkfoder.Forsogslaboratorets Arbog,Copenhagen 1971:271-277.

Jorgensen, G., Brandt, A., Tauson, A-H., Rimeslatten, H., Skrede, A., Juokslahti, T., Valtonen, M., Utne, F. and Carlsson, J. (1987). Vitamins In The Nutrition Of Fur Bearing Animals. F. Hoffman-LaRoche and Co. Ltd. Basle, Switzerland.

Jorgensen, G. and Christensen, R. (1966). Relationship between hemoglobin values and fur properties in mink. Nordisk Veterin Aemedican 18: 166-173.

Jorgensen, G. and Clausen, H. J. (1964). Forsog med tilskud af forsikellige ma ngder af aminosyrerne lysine or methionine til minkhvalpense normalfoder. Bilag Til Mode I. NJF's subsektion for pelsdyr. Ultuna, Sweden.

Jorgensen, G. and Clausen. H. J. and Petersen, F. H. (1962). Fodringsforsog med mink 1960. 1a-24. Experiments with addition of protein and carbohydrate cleaving enzymes in mink kit rations. #322 Beretning Fra Forsogslaboratoriet.

Jorgensen, G. and Eggum, B. O. (1971). Minkskindets opgygning. Dansk Pelsdyrval 34: 261-267.

Jorgensen, G., Enggaard-Hansen, N., Eggum, B.O. and Glem-Hansen, N. (1976). Kemisk sammensaetning og fidervaerdi af torret alca-laseslam or pencillincelium til mink I or II. Meddelse nr. 91 or 92 fra Statens Husdrybrugsforsog, Copenhagen, Denmark, 1976.

Jorgensen, G. and Glem-Hansen, N. (1970). Forsog med forskellig protein concentration kombineret me tilsaetning of methionine. Stenc Bilag Til Forsogslaboratoriets Arsmode, Copenhagen, Denmark.

Jorgensen, G. and Glem-Hansen, N. (1972). Forsog med forskellige proteinmageder til mink. Dansk Pelsyravl 35: 15-23.

Jorgensen. G. and Glem-Hansen, N. (1973a). A cage designed for metabolism and nitrogen balance trials with mink. Acta Agr. Scand. 23(1): 3-4.

Jorgensen, G. and Glem-Hansen, N. (1973b). Fedtsyresammensaetningens Indflydelse pa fedstoffeernes fordojelighed. Forsogslaboratoriets Arbog, Statens Husdyrforsog, Copenhagen: 250-259.

Jorgensen, G. and Glem-Hansen, N. (1973c). Kornarternes fordojelighed efter forskellig behandling. Forsogs Laboratoriets Arbor Husdrybrugsforsog, Copenhagen, 1973: 285-288.

Jorgensen, G. and Glem-Hansen, N. (1973d). Indflydelsen af kogning med eddikesyre pa kornarternes fordojehighest til mink . Forsogslaboratoriets Arbog, Statens Husdyrbrugsforsog, Copenhagen, 1973: 289-292.

Jorgensen, G. and Glem-Hansen, N. (1974). Forsog med mink 1972/1973 Pelsdyrforsogsfarmen "Trollesminde", Danish Pelsdyravl 37(1): 27-29.

Jorgensen, G.. and Glem-Hansen. N. (1975a). Kemisk sammensatning og fordojelighed af majsgluten mel. Meddelse NR 73 fra Statens Husdyrbrugs-Forsog.

Jorgensen, G. and Glem-Hansen, N. (1975b). Traestofindholdet og formalingsgradens indflydelse pa fordojeligheden af sojaskra til mink. Meddelse Nr. 76.1-4.Fra Statens Husdyrbrugsforsog, Copenhagen, Denmark.

Jorgensen, G. and Glem-Hansen, N. (1975c). Fordojelighedforsog med rene stivelsesarter. 422 Beretning fra forsogslaboratriet Statens Husdyrbrugforsog, Kobenhagen. 1975:28-36.

Jorgensen, G. and Glem-Hansen, N. (1975d). Sammensaetning og fordojeligheden af forskellige sajaprodukter. 422 Beretning Fra Forsogslaboratoriet.: 19-27.

Jorgensen, G. and Glem-Hansen, N. (1979). Forsog med forskellig proteinkoncentration kombinet med tilsaetning methionin. Stencileret Bilag Til Statens Husdyrbrugsforsogs Arsmode, Copenhagen, Denmark.

Jorgensen, G., Glem-Hansen, N. and Moller-Jensen, P. (1975). Methanoldyrket bacterieprotein til mink. 422 Beretning fra forsogslaboriet Statens Husdyrbrugforsog, Copenhagen, Denmark 1975: 55-66.

Jorgensen, G. and Hillemann, G. and Clausen, H. J. Forsog med forskellige maegder fedt og kulhydrater i foderet til diegivende mink. 341. Beretning fra forsogslaboratoriet, Statens Husdyrbrugsforsog, Copenhagen: 3-8.

Jorgensen, G., Hillemann, G. and Clausen. H. J. (1963). Forsog med forskellige maegder fedt og kuhlhydrater i foderet til diegivende mink. 341 Beretning fra Forsogslaboratoriet, Statens Husdyrbrugsforsog, Copenhagen : 3-8.

Jorgensen, G., Hjarde, W. and Lieck, H. (1975). Intake, deposition and excretion of thiamine, riboflavin and pyridoxine in mink. 422 Beretning fra Forsogslaboratoriet: 37-54.

Jorgensen. G., Poulsen, J. S. D. and Bindixen, P. (1976). The influence on breeding, production and acid-base balance when mink are fed acidified feed. Nord. Vet. Med. 28: 592-602.

Juokslahti, T., Lindberg, T. D., Pekkanen, J. and Sjogard, B. (1978). Choline chloride in the treatment of fatty liver in mink. Scientifur 2(4): 35-38.

Juokslahti, T., Lindberg, P. and Tyopponen, J. (1980). Organ distribution of some clinically important enzymes in mink. Acta Vet. Scand. 21: 347-353.

Juokslahti, T. and Tanhuanpaa, E. (1984). Fatty acid composition of mink feed. 3rd Inter. Sci.Congr. In Fur Anim. Prod. Versailles, 25-27, April.

Just-Nielsen, A. (1975). Feed evaluation in pigs. Festskrift til Hjalmar Clausen Det Kgls. Danske Landusholdningsselskib, Copenhagen,Denmark : 199-220.

Kacmar. P., Samo, A. and Knezik, J. (1980). Chemical diagnostics of sodium chloride poisoning in thoroughbred fur-bearing animals (fox and coypu) and in turkey and pheasants. Veterinarni Medicina 25 (12): 733-738.

Kainer, R. A. (1954a). The gross anatomy of the digestive system of the mink. I. The headgut and the foregut. Amer. J. of Vet. Res. 15: 82-90.

Kainer, R. A. (1954b). The gross anatomy of the digestive system of the mink. II. The midgut and the hindgut. Amer. J. of Vet. Res. 15: 91-97.

Kakela, R., Polonen, I., Miettinen, M. and Asikainen, J. (2001). Effects of different fat supplements on growth and hepatic lipids and fatty acids of mink. Acta Agr. Scand. 51(1): 217-223.

Kamikawa, E. M., Morris, J. G. and Rogers, Q. R. (1984). Upublished data.

Kangas, J. (1968). Anemia -white underfur. Nat. Fur News 40(2):12-13.

Kangas, J. (1976). Minerals in mink feed and factors affecting the mineral balance. 1st Inter. Sci. Congr. In Fur Anim. Prod. Helsinki, Finland.

Kangas, J. and Makela, J. (1972). The influence of feeding packing plant by-products containing thyroids and parathyroids on the reproduction of rats and mink. Nord. Vet. Med. 24: 162-170.

Kelly, W. H. and Kelly, J. R. (1949). U.S. Patent Office. 2,558,092 in II, 1949 Serial No.70,374. Feed for carnivorous animals and methods for making same.

Kemp, B., Martens, R. P. C. H., Hazeleger, W., Soede, N. M. and Noordhuizen. J. P. T. M. (1993a). The effect of different feeding levels during pregnancy on subsequent breeding results in mink (*Mustela vison*). J. Anim. Physiol. A. Anim. Nutr. 69: 115-119.

Kemp, B., Martens, R. P. C. H., Soede, N. M. and Noordhuizen, J. P. T. M. (1993b). The effect of slimming on embryo surivival rate and the number of corpora lutia at 25 days of pregnancy in minks (*Mustela vison*). Reprod. Dom. Anim. 28: 73-76.

Kennedy, A. H. (1946). Nutritional anemia in minks. Canadian Silver Fox and Fur (April): 12.

Kerminen-Hakkio, M., Dahlman, J., Niemela, P., Jalava, T., Rekila, T. and Syrjala-Quist, L. (2000). Effects of dietary protein levels and quality on growth rate and fur parameters in mink. Scientifur 24(4). IV-A-Nutrition: 7-12. Proc. Of the VIIth Inter. Sci. Congr. In Fur Anim. Prod. Ed. B. D. Murphy and O. Lohi.

Kettelhut, I. C., Foss, M. C. and Migliorini, R. H. (1980). Glucose homeostasis in a carnivorous animal (cat) and in rats fed a high protein diet. Am. J. Physiol. 239: R115-R121.

Kielanowski, J. and Kotarbinska, M. (1070). Further studies on energy metabolism in the pig. Proc. 5th Symp. On Energy Meta. Vitznau: 145-148.

404

Kifer, P.E. (1956). Successful dry rations for use during late growth and maintenance periods of ranch-raised mink. M.S. Thesis. Mich. State University.

Kifer, P.E. and Schaible, P. J.(1955a). Dry diets for ranch mink. I. Suitability for late growth and maintenance. Quart. Bull., Mich. Agr. Exper. Stat. Michigan State College 37(4):592-598.

Kifer, P.E. and Schaible, P. J. (1955b). Dry diets for ranch raised mink. II. Effect on breeding, whelping and early growth of kits. Quart. Bull. Mich. Agr. Exper. Stat. Michigan State College 38(1): 129-138.

Kifer, P.E. and Schaible, P.J. (1955c). A report on an experiment with three dry diets. American Fur Breeder 28 (12): 20-22, 50.

Kifer, P.E. and Schaible, P.J. (1956). Further developments in formulating dry rations for mink. Quart. Bull. Mich. Agr. Exper. Stat., Michigan State College 39(1): 17-24.

Kiiskinen, T. (1984). Apparent and true digestibility of protein and availability of amino acids in some dry protein ingredients. The 3rd Inter. Sc. Congr.in Fur Anim. Prod., April 25-27, Versailles, France.

Kiiskinen, T., Huida, L., Pastuszewska, B. and Berg, H. (1985). Digestibility of nitrogen and amino acids of some dry feedstuffs for mink. Ann. Agric. Fenniae: 107-114.

Kiiskinen, T. and Makela, J. (1975). Om sojamjolets anvandlnings mogligheter som foder til mink. Finsk Palstidskrift 9: 476-485.

Kiiskinen, T. and Makela, J. (1977). The results of mineral supplementation in the feeding of mink. Scientifur 1:41.

Kiiskinen, T. and Makela, J. (1981). Digestibility of pellets on mink and fitch. Scientifur 5(2): 36-41, Finsk Palstidskrift 15: 193-195 and Dansk Pelsdyravl 44: 211-213.

Kiiskinen, T. and Makela, J. (1991). Digestibility of some industrial by-products in mink. Scientifur 15(4): 335.

Kiiskinen, T., Makela, J. and Kangas, J. (1974). Soja som protein tilsats for palsdjur. Finsk Palstdskrift 8: 63-66 and Vara Palsdjur 45: 88-95.

Kilgore, L. (1974). Personal communication. Ohio mink rancher.

Kinsella, J. E. (1971). Biochemical analyses of the lipids of mink milk. Inter. J. Biochem. 2: 6-10.

Kirk, R.. J. (1962a). Raw rough fish and raw wheat in the growing and furring diet for mink kits. Fur Trade J. of Canada 39(3): 10-11.

Kirk, R. J. (1962b). Raw rough fish and uncooked cereal in the mink breeding and whelping diet. Fur Trade J. of Canada 39(7): 10-14.

Kjertnes, T.. Gulbrandsen, K. E., Mundheim, H. and Opstvedt. (1980). Norseamink and Norse-LT -two special qualities of fish meal for fur animals. Biology, Pathology and Genetics of Fur Bearing Animals. Ed. B.D.Murphy and D. B. Hunter. Proc. Of the 4th Inter. Sci. Congr. In Fur Anim. Prod., Rexdale, Ontario, Canada, August 21-24.

Kleiber, M. (1961). "The Fire Of Life: An Introduction to Animal Energetics". Wiley. N.Y.

Kleiber, M. (1975). "The Fire Of Life". Robert E. Krieger Publ. Co. Inc. Huntington, N.Y. p. 453.

Klevesahl, V. (1968, 1970). Personal communication. Georgia mink rancher.

Klingener, D. (1998). Customized Complete Laboratory Anatomy of the Mink 2/e Laboratory Manual. McGraw Hill Companies, Primis Custom Publishing. 56p.

Klingener, D. (2000). 2nd ed. Laboratory Anatomy of the Mink. Wm C. Brown Co., Publishers, Dubuque, Iowa.

Kneuven, C. (2000). Sodium bisulfate, a potential new acidifier for the pet food industry. Petfood Industry. June 2000.

Koch, J. F. (1940). More about bladder stones. Fur-Fish Game 72(5): 41-42.

Kon, S. K. and Cowie, A. T. (1961). Milk, The Mammary Gland and Its Secretions. Vol. 2. Academic Press, N.Y., New York.

Konrad, J., Hanak, J., and Mouka, J. (1973). The clinical and etiological aspects of urolithiasis in mink (Luterola vison). Veterinarni Med. 18(9): 533-540.

Konsulenternes Foderudvalg, Vinterens Fording (1976). Dansk Pelsdyravl 39: 415.

Knox, B. (1977). Nyt fodringsbetinget sygdomssyndrom pa Danish mink farme. Dansk Vet. Tidsskrift 60(5): 196-199.

Koppang, N. (1974). Dimethylnitrosoamine formation in fish meal and toxic effects on pigs. Amer. J. Path. 74(1): 95-108.

Koppang. N. and Helgebostad, A. (1972). Aflatoxin intoxication in mink. Nord. Vet. Med. 24: 213-219.

Koppang, N. and Rimeslatten, H. (1976). Toxic and carcinogenic effects of nitrosodimethylamine in mink. Inter. Agency for Research in Cancer 14: 443-452.

Korhonen, H., Mononen, J., Haapanen, K. and Harri, M. (1991). Factors influencing reproductive performance, kit growth and pre-weaning survival in foxes and mink. Scientifur 15: 43-48.

Krieger, F. (1962). Central Fur Foods Co-Op, Wadsworth, Ill. Personal communication - Wisconsin mink rancher.

405

Kubin, R. and Mason, M. M. (1948). Normal blood and urine values for mink. The Cornell Vet. 38(1): 79-85.

Kumeno, F., Itoyama, K., Hasegawa,.J. and Aoki, S. Nippon Formula Feed Mfg. Co., Japan (1970). Effect of protein and fat levels on complete pelleted diets on the growth of mink kits. J. Anim. Sc. 31(1): 894-899.

Kurhajec,T. (2000). Personal communication. Wisconsin mink rancher.

Kuusi, T. (1963). On the occurrence of thiaminase in Baltic herring. Valtion Teknillinen Tutkimuslaitos. The State Institute for Technical Research, Helsinki, Finland. Tiedotus Sarja IV-Kemia 52.

Kuznecov, G. A. and Diveeva, G. M. (1970). Hereditary predisposition to wet belly in mink. Inst. Puchn. Zverovod Krolokovod 9: 18-24.

Kuznecow, G. A. (1976). Outlines of scientific research work on fur animals in USSR. 1st Inter. Sci. Congr. In Fur Anim. Prod., Helsinki, Finland.

Labas, F. (1971). Composition chimique du lait de lapine evolution au cours de la traite et en function de stade lactation. Ann. Zootech 20: 185-191.

La Due, H. J. (1930). What do mink eat in the wild? Amer. Fur Breeder 3(3):20.

Laerke, H.N., Boisen, S. and Hejlesen, C. (2003). An in vitro method for estimating protein digestibility in mink feed. Annual Report 2002: 65-75. Danish Fur Breeder Research Center, Holstebro, Denmark.

Laplace, J.P. and Rougeot, J. (1976). Mensurations verserales chez le vison selon la sexl et el mode d'alimination. Ann. Zootech 25: 387-396.

Lalor, R. J., Leoschke. W. L. and Elvehjem, C.A. (1951). Yellow fat in mink. J. Nutr. 45: 183-187.

Lalor, R.J., Leoschke, W. L. and Elvehjem, C. A. (1952). Horsemeat blamed for mink disease. Wisc. Agr. Sta. Bul. 496.

Lalor, R.J. (1956).Commentary. Black Fox and Modern Mink Breeder 39(9):18

Langenfeld, E. (1955). Assoc. Fur Farms, Wisc. Personal communication.

Langenfeld, P. (1952). Elcho Mink Ranch, Wisc. Personal communication.

Lau, A., Hejlesen, C. and Borsting, C. F. (1997). Wheat gluten as alternative protein source in mink nutrition. Scientifur 21(4): 302.

Lassen, M. (1994). Requirements of protein and amino acids for mink - the methods for evaluating of protein need. Ph.D. Discourse, 15 pp.

Lauerman, L. H. (1962). Treatment of urinary infection. Amer. Fur Breeder 35(7): 16-17.

Lauerman, L. H. (1965). Diseases affecting the urinary system. Nat. Fur News 37(10): 18, 25. 28, 34.

Lauerman, L.H. and Berman, D. T. (1962a). Urinary tract infection in mink. I. Bacteriology. Am. J. Vet. Res. 23(94): 663-666.

Lauerman, L. H. and Berman, D.T. (1962b). Urinary tract infections in mink. II. Diethylstilbesterol as a predisposing factor. Am. J. Vet. Res. 23(96): 1097-1103.

Layton (nee Collins), H. (1998a). Development of digestive capabilities and improvement of diet utilization by mink pre-weaning and post-weaning with emphasis on gastric lipase. M.S. Thesis. Nova Scotia Agricultural College, Truro and Dalhousie University, Halifax, N. S. 148 pp.

Layton, H. (1998b). Development of digestion capabilities and improvement of diet utilization by mink pre-weaning and post-weaning with emphasis on gastric lipase. Nova Scotia Fur Institute 15th Anniversary Book, 1984-1999: 54-55.

Layton, H. N., Rouvinen-Watt, K. I. And Iverson, S. J. (2000). Body composition in mink (*Mustela vison*) kits during 21-42 days postpartum using estimates of hydrogen isotope dilution and direct carcass analyses. Comparative Biochemistry and Physiology, Part A. 126(2): 295-303.

Leader, R. W. (1955). Pathogenesis and lesions of steatitis. M.S. Thesis. Washington State University.

Lebengartz, (1968) - as reported by Perel'dik, N. S., Milovano, L. V. and Erin, A. T. (1972). Feeding fur bearing animals . Translated from Russian by the Agricultural Research Service, U.S. Dept. of Agr. And Nat. Sci. Foundation, Washington, D.C.

Lee, R.C. (1939). Size and basal metabolism in the rabbit, J. Nutr. 18: 489-500.

Lee, C. F. (1948). Thiaminase in fishery products. A review. Comm. Fish Rev. U.S.D.I., F. W. S. Rep. No. 202.

Lee, C. F., Nilson, H. W. and Clegg, W. (1955). Weight range, proximate composition, and thiaminase content of fish taken in shallow water trawling in Northern Gulf of Mexico. U. S. D. I., F. W. S. Rep. No. 396. Technical Note 31. Comm. Fish. Rev. 17(3):21.

Leekley, J. R. and Cabell, C. A. (1959).Effect of butylated-hydroxy-toluene, biphenyl-para-phenylenediamine and other additives on reproduction and steatitis in mink fed fish diets. J. Anim. Sci. 18: 1534-1535.

Leekley, J. R. and Cabell, C. A. (1961). Anti-oxidants and other feed additives in fish diets for mink. U.S.D.A. Agr. Res. Svc. Prod. Res. Rept. No. 49.

Leekley, J.R., Cabell, C. A. and Damon, R.A. jr.(1962). Anti-oxidants and other additives for improving Alaska fish waste for mink feed. J.Anim. Sci. 21(4): 762-765.

Legrand-Defretin, V. (1994). Differences between cats and dogs: a nutritional review. Proc. Nutr. Soc. 1994; 53:15-24.

Leklem. J. E., Woodford, J. and Brown, R. R. (1969). Comparative tryptophan metabolism in cats and rats. Difference in adaptation of tryptophan oxygenase and in vivo metanbolism of tryptophan, kynurenine and hydroxykynurenine. Comp. Biochemistry and Physiology 31: 95-109.

Leonard, A. (1959). "Wet belly" disease - Albert Leonard answers questions about mink. Amer. Fur Breeder 32(8):10.

Leoschke, W. L. (1952).The vitamin B-12 requirement of the mink (*Mustela vison*). M.S. Thesis, Univ. Wisc., Madison, Wisconsin.

Leoschke, W. L. (1953). Nutrition of the mink. Transactions, American Assoc. of Cereal Chemists. 11(3): 212-216.

Leoschke, W. L. (1954). Ph. D. Thesis. Studies on the nutritional requirements of the mink (*Mustela vison*). Univ. Wisc. Madison, Wisc.

Leoschke, W. L. (1955). University of Wisc. Unpublished data.

Leoschke, W. L. (1956). Phosphoric acid in the mink's diet. Amer. Fur Breeder 29(8): 18.

Leoschke, W. L. (1956). Wet belly research at the Univ. Wisc. Prog. Reports. Mink Farmers Research Foundation.

Leoschke, W. L. (1957-1999). National Research Ranch - Unpubl. Data.

Leoschke, W. L. (1958). Mink research studies VioBin Corp. Unpubl. Data.

Leoschke, W. L. (1958a). Facts about fat in the mink's diet. Nat. Fur News 30(4): 22.

Leoschke, W. L. (1959a). The digestibility of animal fats and proteins by mink. Amer. J. Vet. Res. 20(79): 1086-1089.

Leoschke, W. L. (1959b). Wet belly disease research. Amer. Fur Breeder 32(6): 16-17, 30-31. 34-36.

Leoschke, W. L. (1960). Mink nutrition research at the Univ. Wisc. Univ. Wisc. Research Bul. No. 222.

Leoschke, W. L. (1961). Studies on experimental production of wet belly disease by supplementation of the ranch diet with male hormones. Prog. Rept. Mink Farmers Research Foundation.

Leoschke, W. L. (1962). Nutrition management - protein quality and protein economics. Fur Trade J. of Can. 39(9): 8-13.

Leoschke, W. L. (1962a). Studies on the wet-belly disease of the mink. Proc. of the Indiana Academy of Science for 1962. 72: 136-138.

Leoschke, W. L. (1962b). Studies on the wet-belly disease of the mink. Prog. Rept. Mink Farmers Research Foundation.

Leoschke, W. L. (1962c). Studies on the wet-belly disease of mink. B. Dietary Supplementation with agents designed to increase water intake of the mink. Prog. Repts. Mink Farmers Research Foundation.

Leoschke, W. L. (1963). Digestibility of cereal grain carbohydrates by the mink. Prog. Repts. Mink Farmers Research Foundation.

Leoschke, W. L. (1965a). Dr. W. L. Leoschke on mink nutrition. Diet water intake can help control wet belly. Amer. Fur Breeder 39(9): 26, 36.

Leoschke, W. L. (1965b). Studies on the digestibility of carbohydrates in raw and processed cereal grains by mink. Prog. Rep, Mink Farmers Res. Fd.

Leoschke, W. L. (1965c). Optimum nutrition tied to dietary ash levels. American Fur Breeder 38(7): 22.

Leoschke, W. L. (1966). Caloric density critical during lactation period. American Fur Breeder 39:(4):18.

Leoschke, W. L. (1968). Dry diets for the mink: past, present and future. 1928-1968 Jaar NFE Symposium - Forty Years of the Dutch Fur Breeders Assoc., Rotterdam, Netherlands, September 24, 1968.

Leoschke, W. L. (1969). Nutrition management on the ranch. Section 4. Feeding and Nutrition. Fur Farm Guide Issue. American Fur Breeder 42(12): 85-116.

Leoschke, W. L. (1970a). Field observations on cannibalism - energy balance and mink genetics. U.S. Fur Rancher. 49(9): 13.

Leoschke, W. L. (1970b). Unpublished data, National Research Ranch, Northwood Fur Farms, Cary, Illinois.

Leoschke, W. L. (1972). Key to better feeds: pay attention to mink, not man. U.S. Fur Rancher 51(9): 5-6.

Leoschke, W. L. (1975). Rating feedstuffs based on taste appeal for mink. Blue Book of Fur Farming. 1976 edition: 26-28.

Leoschke, W. L. (1976). Qualitative evaluation of protein feedstuffs for mink. The lst Inter. Sc. Congr. In Fur Anim. Prod. April 1976, Helsinki, Finland.

Leoschke, W. L. (1980). Nutrient Requirements For Mink. Chapter in M. Rechcigl. Ed. Handbook Series in Nutrition and Food. Section D, Volume VI. West Palmbeach, Florida, CRC Press, Inc.

Leoschke, W. L. (1981). Pellets: The mink feed of the future. Deutsche Pelztierzuchter 55(3): 37-39.

Leoschke, W. L. (1984a). Mink Creatinine excretion. The 3rd Inter. Sc. Congr. In Fur Anim. Prod., Versailles, France, April 1984.

Leoschke, W. L. (1984b). Protein nutrition of mink. Blue Book of Fur Farming 1985: 57-61.

Leoschke, W. L. (1985). Fat in modern mink nutrition. 1986 Blue Book of Fur Farming: 28-37.

Leoschke, W. L. (1987a). Carbohydrates in modern mink nutrition. Blue Book of Fur Farming 1987: 39-42.

Leoschke, W. L. (1987b). Kohlenhydrate in der nerzfuterung (Importance of carbohydrates for fur production of mink. Der Deutsche Pelztierzuchter, 11: 163-164 and 12:177-179.

Leoschke, W. L. (1990). Evaluation of protein quality in dehydrated mink feedstuffs. 1991 Blue Book of Fur Farming: 53-60.

Leoschke, W. L. (1992). Use of ferret kits in the assessment of the biological value of protein in dehydrated mink feedstuffs. Norwegian J. Ag. Sc. Supple. No. 9, 1992. Progress in Fur Animal Science. Proceedings from the 5th Inter. Sc. Congr. In Fur Anim. Prod. Oslo, Norway: 232-234.

Leoschke, W. L. (1996). Phosphoric acid in modern mink and fox nutrition. 6th Inter. Sc. Congr. in Fur Anim. Prod.. August 21-23.1996, Warsaw, Poland, Animal Prod. Review, Polish Society of Animal Prod. Applied Science Reports . Progress in Fur Animal Science Reports 28 - Nutrition, Patology And Disease: 77-78.

Leoschke, W. L. (2001). Modern nuitrition of the mink. Recommended protein levels as expressed as percent of metabolic energy. 2002 Blue Book of Fur Farming. November 2001: 34-36.

Leoschke, W. L. (2004). Sodium bisulfate as a mink food preservative. Scientifur 28(3): 140-141.

Leoschke, W. L. and Elvehjem, C. A. (1954). Prevention of urinary calculi Formation in mink by alteration of urinary pH. Proc. Soc. Exp. Biol. Med. 85: 42-44.

Leoschke, W. L. and Elvehjem, C. A. (1956). The importance of arginine and methionine for growth and fur development of mink on purified diets. Fed. Proc. 15: 560.

Leoschke, W. L. and Elvehjem, C. A. (1956).Unpublished data.

Leoschke, W. L. and Elvehjem, C. A. (1959a) The importance of arginine and methionine for the growth and fur production of mink fed purified diets. J. Nutr. 69: 147-150.

Leoschke, W. L. and Elvehjem, C. A. (1959b). The thiamine requirement of the mink for growth and fur development, J. Nutr. 69(3): 211-213.

Leoschke, W. L., Lalor, R. J. and Elvehjem. C. A. (1953). The vitamin B-12 requirement of the mink. J. Nutr. 49: 541-548.

Leoschke, W. L. and Rimeslatten, H. (1954). University of Wisconsin and Royal Agricultural College, Oslo, Norway. Unpublished data.

Leoschke, W. L. and Zawadzke, E. (1988). Urinary orotic acid excretion of mink as an assay of feedstuff arginine availability. Biology, Pathology and Genetics of Fur Bearing Animals. Ed. B.D. Murphy and D. B. Hunter, 4th Inter. Sc. Congr. In Fur Anim. Prod., August 21-24, 1988. Rexdale, Ontario, Canada: 270-277.

Leoschke, W. L., Zikria, E. and Elvehjem,. C. A. (1952). Composition of urinary calculi from mink. Proc. Soc. Exp. Biol. Med. 80: 291-293.

Lewis, U. J. (l951). University of Wisconsin. Unpublished data.

Lisbjerg, S. (2005). Effect of acids in feed on pH in mink urine. Proceedings from NJF-Seminar No. 377. 2005.

Ljokjel, K., Harstad, O.M. and Skrede, A. (2000). Effect of heat treatment of soybean meal and fish meal on amino acid digestibility in mink and dairy cows. Animal Feed Science and Technology 84: 83-95.

Ljokjel, K. and Skrede, A. (2000). Effect of feed extrusion temperatures on digestibility of protein, amino acids and starch in mink. Scientifur 24(4)-IV-A-Nutrition. Proc. Of the 7th Inter. Sci. Congr. In Fur Animal Prod. Ed. B.D. Murphy and O. Lohi, Kastoria. Greece.

Ljokel, K., Sorensen, M., Storebakken, T. and Skrede, A. (2004). Digestibility of protein, amino acids and starch in mink (*Mustela vison*) fed diets processed by different extrusion conditions. Can. J. Anim. Sci. 84:673-680.

Lode, T. (l993). Diet composition and habitat use of sympatric polecat and American mink in Western France. Acta Theriologica 38(2): 161-166.

Loftgard, G.. Moe, A. and Yndestad, M. (1972). Eddikosyrekonservering av karohydrate-grot I pelsdyrfur-produktivoen. Norsk Pelsyrblad 46: 519-521.

Loftgard,G.,Moe, A.and Yndestad,M.(1972)The preservation of cooked cereals in mink feed products by he use of acetic acid. Nord.Vet-Med.24:586-591.

Lohi, O. and Hansen, K. (1990). Heritability of body length and weight in mink. Deutsche Pelztierzuchter 64(1): 4-5.

Lohi, O. (2000). Finish mink scientist. Personal communication.

408

Lohi, O. and Jensen, L. V. (1991). Mineral composition of mink feed and mink hair. Beretning fra Statens Husdyrbrugforsog. Nr. 688: 99-124.

Long, J. B. and Shaw, J.N.(1943). Chastek paralysis produced in Oregon mink and fox by feeding frozen smelts. The North Amer. Vet. 24: 234-237.

Loosli, J. K. and Maynard, L.A. (1939). The digestibility of animal products and cereals by mink. Proc. Amer. Soc. Anim. Prod. 32: 400-03.

Loosli, J.K. and Smith, S.E. (1940). Nutrition experiments with foxes and minks. Amer. Fur Breeder XII (7): 6-12.

Lorek, M. O.,Gugolek, A., Rotkiewicz, T. and Podbielski, M. (1996). Use of animal and plant fats in mink feeding. Proc. 6th Inter. Sc. Congr. In Fur Anim. Prod., August 21-23, Warsaw. Poland. Animal Prod. Review, Polish Society of Anim. Prod. Applied Science Reports 28. Progress in Fur Animal Science. Nutr. Patalogy and Disease, ed. A. Frindt and M. Brzozowski.: 37-46.

Loveridge, G. G. and Rivers, J.P.W. (1989). Body weight changes and energy intakes of cats during pregnancy and lactation. In Nutrition of the Dog and Cat. Waltham Symposium 7, Cambridge University Press, pp. 113-132.

Lund, R.S. (1975a). Fremstilling haantering or brug af sildeensilage. Progress Report to NJF's Subdivision of Fur Bearing Animals. Sweden, September 1975. 6 pages.

Lund, R. S. (1975b). Forsog med majsgluten mel. Dansk Pelsdyravl 38: 386-388.

Lund, R.S. (1979). Experiments with NaCl and other sodium salts in feed for lactating females. Scientifur 3(2): 19-20.

Lund, R.S. (1981). Report from the West experimental farm. Dansk Pelsdyravl 44 (4): 169.

Lund, R.S. (1983). Fodringsforsog med forskellige proteinniveaver til mink. Bilag til Seminar i. NJR's Subsektion for Pelsdyr i Malmo.

Lund, R.S. and Hansen, E. (1987).Forbedring af minkskindenes storrelse og Kvalitet ved hjaelp af aend-ringer af energifordelingen og fording med hojt og laut proteinniveaver. Faglig Arsberetning 1987: 97-106.

Lykken, G., Mahalko, J., Henriksen, L., Canfield, W., Johnson, P. and Sandstead, H. (1982). Effects of browning on zinc retention. Fed. Proc. 41:282.

Lyngs, B. (1992). Taste appeal trials with poultry offal for mink. Prog. In Fur Anim. Sc. Proc. From the 5th Inter. Sc. Congr. In Fur Anim. Prod. Norwegian Agr. Sc. Suppl. No. 9, 1992: 298-307. MacDonald, D. (1992). The velvet claw - a natural history of the carnivores. BBC Books, London, 256 p.

MacDonald, M. L. and Rogers, Q.R. (1984). Nutrition of the domestic cat, a mammalian carnivore. Ann. Rev. Nutr. 4:521-562.

MacDonald, M. L.. Rogers, Q.R. and Morris, J. G. (1983). Role of linoleate as an essential fatty acid for the cat independent of arachidonate synthesis. J. Nutr. 113: 1422-1423.

Makela, J. (1971a). Minkens behov for drikkevand. Dansk Pelsdyravl 34:535-536.

Makela, J. (1971b). Minkin juomaveden trarpeesta (about the need of drinking water for mink). Turkistalous 43: 415-416.

Makela, J. (1979). Resultat fran forsok med vassbuk och mort. NJF's subsektion for palsdjur mole I. Kungalv 10: 10-12.

Makela, J., Huilaja, J. and Kangas, J. (1968). Utfodringsforsok med mjolkpulver. Finsk Palsatidskrift 40: 383-385.

Makela, J., Kangas, J. and Huilaja, J. (1967). Kvark som minkfoder. Finsk Pelstidskrift 39: 252-254.

Makela, J. and Polenen, J. (1978). Unpublished data.

Makela, J. and Valtonen, M. (1982). Water intake and some dietary factors affecting it. Nordiske Jorbruksforskeres Forening, Mote om pelsdyrproduksjon.

Maksimov, A. P. and Nikolaevskii (1986). Moskovskaya Vet. Akademiya, Moscow, USSR: 91-94.

Maran, T., Krunk, H., MacDonald, D.W. and Polma, M. (1998). Diet of two species of mink in Estonia: Displacement of *Mustela Lutreola* by *Mustela vison*. J. Zool. London 245: 218.

March, B.E., Wong, E., Seier, L., Sim, J. and Bielly, J. (1973). Hypervitaminosis E in the chick. J. Nutr. 103: 371-377.

Marcuse, R. (1972). Fettforandringar i livs-och fodered - kemiska och naringsfysilologiska aspekter. Symposium om Forkvalitet, NJF's subsektion for pelsdyr, Roros, Norway, 1972: 28-61.

Marsh. F. (1995). Univ. Wisc. mink researcher, personal communication.

Martin, J. C. (1987). Feeding the mink as it was done half a century ago. Fur Rancher 67(4): 24-25.

Martin, J. P. and Synge, R. L. M. (1945).Analytical chemistry of the proteins. Ad. Prot. Chem. 2: 1-83.

Maskell, I. E. and Johnson, J. V. (1996). Digestion and Absorption. Pp. 25-40 in The Waltham Book of Companion Animal Nutrition, editor, I. Burger. Pergamon Press, New York. 126 pp.

Mason, K. E. and Hartsought, G. R. (1951). "Steatitis" or "Yellow Fat" in mink and its relation to dietary fats and inadequacy of vitamin E. J. Am. Vet. Med. Assoc. 119: 72-75.

May, G. (1970). Aflatoxin poisoning in mink. The British Fur Farmers Gazette. 20(2): 18-19.

Maynard, L. A. (1951). Animal Nutrition. No. 343. McGraw-Hill Book Co., New York, N.Y.

McCarthy, B. H. (1964). Ph.D. Thesis, Cornell University. Pantothenic acid requirements and the histopathology of pantothenic acid requirements in the mink.

McCarthy, B. H., Travis, H. E., Krook, L. and Warner, R. G. (1966). Pantothenic acid deficiency in the mink. J. Nutr. 89(4): 392-398.

McDermid, A.M.and Ott, G. I. (1947). "Yellow Fat" as observed in mink. J. Amer. Vet. Med. Assoc. 110: 34-35.

McWilliams, L. (1990). Personal communication.

Mejborn, H. (1986). Zinc metabolism in mink. Dissertation. Inst. Animal Physiology, Royal Vet. And Agr. Univ., Copenhagen.

Mejborn, H, (1987). The effect of interaction between dietary protein and zinc content on protein utilization in young male mink. Scientifur 11(3): 233-237.

Mejborn, H. (1988). Influence of different zinc intake on zinc absorption, retention and turnover in mink. 4th Inter. Sc. Congr. In Fur Animal Production, Biology, Pathology and Genetics of Fur Bearing Animals. Ed. B.D. Murphy and D.B. Hunter. August 21-24, l988. Rexdale, Ontario, Canada: 247-257.

Mejborn, H. (1989a). Effect of copper addition to mink feed during the growth and moulting period on growth, skin production and copper retention. Scientifur 13(3): 229-233.

Mejborn, H. (1989b). Zinc balances in young and adult mink (*Mustela vison*) in relation to dietary zinc intake. J. Anim. Physiol.,A. Anim. Nutr. 61:187-192.

Mejborn, H. (1990). Endogenous zinc excretion in relation to various levels of dietary zinc intake in mink (*Mustela vison*). J. Nutr. 120: 862-868.

Melnick, D., Hochberg, M. and Oser, B. L. (1945). Physiological availability of the vitamins II. The effect of dietary thiaminase in fish products. J. Nutr. 30(2): 81-88.

Mertin, D., Tocka, I. and Oravcova, E. (1991). Effect of zinc, selenium on some morphological properties of silver foxes in period of fur maturity. Scientifur 15(4): 287-293.

Meyer, G. (1995). Wisconsin mink rancher. Personal communication.

Michels, M. (1980, 1998, 2001, 2002). National Research Ranch. Unpublished data.

Miller, E.L., Carpenter, K. J. and Milner, C. K. (1965). Availability of sulfur amino acids in protein foods. 3. Chemical and nutritional changes in heated cod muscle. Br. J. Nutr. 19: 547-564.

Miller, S. A. and Allison, J. B. (1958). The dietary nitrogen requirement of the cat. J. Nutr. 64: 493-500.

Milovanov, L. V. (1961). Importance of amino acids in rations for young mink. Krolik Zver. No. 9: 18-20. (Nutr. Abstracts and Review, 34; ref. No. 3375.

Mitchel, H. H. (1962). Comparative Nutrition of Man and Domestic Animals. Vol. I. Academic Press, New York, N.Y.

Moller, S. (1986). Digestibility of nutrients and excretion of water and salts in faeces from mink fed different types and levels of fiber. Scientifur 10(1): 62.

Moller,. S. (1988). Temperature preference of drinking water in mink. Biol. Pathology and Genetics of Fur Bearing Animals. Ed. B. D. Murphy and D. B. Hunter. Proc. 4th Inter. Sc. Congr. In Fur Animal Production. Rexdale, Ontario, Canada. April 21-24, 1988.

Moller, S. H. (1997). The importance of weight development and pelting time to the skin length of mink - an example of the development of production management by means of farm experiments. Scientifur 21(4): 309.

Moller, S. H. (1999). Effect of weight development, pelting time, colour type and farm on skin length in mink. Acta Agric. Scand. Sect. A. Anim. Sc.49: 121-126.

Moller-Jensen, D. and Jorgensen, G. (1975). Fremstilling og anvendelse af Fiskeensilage til mink. 427 Beretning fra Statens Husdyrbrugsforsog, Copenhagen, Denmark.

Mosshammer, G. (1985). Indiana mink rancher. Persomnal communication.

Morris, J. G. (2002). Idiosyncratic nutrient requirements of cats appear to be diet induced evolutionary adaptation. Nutr. Res. 15: 153-168.

Morris J. G. and Rogers, Q. R. (1978). Arginine: an essential amino acid for the cat. J. Nutr. 108(12): 1944-1953.

Morris, J. G., Rogers, Q. R., Winterrowd, D. L. and Kamikawa, E. N. (1979). The utilization of ornithine, citrulline by the growing kitten. J. Nutr. 109: 724-729.

410

Morris, J. G., Shiguang, Y. and Rogers, Q. R. (2002). Red hair in black cats is reversed by addition of tyrosine to the diet. J. Nutr. L32: 1646S-1648S.

Moustgaard, J. and Riis, P.M. (1957a). Om minkens proteinov til fosterproduktion og proteinstoffernes biologiske va rdi. Dansk Pelsyravl 20: 11-12, 15-18,21.

Moustgaard, J. and Riis, P.M. (1957b). Protein requirements for growth of mink. The Black Fox Magazine and Modern Mink Breeder 40(8): 8-11.

Mulder, G. J. (1923). The chemistry of animal and vegetable physiology. Quoted in L. B. Mendel, Nutrition. The Chemistry of Life. Yale Univ. Press, New Haven. Conn. 1923. p. 16.

Mullen, A. (1992). Nova Scotia mink rancher. Personal communication.

Muralidhara, K.S. and Hollander, D. (1977). Intestinal absorption of alpha-tocopherol in the anaesthetized rat. The influence of luminal constituents on the absorption process. Lab. Clin. Med. 90: 85-91.

National Academy of Sciences (1978). Nutrient requirements of cats. National Academy of Science, Washington, D.C., USA. 48 pp.

Narasimhalu, P., Belzile, R. J. and LePace, M. (1978). Effects of feeding raw meat and soybean meal on blood composition in mink (*Mustela vison*).Can. J. Anim. Sc. 58: 191-197.

Neil, M. (1992). Effects of diet on water turnover and water requirement in mink. Ph.D. Thesis, Swedish University of Ag. Sc., Dept. Anim. Nutr. And Management. Report 213.

Neilands, J. B. (1947). Thiaminase in aquatic animals of Nova Scotia. J. Fish Res. Bd., Canada 7(2): 94.

Newell, A. I. T. (2001). Nutrient Intake and Excretion in Growing Mink. Nova Scotia Dept. of Ag. And Fisheries Ref. 990027.

Niemimaa, J. and Pokk, J. (1990). Food habits of the mink in the outer archipelago of the Gulf of Finland. Suomen Riista 36: 18-30/

Nordfeldt, S., Melin, G. and Thelander, R.(1954). Experiments with antibiotics in the ration of fur animals. Kungl Lantbrukhogsk Statens Husdjursforsok Sartryck Forhandsmedd. No. 107.

Neseni, R.(1935). Einige anatomische datin von pelztieren prager tierarztle. Archieves 10: 211-216.

Neseni, R. and Piatkokski, B. (1958). Feed passage in the mink. Arch. Tierernahrung 8: 296.

Newell, C. W. (1999). Nutrient flow and manure management in the mink industry. M.S. Thesis. Nova Scotia Ag. College, Truro and Dalhousie Univ. Halifax, N.S. 91 pp.

Ngurjen, H. T., Moreland, A. F. and Shield, R. P. (1979). Urolithiasis in ferrets Mustela putorius). Lab. Anim. Sc. 29: 243-245.

Nielsen, I. M. (1955). Progress in recent research on urinary calculi. The Black Fox Magazine and Modern Mink Breeder 39(4): 13-15, 23-25.

Nielsen, I. M. (1956). Urolithiasis in mink. Pathology, bacteriology and experimental production. J. Urol. 75: 602-614.

Nielsen, T.B., Sandbol, P., Hejlsesen, C.,Hansen, N.E. and Elnif, J.(2004). The amino acid requirements of mink (*Mustela vison*) in the intensive growth period - growth results. Annual Report 2004. Danish Fur Breeders Research Center, Holstebro, Denmark. 89-102.

Noble, P. (1970). Ontario mink rancher. Personal communication.

Oftedal, O. T. (1984). Milk composition, milk yield and energy output at peak lactation. A comparative review. Symp. Zool. Soc. London. No. 5l: 33-85.

Oehm, B. (1977). Ontario mink rancher. Personal communication.

Olesen, C. R., Clausen, T.N. and Wamberg, S. (1992). Compositional changes in mink (*Mustela vison*) milk during lactation. Norwegian J. Ag. Sc. Suppl. No. 9, 1992. Progress in Fur Animal Science. Proc. Of the 5th Int. Sc. Congr. In Fur Animal Prod.: 308-314.

Olson, R.E. (1984). Present Knowledge In Nutrition. The Nutrition Foundation, Inc. 479-505.

Oldfield, J.E. (1949). A study of nitrogen metabolism with special reference to mink. M.S. Thesis. Dept. Anim. Science, Univ. Brit. Columbia, Canada.

Oldfield, J. E., Allen, P. H. and Adair, J. (1956). Identification of cystine calculi in mink. Proc. Soc. Exp. Biol. & Med. 19: 560-562.

Oldfield, J. E., Stout, F.M. and Adair, J. (1960). Experimental studies with mink. Projects Supported by the Mink Farmers Res. Foundation.

Olenik, V.M. and Svetchkina, E. B. (1992). Change in the enzyme spectrum of the digestive tract in mink postnatal ontogeny. Scietifur 16(4): 289-292.

Olenik, V. M. and Svetchkina, E. B. (1993). Some regularities in enzyme spectrum formation in the digestive tract of mink. Scientifur 17(4): 303-305.

411

Olesen, C. R., Clausen, T.N. and Wamberg, S. (1992). Compositional changes in mink (*Mustela vison*) milk during lactation. Norwegian J. Agr. Sci. Supplement No. 9. Progress in Fur Anim. Sc. Proc. 5th Inter. Sc. Congr. In Fur Anim. Product: 308-314.

Opstvedt, J., Miller, R., Hardy, R.W. ad Spinelli, J. (1984). Heat-induced changes in sulfhydryl groups and disulfied bonds in fish protein and their effect on protein and amino acid digestibility in rainbow trout (Salmo Gairdneri). J. Agr. Food Chem. 32: 929-935.

Osborne, T. B. and Mendel, L. B. (1917). The use of soybean as food. J. Biol. Chem. 32: 369.

Ostergard, K. and Mejborn, H. (1989). Effects of heat treatment on the mink digestibility of starch from wheat and barley. Short. Commun. No. 734. National Institute of Animal Sc., Denmark. 4 pp.

Oswald, P. (1970, 1995). Ontario mink rancher. Personal communication.

Ozeki, H., Ito, S., Wakamatsu and Hirobe, T. (1995). Chemical characteristics of hair melanins in various coat-color mutants of mice. J. Investic. Dermatol. 105: 361-366.

Palmer, L. S. (1927). Dietetics and its relationship to fur. American Fox and Fur Farmer 7(2): 22-24.

Pastirnac, N. (1977). The "wet belly" disease in mink (*Mustela vison*). Scientifer (2): 34-39.

Patrick, M. (1985). Wisconsin mink rancher. Personal communication.

Penelaik, P. (1975). Russian Handbook.

Perel'dik, N.S. (1974). Grundlagen der fullerung charakterisierung der futtermittel und zusamensetzung der futterationen.In Edelpelztiere (ed. U.D. Wenzel): 269-422. Veb Deutscher Landwirtschaftsverlag, Berlin. 558 pp.

Perel'dik, M.N. and Titova, M. I. (1950). Experimental determination of feeding standards for adult breeding mink. Karakulevodstvo Zverovodstvo 3(2): 29-35.

Perel' dik, N.S., Milovanov, L. V. and Erwin, A.T. (1972). Feeding fur animals. Translated from the Russian by the Agr. Res. Svc. U.S. Dept. Agr. And Nat. Sc. Foundation. Washington, D.C. 344 pp.

Perel'dik, N.S., Titova, M. I. and Kuznetsova, Y. D. (1970). Reproductive capacity of one-year old female mink depending on the amount of tryptophan and sulfur containing amino acids in rations. Nauch. Tr. Nauch-Issled Inst. Pushnogo Zverovop Krolikovod 9: 175-180. Abstr. In Chemical Abstr. 1971, 75.

Petersen, E. (1982). Personal communication.

Petersen, F.H. (1957).Foringsforsog med avlsmink II. Forsog med tilskud af vitamin E og C til normal foder. 297. Beretning Fra Forsogslaboriet Statens Husdyrbrugsforsog, Copenhagen, Denmark: 11-19.

Petersen, I.M., Sand, O. and Grove-Sorensen, D. (1995). Effects of starvation on the activities of key enzymes of the glycolysis and the glyconeogenesis in the liver of mink. NJF seminar, Oct. 4-6, Gothenberg, Sweden.

Phillips, P.H. and Hart, E. B. (1935). The effect of organic dietary constituents upon chronic fluorine toxicosis in the rat. J. Biol. Chem. 109: 657.

Pierieldik, N.S. and Pierieldik, D. N. (1980). Ferroanemin-prefenting cotton fur. Krolikovodstvo I. Zverodstvo, Moskva 33:33.

Pingel, H., Anke, M. and Salchert, E. (1992). The influence of zinc supplementation on growth and reproduction in mink. Norwegian J. Agr. Sc. Supple. No. 9, Progress in Fur Animal Sc. Proc. Of the 5th Inter. Sc. Congr. in Fur Animal Production.

Pion, P. D., Kittleson, M.D. and Rogers, Q. R. (1987). Myocardial failure in cats associated with low plasma taurine. A reversible cardiomyopathy. Science 237: 764-768.

Polonen, I., Kakela, R., Miettinen, M. and Asikainen, J. (2000). Effects of different fat supplements on liver lipids and fatty acids and growth of mink. Scientifur 24(4): IV-A: Proc. Of the 7th Inter. Sc. Congr. In Fur Anim Prod. Kastoria, Greece. Ed. B. D. Murphy and O. Lohi 92-94.

Polonen, I., Niemala, P., Xiao, Y., Jalkanen, L., Korhonen, H. and Makela, J. (1999). Formic acid, sodium benzoate preserved slaughterhouse offal and supplemental folic acid in mink diets. Anim. Feed Sc.Technol. 78: 39-56.

Polonen, I., Scott, R. and Oldfield, J. E. (1993). Mink diet energy during pre-weaning and early post weaning periods. Scientifur 17: 47-51.

Polonen, I., Vaheristo, L. T. and Tanhuanpaa, E.J. (1999). Effect of folic acid Supplementation on folate status and formate oxidation rate in mink (*Mustela vison*). J. Anim. Sc. 75: 1569-1574.

Poquette, J. (2002). Personal communication.

Porter, D.D. (1965). Transfer gamma globulin from mother to offspring in mink. Proc. Soc. Exp. Biol. & Med. 119 (1): 131-135.

Poulsen, J. S. D. and Jorgensen, G. (1976). Acid-base disorders in mink fed on fish silage. 1st Inter. Sc. Congr. In Fur Anim. Prod. Helsinki, Finland

Poulsen, J.S.D. and Jorgensen, G. (1977). The influence of the pH of feed on the acid-base balance of mink. Nord. Vet-Med. 29: 488-497.

Powell, D.C., Bursian, S. J., Bush, C.R., Napolitano, A.C. and Aulerich, R. J.(1997). Reproduction performance and kit growth in mink fed diets containing copper treated eggs. Scientifur 21(1): 59-66.

Powell, D. C., Bursian, S. J., Bush, C.R., Napolitano, A.C. and Aulerich, R. J. (1997). Reproduction performance and kit growth in mink fed diets containing copper treated eggs. Fur Rancher 78 (fall): 4-6.

Pridham, T. J. (1967). Lead poisoning is insidious killer of ranched mink. Fur Trade J. of Canada 44(7):4.

Pridham,.T.J.(1968).Cause of prolapsed rectum in mink. Fur Trade J.Can.46(6):23.

Quortrup, E. R. and Gorham, J. R. (1949). Susceptibility of fur bearing animals to the toxins of Clostridium botulinum types A, B,C and E. Amer. J. Vet. Res. 10: 269-271.

Quortrup, E..R., Gorham, J. R. and Davis, C. L. (1948). Nonsuppurative paneceilitis (yellow fat) in mink. Vet. Med. 43(6): 228-230.

Quist, A. (1964). Natriumbisulfite-konserverat foder. Dansk Pelsdyravl 27: 412-413.

Quist, A. and Makela, J. (1961). Utfodringsforsok for mink ar l960. Turkistalous 32: 154-162.

Quist, A. and Makela, J. (1963). Konbservering af minkfoder med natriumnbisulfit. Dansk Pelsdyravl 26: 377-380.

Quist, A.and Makela, J.(1964).Praktiska erfarenheter av natriumbisulfite- Konserverat foder och under ar 193 foretagna forsook. Turkistalous 36:192-198.

Rapoport, O. L. and Golushkova, M. A. (1991). Mink's assimilation of various iron combinations. Scientifur 15(4): 316.

Rasmussen, P. V. and Borsting, C. F. (2000). Effects of variations in dietary protein levels on hair growth and pelt quality in mink (Mustela vison). Can. J. Anim. Sci.80: 633-642.

Reich, C. (1975). New York fur buyer. Personal communication.

Restum, J. C., Bush, C. R., Malinczak, R. L., Watson, G. L., Braselton, W. E., Bursian, S.J. and Aulerich, R. J. (1995). Effects of supplemental dietary NaCl and restricted water on mink. Vet. Human Toxicol. 37(1): 4-l0.

Rice, R. P. (1969). Practical mink nutrition. The British Fur Farmers Gazette 19(2): 27-30.

Riedel, W. (1957). Wisconsin mink rancher. Personal communication.

Rietveld, A. A. (1970 & 1973). National Research Ranch, Unpublished data.

Rietveld, A. A. (1976). Applied science in mink ranching. The lst Inter. Sc. Congr. In Fur Anim. Prod. April 27-28th, 1976. Helsinki, Finland.

Rimeslatten, H. (1954 & 1958). Norges Landbrukshogskole, Institute for Fjorfe og Pelsdyr, Vollebekk, Norway. Personal communication.

Rimeslatten, H.(1959). Protein, fett och kulhydrate - minkfoderet. Vara Pelsdjur 30(8): 179-182.

Rimeslatten, H. (1964a). Forsok med vitamin B-6 til mink. Nord Jordrforskn. Suppl. II: 583-587.

Rimeslatten, H. (1964b). Energiforbrucket til mink og rev. Beretning om Nordiske Jordbruksforskeres Forenings Tolvet. XII Kongress, Helsinki, Finland, 1964: 475-482.

Rimeslatten, H. (1966a). Revens og minkens vitamin behov. Kurs i minkfoderteknologi for tilverkare av minkfoderblandninger. Nordiska Jordbruksforskares Forening. Oslo, Norway.

Rimeslatten, H. (1966b). Revens og minkens mineral behov. Kurs i Minkfoderteknologi for tilverkare av minkfoder blandninger. Nordiska Jordbruksforskares Forening, Oslo, Norway.

Rimeslatten, H. (1968). Minkens og revens til a lagre vitamin A i leveren under forskellige fodringforhold. Norsk Pelsdyrblad 42: 542-546.

Rimeslatten, H. (1974). Soyamjol som for til pelsydr. Norsk Pelsdyrblad 48: 219-223.

Rimeslatten, H. and Aam. A. A. (1962). Forsok med torrfiskmjol til solre blarev og mink. Norsk Pelsdyrblad 36: 392-404.

Rimeslatten, H. and Skrede, A. (1957). Personal communication.

Rivers, J. P. W., Sinclair, A. J. and Crawford, M. A. (1975). Inability of the cat to desaturate essential fatty acids. Nature 259: 171-173.

Roberts, W. L. (1953). The ideal or perfect mink food. The Black Fox and Modern Mink Breeder 37(7): 20,31,32.

Roberts, W. L. (1955a). More on nursing sickness and salt. Amer. Fur Breeder 28(7): 38-39.

Roberts, W. L. (1955b). Personal communication.

Roberts, W.L. (1959). Wet belly disease. National Fur News 31(7):16.

Roberts, W.L. (1988). Starch digestion by mink. Fur Rancher 68(7): 4,23.

Roberts, W. L. (1990). The essential amino acid content of proteins. Inter. Mink Show, Madison, WI. January 1990.

Roberts, W. L. (1997). History of pellet development. Inter. Mink Show, Madison, WI. January 1997.

Roberts, W. L. and Kirk, R. J. (1964). Digestibility and N utilization of raw fish and dry meals by mink. Am. J. Vet. Res. 25(109): 1746-1750.

Rogers, Q. R., Baker, D. H., Hayes, K. C. , Kendall, P.J. and Morris, J.C. (1986). Nutrient Requirements of Cats, Revised edition, (1986). Nutrient Requirements of Domestic Animals. National Academy Press, Washington, D.C.

Rogers, Q. R. and Harper, A.E. (1970). Amino acid diets and maximal growth in the rat. J. Nutr. 87: 267-73.

Rogers, Q. R. and Morris, J. G. (1979). Essentiality of amino acids for growing kittens. J. Nutr. 109: 718-723.

Rogers. Q. R. and Morris, J. G. (1980). Why does the cat require a high protein diet? In Nutr. of Dog and Cat. Ed. R.S. Anderson) Oxford, Pergamon Press: 45-66.

Rogers, Q. R. and Morris, J. G. (1982). Do cats really need more protein. J. Small Anim. Practice 23: 521-532.

Rogers, Q. R. and Morris, J. G. (1983). Protein and amino acid nutrition of the cat. AAHA's 50th Annual Meeting Proc.: 333-336.

Rogers, Q. R., Morris, J. G. and Freedland, R. A. (1977). Lack of hepatic enzymatic adaptation to low and high levels of dietary protein in the adult cat. Enzymes 22: 348-396.

Romero, J. J., Castro, E., Diaz, A.M., Reveco, M. and Zaldivar,J. (1994). Evaluation of methods to certify "premium" quality of Chilean fish meals. Aquaculture l24: 351-358.

Rose, J., Stormshak, F., Oldfield, J. E. and Adair, J. (1984). Induction of winter fur growth in mink (*Mustela vison*) with melatonin. J.Anim. Sc. 58(1): 57-61.

Rothenberg, S. and Jorgensen, G. (1971). Some haematological indices in mink. Nord. Vet-Med. 23: 361-366.

Rougeot, J., Melcion, J. P., Charlet-Lery, G. and DeLort-Laval, J. (1971). Possibility of using complete pelleted diets in mink. Ann. Zootech. 20(2): 259-262.

Rouvinen, K. I. (1990. Digestibility of different fats and fatty acids in the mink (Mustela vison). Acta Agr. Scand. 40: 93-99.

Rouvinen, K.I. (1991).Dietary effects of omega-3 PUFA on body fat composition and health status of farm raised blue and silver foxes. Acta Agr. Scand. 41: 401-414.

Rouvinen, K. I. (1991). Effects of dietary fat on production performance, body fat composition and skin storage in farm raised mink and foxes. Ph.D. Dissertation, Publications of University of Kuopio, Finland, Natural Sciences Original Reports 9/91.

Rouvinen, K. I. (1996 & 1998). Personal communication.

Rouvinen-Watt, K. I. (2002). New hypothesis for pathogenesis of nursing sickness in mink. NJF-Seminar No. 347. Vuokalte, Finland, Oct 2-4, 2002: 221-230.

Rouvinen-Watt, K. I. (2003). Nursing sickness in the mink -a metabolic mystery or a familiar foe? The Can. J. Vet. Res. 67: 161-168.

Rouvinen, K. I., Anderson, D.M. and Alward, S. R. (1996). Use of silver hake and herring and the corresponding silages in mink diets during growing furring period. Can. J. Anim. Sc. 76: 127-133.

Rouvinen, K. I., Anderson, D.M. and Alward, S. R. (1997). Effects of high dietary levels of silver hake and Atlantic herring on growing-furring performance and blood clinical-chemistry of mink (*Mustela vison*)/ Can. J. Anim. Sc. 77: 509-517.

Rouvinen, K. I., Archbold, S., Laffin, S. and Harris, M. (1999). Long-term effects of tryptophan on behavior response and growing-furring performance in silver fox (*Vulpes vulpes*). Applied Animal Behavior Science 63: 65-77.

Rouvinen, K. I., Inkinen, R. and Niemela, P. (1991). Effects of slaughter house offal and fish mixture diets on production performance of blue and silver foxes. Acta Agr. Scand. 41(4): 387-399.

Rouvinen-Watt, K. I. and Hynes, A.M. (2004). Mink nursing sickness survey in North America. VIII Inter. Sc. Congr. In Fur Animal Prod. De Ruwenberg, 's-Hertogenbosh, The Netherlands, 15-18 Sept. 2004. Scientifur 28(3): 71-78.

Rouvinen, K. I. and Kiiskinen, T. (1989). Influence of dietary fat source on the body composition of mink (*Mustela vison*) and blue fox (*Alopex lagopus*) Acta Agr. Scand. 39: 279-288.

Rouvinen, K. I. and Kiiskinen, T. (1990). High dietary ash content decreases fat digestibility. Scientifur 14(4): 301.

Rouvinen, K.I. and Kisskinen, T. (1991). High dietary ash content decreases fat digestibility in the mink. Acta Agr. Scand. 41(4): 375-386.

Rouvinen, K. I., Newell, C.W., White, M.B. and Anderson, D.M. (1996). Dietary manipulation to reduce ammonia liberation from mink manure. Anim. Prod. Review, Applied Science Reports 28: 31-36.

Rouvinen, K. I., Niemela, P. and Kiiskinen, T. (1989). Influence of dietary fat on growth and fur quality of mink and fox. Acta Agric. Scand. 39: 269-278.

Rouvinen, K. I., Pitre, I., White, M.B. and Anderson, D. M. (1995). Amino acid supplementation and diet balance in mink. Final Report. Canada's Green Plan. Canada/Nova Scotia Agreement on the Agricultural Component of the Green Plan. 12 pp. http/agri gov. ns cal/rsl/Greenplan/waste/061.html.

Rouvinen, K. I., White, M. B. and Anderson, D. M. (1996). Use of microbial phytase phosphorus excretion in farmed mink. Canada/Nova Scotia Agreement on the Agricultural Component of the Green Plan, Nova Scotia, Canada 13 pp.

Rouvinen-Watt, K.I., White, M. B. and Campbell, R. (2005). Mink Feeds and Feeding, Nova Scotia Agricultural College:49.

Rouvinen-Watt, K. I. (2000). Nutrient management in carnivore fur breeders. Scientifur 24(4): 25-35. Proc. 7th Inter. Sc. Congr. In Fur Anim. Prod. Kastoria,. Greece. Sept. 13-15.

Rouvinen-Watt, K. I. and Harri, M.(2000). Observations on thermoregulatory ontogeny of mink (*Mustela vison*) . J. Thermal Biol. - in press.

Rouvinen-Watt, K.I., White, M., Clarke, N. and Cormier, M.(2002). Impact of feed supplementation on diet palatability by mink. NJF Seminar No. 347. Vuokatt, Finland. Oct 2-4, 2002: 263-269.

Sandberg, P. (1990 & 1996). Oregon mink rancher. Personal communication.

Sandbol, P. and Clausen, T.N. (2004). Sulphur containing amino acids for mink in the period of growth. Ann. Report 2004. Danish Fur Breeders Research Center, Holstebro, Denmark. Pp. 103-109.

Sandbol, P., Clausen, T.N. and Hejlesen, C. (2003a). Methionine and methyl donors for mink (*Mustela vison*) in the furring period. NJF-Seminar No. 354, Lillehammer, Norway, October 8-10, 2003. 6 pp.

Sandbol, P., Clausen, T.N. and Hejlesen, C. (2003b). Amino acid profiles in the furring period of the mink (*Mustela vison*). NJF-Seminar No. 354, Lillehammer, Norway, October 8-10, 2003. 6 pp.

Sandbol, P., Clausen, T.N.and Hejlesen, C. (2004a). Svovlholdige aminosyrer til mink i vaekstperioden. In "Samlet Oversigt Over Fodrings Forsog I Vaekstperioden" 2003: 28-34. Pelsdyre Hvervets Forsogs og Forsknings Center (Intern Raport).

Sandbol, P., Clausen, T.N. and Hejlesen, C. (2004b). Ideal protein for mink (*Mustela vison*) in the growing and furring periods. Ann. Report 2004 Danish Fur Breeders Research Center, Holstebro, Denmark. Pp. 111-119.

Sandbol, P., Clausen, T.N. and Hejlesen, C. (2004c). Ideal protein for mink (*Mustela vison*) in the growing and furring periods. Scientifur 28(3): 120-128.

Sangild, P. T. and Elnif, J. (1996). Intestinal hydrolytic activity in your mink (*Mustela vison*) develops slowly postnatally and exhibits late sensitivity to Glucocorticords. J. Nutr. 126(9): 2061-2068.

Schaefer, A. E., Tove, S. B., Whitehair, C.K. and Elvehjem, C. A. (1947). Use of foxes and minks for studying new B vitamins.International Review of Vitamin Research (International Zeitschrift fur Vitaminforschung). 19: 12-19.

Schaefer, A. E., Tove, S. B.,Whitehair, C. K. and Elvehjem, C. A. (1948). The requirement of unidentified factors for mink. J. Nutr. 35: 157-166.

Schaefer, A. E., Whitehair, C. K. and Elvehjem, C. A. (1946). Purified rations and the importance of folic acid in mink nutrition. Proc. Soc. Exp. Biol. Med. 62: 169-174.

Schaible, P.J. (1961). Unpublished Data.

Schaible, P.J., Aulerich, R.J. and Hartsought, G. R. (1966). An evaluation of the use of high levels of "spent" chickens in mink diets during the late growth and furring. Projects Supported by the Mink Farmers Res. Foundation 1966.

Schaible, P.J., Travis, H. E. and Shelts, G. (1962). Urinary incontinence and "wet belly" in mink. Quarterly Bulletin of the Michigan Agr. Exp. Station. East Lansing, Michigan. 44(3): 466-483.

Schimelman, S., Travis, H. F. and Warner, R. G. (1969). Determination of the biotin requirement, the effects of biotin deficiency and the interaction of antioxidants and oxidized fats with biotin requirements in the growing mink. Progress Reports: Mink Farmers Research Foundation.

Schlough, J. S. (1951). A guide to the dissection of the mink. 2nd ed. NASCO, Ft. Atkinson, WI. 53538.

Schneider, R. R. and Hunter, D. B. (1992). Nursing disease in mink. Scientifur 16(3): 239-242.

Schneider, R. R. and Hunter, D. B. (1993). Nursing disease in mink. Clinical and postmorten findings. Vet. Pathol. 30(6): 512-521.

Schneider, R. R., Hunter, D. B. and Waitner-Tuews, D. (1992). Nursing disease in mink individual-level epidemiology. Prev. Vet. Med. 14:167-179.

Schultz A. (1995). Minnesota mink rancher. Personal communication.

Schwartz, T.M. and Shackelford, R.M. (1973). Pseudodistemper an apparently new aliment of mink. U.S. Fur Rancher 52(8): 6.

Schweigert, F. J., Thomann, E. and Zucker, H. (1951). Vitamin A in the urine of carnivores. Inter. J. Vit. Nutr. Res. 61: 110-113.

Scott, M. I. (1986). Nutrition of Humans and Selected Animal Species. John Wiley & Sons, N.Y. p. 225.

Scott, P.P. (1960). Some aspects of the nutrition of the dog and cat. II. The Cat Vet. Rec. 72:5.

Scott, P. P. (1964). Nutritional requirements and deficiencies. In: Feline Medicine and Surgery. E. J. Catcott, ed. Santa Barbara, Calif. AM Vet. Publ. Inc.

Sealander, J. A. (1943). Winter food habits of mink in Southern Michigan. J. Wildlife Management 7(4): 411-417.

Sealock, R. R., Livermore, A. H. and Evans, C. A. (1943). Thiamine inactivation by the fresh-fish or Chastek-paralysis factor. J. Amer. Chem. Soc. 65: 935-940.

Seier, L.C., Kirk, R.J., Devlin, T. J. and Parker,. R. J. (1970). Evaluation of two dry protein sources in rations for growing-furring mink. Can. J. Anim. Sc. 50: 311-318.

Seimiya, Y., Kikuchi, F., Tanaka, S. and Ohshima. K. (1988). Pathological Observations of nursing sickness in mink. Japan J. Vet. Sci. 50: 255-257.

Semchenko, A. I. (1972). Digestibility of nutrients from the rations in mink of different colours. Trudy Moskovskoi Veterinarnji Akademie 59: 53-54. From Nutr. Abst. & Reviews 43, citation 4176, 1973.

Seton, E.(1929). Lives of game animals. Vol. II, Part II. Doubleday Doran & Co., Inc. Garden City, N.Y.

Shackelford, R. M. (1959). Personal communication.

Shackelford, R.M. and Cochrane, R. L. (1962). Reproduction performance of female mink fed stilbesterol. J. Anim. Sc. 21(2): 226-231.

Shacklady, C. A. (1972). Yeasts grown on hydrocarbons as new sources of protein. World Review of Nutrition and Dietetics 14: 154-179.

Sharma, C., Bursian, S. J. and Aulerich, R. J. (2000). Reproductive toxicity of ergot-contaminated wheat to mink. 2001 Blue Book of Fur Farming: 22-29.

Sherman, H.C. (1941). Chemistry of Foods and Nutrition. 7th Edition, MacMillan, N.Y. p. 233.

Shupe, J. L., Larsen, A.E. and Olsen, A. E. (1987). Effects of diets containing sodium fluoride on mink. J. Wildlife Diseases 23(4): 606-613.

Sibbald, I. R., Sinclair, D. G., Evans, E. V. and Smith, D. L. T. (1962). The rate of passage of feed through the digestive tract of the mink. Canadian J. Biochemistry and Physiology 40: 1391-1394.

Simoes-Nunes, C., Charlet-Lery, G. and Rougeot, J. (1984). Adaptation of the exocrine pancreatic secretion to diet composition in mink. 3rd Inter. Sc. Congr. In Fur Anim. Prod. Versailles, France, April 25-27.

Simova, J., Heger, J., Frydrych, Z. and Vesely, Z. (1982). Changes in nutritional value of potato proteins during potato protein concentration from tuber sap. Die Nahrung 26(9): 789-795.

Sinclair, A. J., McLean, J.G. and Monger,. E. A. (1979). Metabolism of linoleic acid in the cat. Lipids 14: 932-936.

Sinclair, D. G. and Evans, E. V. (1962b). A metabolism cage designed for use with mink. Can. J. Biochem. And Physiol. 40: 1395-1399.

Sinclair, D.G., Evans, E. V. and Sibbald, I. R. (1962a). The influence of apparent digestible energy and apparent digestible nitrogen in the diet on weight gain, feed consumption and nitrogen retention of growing mink. Can. J. Biochem. And Physiol. 40: 1375-1389.

Skrede, A. (1970a). Normal variation in the haemoglobin conc. in mink blood. Acta Agric. Scand. 20(4): 257-264.

Skrede, A. (1970b). Dietary blood in the prevention of fish-induced anemia in mink. I. Iron absorption studies. Acta Agr. Scand. 20(4): 265-274.

Skrede, A. (1970c) Dietary blood in the prevention of fish-induced anemia in mink. II. Feeding experiments. Acta Agr. Scand. 20(4): 275-285.

Skrede, A. (1974). Faktorer av betydning for hemoglobin - dannelsen. Bilag til kursus om vitamin og mineral stoffer i minkens. Foder Deres Indflydelse pa Reproduction, Vaekst og Pelsudvikling. June. Landhoskole, Denmark.

Skrede, A. (1975). Proteinkonsentrasjonen i foret til minkvalper fra avvenning til pelsing. Norsk Pelsdyrblad 49: 270-272.

Skrede, A. (1977). Soybean meal versus fish meal as protein resource in mink diets. Acta Agric. Scand. 27: 145-155.

Skrede, A. (1978a). Utilization of fish and animal by-products in mink nutrition. I. Effect of source and level of protein on nitrogen balance, post weaning growth and characteristics of winter fur quality. Acta Agric. Scand. 28(2): 105-129.

Skrede, A. (1978b). Utilization of fish and animal by-products in mink nutrition. II. Effect of source and level of protein on female reproductive performance and preweaning growth and mortality of progeny. Acta Agric. Scand. 28(2): 130-140.

Skrede, A. (1978c). Utilization of fish and animal by-products in mink nutrition III. Digestibility of diets based on different cod (*Gadus morrhua*) fractions in mink of different ages. Acta Agric. Scand. 28(2): 141-147.

Skrede, A. (1979a). Utilization of fish and animal by-products in mink nutrition IV. Fecal excretion and digestibility of nitrogen and amino acids by mink fed cod (*Gadus morrhua*) fillet and meat-and-bone meal. Acta Agric. Scand. 29: 241-257.

Skrede, A. (1979b). Utilization of fish and animal by-products in mink nutrition V. Content and digestibility of amino acids in cod (*Gadus morrhua*) byproducts. Acta Agric. Scand. 29: 353-362.

Skrede, A. (1980). Amino acid digestibility in mink. The 2nd Inter. Sc. Congr. In Fur Anim. Production, Copenhagen, Denmark, April 8-10.

Skrede, A. (1981a). Utilization of fish and animal by-products in mink nutrition VI. The digestibility of amino acids in fish visceral products and the effects on growth, fur characteristics and reproduction. Acta Agric. Scand. 31(2): 171-198.

Skrede, A. (1981b). Varying fat: carbohydrate ratios in mink diets.I. Effects on reproduction, early kit growth, viability and body composition. Meldinger Fra Norges Landbrukshogskole. Scientific Reports of the Agricultural University of Norway (1981). Melding NR 71, 60 (16): 20 pp.

Skrede, A. (1983). Varying fat: carbohydrate ratios in mink diets. II. Feeding Of kits during weaning to pelting period. Agric. Univ. Norway #74 Vol. 62(15). 20 pp.

Skrede, A. (1984). Evaluation of cepelin oil as an energy source in mink diets. 3rd Inter. Sc. Congr. In Fur Anim. Prod. Versailles, France, April 25-27.

Skrede, A. (1986a). Iron utilization in mink. Scientifur 10(4): 292-293.

Skrede, A. (1986b). Jernutnyttelse hos mink.NJF Meeting 110. 9-11 Sept. Kuipio, Finland.Nordisk Jordbruks Forskning 69:16

Skrede, A. (1987). Effekt av jerntilskudd og cysteine til mink. I. Reproducuks jonsperoden. Norsk Pelsdyrblad 61(8): 14-15.

Skrede, A. (1988).Effects of cysteine on iron absorption in mink. Biol.. Pathology and Genetics of Fur Bearing Animals. Ed. B.D. Murphy and D. B.I Iunter. 4th Inter. Sc. Congr. In Fur Anim. Prod. Rexdale, Ontario, Canada: 258-263.

Skrede, A. and Ahlstrom, O. (1987). Forsok med ulike jern preparat-er og cystein til mink. Norsk Pelsdyrblad 61(3): 3-5.

Skrede, A. and Ahlstrom, O. (2001). Meat-and-Bone meal as feed for the blue fox and mink in the reproduction period. NJF-seminar No. 331- 6 pp.

Skrede, A. and Ahlstrom, O. (2002). Soybean products in fur animal diets. NJF - Seminar No. 347: 17-24. Vuokatti, Finland. Oct 2-4, 2002.

Skrede, A., Birge, G..M., Storebakken, T., Herstad, O., Aarstad, K. G. and Sunstol, F. (1998). Digestibility of bacterial protein grown on natural gas in mink, pigs, chicken and Atlantic salmon. Animal Feed Science and Technology 76:103-116.

Skrede, A. and Herstad, O. (1978). Micronized cereals and soybeans for mink and chickens. Scientific Reports of the Agric. Univ. of Norway.1978.57(9):11

Skrede, A. and Koppang, N. (1982). Gastric ulcers in mink fed raw squid. Norsk Pelsdyrblad 56: 301-303.

Skrede, A. and Krogdahl, A. (1985). Heat affects nutritional characteristics of soybean meal and excretion of proteinases in mink and chicks. Nutr. Reports Int. 32(2): 479-489.

Skrede, A., Krogdahl, A. and Austreng, E. (1980). Digestibility of amino acids in raw fish flesh and meat-and-bone meal for chicken, fox, mink and rainbow trout. Zeitschrift Fir Tierphysiologic, Tierenahrung und Futtermillekunde. Band 43(2): 92-101.

Skrivan, M. (1980). Some problems on amino acid and vitamin nutrition of minks. Agrochemica (Czeckoslovakia). 29: 58-60.

Skrivan, M., Stole, L., Plistova, J. and Plistil, J. (1972b).The effect of calcium pantothenate supplements to rations with higher levels of pantothenate on the fertility of female mink. Biol. Chem. Vet. (Praha) 15: 213-219.

Skrivan, M., Stole, L. and Vorisk, K. (1972a). The effect of calcium pantothenate supplements of rations with low levels of pantothenate on fertility of one year old female mink. Biol. Chem. Vet. (Praha) 15: 209-212.

Smith, G. E. (1927). Progress reports - results of experiments 1926-1927: Dominion Experimental Fox Ranch, Summerside, Prince Edward Island.

Smith, H. (1972). Iowa mink rancher. Personal communication.

Smith, H. J. (1932). Foxes thrive on exclusive Purina Chows. Amer. Fur Breeder V(1): 14-15.

Smith, S. E. and Barnes, L. L. (1941). Experimental rickets in silver foxes (*Vulpes fulva*) and minks (*Mustela vison*). U.S. Bur. Anim. Ind. and CornellUniv. Ithaca, New York. Unpublished data.

Smith, S.E. and Hodson, A. Z. (1941). The composition of vesical calculi of fur animals. The Cornel Vet. 31(1): 30-34.

Sompolinsky, D. (1950). Urolithiasm in mink. The Cornell Vet. XL(4): 367-377.

Sonnenberg, R. (2001). Minnesota mink rancher. Personal communication.

Sopropeche, B. P. (1991)-62204 Boulogne - Sur-Mer Cedex - France.

Sorensen, P. G., Petersen, I.M. and Sand, O. (1995).Activities of carbohydrate and amino acid metabolizing enzymes from the liver of mink (*Mustela vison*) and preliminary observations on steady state kinetics of the enzymes. Comp. Biochem. Physiology 112B: 59-64.

417

Sorfleet, J. L. and Chavez, E.R. (1980). Comparative biochemical profiles in blood and urine of two strains of mink and changes associated with the incidence of wet-belly disease. Can. J. Physiol. Pharm. 58(5): 499-503.

Spencer, R. P. and Brody, K. K. (1964). Biotin transport by small intestine of the rat, hamster and other species. Am. J. Physiology 206: 653-655.

Spitzer, E.H., Coombs, A. I., Elvehjem, C. A. and Wisnicky, W. (1941). Inactivation of vitamin B-1 by raw fish, Soc. Exp. Biol. Med. Proc. 48: 376-379.

Steger, H. and Piatkowski, B. (1959). Nutrient content and digestibility of silkworm pupa meal in studies wituh silver foxes and mink. Arch Tierernahrung, 9:463.

Stevens, C.E. (1977). Comparative physiology of the digestive systems. In Swenson, M.J. (ed.). Dukes' Physiology of Domestic Animals. 9th ed. Cornell University Press, Ithaca, New York. USA 14850: 216-232.

Steenboch, H. (1939). Rations low in phosphorus may cause urinary calculi in minks. Amer. Nat. Fur and Market Journal 18(2): 15,24.

Stout, F.M. and Adair, J. (1969). Biotin deficiency in mink fed poultry by-products. Amer. Fur Breeder 42(6): 10-12.

Stout, F.M. and Adair, J. (1970). Factors influencing fur color and quality in mink. Projects Supported by the mink Farmers Research Foundation.

Stout, F.M., Adair, J., Costley, G..E. and Oldfield, J.E. (1967). Nutrition and other factors influencing fur color in mink. Amer. Fur Breeder 40(6): 10-12.

Stout, F.M., Adair, J. and Oldfield, J.E.(1966). Turkey waste graying - an Answer. Fur Trade J. of Canada, May 1966: 14-15.

Stout, F.M., Adair, J. and Oldfield, J. E. (1968). Hepatotoxicosis in mink associated with feeding toxic herring meal. Amer. Fur Breeder 41(6):12,14,18 & 19.

Stout, F.M., Adair,J., Oldfield, J. E. and Brow, W. G. (1963). Influence of protein and energy levels on mink growth and furring. J. Anim. Sc. 22: 847-855.

Stout, F.M., Baily, D.E., Adair, J. and Oldfield, J.E.(1968). Iron - a chromomeric nutrient. J. Anim. Sc. 27(4): 1157.

Stout, F.M., Oldfield, J.E. and Adair, J. (1960a). Nature and cause of "cotton fur" abnormality in mink. J. Nutr. 70: 421-426.

Stout, F.M., Oldfield, J.E. and Adair, J. (1960b). Aberrant iron metabolism and the "cotton fur" abnormality in mink. J. Nutr. 72: 46-52.

Stout, F.M., Oldfield, J.E. and Adair, J. (1962). Increasing production efficiency - new concepts in dry rations. Nat. Fur News 34(5): 8.

Stout, F.M., Oldfield, J.E. and Adair, J. (1963). A secondary induced thiamine deficiency in mink. Nature 197, No. 4869: 810-811.

Stowe, H.D., Schmidt, D.A., Travis, H. F. and Orentas, S. (1959). Pathology of tocopherol-deficient mink. Projects Supported by the Mink Farmers Research Foundation.

Stowe, H.D. and Whitehair, C. K. (1963). Gross and microscopic pathology of tocopherol deficiency mink. J. Nutr. 81: 287-300.

Stowe, H.D. and Whitehair, C. K. (1965). Clinical pathology of tocopherol deficient mink. Am. J. Vet. Res. 25(108): 1542-1549.

Stowe, H.D., Whitehair, C.K. and Travis, H. E. (1960). Pathology of vitamin E deficiency in the mink. Fed. Proc. 19(1): 420.

Struthers, B. J. and MacDonald, J. R. (1983). Comparative inhibition of trypsins from several species by soybean trypsin inhibitors. J. Nutr. 113(4): 800-804.

Sturdy, C. (1980). Wisconsin mink rancher. Personal communication.

Stryer, L. (1988). Biochemistry. 3rd Ed. Freeman and Co., San Francisco, USA, 1089 pp.

Sudadolnik, R.J., Stevens, C.O., Dechner, R.H., Henderson, L.M. and Hankes, L. V. (1957). Species variations in the metabolism of 3-hydroxy- anthranilate to pyridine carboxylic acid. J. Biol. Chem. 228: 973-982.

Sudenko, V. I., Groma, L. T. and Podgorsky, V. S. (1995). Studies on the microform of the gastrointestinal tract of mink kept in the zone of the Chenobyl NPP and outside it. Mikrobiology 57(2): 54-60.

Svenden, A. (1995). Palatability of ground alfalfa in the feed for grown male mink. Scientifur 19(3): 225.

Szymecko, R., Jorgensen, G., Bieguszewski, H. and Borsting, C. (1992). The effect of protein source on digestive passage and nutrient digestibility in polar foxes. Norwegian J. of Agr. Sc. Supplement No. 9, 1992. Progress in Fur Animal Science. Proc. Of the 5th Inter. Sc. Congr. in Fur Anim. Prod.

Szymecko, R. and Skrede, A. (1990). Protein digestion in mink. Acta Agr. Scand. 40(2): 189-200.

Taranov, G. S. (1977). Biology and pathology of farm bred fur-bearing animals. Abstracts of papers presented at the Second All-Union Scientific Conference. Kiro, USSR: 248-249.

Tauson, A-H. (1988a). Varied energy concentrations in mink diets. Apparent Digestibility of the experimental diets. Acta Agr. Scand. 38: 223-229.

Tauson, A-H (1988b). Varied energy concentrations in mink diets. II. Effects on kit growth performance, female weight changes and water turnover in the lactation period. Acta Agric.Scand. 38: 231-242.

Tauson, A-H.(1993a). High dietary levels of polyunsaturated fatty acids and varied vitamin E supplementation in the reproduction period of mink. J. Anim. Physiol. and Anim.Nutr. 72: 1-13.

Tauson, A-H. (1993b). Effect of body condition and dietary energy supply on reproduction processes in the female mink (*Mustela vison*). J. Reprod. Fert., Suppl. 47: 37-45.

Tauson, A-H. (1994). Postnatal development in mink kits. Acta Agric. Scand. Sect. A. Animal Sc. 44: 177-184.

Tauson, A-H. (1999a). Water intake and excretion, urinary solute excretion and some stress indicators in mink (*Mustela vison*). I. Effect of ambient temperature and quantitative water supply to adult males. Anim. Sc. 69: 171-181.

Tauson, A-H. (1999b). Water intake and excretion, urinary solute excretion and some stress indicators in mink (*Mustela vison*). 2. Short term response of adult males to changes in ambient temperatures 5 degrees and 20 degrees Celsius. Anim. Sc. 69: 183-190.

Tauson, A-H. and Alden, E. (1980). Potato protein for growing mink. Vara Palsdjur 51 (7): 176-185.

Tauson, A-H. and Alden, E. (1984). Pre-mating body weight changes and reproduction performance in female mink. Acta Agr. Scand. 34: 177-182.

Tauson, A-H. and Elnif, J. (1994). The pregnant mink (*Mustela vison*) - energy metabolism, nutrient oxidation and metabolic hormones. Proceedings of the 13th Symposium on Energy Metabolism. EAAP Publ. No. 76: 79-82.

Tauson, A-H., Elnif, J. and Enggaard Hansen, N. (1992). Energy metabolism and foetal growth in the pregnant mink (*Mustela vison*). Norwegian J. of Agr. Suppl. No. 9. Progress in Fur Anim. Sc., Proc.of the 5th Inter. Sc. Congr. In Fur Anim. Prod., Oslo, Norway, 1992: 261-267.

Tauson, A- H., Elnif, J. and Enggaard-Hansen, N. (1994). Energy metabolism and nutrient oxidation in the pregnant mink (*Mustela vison*) as a model for other carnivores. J. Nutr. 124: 26095-26135.

Tauson, A-H. and Englund, L. (1989). Prenatal mortality in farmed fur-bearing animals. NJS-Seminar 170. Stockholm.

Tauson, A-H. and Neil, M. (1991a). Fish oil and rapeseed oil as the main fat sources in mink diets in the growing-furring period. J. Anim. Physiol. And Anim. Nutr. 65(2): 84-95.

Tauson, A-H. and Neil, M. (1991b). Varied dietary levels of biotin for mink in the growing furring period. J. Anim. Physiol. and Anim. Nutr.(65): 235-243.

Tauson, A-H. and Neil, M. (1993a). Vitamin B-12 supplementation to mink (*Mustela vison*) in the prevention of feed-induced iron deficiency anemia. I. Effect on growth performance and fur quality characteristics. Acta Agr. Scand. Sect. A. Anim. Sc. 43: 116-122.

Tauson, A-H. and Neil, M. (1993b). Vitamin B-12 supplementation in mink (*Mustela vison*) in the prevention of feed-induced iron deficiency anaemia. II. Effect on hematological parameters and mineral content of the liver. Acta Agr. Scand., Sect. A. Anim. Sc. 43: 123-128.

Tauson, A-H., Sorensen, H.J., Wamberg, S. and Chwalibog, A. (1998). Energy metabolism, nutrient oxidation and water turnover in the lactating mink (*Mustela vison*). J. Nutr. 128: 2615S-2617S.

Taylor, R. J. (2000). Ontario, Canadian mink rancher, Personal Communication.

Taylor, P. G., Marinez-Torres, C., Romano, E. L. and Layrisse, M. (1986). The effects of cysteine containing peptides released during meat digestion on iron absorption in humans. Am. J. Clin. Nutr. 43: 68-71.

Teeter, R. G., Baker, D.H. and Corbin, J.E. (1978a).Methionine essentiality for the cat. J. Anim. Sc. 46: 1287-1292.

Teeter, R. G., Baker, D.H. and Corbin, J. E. (1978b). Methionine and cystine Requirements of the cat. J. Nutr. 108: 291.

Thorbek, G. (1975). Studies on energy metabolism in growing pigs. 424. Beretning Fra Statens Husdyrbrugsforsog, Copenhagen, Denmark. Pp. 159.

Thorbek, G. and Henckel, S. (1976). Studies on energy requirements for maintenance in farm animals. Proc. 7th Symposium of Energy Metabolism. Vichy, France. Pp. 4.

Tjurina, N.W. and Tjutunnik, N.N. (1982). Ergebnisse der utersuchung der mineralzusammenet der haare von fuchen und nerzen bei kafighaltung. In V.A. Berestov: Fiziologiceskoe Sostojanie Putnych Zwerej I Put. Ego-regulajacil. Petrosawodsk.

Tolonen, A. (1982). Summary: The food of the mink (*Mustela vison*) in Northeastern Finnish Lapland in 1967-1976. Suomen Riista 29: 61-65.

Tomlinson, M.J., Perman, V. and Westlake, R. L. (1982). Urate nephrolithiasis in ranch mink. J. of the Amer. Vet. Med. Assoc. 180(6):622-626.

Tomlinson, M. J., Perman, V. and Westlake, R. L. (1987). Urate nephrolithiasis in ranch mink is explained. Fur Rancher 67(3): 4.

Tove, S. B. (1950). Ph.D. Thesis, University of Wisconsin.

Tove, S. B., Lalor, R. J. and Elvehjem, C. A. (1950a). Properties of the methanol soluble factor required by the mink. Soc. Exp. Biol. Med. 75: 71-74.

Tove, S. B., Lalor, R.J. and Elvehjem, C. A. (1950b). An unidentified factor present in hog intestinal mucosa required by the mink. J. Nutr. 42: 433-441.

Tove, S. B., Schaefer, A. E. and Elvehjem, C. A. (1949). Folic acid studies in the mink and fox. J. Nutr. 38: 469-478.

Travis, H.F. Howell, R.E., Groschke, A. G. and Card, C. G.(1949). Mink feeding experiments: Report I. Quarterly Bulletin, Michigan Agri. Exper. Stat., Michigan State College 32(1): 64-69.

Travis, H. F. (1956). A study of the amino acid requirements of growing mink. Projects Supported by the Mink Ranchers Res. Foundation.

Travis. H. F. (1977). The effects of high levels of vitamin A upon the reproductive performance of mink. Projects Supported by the Mink Farmers Research Foundation.

Travis, H.F., Bassett, C. F., Loosli, J. K. and Warner, R. G. (1955). Stilbesterol tolerance of mink for reproduction, a progress report. The Black Fox and Modern Mink Breeder 39(3): 17-21.

Travis, H.F.,. Basset, C. F. and Warner, R. G. (1964). Gullet trimmings can be disastrous to reproduction. Amer. Fur Breeder 37(2): 14,34.

Travis, H.F., Bassett, C. F., Warner, R. G. and Loosli, J. K. (1956). Influence of DES on growth and reproduction in mink. Fed. Proc. 15:575.

Travis, H. F., Bassett, C.F., Warner, R.G. and Reineke, E.P. (1966). Some effects of feeding products high in naturally occurring thyro-active compounds upon reproduction of mink. Am. J. Vet. Res. 27(118): 81005-8170005.

Travis, H. F. and Duby, R. T. (1969). Are you feeding thyroids to your mink? Amer. Fur Breeder 42(9): 11.

Travis, H. F., Evans, E.V., Jorgensen, G.. Aulerich, R. J., Leoschke, W. L. and Oldfield, J. E. (1982). Nutrient Requirements of Domestic Animals. Number 7. Nutrient Requirements of Mink and Foxes. Second Revised Ed. 1982. National Academy Press, Washington, D.C.

Travis, H. F., Martin, T. F. Jr. and Pilbean, J. E. (1977). Manage waste to protect environment. Blue Book of Fur Farming 1978: 37-41.

Travis. H. F. and Pilbean, J. E. (1978a). Effects of storage on the vitamin E and oxidative rancidity levels in feeds. Fur Rancher 58(2): 10-11.

Travis, H. F. and Pilbean, J. E. (1978b). Effects of supplementation of vitamin E on reproduction and growth of mink kits from birth to six weeks. Fur Rancher 58(6): 17-18.

Travis, H. F., Ringer, R. K. and Schaible, P.J. (1961). Vitamin K in the nutrition of mink. J. Nutr. 74:181-184.

Travis, H. F. and Schaible, P.J. (1961a). Investigation of the safety of high levels of antioxidants in mink food. Nat. Fur News 33(5): 15-16.

Travis, H. F. and Schaible, P. J. (1961b). Effect of dietary fat levels upon reproduction performance of mink. Mich. Ag. Exp. Stat. Quart. Bull. 45(3): 518-521.

Travis, H.F. and Schaible, P.J. (1961c). Studies of dry rations for the growth and furring of mink. Mich. Ag. Exp. Stat. Quart. Bull. 44(1):32-44.

Travis, H.F. and Schaible, P.. J. (1962). Effects of DES fed periodicly during gestation of female mink upon reproduction and kit performance. Amer. J. of Vet. Res. 23(93): 359-361.

Travis, H.F. and Schaible, P.J. (1962a). Dry diets for mink during maintenance, reproduction and early kit growth. Mich. Agr. Exp. Stat. Quart. Bull. 44(1): 45-51.

Travis, H.F., Warner, R. G. and Schimelman, S. (1967). Progress in dry diet research. Amer. Fur Breeder 40(6): 16-17.

Travis, H. F., Warner, R. G. and Schimelman, S. (1968). A study of the need for certain vitamins by mink using purified diets. Projects Supported by the Mink Farmers Res. Foundation.

Treuhardt, J. (1992). Hematology, anti-oxidant trace elements, the related enzyme activities and vitamin E in growing mink on normal and anaemiogenic fish feeding. Ph.D. Dissertation, Abo Academi, Finland. Acta Academiae Aboensis. Ser. B. Mathematica et Physics. 52(4).

Trimberger, G. (1993). Wisconsin mink rancher. Personal Communication.

Trouw of Canada.(1970) Advertisement. Fur Trade Journal of Canada. 48(3):11.

Tyopponen, J, Berg, H. and Valtonen, M. (1987). Effects of dietary supplement of methionine and lysine on blood parameters and fur quality in mink fed with low protein diets. Acta Agr. Scand. 37(4):487-494.

Tyopponen, J., Hakkarainen, J., Juokslahti, J. and Lindberg, P. (1984). Vitamin E requirement of mink with special reference to tocopherol composition in plasma, liver and adipose tissue. Amer. J. Vet. Res. 45(9): 1790-1794.

Tyopponen, J., Polonen, I. and Valtonen, M. (1988). Effect of heat processing on amino acid availability from mink feed. Possible effect on iron absorption. Biology, Pathology and Genetics of Fur Bearing Animals. Ed. B. D. Murphy and D.B.Hunter. Proc. Of the 4th Inter. Sc. Congr. In Fur Anim. Prod., Rexdale, Ontario, Canada: 288-297.

420

Tyopponen, J., Valtonen, M. and Berg, H. (1986). Low protein feeding in mink. Effects of plasma free amino acids, clinical blood parameters and fur Quality. Acta Agr. Scand. 36(4): 421-428.

Ugletveit, S. (1975. Sildemel tilpelsdyr. Norsk Pelsdyrblad. 49: 159-161.

Utah Rancher (1996). Personnel communication.

Utne, F. (1974). B-vitaminer i minkfodringen. Bilag til kursus om vitamin og mineralstoffer I. Minkens foder deros indflydelse pa reproduction, vae ket og. Pelsudvikling. Tune Landboskole, Denmark.

Van Campen, D. R. (1969). Copper interference with the intestinal absorption of zinc-65 by rats. J. Nutr. 97: 164-168.

Van Campen, D.R. and Scaifi, P.U. (1967). Zinc interference with copper absorption in rats. J. Nutr. 91: 473-476.

Van Limborgh, G. I., Van Der Wind, J.J. and Blommaert, M. (1969). Successful complete dry feeding with Pelsifood. Nat. Fur News 42(1): 20-28

Varnish, S. A. and Carpenter, K. J. (1975). Mechanisms of heat damage in protein, the nutritional values of heat-damaged and propionylated proteins as sources of lysine, methionine and tryptophan. Br. J. Nutr. 34: 325-328.

Vermeulen, C. W., Ragens, H.D., Grove, W. J. and Goetz, R. (1951). Experimental urolithasis III. Prevention and dissolution of calculi by alteration of urinary pH. J. Urology 66(1): 1-5.

Vitamins, Inc. Chicago (1960). Food preservation with phosphoric acid. Fish-Pak. Data Sheet #1018.

Waanachek, J. (1966). Washington mink rancher. Personal communication.

Walker, B.L. and Lischenko, V. F. (1966). Fatty acid composition of normal mink tissues. Can. J. Biochem. 44: 179-185.

Wamberg, S. (1994). Rates of heat and water loss in female mink (Mustela vison) measured by direct calorimetry. Comp. Biochem. & Physiol. 107A: 451-458.

Wamberg, S., Clausen, T. N., Olesen, C. R. and Hansen, O. (1992). Nursing sickness in lactating mink (Mustela vison). II. Pathophysiology and changes in body fluid composition. Can. J. Vet. Res. 56: 95-101.

Wamberg, S., Olesen, C. R. and Hansen,. H.O. (1992). Influence of dietary source of fat on lipid synthesis in the mink (Mustela vison) mammary tissue. Comp. Biochem. Physiol. 103A (1):199-204.

Wamberg, S., Tauson, A-H and Elnif, J. (1996). Effects of feeding and short term fasting on water and electrolyte turnover in female mink (Mustela vison). Br. J. Nutr. 76: 711-725.

Wamberg, S. and Tauson, A-H. (1998a). The measurement of daily milk intake in suckling mink (Mustela vison). J. Nutr. 128: 2615S-2617S.

Wamberg, S. and Tauson, A-H. (1998b). Accuracy of quantitative collection of urine in carnivores. J. Nutr. 128: 2758S-2760S.

Wamberg, S. and Tauson, A-H. (1998c). Daily milk intake and body water turnover in suckling mink (Mustela vison) kits. Comp. Biochem. Physiol. 119A: 931-939.

Wamberg, S. and Tauson, A-H. (1998d). Accurate measurements of daily milk intake in suckling mink (Mustela vison) . Scientifur 22(1): 53.

Wambergm S. and Tauson, A-H. (1998e). Direct measurement of daily milk intake of suckling mink (Mustela vison) kits. J. Nutr. 128: 2620S-2622S.

Warner, R.G. (1966). Personal communication.

Warner, R.G. and Bassett, C.F. (1962). Purified diet experiments today will make cheaper diets in 1970. Fur Trade J. of Canada 39(11): 13-15.

Warner, R. G. and Travis, H. F. (1966). Availability of beta-carotene for growing mink. Projects of the Mink Farmers Research Foundation.

Warner, R.G., Travis, H. F. and Bassett, C.F. (1964). The determination of the minimum nutrient requirements of the mink using purified diets. Projects of the Mink Farmers Research Foundation.

Warner, R.G., Travis, H.F., Bassett, C.F., Krook, L. and McCarthy, B.(1963). Utilization of carotene by growing mink kits. Projects of the Mink Farmers Research Foundation.

Warner, R. G., Travis, H.F., Bassett, C.F., McCarthy, B. and Abernathy, R. P. (1968). Niacin requirement of growing mink. J. Nutr. 95(4): 563-568.

Watkins, B.F., Adair, J. and Oldfield, J.E.. Evaluation of shrimp and king crab processing by-products as feed supplements for mink. J. Anim. Sc. 55(3): 578-589.

Watt, P. R. (1951). Report on fur farming experiments. Oregon State College, Report 1950B.

Watt, P. R. (1952). DL-methionine supplementation of high fish diets. Mink Nutrition Research Progress Report Number 2. Dept. Fish and Game Management, Oregon Exp. Stn., Corvallis, Oregon.

Watt, P. R. (1953). Mink nutrition Research Progress Report. No. 3. Oregon Exper. Sta.

Wehr, N. B., Adair, J. and Oldfield, J.E. (1980). Biotin deficiency in mink fed spray dried eggs. J. Anim. Sc. 50(5): 877-884.

Wehr, N. B., Adair, J. and Oldfield, J.E. (1983). Studies on the nature and cause of wet belly disease in mink - a review. Experimental Fur Farm, Dept. of Anim. Sc., Oregon State Universitym Corvallis, Oregon.

Wehr, N. B., Adair, J., Scott, R.,. Thomson, C. and Oldfield, J.E. (1976 & 1977). Effects of Hemax and sodium bisulfite on prevention of formaldehyde or hake-induced anemia in mink fed fish and non-fish rations. Projects of the Mink Farmers Research Foundation.

Wehr, N. B., Oldfield, J. E. and Adair, J. (1981). Fur growth and development: Nutritional implications. Blue Book of Fur Farming -1982 Edition: 43-48.

Weis, D. J., Perman, V., Wuestenberg, W. and Bucci, T.J. (1994). Hematologioc response of adult brown mink to oxidative stress. Vet. Human Toxicol. 36(2): 109-111.

Weiss, V. (1981). Royal Vet and Ag. Univ., 1-56 Res. Farm West Herningvej 112. Denmark.

Weis, W. (1998). Personal communication.

Wenzel, U. D. and Arnold, P. (1971). Skeletal diseases of farm raised minks (*Mustela vison*). J. Bruhl 12(6): 10-12.

Westlake, R. (1988). Personal communication.

Westlake, R. and Newman, A. (1985). Veterinarian and a Minnesota mink rancher. Personal communication.

Westlake, R., Detroit Lakes Animal Clinic and Poppenca, R., Toxicologist, Diagnostic Laboratory, Michigan State Univ., East Lansing, Mich. (1986). Personal communication.

White, M.B., Egan, L. A. and Anderson, D.M. (1992). The effect of dietary acidifiers in diets of mature ranched foxes with a history of chronic urolithiasis. Norwegian J. Agr. Sc. Supplement No. 9: 3226-3231. ISSN 0801 5341.

White, M.B., Rouvinen-Watt, K. I., , Boudreau, D., Longmire, L. and Johnson. M. (2002). Apparent digestibility and storage stability of amino acids in feedstuffs prepared from end-of-cycle laying hens by formic acid preservation. NJF-Seminar No. 347. Vuokatti, Finland, Oct. 2-4, 2002:33-46.

Whitehair, C. K., Schaefer, A..E. and Elvehjem, C.A. (1949). Nutritional disease in mink with special reference to hemorrhagic gastroenteritis, "yellow fat" and anemia. Am. J. Vet. Med. Assoc. ll5: 54-58.

Williams. C., Elnif, J. and Buddington, R. K. (1998). The gastrointestinal bacteria of mink (*Mustela vison*). Influence of age and diet. Acta Vet. Scand. 39: 473-482.

Wilson, H.C. (1983). Wilson and Partners Veterinary Surgeons,136 Bonnygate, Cupar, Fife, KY15 4LF, Scotland. Personal communication.

Wisconsin Analytical, Inc. (1987). Watertown, WI.

Witz, W. (1984). Field feeding trials. Blue Book of Fur Farming: 93-97.

Wobeser, G., Nielsen, N.O. and Schiefer, B. (1975a). Mercury and mink. The use of mercury contaminated fish as food for ranch mink. Can.. J. Comp. Med. 40: 30-33.

Wobeser, G., Nielsen, N. O. and Schiefer, B. (1975b). Mercury and mink. II. Experimental methyl mercury intoxication. Can. J. Comp. Med. 40: 34-38.

Wobeser, G. and Swift, M. (1976). Mercury poisoning in a wild mink. J. Wildlife Disease, 12: 335-338.

Woller, J. (1977). Investigation of the effect of increasing amounts of histamine in feed for mink kits. Scientifur 1(4):19-27. Wolf(1942).

Wood, A. J. (1956a). Time of passage of food in mink. The Black Fox Magazine and Modern Mink Breeder 39(9): 12-13.

Wood, A.J. (1956b). Time of passage of food in the mink. Fur Trade J.l of Canada 33(6): 13, 18.

Wood, A. J. (1958). Whale meat for mink. Fur Trade J. of Canada 35(11):12-14.

Wood, A. J. (1962). Nutrient requirements of the mink. Western Fur Farmer April: 10.

Wood, A. J. (1964). Meeting the mink's protein requirements in practice. Fur Trade J. of Canada (6): 10-12.

Worthy, G. A. J., Rose, J. and Stormshak, F. (1987). Anatomy and physiology of fur growth: The pelage priming process. In: Wild Fur Bearer Management and Conservation in North America (Eds. M. Novak and J. Baker). Ontario Ministry of Natural Resources and Ontario Trappers Assocation. Pp. 827-841.

Wretlind, A. J. and Rose, C. (1950). Methionine requirements for growth and utilization of its optical isomer. J. Biol. Chem. 187: 697-704.

Yeager, L. E. (1943). Storing of muskrats and other foods by minks. J. Mammalogy 24(1): 100-101.

Yu, T. C. and Sinnhuber, R. O. (1967). Development of a fat quality test for application to mink feeds.Projects of the Mink Farmers Research Foundation.

Zawadke, E. (1988). Valparaiso University student. Personal communication.

Zellen, G. (1996). Urinary system of the mink. 1-13: l6. In Mink - Biology, Health and Disease eds. B.D. Hunter and N. Lemieux. Graphic and Print Svc. Univ. Guelph, Guelph, Ontario, Canada.

Zimbal, R. (1975, 1998, 2000 and 2001). Wisconsin mink rancher. Personal communication.

Zimbal, R., Jr. (1996). Raw egg preservation. Nova Scotia Agr. College Seminar.

Zimbal, R., Jr. and Zimbal, L. (1995). Wisconsin mink ranchers. Personal communication.

Zimmerman, H. (1976). Observations on wet belly disease in male mink. (Beobachtungen zur bachnasse der nerzruden).J. Bruhl 17(6): l0.

Zimmerman, H. (1981). Vitamin B-1-mangel bei tragenden. Nerzfahen Mh. Vet. Med. 36:508-511.

Zimmerman, H. (1992). Vitamin E disturbance in mink after rabbit offal feeding. Der Deutsche Pelztierzuchter. 66(6): 6.

Zong, D. Q. (1986). Fishmeal poisoning in mink. Scientifur 10(2): 127.

Index

Nutrition and Nutritional Physiology Of The Mink–A Historical Perspective

425